STRUCTURE REPORTS

for 1982

Volume 49A

Structure Reports is prepared under the guidance of a Commission of the International Union of Crystallography. The members of the Commission sometime concerned with the preparation of this volume are listed below.

STRUCTURE REPORTS

for 1982

Volume 49A

METALS AND INORGANIC SECTIONS

General editor

G. Ferguson

Section editor

J. Trotter

Published for the

INTERNATIONAL UNION OF CRYSTALLOGRAPHY

by

KLUWER ACADEMIC PUBLISHERS

DORDRECHT / BOSTON / LONDON

First published in 1989

ISSN 0166–6983

printed on acid free paper

ISBN 0–7923–0239–7

Printed in the Netherlands

TABLE OF CONTENTS

INTRODUCTION

The present volume continues the aim of Structure Reports to present critical accounts of all crystallographic structure determinations. Details of the arrangement in the volumes, symbols used etc. are given in previous volumes (e.g. 41B or 42A, pages vi–viii).

University of Guelph,
Guelph, Ontario, Canada

G. FERGUSON

22 January 1989

STRUCTURE REPORTS

SECTION I

METALS

Edited by

J. Trotter
(University of British Columbia)

with the assistance of

L. D. Calvert

ARRANGEMENT

As in previous volumes the arrangement in the Metals section is approximately, but not strictly, alphabetical, and to find particular substances the subject index or formula index should be used.

ALUMINUM ARSENIC CALCIUM
$AlCa_3As_3$

ALUMINUM ANTIMONY BARIUM
$AlBa_3Sb_3$

G. CORDIER, G. SAVELSBERG and H. SCHÄFER, 1982, Z. Naturforsch., **37B**, 975-980.

$AlCa_3As_3$, orthorhombic, Pnma, a = 12.212, b = 4.201, c = 13.434 A, Z = 4. Mo radiation, R = 0.093 for 1074 reflexions.

$AlBa_3Sb_3$, orthorhombic, Cmca, a = 21.133, b = 7.194, c = 14.069 A, Z = 8. Mo radiation, R = 0.059 for 978 reflexions.

$AlCa_3As_3$ contains infinite chains of corner-sharing $AlAs_4$ tetrahedra, while $AlBa_3Sb_3$ contains isolated Al_2Sb_6 groups of two edge-sharing tetrahedra (Fig. 1). Al-As = 2.50-2.54, Al-Sb = 2.71, 2.72, Ca-As = 2.91-3.38, Ba-Sb = 3.48-3.80 A.

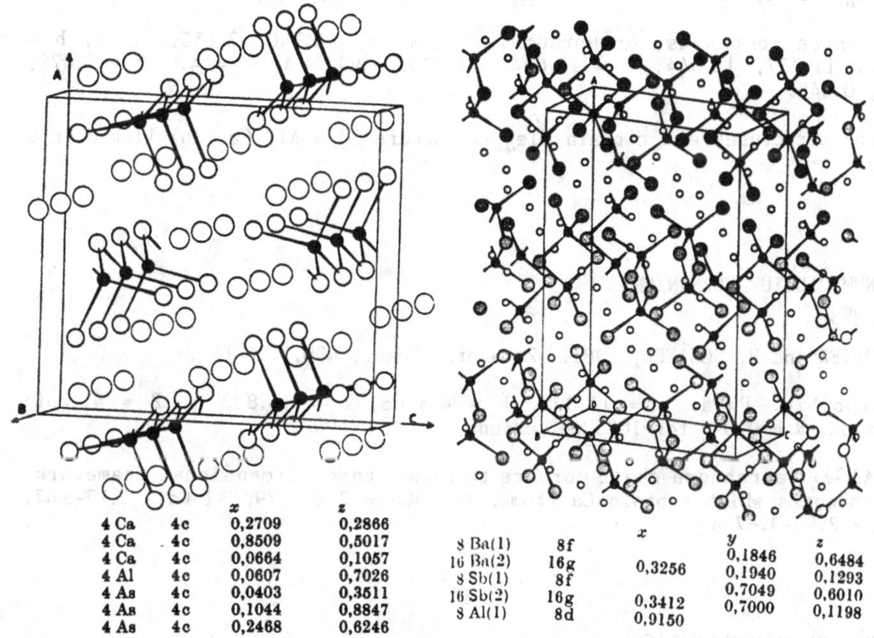

		x	z
4 Ca	4c	0,2709	0,2866
4 Ca	4c	0,8509	0,5017
4 Ca	4c	0,0664	0,1057
4 Al	4c	0,0607	0,7026
4 As	4c	0,0403	0,3511
4 As	4c	0,1044	0,8847
4 As	4c	0,2468	0,6246

		x	y	z
8 Ba(1)	8f		0,1846	0,6484
16 Ba(2)	16g	0,3256	0,1940	0,1293
8 Sb(1)	8f		0,7049	0,6010
16 Sb(2)	16g	0,3412	0,7000	0,1198
8 Al(1)	8d	0,9150		

Fig. 1. Structures of $AlCa_3As_3$ (left, y = 1/4) and $AlBa_3Sb_3$ (right).

ALUMINUM BARIUM SULPHUR
$BaAl_2S_4$

BARIUM GALLIUM SULPHUR
$BaGa_2S_4$

B. EISENMANN, M. JAKOWSKI and H. SCHÄFER, 1982. Mater. Res. Bull., **17**,1169-1175.

Cubic, Pa3, a = 12.650, 12.685 A, Z = 12. Mo radiation, R = 0.112, 0.084 for 769, 956 reflexions. Ba(1) in 4(a): 0,0,0; Ba(2) in 8(c): x,x,x, x = 0.3736, 0.3722; Al or Ga in 24(d): 0.1966,0.3571,0.1253 or 0.1959,0.3568,0.1263; S(1) in 24(d): 0.3599,0.2899,0.1299 or 0.3627,0.2935,0.1268; S(2) in 24(d): 0.2263,0.5270,0.0797 or 0.2291,0.5267,0.0773.

AlS_4 (or GaS_4) tetrahedra share corners to form a three-dimensional framework, with Ba in holes with 12- and 6-coordinations. Al-S = 2.20-2.26, Ga-S = 2.25-2.30, Ba-S = 3.16-3.62 A.

ALUMINUM BARIUM TELLURIUM
Al_2BaTe_4

GALLIUM STRONTIUM TELLURIUM
INDIUM STRONTIUM TELLURIUM
BARIUM INDIUM TELLURIUM
Ga_2SrTe_4, In_2SrTe_4, $BaIn_2Te_4$

B. EISENMANN, M. JAKOWSKI and H. SCHÄFER, 1982. Rev. Chim. Minér., **19**, 263-273.

Al_2BaTe_4, tetragonal, P4/nbm, a = 8.516, c = 6.713 A, Z = 2. R = 0.051.

Other three compounds, orthorhombic, Cccm, a = 6.740, 7.155, 7.127, b = 11.621, 11.737, 12.044, c = 11.634, 11.770, 12.115 A, Z = 4. R = 0.079, 0.099, 0.046.

The structures all contain MTe_4 tetrahedra (M = Al, Ga, In) linked into chains.

ALUMINUM CALCIUM GERMANIUM
$Al_2Ca_3Ge_3$

G. CORDIER and H. SCHÄFER, 1982. Z. anorg. Chem., **490**, 136-140.

Orthorhombic, Pnma, a = 11.395, b = 4.3468, c = 14.833 A, Z = 4. Mo radiation, R = 0.093 for 1025 reflexions.

$AlGe_4$ tetrahedra share corners to form a three-dimensional framework, with channels which contain Ca atoms. Al-4Ge = 2.55-2.79, Al-Ca = 3.15-3.47, Ge-Ca = 2.99-3.47 A

ALUMINUM GADOLINIUM SILICON
Al_2GdSi_2

R. NESPER, H.G. von SCHNERING and J. CURDA, 1982. Z. Naturforsch., **37**B, 1514-1517.

Trigonal, $P\bar{3}m1$, a = 4.194, c = 6.651 A, Z = 1. Mo radiation, R = 0.034 for 420 reflexions. Al in 2(d): 1/3,2/3,0.6432; Gd in 1(a): 0,0,0; Si in 2(d): 1/3,2/3,0.2617. Al_2CaSi_2-type (**1**). Al-Si = 2.50, 2.54, Gd-Si = 2.98 A.

1. Structure Reports, **32**A, 5.

ALUMINUM HYDROGEN LANTHANUM NICKEL
$AlLaNi_4D_{4.1}$

V.A. JARTYS', V.V. BURNAŠEVA, S.E. CIRKUNOVA, E.N. KOZLOV and K.N. SEMENENKO,

1982. Kristallografija, **27**, 242-246 [Soviet Physics - Crystallography, **27**, 148-151].

Trigonal, P321, a = 5.300, c = 4.242 A, Z = 1. Neutron powder data. 1 La in 1(a): 0,0,0; 2 Ni in 2(d): 1/3,2/3,1.003; 2Ni + 1Al in 3(f): 0.501,0,1/2; 2.7 D(1) in 6(g): 0.484,0.457,0.098; 1.4 D(2) in 6(g): 0.181,0.285,0.463. $CaCu_5$-type structure (**1**), with D in two tetrahedral sites, and lowering of symmetry from P6/mmm to P321.

$LaNi_{4.5}Al_{0.5}D_{4.5}$

C. CROWDER, W.J. JAMES and W. YELON, 1982. J. Appl. Phys., **53**, 2637-2639.

Hexagonal, P6/mmm, a = 5.341, 5.331, c = 4.236, 4.236 A, at 298, 77K, Z = 1. Neutron powder data. $LaNi_5$-type (**1**), with Al in the 3(g) Ni site, and D in 6(i) and 6(m) sites [compare lower-symmetry descriptions, above and following report].

1. Structure Reports, **11**, 59; **46A**, 70.

HYDROGEN LANTHANUM NICKEL
$LaNi_5D_{6.0}$

V.V. BURNAŠEVA, V.A. JARTYS', N.V. FADEEVA, S.P. SOLOV'EV and K.N. SEMENENKO, 1982. Vest. Mosk. Univ., Ser. 2: Khim., **23**, 163-167.

Trigonal, P321, a = 5.43, c = 4.28 A, Z = 1. Neutron radiation, R = 0.083. [Compare, e.g., preceding report.]

ALUMINUM NICKEL ZIRCONIUM
Al_5Ni_2Zr

V.Ja. MARKIV, P.I. KRIPJAKEVIČ and N.M. BELJAVINA, 1982. Dopov. Akad. Nauk Ukr.RSR, Ser. A: Fiz.-Mat. Tekh. Nauki, No. 3, 76-79.

Tetragonal, I4/mmm, a = 4.023, c = 14.44 A, Z = 2. R = 0.12 for 54h0ℓ reflexions. Zr in 2(a): 0,0,0; Ni in 4(e): 0,0,0.238; Al(1) in 8(g): 0,1/2,0.149; Al(2) in 2(b): 0,0,1/2.

New structure type; coordination numbers are: Zr = 12+2, Ni = 12+2, Al = 12, 13 (Fig. 1).

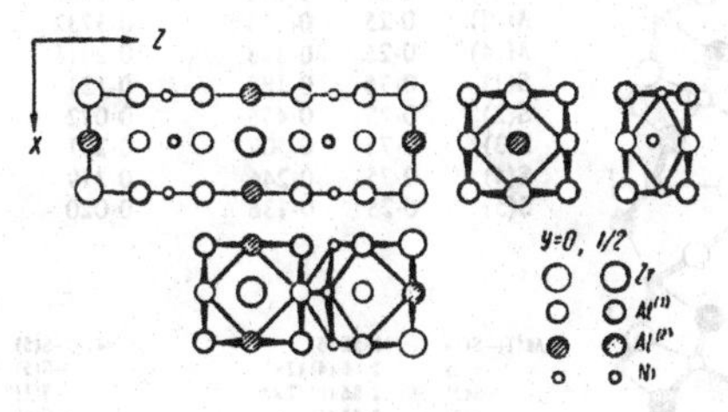

Fig. 1. Structure of Al_5Ni_2Zr.

ALUMINUM STRONTIUM
Al_2Sr

G. CORDIER, E. CZECH and H. SCHÄFER, 1982. Z. Naturforsch., **37**B, 1442-1445.

Normal-pressure, orthorhombic, Imma, a = 7.905, b = 4.801, c = 7.974 A, Z = 4. Mo radiation, R = 0.036 for 264 reflexions. Sr in 4(e): 0,1/4,0.6995; Al in 8(i): 0.1823,1/4,0.0889. Hg_2K-type (**1**), as previously described (**2**).

High-pressure, cubic, Fd3m, a = 8.325 A, Z = 8. Powder data. Sr in 8(a); Al in 16(d). Cu_2Mg-type (**3**).

1. Structure Reports, **19**, 231.
2. Ibid., **41**A, 113.
3. Strukturbericht, **1**, 490.

ANTIMONY ARSENIC SULPHUR THALLIUM (REBULITE)
$Tl_5Sb_5As_8S_{22}$

T. BALIĆ-ŽUNIĆ, S. ŠĆAVNIČAR and P. ENGEL, 1982. Z. Kristallogr., **160**, 109-125.

Monoclinic, $P2_1/c$, a = 17.441, b = 7.363, c = 32.052 A, β = 105.03°, Z = 4. Mo radiation, R = 0.057 for 5665 reflexions.

Three-dimensional framework of connected AsS_3 trigonal pyramids and Sb coordination polyhedra (3+4 or 4+3 S neighbours), with Tl ions having (6+2)-, (6+3)-, and 8-coordinations. Some As and Sb sites have partial occupancies.

ANTIMONY BISMUTH SULPHUR TIN
$(Bi,Sb)_2Sn_2S_5$

V. KUPČIK and M. WENDSCHUH, 1982. Acta Cryst., B**38**, 3070-3071.

Orthorhombic, Pmcn, a = 3.95, b = 11.26, c = 19.49 A, Z = 4. Mo radiation, R = 0.101 for 136 reflexions.

Chains along **a** of slightly-distorted MS_5 square pyramids, with chains connected by longer M-S bonds which complete distorted octahedral coordination geometry (7-coordination for M(4)) (Fig. 1).

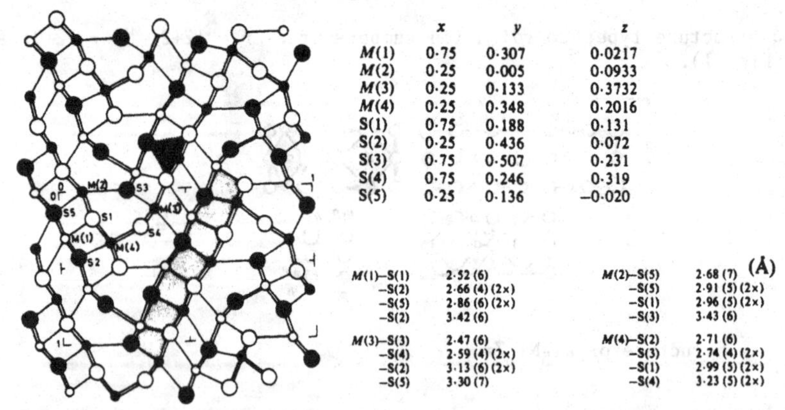

	x	y	z
M(1)	0·75	0·307	0·0217
M(2)	0·25	0·005	0·0933
M(3)	0·25	0·133	0·3732
M(4)	0·25	0·348	0·2016
S(1)	0·75	0·188	0·131
S(2)	0·25	0·436	0·072
S(3)	0·75	0·507	0·231
S(4)	0·75	0·246	0·319
S(5)	0·25	0·136	−0·020

			(Å)
M(1)–S(1)	2·52 (6)	M(2)–S(5)	2·68 (7)
–S(2)	2·66 (4) (2×)	–S(5)	2·91 (5) (2×)
–S(5)	2·86 (6) (2×)	–S(1)	2·96 (5) (2×)
–S(2)	3·42 (6)	–S(3)	3·43 (6)
M(3)–S(3)	2·47 (6)	M(4)–S(2)	2·71 (6)
–S(4)	2·59 (4) (2×)	–S(3)	2·74 (4) (2×)
–S(2)	3·13 (6) (2×)	–S(1)	2·99 (5) (2×)
–S(5)	3·30 (7)	–S(4)	3·23 (5) (2×)

Fig. 1. Structure of $(Bi,Sb)_2Sn_2S_5$.

ANTIMONY CERIUM LITHIUM
ANTIMONY LITHIUM PRASEODYMIUM
ANTIMONY LITHIUM NEODYMIUM
$CeLi_2Sb_2$, $PrLi_2Sb_2$, $NdLi_2Sb_2$

H.-O. FISCHER and H.-U. SCHUSTER, 1982. Z. anorg. Chem., **491**, 119-123.

Tetragonal, P4/nmm, a = 4.335, 4.329, 4.280, c = 10.960, 10.999, 10.910 A, Z = 2. Mo radiation, R = 0.058, 0.058, 0.053 for 177, 287, 344 reflexions. Ln in 2(c): 0,1/2,z, z = 0.2657, 0.2658, 0.2668; Li(1) in 2(a): 0,0,0; Li(2) in 2(c): z = 0.599, 0.620, 0.598; Sb(1) in 2(b): 0,0,1/2; Sb(2) in 2(c): z = 0.8385, 0.8378, 0.8353. $CaBe_2Ge_2$-type structures (1).

1. Structure Reports, **43**A, 28.

ANTIMONY LEAD SILVER SULPHUR (SYNTHETIC ANDORITE)
$Pb_{0.61}Ag_{1.17}Sb_{3.05}S_6$

N.I. ORGANOVA, O.V. KUZ'MINA, N.S. BORTNIKOV and N.N. MOZGOVA, 1982. Dokl. Akad. Nauk SSSR, **267**, 939-942.

Orthorhombic, Bbmm, a = 13.00, b = 19.17, c = 4.246 A, Z = 4. Mo radiation, R = 0.102 for 738 reflexions.

Atomic positions

			x	y	z
4	M(1)	in 4(c)	0.3346	3/4	0
8	M(2)	8(f)	0.1287	0.4477	0
8	M(3)	8(f)	0.4042	0.1338	0
8	S(1)	8(f)	0.068	0.1619	0
8	S(2)	16(h)	0.2565	0.9003	0.077
4	S(3)	4(c)	0.312	1/4	0
4	S(4)	8(e)	1/2	0	0.019

M(1) = 0.6 Pb + 0.3 Sb
M(2) = Sb
M(3) = 0.35 Sb + 0.6 Ag

Isostructural with the natural mineral (andorite-VI = ramdohrite) (1).

1. Structure Reports, **38**A, 18.

ANTIMONY STRONTIUM SULPHUR
$Sr_3Sb_4S_9$

G. CORDIER, C. SCHWIDETZKY and H. SCHÄFER, 1982. Rev. Chim. Minér., **19**, 179-186.

Orthorhombic, $Pna2_1$, a = 16.579, b = 24.000, c = 4.090 A, Z = 4. R = 0.075.

Chains along **b** of edge-sharing SbS_5E octahedra and of corner-sharing SbS_3E tetrahedra, arranged so that the lone pairs (E) point towards each other.

ANTIMONY SULPHUR THALLIUM
$TlSb_3S_5$

M. GOSTOJIĆ, W. NOWACKI and P. ENGEL, 1982. Z. Kristallogr., **159**, 217-224.

Monoclinic, $P2_1/c$, a = 7.225, b = 15.547, c = 8.946 A, β = 113.55°, Z = 4. Mo radiation, R = 0.067 for 2700 reflexions.

SbS_3 trigonal pyramids share corners to form Sb_3S_5 (010) layers, which are linked by one longer Sb-S bond; within the layers are chains along **c** of face-sharing TlS_8 bicapped trigonal prisms. Sb-S = 2.43-2.64, 2.85, Tl-S = 3.04-3.56 A.

ARSENIC CALCIUM COPPER
$CaCu_4As_2$

M. PFISTERER and G. NAGORSEN, 1982. Z. Naturforsch., **37B**, 420-422.

Rhombohedral, $R\bar{3}m$, a = 4.173, c = 22.62 A, Z = 3 (rhombohedral, a = 7.91 A, α = 30.6°, Z = 1). Mo radiation, R = 0.11 for 297 reflexions. Ca in 3(a): 0,0,0; As, Cu(1), Cu(2) in 6(c): 0,0,z, z = 0.252, 0.143, 0.436. Isostructural with $CaCu_4P_2$ (**1**). Ca-6As = 3.03, Ca-8Cu = 3.23, 3.34, Cu-3 or 4As= 2.45-2.78 A.

1. Structure Reports, **46A**, 50.

ARSENIC CALCIUM SILICON
ARSENIC CALCIUM GERMANIUM
ARSENIC GERMANIUM STRONTIUM
$Ca_3Si_2As_4$, $Ca_3Ge_2As_4$, $Sr_3Ge_2As_4$

ARSENIC SILICON STRONTIUM
$Sr_3Si_2As_4$

B. EISENMANN and H. SCHÄFER, 1982. Z. anorg. Chem., **484**, 142-152.

First three compounds, monoclinic, $P2_1/c$, a = 7.303, 7.308, 7.625, b = 17.730, 17.886, 18.514, c = 7.155, 7.239, 7.470 A, β = 111.66, 111.75, 111.76°, Z = 4. Mo radiation, R = 0.073, 0.088, 0.107 for 2094, 2198, 1383 reflexions.

$Sr_3Si_2As_4$, monoclinic, C2/c, a = 9.205, b = 16.832, c = 7.376 A, β = 122.46°, Z = 4. Mo radiation, R = 0.056 for 836 reflexions.

Both structure types contain As_3M-MAs_3 groups (M = Si, Ge) connected into chains in different ways [see this volume, p. **19** for illustrations]: in the first three structures, AsS_3 groups share 2 S by spanning neighbouring S-As-As-S groupings to give five-membered As_3S_2 rings; in $Sr_3Si_2As_4$, AsS_3 groups share one edge to give four-membered As_2S_2 rings. Chains are arranged in sheets which are linked by the alkali-metal cations.

ARSENIC COBALT SULPHUR (COBALTITE)
CoAsS

P. BAYLISS, 1982. Amer. Min., **67**, 1048-1057.

Orthorhombic, $Pca2_1$, a = 5.592, b = 5.587, c = 5.567 A, Z = 4. R = 0.049-0.128 for 284-303 reflexions for four samples (twinned crystals). Co: 0.9955,0.2596,0; As: 0.6195,0.8694,0.6168; S: 0.3845,0.6314,0.3809. All the samples show some deviation from pyrite-like Pa3.

ARSENIC NICKEL SULPHUR (GERSDORFFITE)
NiAsS

P. BAYLISS, 1982. Amer. Min., **67**, 1058-1064.

Cubic, Pa3 (pyrite-type); cubic, $P2_13$ (ullmannite-type); and orthorhombic, $Pca2_1$ (cobaltite-type (see preceding report)) were found.

ARSENIC MANGANESE SULPHUR THALLIUM
$Tl_2MnAs_2S_5$

M. GOSTOJIĆ, A. EDENHARTER, W. NOWACKI and P. ENGEL, 1982. Z. Kristallogr., **158**, 43-51.

Orthorhombic, Cmca, a = 15.340, b = 7.608, c = 16.651 A, Z = 8. Mo radiation, R = 0.078 for 1392 reflexions.

As_2S_5 groups of two corner-sharing AsS_3 trigonal pyramids, linked into (001) slabs by MnS_6 octahedra and Tl(1)-(4+2)S polyhedra; slabs are connected by Tl(2) ions which also have (4+2)-coordination. As-S = 2.22-2.29, Mn-S = 2.57-2.65, Tl-S = 2.93-3.25, 3.42-3.89 A.

ARSENIC MERCURY SULPHUR THALLIUM (SIMONITE)
$TlHgAs_3S_6$

P. ENGEL, W. NOWACKI, T. BALIĆ-ŽUNIĆ and S. ŠĆAVNIČAR, 1982. Z. Kristallogr., **161**, 159-166.

Monoclinic, $P2_1/n$, a = 5.948, b = 11.404, c = 15.979 A, β = 90.15°, Z = 4. Mo radiation, R = 0.05 for 2462 reflexions.

AsS_2 chains along [101] of corner-sharing AsS_3 trigonal pyramids, double chains along [100] of distorted HgS_6 octahedra, and (7+2)-coordinate Tl ions; the coordination polyhedra form layers parallel to (001). As-S = 2.18-2.34, Hg-S = 2.40-2.73, 3.06-3.41, Tl-S = 3.20-3.47, 3.62, 3.84 A.

ARSENIC TELLURIUM
As_2Te_3

A.S. KANIŠČEVA, Ju.N. MIKHAILOV and A.P. ČERNOV, 1982. Izv. Akad. Nauk SSSR, Neorg. Khim., **18**, 949-952.

Monoclinic, C2/m, a = 14.364, b = 4.025, c = 9.889 A, β = 95.14°, Z = 4. R = 0.094

The structure contains columns of edge-sharing $AsTe_6$ octahedra.

BARIUM BISMUTH CADMIUM
$Ba_2Bi_4Cd_3$

G. CORDIER, P. WOLL and H. SCHÄFER, 1982. J. Less-Common Metals, **86**, 129-136.

Orthorhombic, Cmca, a = 7.037, b = 17.438, c = 9.267 A, Z = 4. Mo radiation, R = 0.121 for 892 reflexions.

Framework of edge- and corner-sharing $CdBi_6$ octahedra and $CdBi_5$ trigonal bipyramids, with Ba coordinated to 8 Bi (Fig. 1). Cd-Bi = 2.94-3.40, Ba-Bi = 3.60-3.75 A.

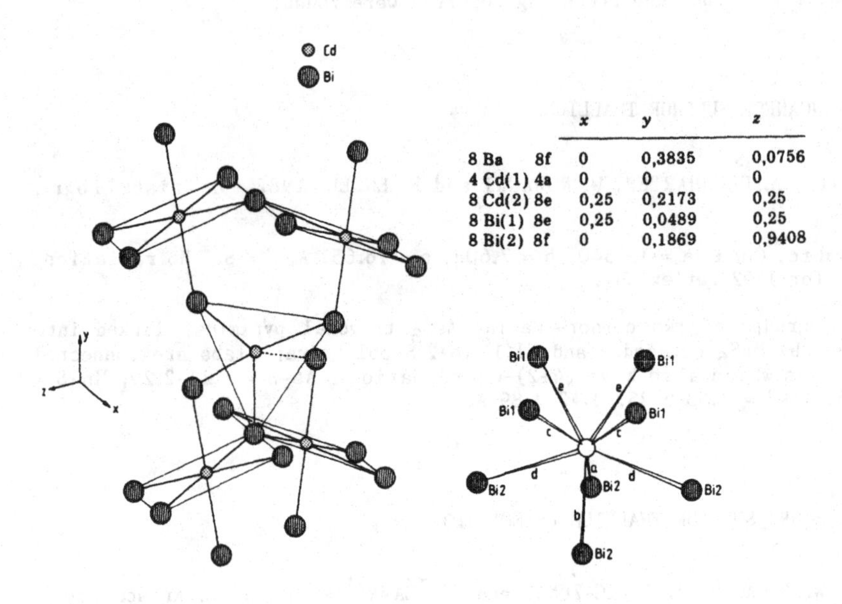

		x	y	z
8 Ba	8f	0	0,3835	0,0756
4 Cd(1)	4a	0	0	0
8 Cd(2)	8e	0,25	0,2173	0,25
8 Bi(1)	8e	0,25	0,0489	0,25
8 Bi(2)	8f	0	0,1869	0,9408

Fig. 1. Structure of $Ba_2Bi_4Cd_3$.

BARIUM GERMANIUM PHOSPHORUS
Ba_2GeP_2

ARSENIC BARIUM GERMANIUM
Ba_2GeAs_2

B. EISENMANN, H. JORDAN and H. SCHÄFER, 1982. Z. Naturforsch., **37**B, 1221-1224.

Monoclinic, $P2_1/c$, a = 8.553, 8.757, b = 9.515, 9.700, c = 7.481, 7.667 A, β = 105.95, 106.03°, Z = 4. Mo radiation, R = 0.044, 0.064 for 1632, 1657 reflexions.

Atomic positions

Ba_2GeP_2	x	y	z
Ba(1)	0.9926	0.3574	0.7384
Ba(2)	0.4806	0.1631	0.0994
Ge	0.6427	0.4722	0.0274
P(1)	0.7631	0.9020	0.3408
P(2)	0.7542	0.4029	0.3274
Ba_2GeAs_2			
Ba(1)	0.9972	0.3585	0.7422
Ba(2)	0.4823	0.1626	0.1026
Ge	0.6397	0.4698	0.0257
As(1)	0.7632	0.9003	0.3350
As(2)	0.7556	0.3991	0.3334

The structures (Fig. 1) contain sheets of isolated $Ge_2X_4^{8-}$ anions, linked by Ba ions coordinated to 6 P or As, Ba-P = 3.20-3.59, Ba-As = 3.28-3.64 A; there is a short Ba(2)-Ba(2) contact in each structure, 3.50 and 3.58 A.

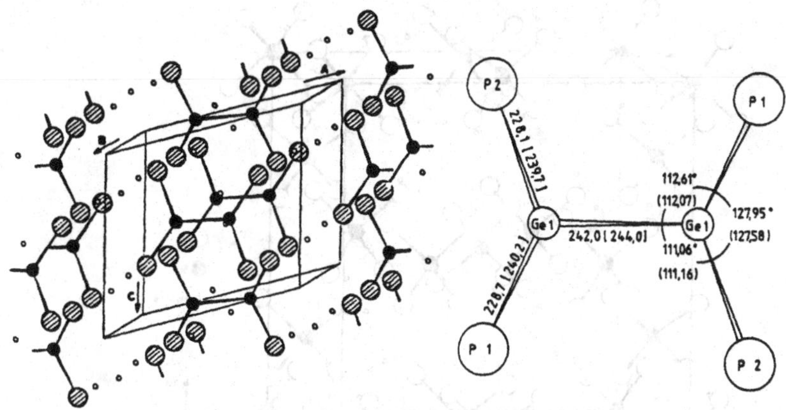

Fig. 1. Structures of Ba_2GeP_2 and Ba_2GeAs_2 (bond distances in A x 10^2).

BARIUM IRON SULPHUR
$Ba_2Fe_4S_5$

J.S. SWINNEA, G.A. EISMAN, T.P. PERNG, N. KIMIZUKA and H. STEINFINK, 1982. J. Solid State Chem., **41**, 104-108.

Orthorhombic, Pmmn, a = 4.016, b = 9.616, c = 6.514 A, Z = 1. Mo radiation, R = 0.036 for 432 reflexions. Ba in 2(a): 1/4,1/4,0.5279; Fe in 4(e); 1/4,0.9178,0.8792; S(1) in 4(e): 1/4,0.9613,0.2354; 1 S(2) in 2(b): 1/4,3/4,0.7318.

BaS_6 trigonal prisms share triangular faces to form columns along **c**, with zigzag chains of Fe along **b** formed by filling tetrahedral interstices; one S has only 50% occupancy. Fe-S = 1.879(1), Ba-S = 2.626-3.367 A.

BARIUM PHOSPHORUS SILICON
GERMANIUM PHOSPHORUS STRONTIUM
BARIUM GERMANIUM PHOSPHORUS
Ba_4SiP_4, Sr_4GeP_4, Ba_4GeP_4

B. EISENMANN, H. JORDAN and H. SCHÄFER, 1982. Mater. Res. Bull., **17**, 95-99.

Cubic, $P\bar{4}3n$, a = 13.023, 12.506, 13.044 A, Z = 8. Mo radiation, R = 0.078, 0.108, 0.064 for 1079, 537, 605 reflexions. Ba or Sr(1) in 8(e): x,x,x, x = 0.6463, 0.6455, 0.6457; Ba or Sr(2) in 24(i): 0.4069,0.1464,0.3671; 0.4056,0.1459,0.3679; 0.4059,0.1456,0.3674; Si or Ge(1) in 2(a): 0,0,0; Si or Ge(2) in 6(d): 1/4,0,1/2; P(1) in 8(e): x = 0.3996, 0.3904, 0.3976; P(2) in 24(i): 0.1546,0.1082,0.4000; 0.1487,0.1172,0.3954; 0.1534,0.1107,0.3984.

Isolated SiP_4 or GeP_4 tetrahedra, linked by Ba or Sr coordinated to six P (Fig. 1). Ca_4SiP_4 and Sr_4SiP_4 are isostructural.

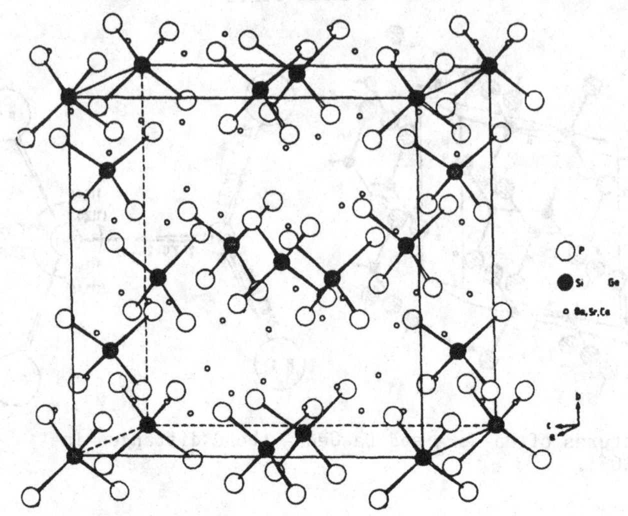

Fig. 1. Structure of Ba_4SiP_4 and isotypes.

BARIUM PLATINUM SULPHUR
$BaPt_2S_3$

J. HUSTER, 1982. J. Less-Common Metals, **84**, 125-131.

Tetragonal, $P4_12_12$, a= 6.7648, c = 12.3489 A, Z = 4. Ag radiation, R = 0.028 for 235 reflexions.

Three-dimensional linkage of PtS_4 square planes (Fig. 1), with Ba coordinated to six S. Pt-S = 2.28-2.34, Ba-S = 3.17-3.51 A.

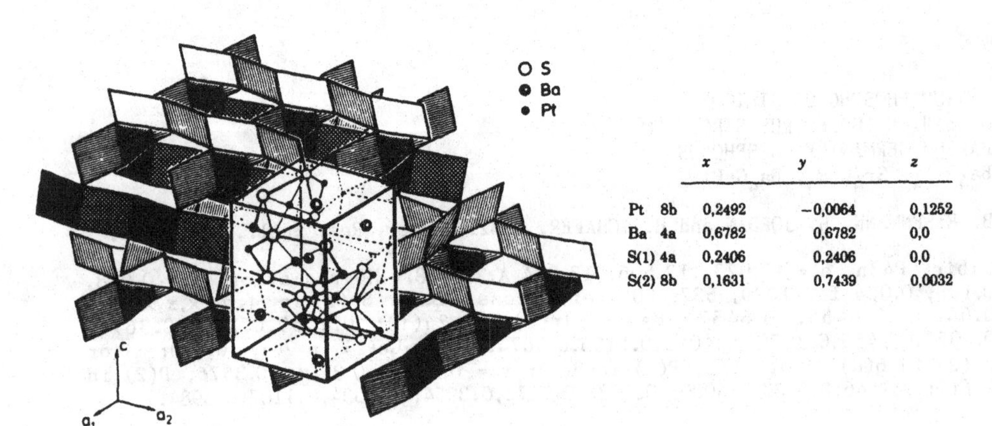

	x	y	z
Pt 8b	0,2492	-0,0064	0,1252
Ba 4a	0,6782	0,6782	0,0
S(1) 4a	0,2406	0,2406	0,0
S(2) 8b	0,1631	0,7439	0,0032

Fig. 1. Structure of $BaPt_2S_3$.

BARIUM RHENIUM SULPHUR
$Ba_2Re_6S_{11}$

W. BRONGER and H.-J. MIESSEN, 1982. J. Less-Common Metals, **83**, 29-38.

Monoclinic, C2/c, a = 15.860, b = 9.158, c = 11.916 A, β = 116.32°, Z = 4. Ag radiation, R = 0.04 for 1968 reflexions.

Re_6S_8 clusters are linked via additional S atoms to form a three-dimensional framework, with Ba ions in holes (Fig. 1). Re-Re = 2.610-2.638 A. The Sr compound is isostructural.

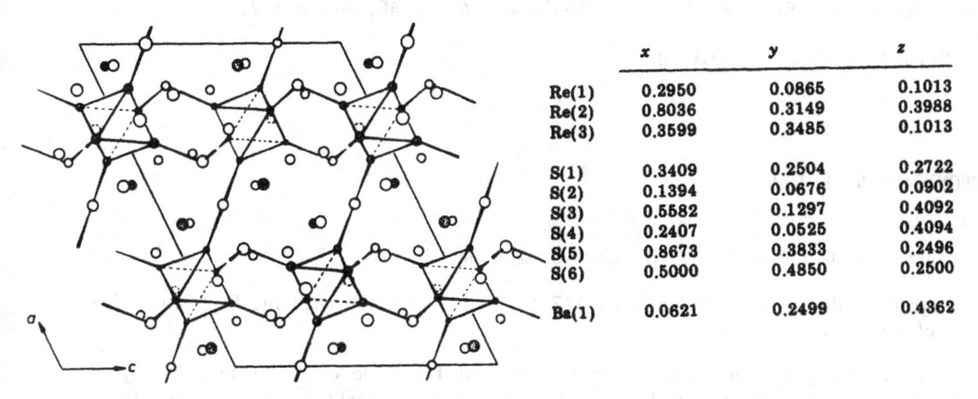

	x	y	z
Re(1)	0.2950	0.0865	0.1013
Re(2)	0.8036	0.3149	0.3988
Re(3)	0.3599	0.3485	0.1013
S(1)	0.3409	0.2504	0.2722
S(2)	0.1394	0.0676	0.0902
S(3)	0.5582	0.1297	0.4092
S(4)	0.2407	0.0525	0.4094
S(5)	0.8673	0.3833	0.2496
S(6)	0.5000	0.4850	0.2500
Ba(1)	0.0621	0.2499	0.4362

Fig. 1. Structure of $Ba_2Re_6S_{11}$.

BISMUTH EUROPIUM SULPHUR
$BiEu_2S_4$

P. LEMOINE, D. CARRÉ and M. GUITTARD, 1982. Acta Cryst., **B38**, 727-729.

Orthorhombic, Pnam, a = 11.579, b = 14.523, c = 4.089 A, Z = 4. Mo radiation, R = 0.05 for 795 reflexions.

Eu(II) and Eu(III) ions have 8- and 7-coordinations, respectively, and Bi has octahedral coordination (Fig. 1). [The structure is basically Bi_2PbS_4-type (1).]

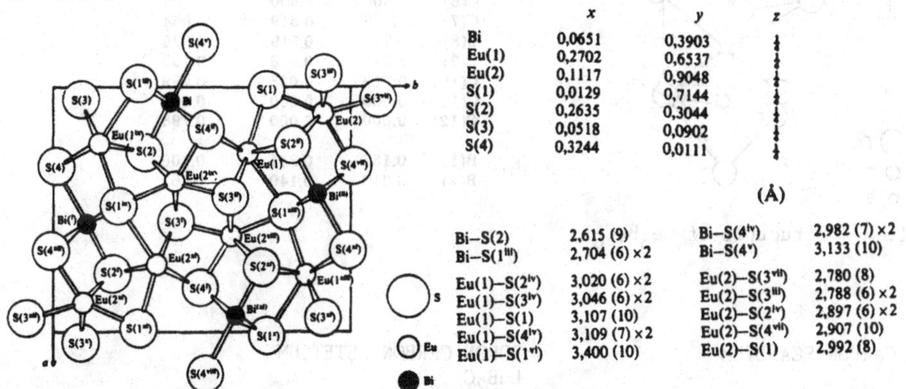

	x	y	z
Bi	0,0651	0,3903	$\frac{1}{4}$
Eu(1)	0,2702	0,6537	$\frac{1}{4}$
Eu(2)	0,1117	0,9048	$\frac{1}{4}$
S(1)	0,0129	0,7144	$\frac{1}{4}$
S(2)	0,2635	0,3044	$\frac{1}{4}$
S(3)	0,0518	0,0902	$\frac{1}{4}$
S(4)	0,3244	0,0111	$\frac{1}{4}$

(Å)

Bond	Distance	Bond	Distance
Bi–S(2)	2,615 (9)	Bi–S(4^{iv})	2,982 (7) ×2
Bi–S(1^{iii})	2,704 (6) ×2	Bi–S(4^{v})	3,133 (10)
Eu(1)–S(2^{iv})	3,020 (6) ×2	Eu(2)–S(3^{vii})	2,780 (8)
Eu(1)–S(3^{iv})	3,046 (6) ×2	Eu(2)–S(3^{iii})	2,788 (6) ×2
Eu(1)–S(1)	3,107 (10)	Eu(2)–S(2^{iv})	2,897 (6) ×2
Eu(1)–S(4^{iv})	3,109 (7) ×2	Eu(2)–S(4^{vii})	2,907 (10)
Eu(1)–S(1^{vi})	3,400 (10)	Eu(2)–S(1)	2,992 (8)

Fig. 1. Structure of $BiEu_2S_4$.

1. Structure Reports, **15**, 239; **27**, 81.

BISMUTH SELENIUM STRONTIUM
$BiSrSe_3$

R. COOK and H. SCHÄFER, 1982. Rev. Chim. Minér., **19**, 19-27.

Orthorhombic, $P2_12_12_1$, a = 33.552, b = 15.755, c = 4.2609 A, Z = 16. Mo radiation, R = 0.053 for 2483 reflexions [coordinates for Se(5) are misprinted; correct values are 0.3462,0.3571,0.0188].

$BiSe_6$ octahedra link to form infinite $Bi_8Se_{18}^{12-}$ strands which are connected by Se_3^{2-} anions (similar arrays occur in $BaBiSi_3$ (1)); Sr ions have 9-coordination. Bi-Se = 2.65-3.43, Se-Se = 2.40, 2.48, Sr-Se = 2.90-3.52 A.

1. Structure Reports, **46**A, 8.

BORON CARBON CERIUM
$Ce_5B_2C_6$

J. BAUER and O. BARS, 1982. J. Less-Common Metals, **83**, 17-27.

Tetragonal, P4, a = 8.418, c = 12.077 A, Z = 4. Mo radiation, R = 0.052 for 640 reflexions.

New structure type, which can be derived from the CaC_2 and $Sc_{15}C_{19}$ types by filling the distorted octahedral voids alternately with C atoms and C_2 pairs; B atoms are needed to stabilize the structure (Fig. 1).

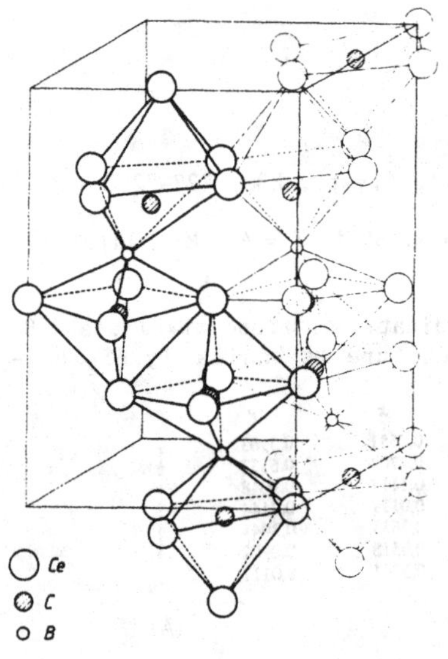

Fig. 1. Structure of $Ce_5B_2C_6$.

	x	y	z
Ce(1)	0.4029	0.2072	−0.0304
Ce(2)	0.2053	0.4036	0.4631
Ce(3)	0.2941	0.0976	0.2494
Ce(4)	0.0971	0.2960	0.7561
Ce(5)	0.5000	0.5000	0.2300
Ce(6)	0.5000	0.5000	0.7390
Ce(7)	0.0000	0.0000	0.4913
Ce(8)	0.0000	0.0000	0.0000
C(1)	0.090	0.291	−0.024
C(2)	0.288	0.103	0.445
C(3)	0.191	0.386	0.224
C(4)	0.379	0.187	0.689
C(5)	0.500	0.500	0.429
C(6)	0.500	0.500	0.538
C(7)	0.538	0.519	−0.064
C(8)	0.538	0.519	0.025
C(9)	0.019	0.038	0.199
C(10)	0.019	0.038	0.289
C(11)	0.000	0.000	0.691
C(12)	0.000	0.000	0.798
B(1)	0.153	0.341	0.106
B(2)	0.310	0.140	0.563

BORON CARBON SCANDIUM
ScB_2C

BORON CARBON LUTETIUM
LuB_2C

J. BAUER, 1982. J. Less-Common Metals, **87**, 45-52.

Tetragonal, $P4_2/mbc$, a = 6.651, 6.7546, c = 6.763, 7.1781 A, Z = 8. Powder data for the Sc compound. Sc in 8(g): x,1/2+x,1/4, x = 0.313; B(1), B(2), and C in 8(h): x,y,0, x = 0.095, 0.140, 0.456, y = 0.595, 0.035, 0.322 [parameters for YB_2C]. YB_2C-type (1).

1. Structure Reports, 37A, 33; **44A**, 95.

BORON LANTHANUM RHENIUM
$La_2Re_3B_7$

Ju.B. KUZ'MA, S.I. MIKHALENKO, B.Ja. KOTUR and Ja.P. JARMOLJUK, 1982. Dopov. Akad. Nauk Ukr.RSR, Ser. B: Geol., Khim. Biol. Nauki, No. 3, 24-27.

Orthorhombic, Pcca, a = 7.681, b = 6.773, c = 11.658 A, Z = 4. Mo radiation, R = 0.098 for 874 reflexions.

The structure (Fig. 1) contains zigzag chains of B atoms. Coordination numbers are: La = 23, Re = 13, 14, B = 7, 8, 9.

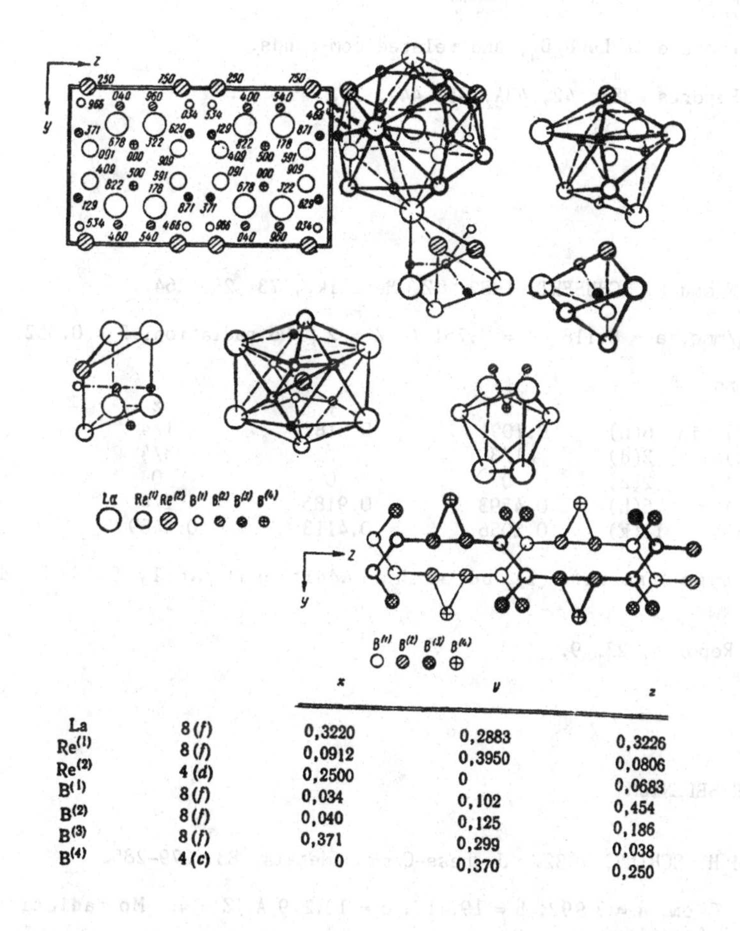

		x	y	z
La	8(f)	0,3220	0,2883	0,3226
Re(1)	8(f)	0,0912	0,3950	0,0806
Re(2)	4(d)	0,2500	0	0,0683
B(1)	8(f)	0,034	0,102	0,454
B(2)	8(f)	0,040	0,125	0,186
B(3)	8(f)	0,371	0,299	0,038
B(4)	4(c)	0	0,370	0,250

Fig. 1. Structure of $La_2Re_3B_7$.

BORON LUTETIUM RHODIUM
$LuRh_4B_4$

K. YVON and D.C. JOHNSTON, 1982. Acta Cryst., **B38**, 247-250.

Orthorhombic, Ccca, a = 7.410, b = 22.26, c = 7.440 A, Z = 12. Mo radiation, R = 0.05 for 388 reflexions.

NaCl-type arrangement of tetrahedral Rh_4 clusters and Lu atoms, with B-B pairs (Fig. 1), as in related compounds (**1**). Rh-Rh = 2.662-3.110(5), Lu-Rh = 2.862-3.207(4), B-B = 1.4(1) A.

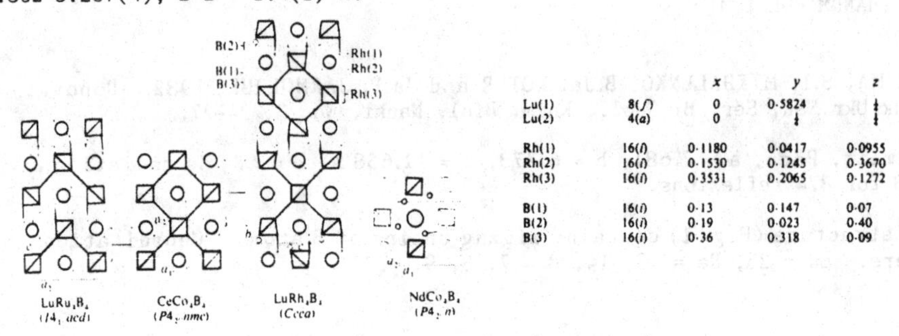

		x	y	z
Lu(1)	8(f)	0	0·5824	1/4
Lu(2)	4(a)	0	1/4	1/4
Rh(1)	16(i)	0·1180	0·0417	0·0955
Rh(2)	16(i)	0·1530	0·1245	0·3670
Rh(3)	16(i)	0·3531	0·2065	0·1272
B(1)	16(i)	0·13	0·147	0·07
B(2)	16(i)	0·19	0·023	0·40
B(3)	16(i)	0·36	0·318	0·09

Fig. 1. Structure of $LuRh_4B_4$, and related compounds.

1. Structure Reports, **38A**, 42; **43A**, 35; **44A**, 36.

CADMIUM COPPER
$Cd_{10}Cu_3$

T. RAJASEKHARAN and K. SCHUBERT, 1982. Z. Metallk., **73**, 262-264.

Hexagonal, $P6_3/mmc$, a = 8.118, c = 8.751 A, Z = 2. Mo radiation, R = 0.052.

Atomic positions

			x	y	z
6	Cu(1)	in 6(h)	0.1091	0.2182	1/4
0.8	Cu(2)	2(d)	1/3	2/3	3/4
1.6	Cd(1)	2(a)	0	0	0
6	Cd(2)	6(h)	0.4593	0.9185	1/4
12	Cd(3)	12(k)	0.2056	0.4113	0.9469

$Al_{10}Mn_3$-type structure (**1**), but with an additional partly-filled 2(d) site.

1. Structure Reports, **23**, 9.

CAESIUM COPPER SELENIUM
$Cs_2Cu_5Se_4$

W. BRONGER and H. SCHILS, 1982. J. Less-Common Metals, **83**, 279-285.

Orthorhombic, Cmcm, a = 3.992, b = 19.417, c = 13.219 A, Z = 4. Mo radiation, R = 0.044 for 265 reflexions.

The structure (Fig. 1) is related to that of $Cs_3Cu_8Se_6$ (1), with close-packed layers of Se and Cs, with Cu in Se holes; Cu-Cu distances (2.46-2.84 A) are only a little longer than in Cu metal.

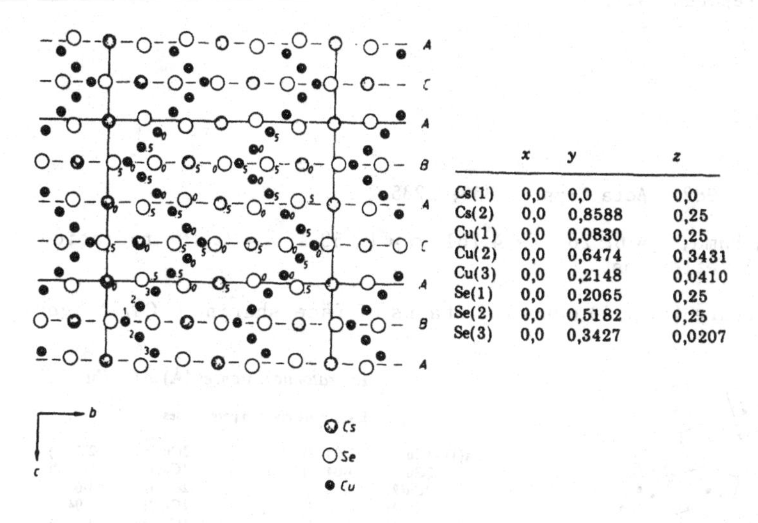

	x	y	z
Cs(1)	0,0	0,0	0,0
Cs(2)	0,0	0,8588	0,25
Cu(1)	0,0	0,0830	0,25
Cu(2)	0,0	0,6474	0,3431
Cu(3)	0,0	0,2148	0,0410
Se(1)	0,0	0,2065	0,25
Se(2)	0,0	0,5182	0,25
Se(3)	0,0	0,3427	0,0207

Fig. 1. Structure of $Cs_2Cu_5Se_4$.

1. Structure Reports, **45**A, 47.

CAESIUM SULPHUR
Cs_2S_5

P. BÖTTCHER and K. KRUSE, 1982. J. Less-Common Metals, **83**, 115-125.

Orthorhombic, $P2_12_12_1$, a = 7.148, b = 18.48, c = 6.780 A, Z = 4. Ag and Mo radiations, R = 0.066, 0.037 for 2055, 627 reflexions.

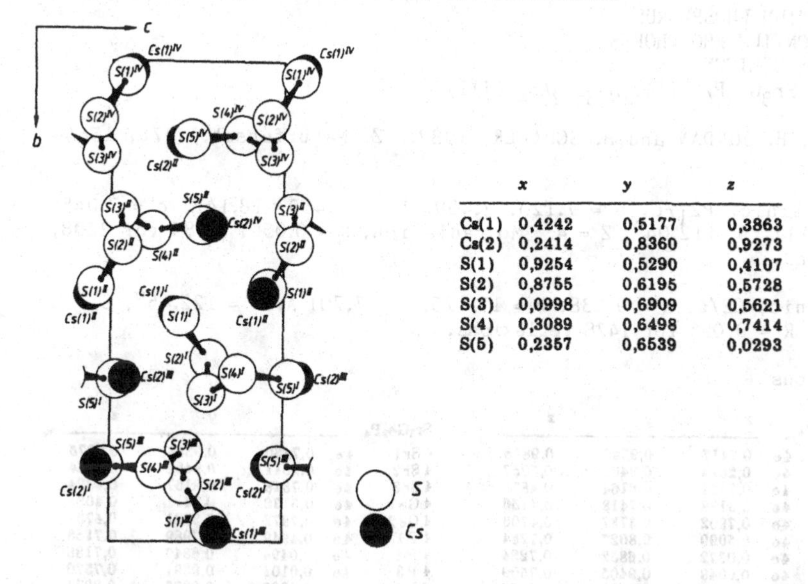

	x	y	z
Cs(1)	0,4242	0,5177	0,3863
Cs(2)	0,2414	0,8360	0,9273
S(1)	0,9254	0,5290	0,4107
S(2)	0,8755	0,6195	0,5728
S(3)	0,0998	0,6909	0,5621
S(4)	0,3089	0,6498	0,7414
S(5)	0,2357	0,6539	0,0293

Fig. 1. Structure of Cs_2S_5.

Isostructural with K_2S_5 (**1**), with S_5^{2-} chains (Fig. 1).

1. Structure Reports, **42**A, 120.

CALCIUM COPPER
Ca_2Cu

M.L. FORNASINI, 1982. Acta Cryst., B**38**, 2235-2236.

Orthorhombic, Pnma, a = 6.126, b = 4.161, c = 14.53 A, Z = 4. Mo radiation, R = 0.082 for 118 reflexions.

The structure (Fig. 1) contains chains of face-sharing $CuCa_6$ trigonal prisms.

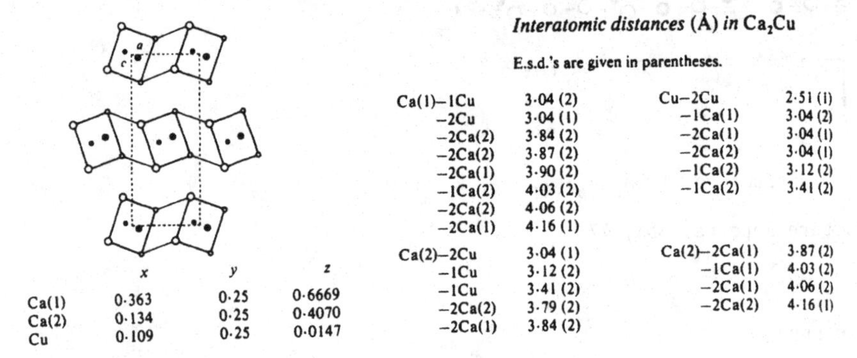

	x	y	z
Ca(1)	0·363	0·25	0·6669
Ca(2)	0·134	0·25	0·4070
Cu	0·109	0·25	0·0147

Interatomic distances (Å) *in* Ca_2Cu

E.s.d.'s are given in parentheses.

Ca(1)–1Cu	3·04 (2)	Cu–2Cu	2·51 (1)
–2Cu	3·04 (1)	–1Ca(1)	3·04 (2)
–2Ca(2)	3·84 (2)	–2Ca(1)	3·04 (1)
–2Ca(2)	3·87 (2)	–2Ca(2)	3·04 (1)
–2Ca(1)	3·90 (2)	–1Ca(2)	3·12 (2)
–1Ca(2)	4·03 (2)	–1Ca(2)	3·41 (2)
–2Ca(2)	4·06 (2)		
–2Ca(1)	4·16 (1)		
Ca(2)–2Cu	3·04 (1)	Ca(2)–2Ca(1)	3·87 (2)
–1Cu	3·12 (2)	–1Ca(1)	4·03 (2)
–1Cu	3·41 (2)	–2Ca(1)	4·06 (2)
–2Ca(2)	3·79 (2)	–2Ca(2)	4·16 (1)
–2Ca(1)	3·84 (2)		

Fig. 1. Structure of Ca_2Cu.

CALCIUM GERMANIUM PHOSPHORUS
GERMANIUM STRONTIUM PHOSPHORUS
ARSENIC BARIUM SILICON
$Ca_3Ge_2P_4$ (I), $Sr_3Ge_2P_4$ (II), $Ba_3Si_2As_4$ (III)

B. EISENMANN, H. JORDAN and H. SCHÄFER, 1982. Z. Naturforsch., **37**B, 1564-1568.

I, II, monoclinic, $P2_1/c$, a = 7.123, 7.459, b = 17.459, 18.142, c = 7.045, 7.259 A, ß = 111.79, 112.04°, Z = 4. Mo radiation, R = 0.050, 0.085 for 2208, 2365 reflexions.

III, monoclinic, C2/c, a = 9.538, b = 17.425, c = 7.701 A, ß = 122.45°, Z = 4. Mo radiation, R = 0.056 for 1478 reflexions.

Atomic positions

$Ca_3Ge_2P_4$		x	y	z
4 Ca 1	4e	0,7417	0,9390	0,9878
4 Ca 2	4e	0,2624	0,9405	0,5057
4 Ca 3	4e	0,7857	0,8164	0,4879
4 Ge 1	4e	0,3188	0,7418	0,4156
4 Ge 2	4e	0,7902	0,3737	0,4706
4 P 1	4e	0,5099	0,8027	0,7264
4 P 2	4e	0,0372	0,6852	0,7224
4 P 3	4e	0,0048	0,9405	0,7509
4 P 4	4e	0,4995	0,9329	0,2455

$Sr_3Ge_2P_4$		x	y	z
4 Sr 1	4e	0,7430	0,9358	0,9875
4 Sr 2	4e	0,2611	0,9422	0,5054
4 Sr 3	4e	0,7848	0,8157	0,4894
4 Ge 1	4e	0,3139	0,7411	0,4084
4 Ge 2	4e	0,7872	0,3721	0,4734
4 P 1	4e	0,4949	0,7989	0,7168
4 P 2	4e	0,0494	0,6846	0,7188
4 P 3	4e	0,0101	0,9391	0,7570
4 P 4	4e	0,4906	0,9307	0,2374

$Ba_3Si_2As_4$		x	y	z
4 Ba1	4e	0,5	0,3098	0,25
4 Ba2	4e	0,5	0,0659	0,25
4 Ba3	4e	0,5	0,4436	0,75
8 Si	8f	0,1483	0,3185	0,3435
8 As1	8f	0,2465	0,1983	0,2960
8 As2	8f	0,2602	0,0726	0,7458

Both types of structure contain X_3M-MX_3 units (M = Ge, Si; X = P, As), linked into different kinds of $M_2X_4^{6-}$ chains (Fig. 1), with chains connected by alkali-metal cations. [See this volume, p. 8 for isostructural materials.]

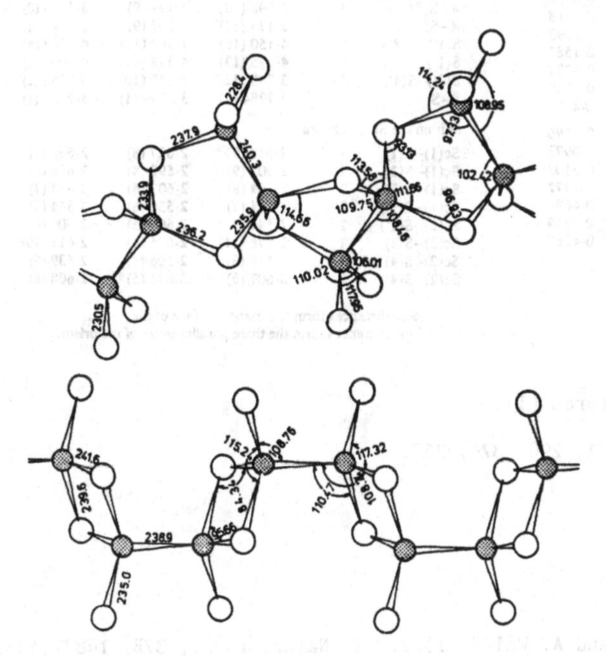

Fig. 1. $M_2X_4^{6-}$ chains in $Ca_3Ge_2P_4$ ($Sr_3Ge_2P_4$) and $Ba_3Si_2As_4$.

CALCIUM SCANDIUM SULPHUR
STRONTIUM SCANDIUM SULPHUR
LEAD SCANDIUM SULPHUR
MSc_2S_4 (M = Ca, Sr, Pb)

D.J.W. IJDO, 1982. Acta Cryst., B38, 1549-1551.

Orthorhombic, Pnam, a = 11.5014, 11.6352, 11.6595, b = 13.4695, 13.6523, 13.6933, c = 3.7284, 3.7799, 3.7531 A, Z = 4. Neutron powder data.

Atomic positions (z = 1/4) and interatomic distances (A)

		x	y
Ca	4(c)	0·7602	0·6658
Sc(1)	4(c)	0·4149	0·0972
Sc(2)	4(c)	0·4413	0·6110
S(1)	4(c)	0·2021	0·1657
S(2)	4(c)	0·1281	0·4673
S(3)	4(c)	0·5308	0·7864
S(4)	4(c)	0·4065	0·4263
Sr	4(c)	0·7609	0·6655
Sc(1)	4(c)	0·4185	0·0988
Sc(2)	4(c)	0·4394	0·6093
S(1)	4(c)	0·2037	0·1583
S(2)	4(c)	0·1255	0·4727
S(3)	4(c)	0·5262	0·7868
S(4)	4(c)	0·4118	0·4272
Pb	4(c)	0·7605	0·6696
Sc(1)	4(c)	0·4178	0·0973
Sc(2)	4(c)	0·4412	0·6103
S(1)	4(c)	0·2029	0·1572
S(2)	4(c)	0·1245	0·4694
S(3)	4(c)	0·5285	0·7859
S(4)	4(c)	0·4097	0·4268

Primed atoms have $z = \frac{3}{4}$ or $z = -\frac{1}{4}$, unprimed atoms have $z = \frac{1}{4}$. The second column in each grouping indicates the bond multiplicity.

		$CaSc_2S_4$	$SrSc_2S_4$	$PbSc_2S_4$
Within the *A* polyhedron				
A–S(2′)	2	2·888 (9)	2·980 (7)	2·990 (8)
A–S(4′)	2	2·947 (8)	3·036 (6)	3·033 (7)
A–S(1′)	2	2·969 (9)	3·087 (8)	3·055 (8)
A–S(3)	1	3·098 (12)	3·194 (9)	3·139 (10)
A–S(3)	1	3·179 (12)	3·154 (9)	3·184 (11)
S(1′)–S(2′)	*	4·150 (16)	4·389 (13)	4·372 (14)
S(1′)–S(4′)	*	4·225 (13)	4·398 (11)	4·409 (13)
S(2′)–S(4′)	*	3·248 (12)	3·389 (10)	3·375 (12)
S–S	†	3·7284 (1)	3·7799 (1)	3·7531 (1)
Within the Sc octahedra				
Sc(1)–S(1)	1	2·615 (9)	2·627 (8)	2·636 (8)
Sc(1)–S(2)	1	2·603 (9)	2·599 (8)	2·678 (9)
Sc(1)–S(2′)	2	2·604 (8)	2·607 (6)	2·614 (7)
Sc(1)–S(3′)	2	2·514 (7)	2·536 (6)	2·544 (7)
Sc(2)–S(1′)	2	2·595 (7)	2·606 (5)	2·600 (6)
Sc(2)–S(3)	1	2·578 (11)	2·625 (8)	2·611 (10)
Sc(2)–S(4)	1	2·519 (9)	2·506 (8)	2·539 (8)
Sc(2)–S(4′)	2	2·607 (6)	2·611 (5)	2·608 (6)

* These distances form the triangular face of the prism.
† These distances form the three parallel edges of the prism.

$CaFe_2O_4$-type structures (1).

1. Structure Reports, **21**, 290; **37**A, 257.

CALCIUM SILICON
$CaSi_2$

J. EVERS, G. OEHLINGER and A. WEISS, 1982. Z. Naturforsch., **37**B, 1487-1488.

Tetragonal, $I4_1/amd$, a = 4.283, c = 13.53 A, Z = 4. Mo radiation, R = 0.048 for 218 reflexions. Ca in 4(a): 0,0,0; Si in 8(e): 0,0,0.41505. High-temperature, high-pressure phase, with α-$ThSi_2$-type structure (1). Ca-Si = 3.09, 3.24, Ca-Ca = 4.00, 4.28, Si-Si = 2.30, 2.40 A.

1. Structure Reports, **9**, 121.

CALCIUM SILVER
CALCIUM GOLD
AgCa, AuCa

STRONTIUM ZINC
SrZn

F. MERLO, 1982. J. Less-Common Metals, **86**, 241-246.

AgCa, AuCa, orthorhombic, Cmcm, a = 4.058, 3.961, b = 11.457, 11.075, c = 4.654, 4.576 A, Z = 4. Mo radiation, R = 0.013 for 122 reflexions for AgCa; powder data for AuCa. Atoms in 4(c): 0,y,1/4, y(Ag) = 0.4260, y(Ca) = 0.1424; y(Au) = 0.42, y(Ca) = 0.14. CrB-type (1). Ag-Ag = 2.88, Ag-Ca = 3.19-3.25, Ca-Ca = 3.95-4.65 A.

SrZn, orthorhombic, Pnma, a = 8.724, b = 4.607, c = 6.417 A, Z = 4. Mo radiation, R = 0.026 for 416 reflexions. Atoms in 4(c): x,1/4,z, x = 0.1784, 0.0279, z = 0.1172, 0.6008, for Sr, Zn. FeB-type (2). Zn-Zn = 2.69, Zn-Sr = 3.35-3.56, Sr-Sr = 4.14-4.68 A.

1. Structure Reports, **12**, 30.
2. Strukturbericht, **2**, 7, 241; **3**, 12, 619.

CARBON CERIUM NICKEL
$Ce_2Ni_{22}C_3$

O.I. BODAK, E.P.MARUSIN, V.S. FUNDAMENSKIJ and V.A. BRUSKOV, 1982, Kristallografija, **27**, 1098-1101 [Soviet Physics - Crystallography, **27**, 657-659].

Orthorhombic, Cmca, a = 11.384, b = 15.024, c = 14.671 A, Z = 8. Mo radiation, R = 0.082 for 890 reflexions.

The structure is shown in Fig. 1.

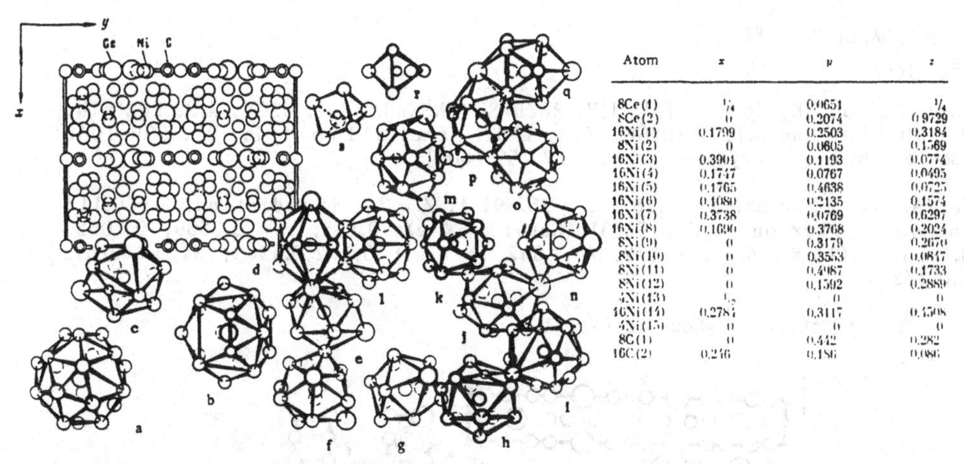

Atom	x	y	z
8Ce(1)	1/4	0.0651	1/4
8Ce(2)	0	0.2074	0.9729
16Ni(1)	0.1799	0.2503	0.3184
8Ni(2)	0	0.0605	0.1569
16Ni(3)	0.3901	0.1193	0.0774
16Ni(4)	0.1747	0.0767	0.0495
16Ni(5)	0.1765	0.4638	0.0725
16Ni(6)	0.1080	0.2135	0.1574
16Ni(7)	0.3738	0.0769	0.6297
16Ni(8)	0.1690	0.3768	0.2024
8Ni(9)	0	0.3179	0.2670
8Ni(10)	0	0.3553	0.0847
8Ni(11)	0	0.4987	0.1733
8Ni(12)	0	0.1592	0.2889
4Ni(13)	1/2	0	0
16Ni(14)	0.2784	0.3117	0.4508
4Ni(15)	0	0	0
8C(1)	0	0.442	0.282
16C(2)	0.246	0.186	0.086

Projection of structure of $Ce_2Ni_{22}C_3$ on xy plane, and coordination polyhedra of atoms. a) Ce (1); b) Ce (2); c) Ni (1); d) Ni (2); e) Ni (3); f) Ni (4); g) Ni (5); h) Ni (6); i) Ni (7); j) Ni (8); k) Ni (9); l) Ni (10); m) Ni (11); n) Ni (12); o) Ni (13); p) Ni (14); q) Ni (15); r) C (1); s) C (2).

Fig. 1. Structure of $Ce_2Ni_{22}C_3$.

CARBON CHROMIUM VANADIUM
BORON CARBON CHROMIUM
CARBON CHROMIUM NITROGEN
$(Cr_{0.54}V_{0.46})(Cr_{0.78}V_{0.22})_2C_{1.44}$, $Cr_3(B_{0.44}C_{0.56})C_{0.85}$, $Cr_3C(C_{0.52}N_{0.48})$

I. H. NOWOTNY, P. ROGL and J.C. SCHUSTER, 1982. J. Solid State Chem., **44**, 126-133.
II. P. ROGL, B. KUNSCH, P. ETTMAYER, H. NOWOTNY and W. STEURER, 1982. Z. Kristallogr., **160**, 275-284.

Orthorhombic, Cmcm, a = 2.876, 2.857, 2.833, b = 9.310, 9.233, 9.249, c = 6.987, 6.967, 6.937 A, Z = 4. Neutron powder data. Filled Re_3B-type, as previously described (1), with the smaller non-metal atoms preferring octahedral holes in the metal lattice.

1. Structure Reports, **31A**, 26.

CARBON ERBIUM SILICON
$Er_5Si_3C_{0.7}$, $Er_5Si_3C_{1.0}$

G.M.Y. AL-SHAHERY, D.W. JONES, I.J. McCOLM and R. STEADMAN, 1982. J. Less-Common Metals, **87**, 99-108.

$Er_5Si_3C_{0.7}$, hexagonal, $P\bar{6}m2$, a = 14.41, c = 6.322 A, cell contents = $Er_{28}Si_{16}C_4$. R = 0.12 for 594 reflexions.

$Er_5Si_3C_{1.0}$, trigonal, $P\bar{3}m1$, a = 14.4, c = 18.0 A, Z = 18. R = 0.19 for 684 reflexions.

Superstructure variants of the $D8_8$ type of Er_5Si_3 (1), with **a**$\sqrt{3}$ for both phases, and tripled **c** for the second phase. C occupies octahedral sites.

1. Strukturbericht, **4**, 24.

CERIUM GALLIUM NICKEL
$Ce_2Ga_{10}Ni$

Ja.P. JARMOLJUK, Ju.N. GRIN', I.V. ROŽDESTVENSKAJA, O.A. USOV, A.M. KUZ'MIN, V.A. BRUSKOV and E.I. GLADYŠEVSKIJ, 1982. Kristallografija, **27**, 999-1001 [Soviet Physics - Crystallography, **27**, 599-600].

Tetragonal, I4/mmm, a = 4.262, c = 26.391 A, Z = 2. Mo radiation, R = 0.113 for 408 reflexions. Ce, Ga(3), Ga(4) in 4(e): 0,0,z, z = 0.3539, 0.1065, 0.1968; Ga(1) in 4(d): 0,1/2,1/4; Ga2 in 8(g): 0,1/2,0.4509; Ni in 2(b): 0,0,1/2.

The structure is shown in Fig. 1.

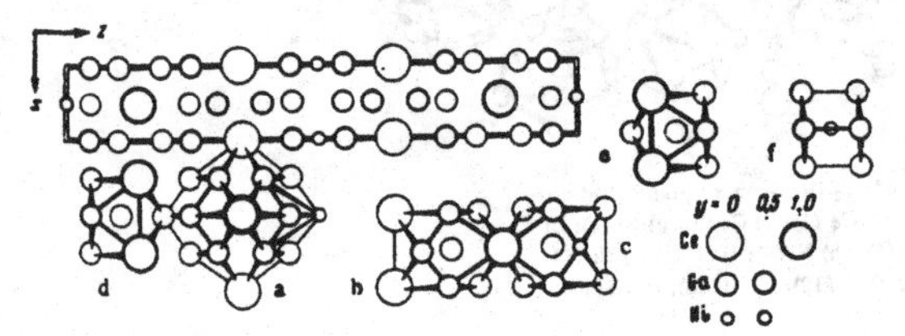

Projections of structure of $Ce_2Ga_{10}Ni$ and of coordination polyhedra of Ce (a), Ga (b-e), and Ni (f) atoms on xz plane.

		C.n.			C.n.
Ce–4Ce	4.262(3)		Ga^{2}–2Ce	3.331(4)	
–$4Ga^{1}$	3.473(3)		–$4Ga^{2}$	3.014(0)	
–$4Ga^{2}$	3.331(4)	20	–$2Ga^{3}$	2.614(5)	12
–$4Ga^{4}$	3.297(3)		–$2Ga^{2}$	2.591(6)	
–$4Ga^{3}$	3.190(3)		–2Ni	2.494(2)	
Ga^{1}–4Ce	3.473(3)		Ga^{3}–4Ce	3.190(3)	
–$4Ga^{1}$	3.014(0)	12	–$4Ga^{1}$	2.552(4)	9
–$4Ga^{4}$	2.552(4)		–Ga^{3}	2.383(11)	
			Ni–$8Ga^{2}$	2.494(2)	8

Fig. 1. Structure of $Ce_2Ga_{10}Ni$.

CERIUM HYDROGEN
$CeD_{2.26}$ (I), $CeD_{2.43}$ (II)

V.K. FEDOTOV, V.G. FEDOTOV, M.E. KOST and E.G. PONJATOVSKIJ, 1982. Fiz. Tverd. Tela, **24**, 2201-2208 [Soviet Physics - Solid State, **24**, 1252-1257].

Tetragonal, I4/mmm (I) and $I4_1md$ (II) superstructures (**a** x **a** x 2**a**) of the f.c.c. metal structure. D sites proposed from neutron powder data.

CHROMIUM COPPER PHOSPHORUS SULPHUR
$Cr_{0.50}Cu_{0.50}PS_3$

P. COLOMBET, A. LEBLANC, M. DANOT and J. ROUXEL, 1982. J. Solid State Chem., **41**, 174-184.

Monoclinic, C2/c, a = 5.916, b = 10.246, c = 13.415 A, β = 107.09°, Z = 8. Mo radiation, R = 0.056 for 1664 reflexions.

Atomic positions

	x	y	z
S(A)	0.2471	0.1823	0.3736
S(B)	0.2656	0.1724	0.8736
S(C)	0.7240	0.9956	0.3757
P	0.0533	0.3319	0.8344
Cr	0	0.3351	1/4
2.6 Cu(A)	0.0607	0.0021	0.3482
1.4 Cu(B)	0.4966	0.5020	0.2670

The structure can be derived from that of $FePS_3$ (**1**) by doubling **c**; it has as ABC sulphur stacking, with the van der Waals gap empty, and the octahedral sites of the filled layer occupied in an ordered manner by Cr, P_2 pairs, and Cu (Cu is distributed in two off-centre sites).

1. Structure Reports, **35**A, 138; **39**A, 76.

CHROMIUM IRON MOLYBDENUM NICKEL (σ PHASE)
$Cr_{11}Fe_{13}Mo_3Ni_3$

I. HJERTÉN, B.-O. MARINDER, A. SALWÉN and P.-E. WERNER, 1982. Acta Chem. Scand., A**36**, 203-206.

Monoclinic, P2 (pseudo-P4/mnm), a = 8.8915, b = 4.6088, c = 8.6818 A, β = 90.15°, Z = 1. Powder data. Slight distortion of the tetragonal σ-phase structure (**1**).

1. Structure Reports, **18**, 104.

CHROMIUM SCANDIUM SILICON
$Cr_4Sc_2Si_5$

B.Ya. KOTUR and M. SIKIRICA, 1982. J. Less-Common Metals, **83**, L29-L31.

Orthorhombic, Ibam, a = 7.585, b = 16.138, c = 4.932 A, Z = 4. Cu radiation, R = 0.060 for 252 reflexions.

Atomic positions and interatomic distances (A)

		x	*y*	*z*
(T_1) Sc	8j	0.1181	0.1421	0
(T_2) Cr(1)	8j	0.2460	0.4380	0
(T_3) Cr(2)	8g	0	0.3093	0.25
Si(1)	8j	0.2947	0.2888	0
Si(2)	8j	0.4223	0.0638	0
Si(3)	4a	0	0	0.25

Sc-Cr(1)	3.451(5)	Cr(1)-Sc	3.451(5)	Si(1)-2Si(1)	2.848(5)
-2Cr(2)	3.244(4)	-2Cr(1)	3.176(3)	-2Sc	2.786(4)
-Cr(1)	3.104(5)	-Sc	3.104(5)	-Sc	2.720(7)
-2Cr(2)	3.099(4)	-2Cr(2)	3.052(4)	-Sc	2.695(7)
-2Sc	3.048(3)	-2Sc	2.969(3)	-2Cr(2)	2.574(5)
-2Cr(1)	2.969(3)	-2Si(2)	2.777(3)	-2Cr(2)	2.540(6)
-2Si(1)	2.786(4)	-2Si(3)	2.497(3)	-Cr(1)	2.436(7)
-2Si(3)	2.753(3)	-Si(2)	2.455(7)		
-Si(1)	2.720(7)	-Si(1)	2.436(7)	Si(2)-2Cr(1)	2.777(3)
-Si(1)	2.695(7)	-Si(2)	2.398(7)	-2Si(2)	2.733(4)
-Si(2)	2.631(7)			-Sc	2.631(7)
		Cr(2)-2Sc	3.244(4)	-2Cr(2)	2.462(6)
		-2Sc	3.099(4)	-Cr(1)	2.455(7)
		-2Cr(1)	3.052(4)	-Cr(1)	2.398(7)
		-2Si(1)	2.574(5)	-Si(2)	2.373(9)
		-2Si(1)	2.540(6)		
		-2Cr(2)	2.466(1)	Si(3)-4Sc	2.753(3)
		-2Si(2)	2.462(6)	-4Cr(1)	2.497(3)
				-2Si(3)	2.466(1)

Ordered V_6Si_5-type (**1**) ($Cr_4Nb_2Si_5$-type (**2**)) [note that the space group settings are different in **1** and **2**].

1. Structure Reports, **38**A, 101.
2. Ibid., **33**A, 64.

CHROMIUM SILICON
Cr_3Si

J.-E. JØRGENSEN and S.E. RASMUSSEN, 1982. Acta Cryst., B**38**, 346-347.

Cubic, Pm3n, a = 4.5599 A, Z = 2. Neutron radiation, R = 0.018 for 21 reflexions. Cr in 6(c): 0,1/2,1/4; Si in 2(a): 0,0,0. A15-type, as previously described (1).

1. Strukturbericht, **3**, 628; Structure Reports, **20**, 78; **45**A, 118.

CHROMIUM SILICON ZIRCONIUM
$CrZrSi_2$

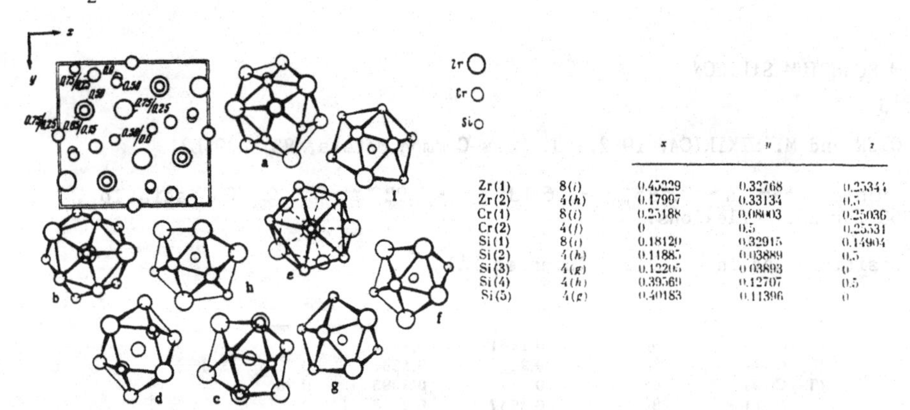

		x	y	z
Zr(1)	8(i)	0.45229	0.32768	0.25344
Zr(2)	4(h)	0.17997	0.33134	0.5
Cr(1)	8(i)	0.25188	0.08003	0.25036
Cr(2)	4(f)	0	0.5	0.25531
Si(1)	8(i)	0.18120	0.32915	0.14904
Si(2)	4(h)	0.11885	0.03889	0.5
Si(3)	4(g)	0.12205	0.03893	0
Si(4)	4(h)	0.39569	0.12707	0.5
Si(5)	4(g)	0.40183	0.11396	0

Projection of $ZrCrSi_2$ structure on xy plane and coordination polyhedra of atoms. a) Zr(1); b) Zr(2); c) Cr(1); d) Cr(2); e) Si(1); f) Si(3); g) Si(4); h) Si(4); i) Si(5).

Fig. 1. Structure of $CrZrSi_2$.

Ja.P. JARMOLJUK, M. SIKIRICA, L.G. AKSEL'RUD, L.A. LYSENKO and E.I. GLADYŠEVSKIJ, 1982. Kristallografija, **27**, 1090-1093 [Soviet Physics - Crystallography, **27**, 652-653].

Orthorhombic, Pbam, a = 9.874, b = 9.144, c = 7.998 A, Z = 12. Mo radiation, R = 0.061 for 1790 reflexions.

The structure is shown in Fig. 1.

CHROMIUM SILVER SULPHUR
$Ag_{0.37}Cr_{1.21}S_2$

K.D. BRONSEMA and G.A. WIEGERS, 1982. Acta Cryst., B**38**, 2229-2232.

Rhombohedral, $R\bar{3}m$, a = 3.4325, c = 37.190 A, Z = 6. Mo radiation, R = 0.102 for 908 reflexions.

The structure (Fig. 1) contains $(hcch)_3$ S layers forming CrS_2 sandwiches with Cr in octahedral coordination. Gaps between these sandwiches are occupied by partially-occupied Cr in octahedral and Ag in tetrahedral holes.

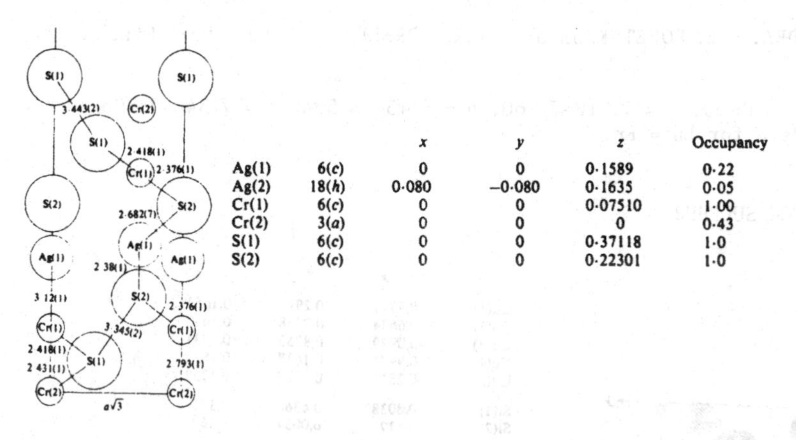

		x	y	z	Occupancy
Ag(1)	6(c)	0	0	0·1589	0·22
Ag(2)	18(h)	0·080	−0·080	0·1635	0·05
Cr(1)	6(c)	0	0	0·07510	1·00
Cr(2)	3(a)	0	0	0	0·43
S(1)	6(c)	0	0	0·37118	1·0
S(2)	6(c)	0	0	0·22301	1·0

Fig. 1. Structure of $Ag_{0.37}Cr_{1.21}S_2$.

COBALT GERMANIUM ZIRCONIUM
$Co_7Ge_6Zr_4$

O.I. ZJUBRIK, R.R. OLENIČ and Ja.P. JARMOLJUK, 1982. Dopov. Akad. Nauk Ukr.RSR, Ser. A: Fiz.-Mat. Tekh. Nauki, No. 11, 74-76.

Cubic, Im3m, a = 7.886 A, Z = 2. Powder data. Zr in 8(c): 1/4,1/4,1/4; Co(1) in 12(d): 1/4,0,1/2; Co(2) in 2(a): 0,0,0; Ge in 12(e): 0.317,0,0. The Hf compound is isostructural (a = 7.831 A).

$Re_7Si_6U_4$-type (**1**) (Fig. 1).

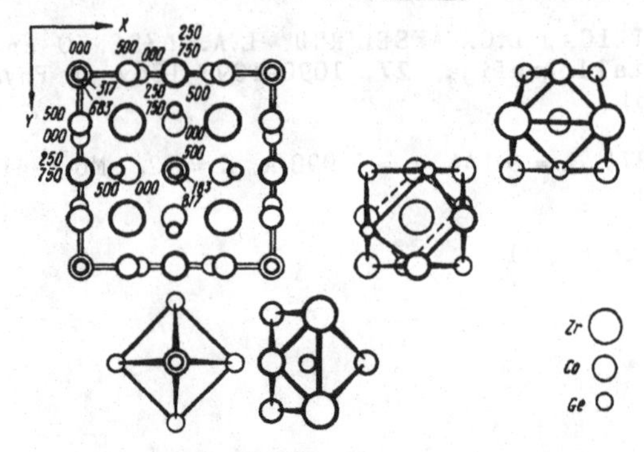

Fig. 1. Structure of $Co_7Ge_6Zr_4$.

1. Structure Reports, **44**A, 96.

COBALT LANTHANON TIN
CoLnSn

R.V. SKOLOZDRA, O.E. KORETSKAJA and Ju.K. GORELENKO, 1982. Ukr. Fiz. Ž., **27**, 263-266.

Orthorhombic, Pnma, a = 7.018-7.260, b = 4.456-4.534, c = 7.348-7.503 A, Z = 4. Powder data for Ln = Er.

COPPER SILICON SULPHUR
$Cu_5Si_2S_7$

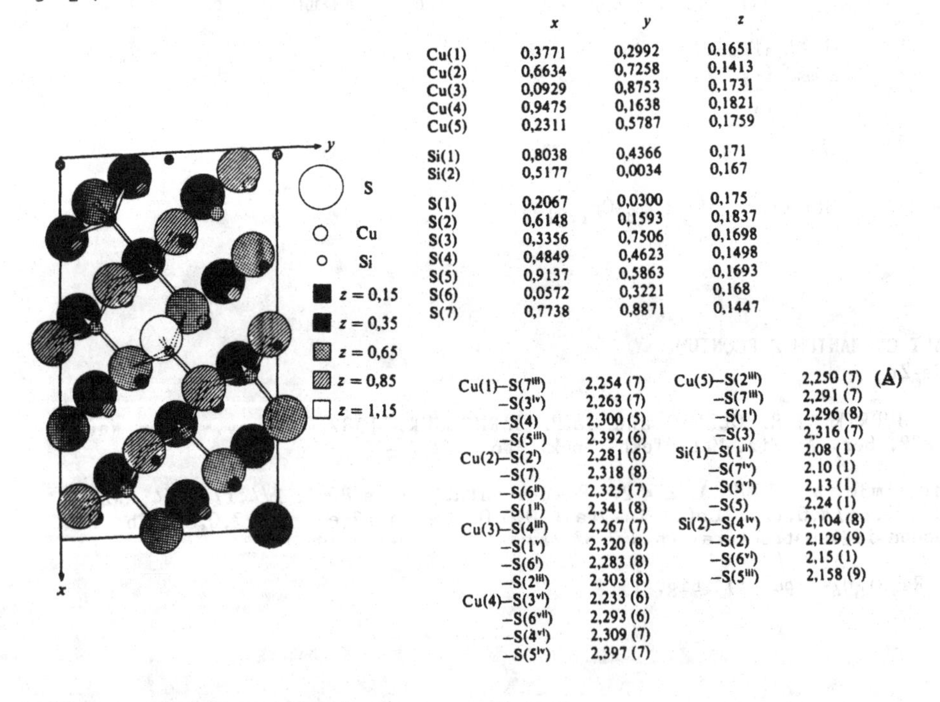

	x	y	z
Cu(1)	0,3771	0,2992	0,1651
Cu(2)	0,6634	0,7258	0,1413
Cu(3)	0,0929	0,8753	0,1731
Cu(4)	0,9475	0,1638	0,1821
Cu(5)	0,2311	0,5787	0,1759
Si(1)	0,8038	0,4366	0,171
Si(2)	0,5177	0,0034	0,167
S(1)	0,2067	0,0300	0,175
S(2)	0,6148	0,1593	0,1837
S(3)	0,3356	0,7506	0,1698
S(4)	0,4849	0,4623	0,1498
S(5)	0,9137	0,5863	0,1693
S(6)	0,0572	0,3221	0,168
S(7)	0,7738	0,8871	0,1447

Bond	(Å)	Bond	(Å)
Cu(1)–S(7^{iii})	2,254 (7)	Cu(5)–S(2^{iii})	2,250 (7)
–S(3^{iv})	2,263 (7)	–S(7^{iii})	2,291 (7)
–S(4)	2,300 (5)	–S(1^{i})	2,296 (8)
–S(5^{iii})	2,392 (6)	–S(3)	2,316 (7)
Cu(2)–S(2^{i})	2,281 (6)	Si(1)–S(1^{ii})	2,08 (1)
–S(7)	2,318 (8)	–S(7^{iv})	2,10 (1)
–S(6^{ii})	2,325 (7)	–S(3^{vi})	2,13 (1)
–S(1^{ii})	2,341 (8)	–S(5)	2,24 (1)
Cu(3)–S(4^{iii})	2,267 (7)	Si(2)–S(4^{iv})	2,104 (8)
–S(1^{v})	2,320 (8)	–S(2)	2,129 (9)
–S(6^{i})	2,283 (8)	–S(6^{vi})	2,15 (1)
–S(2^{iii})	2,303 (8)	–S(5^{iii})	2,158 (9)
Cu(4)–S(3^{vi})	2,233 (6)		
–S(6^{vii})	2,293 (6)		
–S(4^{vi})	2,309 (7)		
–S(5^{iv})	2,397 (7)		

Fig. 1. Structure of $Cu_5Si_2S_7$.

M. DOGGUY, S. JAULMES, P. LARUELLE and J. RIVET, 1982. Acta Cryst., B**38**, 2014-2016.

Monoclinic, Bb, a = 16.216, b = 9.597, c = 6.317 A, γ = 92.38°, Z = 4. Mo radiation, R = 0.066 for 1426 reflexions.

The structure (Fig 1) contains a compact hexagonal arrangement of S, with Si and Cu in tetrahedral holes; SiS_4 tetrahedra are linked to form Si_2S_7 groups.

COPPER SULPHUR TIN
$CuSn_{3.75}S_8$

S. JAULMES, M. JULIEN-POUZOL, J. RIVET, J.C. JUMAS and M. MAURIN, 1982. Acta Cryst., B**38**, 51-54.

Cubic, $F\bar{4}3m$, a = 10.393 A, Z = 4. Mo radiation, R = 0.060 for 115 reflexions. Cu in 4(c): 1/4,1/4,1/4; Sn, S(1), S(2) in 16(e): x,x,x, x = 0.6250, 0.383, 0.8742.

Defect spinel with Cu in 1/2 of the tetrahedral sites and Sn in 15/16 of the octahedral sites. Cu-S = 2.39(2), Sn-S = 2.52, 2.59(1) A.

DYSPROSIUM RHODIUM SILICON
$Dy_2Rh_3Si_5$

B. CHEVALIER, P. LEJAY, J. ETOURNEAU, M. VLASSE and P. HAGENMULLER, 1982. Mater. Res. Bull., **17**, 1211-1220.

Orthorhombic, Ibam, a = 9.80, b = 11.68, c not given [5.68] A, Z = 4. Powder data. Dy in 8(j): 0.267,0.135,0; Rh(1) in 4(a): 0,0,1/4; Rh(2) in 8(j): 0.107,0.358,0; Si(1) in 4(b): 1/2,0,1/4; Si(2) in 8(g): 0,0.217,1/4; Si(3) in 8(j): 0.340,0.396,0. $U_2Co_3Si_5$-type (**1**).

1. Structure Reports, **43A**, 50.

EUROPIUM SULPHUR TIN
Eu_2SnS_5

S. JAULMES, M. JULIEN-POUZOL, P. LARUELLE and M. GUITTARD, 1982. Acta Cryst., B**38**, 79-82.

Orthorhombic, Pmcb, a = 4.100, b = 15.621, c = 11.507 A, Z = 4. Mo radiation, R = 0.051 for 1086 reflexions.

Disordered Sn atoms have tetrahedral coordinations and Eu atoms are 9-coordinate (Fig. 1); three S atoms form a polysulphide anion, S_3^{2-}.

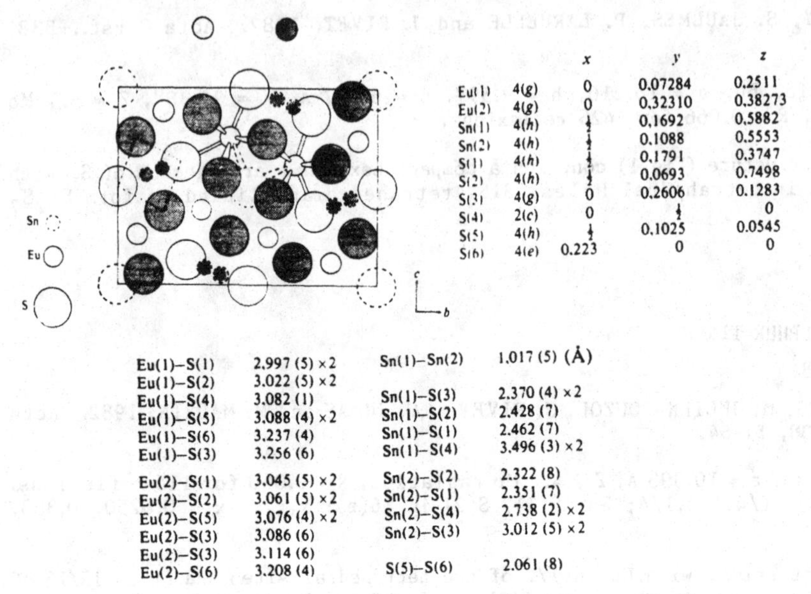

		x	y	z
Eu(1)	4(g)	0	0.07284	0.2511
Eu(2)	4(g)	0	0.32310	0.38273
Sn(1)	4(h)	½	0.1692	0.5882
Sn(2)	4(h)	½	0.1088	0.5553
S(1)	4(h)	½	0.1791	0.3747
S(2)	4(h)	½	0.0693	0.7498
S(3)	4(g)	0	0.2606	0.1283
S(4)	2(c)	0	½	0
S(5)	4(h)	½	0.1025	0.0545
S(6)	4(e)	0.223	0	0

Eu(1)–S(1)	2.997 (5) ×2	Sn(1)–Sn(2)	1.017 (5) (Å)
Eu(1)–S(2)	3.022 (5) ×2		
Eu(1)–S(4)	3.082 (1)	Sn(1)–S(3)	2.370 (4) ×2
Eu(1)–S(5)	3.088 (4) ×2	Sn(1)–S(2)	2.428 (7)
Eu(1)–S(6)	3.237 (4)	Sn(1)–S(1)	2.462 (7)
Eu(1)–S(3)	3.256 (6)	Sn(1)–S(4)	3.496 (3) ×2
Eu(2)–S(1)	3.045 (5) ×2	Sn(2)–S(2)	2.322 (8)
Eu(2)–S(2)	3.061 (5) ×2	Sn(2)–S(1)	2.351 (7)
Eu(2)–S(5)	3.076 (4) ×2	Sn(2)–S(4)	2.738 (2) ×2
Eu(2)–S(3)	3.086 (6)	Sn(2)–S(3)	3.012 (5) ×2
Eu(2)–S(3)	3.114 (6)		
Eu(2)–S(6)	3.208 (4)	S(5)–S(6)	2.061 (8)

Fig. 1. Structure of Eu_2SnS_5 (Sn atoms have occupancy = 0.5).

GALLIUM IRON SULPHUR (3R)
$Fe_2Ga_2S_5$

L. DOGGUY-SMIRI and NGUYEN-HUY-DUNG, 1982. Acta Cryst., B38, 372-375.

Rhombohedral, $R\bar{3}m$, a = 3.6508, c = 44.843 A, Z = 3. Mo radiation, R = 0.048 for 324 reflexions. Fe, Ga, S(2), S(3) in 6(c): 0,0,z, z = 0.13305, 0.27717, 0.22748, 0.36811; S(1) in 3(b): 0,0,1/2.

15 S-atom plane polytype, with a double layer of GaS_4 tetrahedra; the fifth plane of S atoms has metal vacancies (Fig. 1).

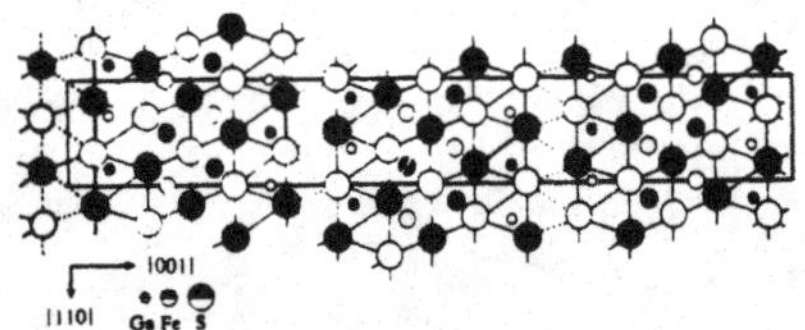

Fig. 1. Structure of 3R-$Fe_2Ga_2S_5$.

GALLIUM LANTHANUM NICKEL
$Ga_6LaNi_{0.6}$

I. Ju.N. GRIN', Ja.P. JARMOLJUK and E.I. GLADYŠEVSKIJ, 1982. Kristallographija, 27, 686-692 [Soviet Physics - Crystallography, 27, 413-417].
II. Ju.N. GRIN', Ja.P. JARMOLJUK, I.V. ROŽDESTVENSKAJA and E.I. GLADYSEVŠKIJ, 1982. Ibid., 27, 693-696 [Ibid., 27, 418-419].

Tetragonal, P4/mmm, a = 4.300, c = 15.632 A, Z = 2. Mo radiation, R = 0.092 for 142 reflexions.

Atomic positions

			x	y	z
	La	in 2(g)	0	0	0.2534
	Ga(1)	2(h)	1/2	1/2	0.321
	Ga(2)	4(i)	0	1/2	0.4179
	Ga(3)	2(h)	1/2	1/2	0.161
	Ga(4)	4(i)	0	1/2	0.078
	Ni(1)	1(b)	0	0	1/2
0.2	Ni(2)	1(a)	0	0	0

The structure (Fig. 1) contains fragments of the Al_4Ba and CaF_2 types. Coordination numbers are Ga = 9-11, La = 20, Ni = 8.

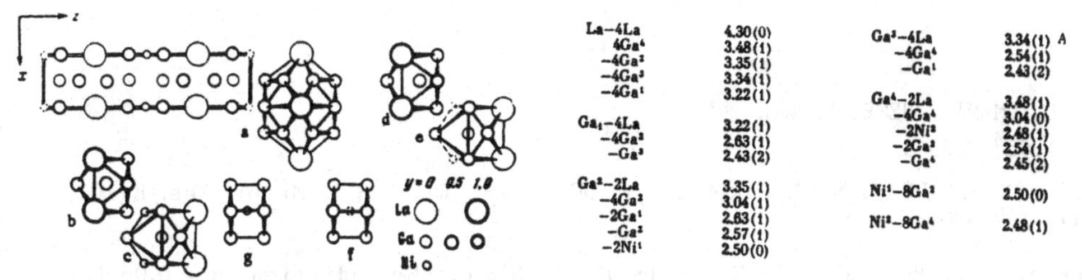

Projection of $LaGa_6Ni_{1-x}$ structure on xy plane, and coordination polyhedra of atoms: a) La; b) Ga[1]; c) Ga[2]; d) Ga[3]; e) Ga[4]; f) Ni[1]; g) Ni[2].

Fig. 1. Structure of $Ga_6LaNi_{0.6}$.

GALLIUM LANTHANUM SULPHUR

$GaLaS_3$

M. JULIEN-POUZOL, S. JAULMES and C. DAGRON, 1982. Acta Cryst., B38, 1566-1568.

Monoclinic, $P2_1/b$, a = 10.33, b = 12.82, c = 10.56 A, γ = 98.90°, Z = 12. Mo radiation, R = 0.057 for 3185 reflexions.

Atomic positions and interatomic distances (A)

	x	y	z
La(1)	0,05478	0,32670	0,02841
La(2)	0,48596	0,32592	0,05355
La(3)	0,81200	0,94336	0,14633
Ga(1)	0,2580	0,8556	0,0832
Ga(2)	0,1995	0,5764	0,2312
Ga(3)	0,6844	0,1686	0,2526
S(1)	0,0638	0,8945	0,0201
S(2)	0,5990	0,0846	0,0769
S(3)	0,2481	0,6775	0,0505
S(4)	0,2456	0,1748	0,0922
S(5)	0,6706	0,5428	0,0915
S(6)	0,9959	0,5134	0,1629
S(7)	0,5625	0,7941	0,1885
S(8)	0,3033	0,4291	0,2237
S(9)	0,8940	0,2492	0,2631

La(1)–S(6)	2,93 (1)	La(1)–S(9)	3,06 (1) (Å)
S(1)	2,95 (1)	S(3)	3,23 (3)
S(6)	2,982 (9)	S(5)	3,31 (2)
S(9)	3,04 (1)	S(8)	3,39 (2)
S(4)	3,05 (2)		
La(2)–S(7)	2,88 (1)	La(2)–S(7)	2,99 (1)
S(4)	2,93 (2)	S(8)	3,05 (1)
S(5)	2,94 (1)	S(5)	3,14 (1)
S(3)	2,97 (2)	S(2)	3,48 (2)
La(3)–S(6)	2,82 (1)	La(3)–S(7)	3,00 (2)
S(1)	2,869 (9)	S(1)	3,07 (2)
S(9)	2,91 (1)	S(2)	3,14 (2)
S(4)	2,95 (1)	S(5)	3,46 (1)
Ga(1)–S(1)	2,24 (2)	Ga(1)–S(3)	2,30 (1)
S(8)	2,265 (8)	S(2)	2,30 (1)
Ga(2)–S(6)	2,25 (2)	Ga(2)–S(8)	2,31 (1)
S(4)	2,261 (8)	S(3)	2,318 (8)
Ga(3)–S(9)	2,25 (2)	Ga(3)–S(7)	2,28 (1)
S(2)	2,255 (7)	S(5)	2,294 (8)

The structure contains GaS_4 tetrahedra, and 9-, 8-, and 8-coordinated La.

GALLIUM LITHIUM
$Ga_{14}Li_3$

I. C. BELIN and R.G. LING, 1982. J. Solid State Chem., **45**, 290-292.
II. J. STOEHR and H. SCHÄFER, 1982. Rev. Chim. Minér., **19**, 122-127.

Rhombohedral, $R\bar{3}m$, a = 8.441, c = 16.793 A, Z = 3. Mo radiation, R = 0.082 for 325 reflexions. Ga(1), Ga(2), 0.5 Li in 18(h): $x,\bar{x},z$, x = 0.1608, 0.5611, 0.8229, z = -0.1354, 0.9485, 0.9224; Ga(3) in 6(c): 0,0,0.9224.

Open stacking of interconnected Ga_{12} icosahedra, with half-occupied Li sites in cavities. Ga-Ga = 2.49-2.73, Ga-Li = 2.61-3.45 A.

GALLIUM MANGANESE SULPHUR
$Ga_{1.85}Mn_{0.23}S_3$

NGUYEN-HUY DUNG, M.-P. PARDO and L. DOGGUY-SMIRI, 1982. Mater. Res. Bull., **17**, 293-300.

Hexagonal, $P6_1$, a = 6.397, c = 18.027 A, Z = 6. Mo radiation, R = 0.06 for 323 reflexions. S(1): 0.325,-0.012,0; S(2): 0.012,0.338,-0.0052; S(3): 0.672,0.681,-0.0056; Ga(1): -0.003,0.338,0.1213; 5.1 Ga(2): 0.654,0.659, 0.1183; 1.38 Mn: 0.306,0.973,0.113. Wurtzite superstructure, as described for α'-Ga_2S_3 (**1**), with Mn in a vacant metal site of Ga_2S_3.

1. Structure Reports, **19**, 405.

GALLIUM NEODYMIUM NICKEL
Ga_2NdNi

Ju.M. GRIN' and Ja.P. JARMOLJUK, 1982. Dopov. Akad. Nauk Ukr.RSR, Ser. A: Fiz.-Mat. Tekh. Nauki, No. 3, 69-72.

Orthorhombic, Cmmm, a = 4.192, b = 17.564, c = 4.133 A, Z = 4. Powder data. Nd in 4(i): 0,0.136,0; Ga(1) in 4(j): 0,0.288,1/2; Ga(2) in 2(b): 1/2,0,0; Ga(3) in 2(d): 0,0,1/2; Ni in 4(j): 0,0.416,1/2.

New structure type, with fragments of the Al_4Ba and AlB_2 types (Fig. 1).

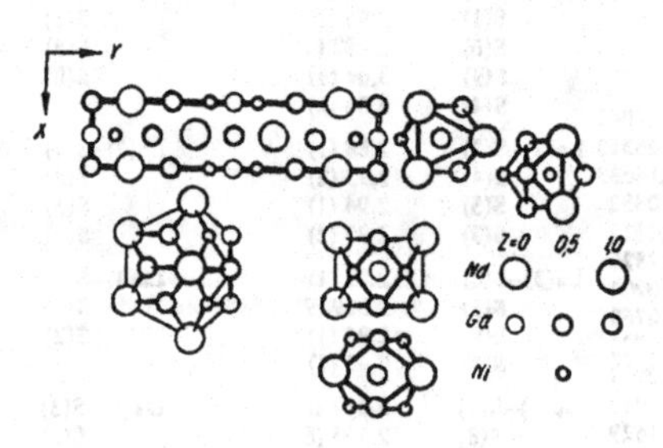

Fig. 1. Structure of Ga_2NdNi.

GALLIUM NICKEL ZINC
$Ga_4Ni_3Zn_6$

T. RAJASEKHARAN, N. SARAH and K. SCHUBERT, 1982. Z. Metallk., **73**, 526-529.

Monoclinic, C2/m, a = 15.392, b = 8.033, c = 12.426 A, β = 108.05°, Z = 8. Mo radiation, R = 0.064.

The structure can be derived from that of $Al_{13}Fe_4$ (**1**), by replacing Al by Ga and Zn, and Fe by Ni.

1. Structure Reports, **19**, 22.

GALLIUM POTASSIUM
Ga_3K

C. BELIN and R.G. LING, 1982. C.R. Acad. Sci., Ser. II, **294**, 1083-1086.

Tetragonal, I$\bar{4}$m2, a = 6.278, c = 14.799 A, Z = 6. R = 0.059 for 142 reflexions. K(1) in 2(a): 0,0,0; K(2) in 4(f): 0,1/2,0.3725; Ga(1) in 8(i): 0.2063,0,0.2217; Ga(2) in 8(i): 0.3054,0,0.3931; Ga(3) in 2(b): 0,0,1/2.

Isostructural with Ga_3Rb (**1**), with a loose stacking of Ga dodecahedra, space for K atoms, and short K-K distances (3.63-3.66 A).

1. Structure Reports, **48A**, 57.

GALLIUM SELENIUM
β-Ga_2Se_3

D. LÜBBERS and V. LEUTE, 1982. J. Solid State Chem., **43**, 339-345.

Monoclinic, Cc, a = 6.6608, b = 11.6516, c = 6.6491 A, β = 108.840°, Z = 4. Powder data. Ga: 0,7/12,0 and 0,3/12,0; Se(1): 3/8,7/12,1/8; Se(2): 3/8,3/12, 1/8 and 3/8,11/12,1/8. Superstructure of α-Ga_2Se_3 zincblende type. Previous studies in **1**.

1. Structure Reports, **12**, 178; **30A**, 140.

GALLIUM SODIUM
$Ga_{13}Na_7$ (forms I and II), Ga_4Na

I. U. FRANK-CORDIER, G. CORDIER and H. SCHÄFER, 1982. Z. Naturforsch., **37B**, 119-126.
II. Idem, 1982. Ibid., **37B**, 127-135.

Form I, rhombohedral, R$\bar{3}$m, a = 14.965, c = 3.893 A, Z = 2. Mo radiation, R = 0.112 for 802 reflexions.

Form II, orthorhombic, Pnma, a = 15.625, b = 14.979, c = 21.678 A. Z = 12. Mo radiation, R = 0.117 for 1875 reflexions. [Perhaps the same phase described below as $Ga_{39}Na_{22}$.]

Ga_4Na, tetragonal, I4/mmm, a = 4.230, c = 11.272 A, Z = 2. Mo radiation, R = 0.096 for 104 reflexions. Na in 2(a): 0,0,0; Ga(1) in 4(e): 0,0,0.3888; Ga(2) in 4(d): 0,1/2,1/4. Al_4Ba-type (**1**), as previously described (**2**).

Both forms of $Ga_{13}Na_7$ contain Ga_{12} icosahedra and Ga_{15} clusters connected into a three-dimensional framework in an $MgCu_2$-type arrangement (Fig. 1).

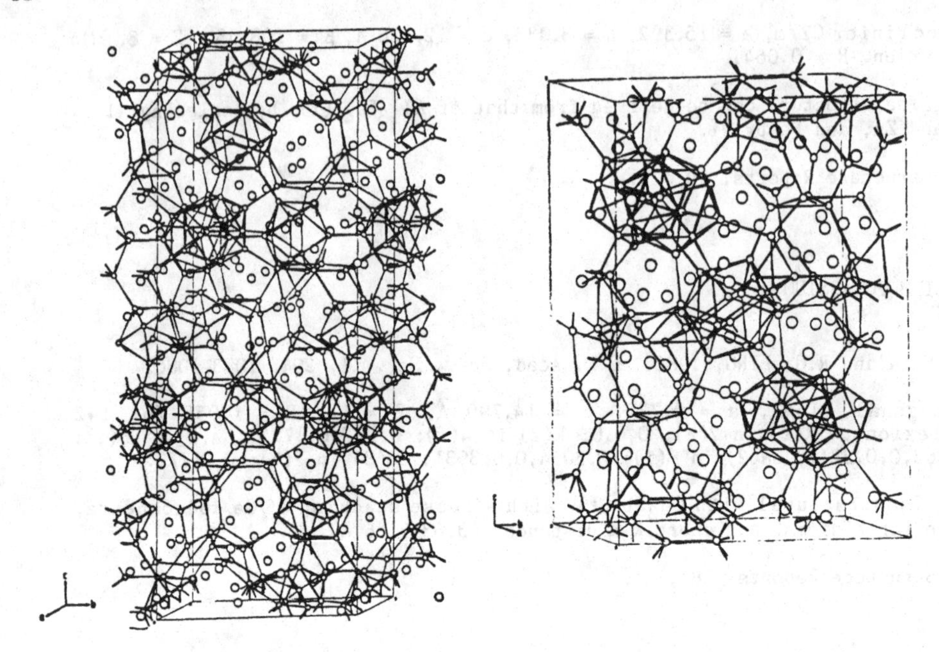

Fig. 1. Structures of $Ga_{13}Na_7$, forms I (left) and II (right); small circles = Ga; large circles = Na.

$Ga_{39}Na_{22}$

III. R.G. LING and C. BELIN, 1982. Acta Cryst., B38, 1101-1104.

Orthorhombic, Pnma, a = 15.585, b = 14.948, c = 21.632 A, Z = 4. Mo radiation, R = 0.039 for 1722 reflexions. [Perhaps the same phase described above as $Ga_{13}Na_7$ = $Ga_{39}Na_{21}$.]

Atomic positions

		x (×10⁴)	y	z
Ga(1)	8(d)	8563	4211	438
Ga(2)		1846	6610	1660
Ga(3)		1896	5955	2817
Ga(4)		5694	5926	−73
Ga(5)		8366	5959	2934
Ga(6)		116	4245	1124
Ga(7)		961	5777	809
Ga(8)		8391	6600	1754
Ga(9)		7079	3444	859
Ga(10)		9225	5740	881
Ga(11)		7101	6500	−573
Ga(12)		1446	4264	288
Ga(13)		5694	4443	621
Ga(14)		1202	4400	3178
Ga(15)		293	3486	2276
Ga(16)		9976	3357	42
Ga(17)	4(c)	4241	7500	2378
Ga(18)		6915	7500	1452
Ga(19)		3332	7500	1333
Ga(20)		5955	7500	2445
Ga(21)		6236	7500	345
Ga(22)		1006	7500	7429
Ga(23)		6933	7500	−1522
Na(1)	8(d)	3306	4338	3055
Na(2)		4182	4237	1583
Na(3)		7187	5613	786
Na(4)		3112	5636	563
Na(5)		5147	6265	1301
Na(6)		107	5593	2225
Na(7)		2017	4398	1769
Na(8)	4(c)	4774	7500	8736
Na(9)		103	7500	1283
Na(10)		3280	7500	4592
Na(11)		7988	7500	7227
Na(12)		1948	7500	285
Na(13)		8790	7500	8863
Na(14)		6064	7500	6199
Na(15)		9254	7500	4914

Complex structure (Fig. 2), with most Ga in a non-compact framework of linked icosahedra, and a few less-coordinated satellite Ga atoms; the framework contains spaces for Na atoms.

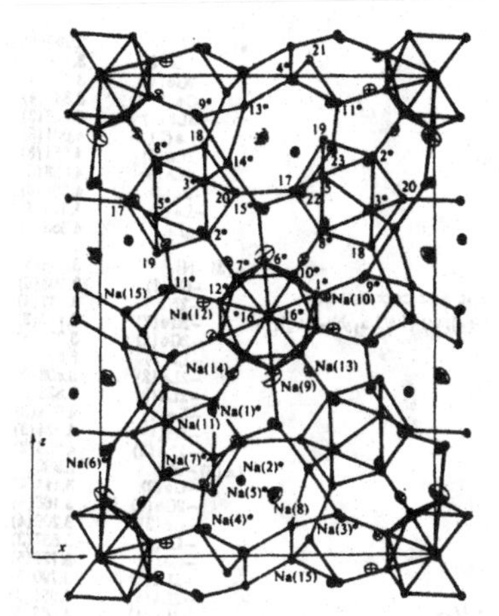

Fig. 1. Structure of $Ga_{39}Na_{22}$.

1. Strukturbericht, **3**, 45, 330.
2. Structure Reports, **34**A, 140.

GALLIUM THULIUM
Ga_2Tm

H. LÜSCHER, K. GIRGIS and P. FISCHER, 1982. J. Less-Common Metals, **83**, L23-L25.

Orthorhombic, Imma, a = 4.200, b = 6.879, c = 8.067 A, Z = 4. Neutron powder data. Ga in 8(h): 0,0.0498,0.1611; Tm in 4(e): 0,1/4,0.5617. $CeCu_2$-type (**1**). Tm-Tm = 3.56-3.67, Tm-Ga = 3.03-3.49, Ga-Ga = 2.53-2.73 A.

1. Structure Reports, **26**, 107.

GERMANIUM LANTHANUM NICKEL
Ge_2La_3Ni

O.I. BODAK, V.A. BRUSKOV and V.K. PEČARSKIJ, 1982. Kristallografija, **27**, 896-899 [Soviet Physics - Crystallography, **27**, 538-539].

Orthorhombic, Pnma, a = 12.041, b = 4.358, c = 11.871 A, Z = 4. Mo radiation, R = 0.053 for 631 reflexions.

The structure (Fig. 1) is a filled analogue of the Hf_3P_2-type, with 17-coordination for La atoms and tricapped trigonal prismatic coordination for Ge and Ni atoms.

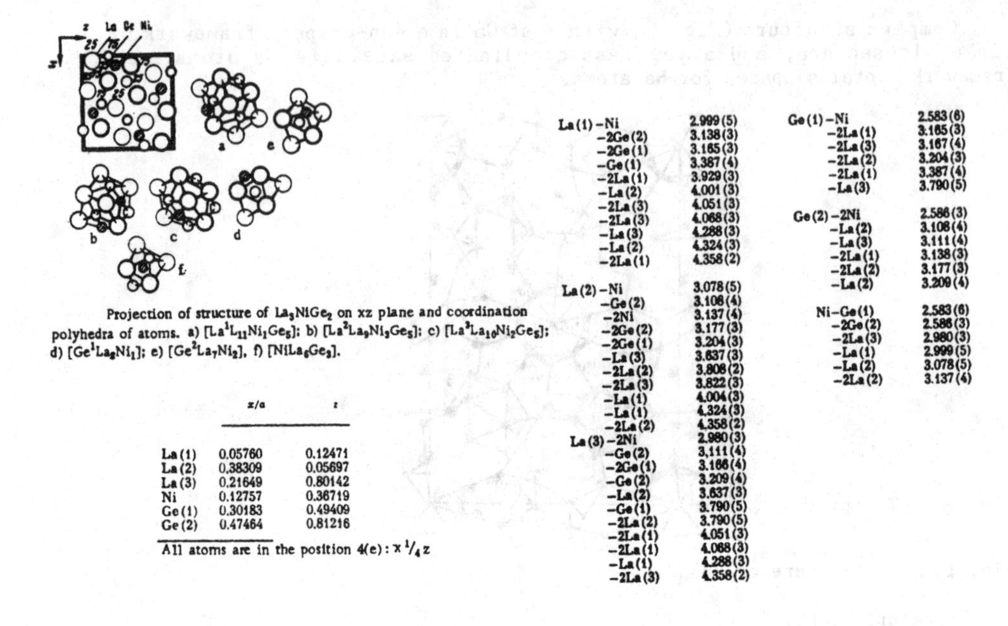

Projection of structure of La_3NiGe_2 on xz plane and coordination polyhedra of atoms. a) $[La^1L_{11}Ni_1Ge_5]$; b) $[La^2La_9Ni_3Ge_5]$; c) $[La^3La_{10}Ni_2Ge_5]$; d) $[Ge^1La_8Ni_1]$; e) $[Ge^2La_7Ni_2]$. f) $[NiLa_6Ge_3]$.

	x/a	z
La(1)	0.05760	0.12471
La(2)	0.38309	0.05697
La(3)	0.21649	0.80142
Ni	0.12757	0.36719
Ge(1)	0.30183	0.49409
Ge(2)	0.47464	0.81216

All atoms are in the position 4(e): x $^1/_4$ z

La(1)–Ni	2.999(5)
–2Ge(2)	3.138(3)
–2Ge(1)	3.165(3)
–Ge(1)	3.387(4)
–2La(1)	3.929(3)
–La(2)	4.001(3)
–2La(3)	4.051(3)
–2La(3)	4.068(3)
–La(3)	4.288(3)
–La(2)	4.324(3)
–2La(1)	4.358(2)
La(2)–Ni	3.078(5)
–Ge(2)	3.108(4)
–2Ni	3.137(4)
–2Ge(2)	3.177(3)
–2Ge(1)	3.204(3)
–La(3)	3.637(3)
–2La(2)	3.808(2)
–2La(3)	3.822(3)
–La(1)	4.004(3)
–La(1)	4.324(3)
–2La(2)	4.358(2)
La(3)–2Ni	2.980(3)
–Ge(2)	3,111(4)
–2Ge(1)	3.166(4)
–Ge(2)	3.209(4)
–La(2)	3.637(3)
–Ge(1)	3.790(5)
–2La(2)	3.790(5)
–2La(1)	4.051(3)
–2La(1)	4.068(3)
–La(1)	4.288(3)
–2La(3)	4.358(2)
Ge(1)–Ni	2.583(6)
–2La(1)	3.165(3)
–2La(3)	3.167(4)
–2La(2)	3.204(3)
–2La(1)	3.387(4)
–La(3)	3.790(5)
Ge(2)–2Ni	2.586(3)
–La(2)	3.108(4)
–La(3)	3.111(4)
–2La(1)	3.138(3)
–2La(2)	3.177(3)
–La(2)	3.209(4)
Ni–Ge(1)	2.583(6)
–2Ge(2)	2.586(3)
–2La(3)	2.980(3)
–La(1)	2.999(5)
–La(2)	3.078(5)
–2La(2)	3.137(4)

Fig. 1. Structure of Ge_2La_3Ni.

GERMANIUM LANTHANUM RHODIUM
$Ge_4La_3Rh_4$

E. HOVESTREYDT, K. KLEPP and E. PARTHÉ, 1982. Acta Cryst., B38, 1803-1805.

Orthorhombic, Immm, a = 4.1746, b = 4.2412, c = 25.234 A, Z = 2. Mo radiation, R = 0.035 for 271 reflexions.

The structure (Fig. 1) is of $U_3Ni_4Si_4$-type (1).

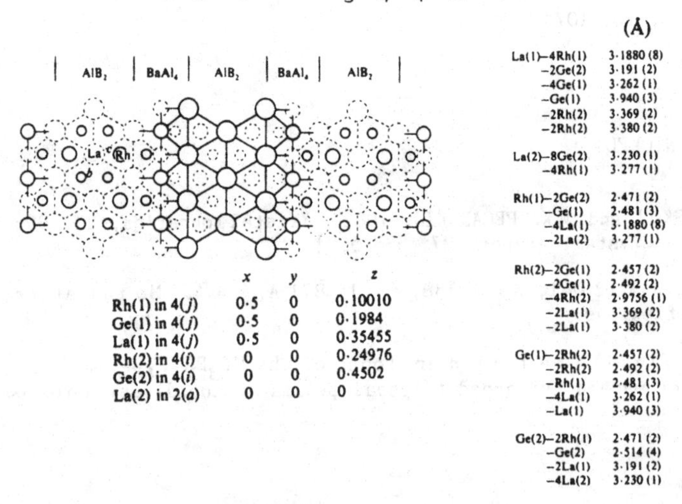

	x	y	z
Rh(1) in 4(j)	0·5	0	0·10010
Ge(1) in 4(j)	0·5	0	0·1984
La(1) in 4(j)	0·5	0	0·35455
Rh(2) in 4(i)	0	0	0·24976
Ge(2) in 4(i)	0	0	0·4502
La(2) in 2(a)	0	0	0

	(Å)
La(1)–4Rh(1)	3·1880 (8)
–2Ge(2)	3·191 (2)
–4Ge(1)	3·262 (1)
–Ge(1)	3·940 (3)
–2Rh(2)	3·369 (2)
–2Rh(2)	3·380 (2)
La(2)–8Ge(2)	3·230 (1)
–4Rh(1)	3·277 (1)
Rh(1)–2Ge(2)	2·471 (2)
–Ge(1)	2·481 (3)
–4La(1)	3·1880 (8)
–2La(2)	3·277 (1)
Rh(2)–2Ge(1)	2·457 (2)
–2Ge(1)	2·492 (2)
–4Rh(2)	2·9756 (1)
–2La(1)	3·369 (2)
–2La(1)	3·380 (2)
Ge(1)–2Rh(2)	2·457 (2)
–2Rh(2)	2·492 (2)
–Rh(1)	2·481 (3)
–4La(1)	3·262 (1)
–La(1)	3·940 (3)
Ge(2)–2Rh(1)	2·471 (2)
–Ge(2)	2·514 (4)
–2La(1)	3·191 (2)
–4La(2)	3·230 (1)

Fig. 1. Structure of $Ge_4La_3Rh_4$.

1. Structure Reports, 45A, 96.

GERMANIUM NICKEL SCANDIUM
$Ge_4Ni_4Sc_3$

B.Ja. KOTUR and M. SIKIRICA, 1982. Acta Cryst., B**38**, 917-918.

Orthorhombic, Immm, a = 12.910, b = 6.598, c = 3.908 A, Z = 2. Mo radiation, R = 0.042 for 205 reflexions.

Atomic positions (x 10^4) and interatomic distances (A)

		x	y	z
Sc(1)	2(*d*)	0	½	0
Sc(2)	4(*e*)	1266	0	0
Ni	8(*n*)	3231	1960	0
Ge(1)	4(*f*)	2189	½	0
Ge(2)	4(*h*)	0	1941	½

Sc(1)–4Sc(2)	3·682 (3)
8Ni	3·272 (2)
2Ge(1)	2·826 (3)
4Ge(2)	2·809 (2)
Sc(2)–2Sc(1)	3·682 (3)
2Ge(1)	3·508 (3)
Sc(2)	3·269 (6)
4Ni	2·875 (4)
4Ge(2)	2·851 (3)
2Ni	2·847 (5)
2Ge(1)	2·792 (4)
Ni–2Sc(1)	3·272 (2)
2Sc(2)	2·875 (4)
Sc(2)	2·847 (5)
2Ni	2·809 (3)
Ni	2·586 (3)
Ge(1)	2·415 (3)
2Ge(1)	2·405 (2)
Ge(2)	2·396 (3)
Ge(1)–2Sc(2)	3·508 (3)
Sc(1)	2·826 (3)
2Sc(2)	2·792 (4)
2Ni	2·415 (3)
4Ni	2·405 (2)
Ge(2)–4Sc(2)	2·851 (3)
2Sc(1)	2·809 (2)
Ge(2)	2·561 (4)
2Ni	2·396 (3)

$Gd_3Cu_4Ge_4$-type structure (**1**).

1. Structure Reports, **35**A, 54.

GERMANIUM PALLADIUM URANIUM
PALLADIUM SILICON URANIUM
RHODIUM SILICON URANIUM
UPd_2Ge_2, UPd_2Si_2, URh_2Si_2

GERMANIUM RHODIUM URANIUM
URh_2Ge_2

H. PTASIEWICZ-BAK, J. LECIEJEWICZ and A. ZYGMUNT, 1981. J. Phys. F: Metal Phys., **11**, 1225-1235.

First three compounds, tetragonal, I4/mmm, a = 4.200, 4.097, 4.009, c = 10.230, 10.046, 10.025 A, Z = 2. Neutron powder data. U in 2(a): 0,0,0; Pd or Rh in 4(d): 1/2,1/4,0; Ge or Si in 4(e): 0,0,z, z = 0.3830, 0.3816, 0.3832. $ThCr_2Si_2$-type (**1**).

URh_2Ge_2, tetragonal, P4/mmm, a = 4.155, c = 9.771 A, Z = 2. Neutron powder data. U(1) in 1(a): 0,0,0; U(2) in 1(d): 1/2,1/2,1/2; Rh + Ge in 2(g): 0,0,0.4041, 2(h): 1/2,1/2,0.1258, and 4(i): 0,1/2,0.2500. [Some interatomic distances are very short.]

1. Structure Reports, **43**A, 99.

GERMANIUM TANTALUM
$GeTa_3$

I. J.-O. WILLERSTRÖM, 1982. J. Less-Common Metals, **86**, 85-104.
II. Idem, 1982. Ibid., **86**, 105-114.

High-temperature form, tetragonal, $I\bar{4}$, a = 10.342, c = 5.153 A, Z = 8. Powder data. Fe_3P-type (1).

Low-temperature form, tetragonal, $P4_2/n$, a = 10.271, c = 5.215 A, Z = 8. Powder data. Ta(1), Ta(2), Ta(3), Ge in 8(g): x,y,z, x = 0.7874, 0.8520, 0.8986, 0.7909, y = 0.3126, 0.9843, 0.4152, 0.9768, z = 0.0093, 0.2294, 0.5306, 0.7293. Ti_3P-type (2).

1. Structure Reports, **27**, 97.
2. Ibid., **32**A, 111.

GOLD INDIUM
$Au_{10}In_3$

M.Z. JANDALI, T. RAJASEKHARAN and K. SCHUBERT, 1982. Z. Metallk., **73**, 463-467.

Hexagonal, $P6_3/m$, a = 10.5387, c = 4.7862 A, Z = 2. Mo radiation, R = 0.061 for 812 reflexions.

Atomic positions

		x	y	z
Au(1)	in 6(h)	0.1774	0.1269	1/4
Au(2)	6(h)	0.4107	0.4343	1/4
Au(3)	6(h)	0.4862	0.2070	1/4
Au(4)	2(c)	2/3	1/3	3/4
In	6(h)	0.1050	0.3640	1/4

$Cu_{10}Sb_3$-type structure (1).

1. Structure Reports, **22**, 32.

HAFNIUM MOLYBDENUM PHOSPHORUS
$Hf_{9.29}Mo_{3.71}P$

A. HÅRSTA, 1982. Acta Chem. Scand., A**36**, 535-539.

Hexagonal, $P6_3/mmc$, a = 8.6236, c = 8.6101 A, Z = 2. Mo radiation, R = 0.12 for 1130 reflexions. 12 Hf(1) in 12(k): 0.19879,0.39758,0.40836; 6 Hf(2) in 6(h): 0.53992,0.07984,1/4; 5.598 Mo(1) + 0.402 Hf(3) in 6(h): 0.88991,0.77982, 1/4; 1.832 Mo(2) + 0.168 Hf(4) in 2(a): 0,0,0; 2 P in 2(c): 1/3,2/3,1/4.

The material is described as isostructural with κ-Hf_9Mo_4Ni (1) [but site 6(g) is not occupied, so that the structure should perhaps be described as Al_5Co_2-type (2), as for Hf_9Mo_4B (3)].

1. Structure Reports, **48**A, 64.
2. Ibid., **26**, 6.
3. Ibid., **39**A, 37; **43**A, 43.

HOLMIUM HYDROGEN NICKEL
$HoNi_3D_{1.8}$

V.V. BURNAŠEVA, V.A. JARTYS', S.P. SOLOV'EV, N.V. FADEEVA and K.N. SEMENENKO, 1982. Kristallografija, **27**, 680-685 [Soviet Physics - Crystallography, **27**, 409-413].

Rhombohedral, R$\bar{3}$m, a = 4.99, c = 26.12 A, Z = 9. Neutron powder data. $PuNi_3$-type structure (1), with D distributed in two tetrahedral sites.

1. Structure Reports, **23**, 110.

HOLMIUM RHODIUM
Ho_5Rh_3

J. LE ROY, J.M. MOREAU and D. PACCARD, 1982. J. Less-Common Metals, **86**, 63-67.

Hexagonal, $P6_3/mcm$, a = 8.100, c = 6.337 A, Z = 2. Mo radiation, R = 0.11 for 114 reflexions. Ho(1) in 6(g): 0.2457,0,1/4; Ho(2) in 4(d): 1/3,2/3,0; Rh in 6(g): 0.599,0,1/4. Mn_5Si_3-type (1).

1. Strukturbericht, **4**, 24, 137, 246; Structure Reports, **32A**, 102.

HOLMIUM SULPHUR TELLURIUM
$(HoS)_2Te_{1.34}$

I. G. GHÉMARD, J. ETIENNE and J. FLAHAUT, 1982. J. Solid State Chem., **45**, 140-145.
II. G. GHÉMARD, J. ETIENNE, G. SCHIFFMACHER and J. FLAHAUT, 1982. Ibid., **45**, 146-153.

Subcell, orthorhombic, Immm, a = 4.158, b = 5.199, c = 13.657 A, Z = 2. Mo radiation, R = 0.061 for 246 reflexions. 4 Ho in 4(j): 1/2,0,0.1608; 4 S in 4(i): 0,0,0.2868; 0.84 Te in 2(a): 0,0,0; 1.84 Te in 4(g): 0,0.2664,0.

Supercell, monoclinic, B2/m, a= 6.658, b = 6.658, c = 13.657 A, γ = 102.7°, Z = 4. Mo radiation, R = 0.043 for 165 reflexions. 8 Ho in 8(j): 0.2596,0.2557,0.1606; 4 S(1) in 4(h): 1/2,1/2,0.2824; 4 S(2) in 4(g): 0,0,0.2879; 1.68 Te(1) in 2(a): 0,0,0; 3.68 Te(2) in 4(i): 0.6330,0.3663,0.

The structure contains (001) layers of Ho_4S tetrahedra, separated by partially-occupied Te sites. Ordering (but twinning) occurs in the superstructure.

HYDROGEN IRON MOLYBDENUM ZIRCONIUM
$FeMoZrD_{2.6}$

V.A. JARTYS', V.V. BURNAŠEVA, N.V. FADEEVA, S.P. SOLOV'EV and K.N. SEMENENKO, 1982. Kristallografija, **27**, 900-904 [Soviet Physics - Crystallography, **27**, 540-543].

Hexagonal, $P6_3/mmc$, a = 5.420, c = 8.826, Z = 4. Neutron powder data. Zr in 4(f): 1/3,2/3,0.069; Fe,Mo(1) in 2(a): 0,0,0; Fe,Mo(2) in 6(h): x,2x,1/4, x = 0.837; 3.8 D(1) in 6(h): x = 0.463; 6.6 D(2) in 24(ℓ): 0.292,0.350,0.964.

$MgZn_2$-type (1), with D in two types of tetrahedral holes.

1. Strukturbericht, **1**, 180.

HYDROGEN LANTHANUM NICKEL
$LaNiD_{3.7}$

V.V. BURNAŠEVA, V.A. JARTYS', N.V. FADEEVA, S.P. SOLOV'EV and K.N. SEMENENKO, 1982. Ž. Neorg. Khim., **27**, 1112-1116 [Russian J. Inorg. Chem., **27**, 625-627].

Orthorhombic, Cmcm, a = 3.982, b = 11.94, c = 4.873 A, Z = 4. Neutron powder data.

Atomic positions

			x	y	z
4	La	in 4(c)	0	0.131	1/4
4	Ni	4(c)	0	0.418	1/4
4	D(1)	4(c)	0	0.912	1/4
7.6	D(2)	8(f)	0	0.309	0.483
3.2	D(3)	4(b)	0	1/2	0

LnNi (CrB)-type structure (**1**), with D in three types of cavity with coordination numbers 5 (trigonal bipyramidal), 4 (tetrahedral), and 2 (linear).

1. Structure Reports, **29**, 51.

HYDROGEN NICKEL ZIRCONIUM
NiZrD

D.G. WESTLAKE, H. SHAKED, P.R. MASON, B.R. McCART, M.H. MUELLER, T. MATSUMOTO and M. AMANO, 1982. J. Less-Common Metals, **88**, 17-23.

Orthorhombic, Cmcm, a = 3.28, b = 10.12, c = 4.05 A, Z = 4. Neutron powder data. Atoms in 4(c): 0,y,1/4, y = -0.140, -0.429, 0.215 for Zr, Ni, D. D is tetrahedrally surrounded by four Zr, and occupation of this site causes a triclinic distortion.

INDIUM SELENIUM
InSe

Y. WATANABE, H. IWASAKI, N. KURODA and Y. NISHINA, 1982. J. Solid State Chem., **43**, 140-150.

Monoclinic, P2/m (**a** unique), a = 4.11, b = 4.61, c = 11.02 A, α = 87.2°, Z = 4. Powder data. In(1) in 2(m): 0,-0.121,0.116; In(2) in 2(n): 1/2,0.621,0.384; Se(1) in 2(m): 0,0.018,0.340; Se(2) in 2(n): 1/2,0.482,0.160. The material is synthesized at high-pressure from the 3R polytype (**1**), and the structure is a modified version of the InS-type; there is some disorder in the layer stacking.

1. Structure Reports, **46A**, 88.

IRIDIUM LANTHANUM SILICON
IrLaSi

I. K. KLEPP and E. PARTHÉ, 1982. Acta Cryst., B**38**, 1541-1544.
II. B. CHEVALIER, P. LEJAY, A. COLE, M. VLASSE and J. ETOURNEAU, 1982. Solid State Comm., **41**, 801-804.

Cubic, $P2_13$, a = 6.363, 6.337 A, Z = 4. I, Mo radiation, R = 0.056 for 108 reflexions; II, powder data. Atoms in 4(a): x,x,x, x = 0.4320, 0.1325, 0.839 for Ir, La, Si in I: x-1/2 = 0.423, 0.135, 0.826 in II.

I describes the structure as an ordered $SrSi_2$ (1) derivative structure (Fig. 1); II describes it as ZrOS-type (2). The two sets of results are essentially identical.

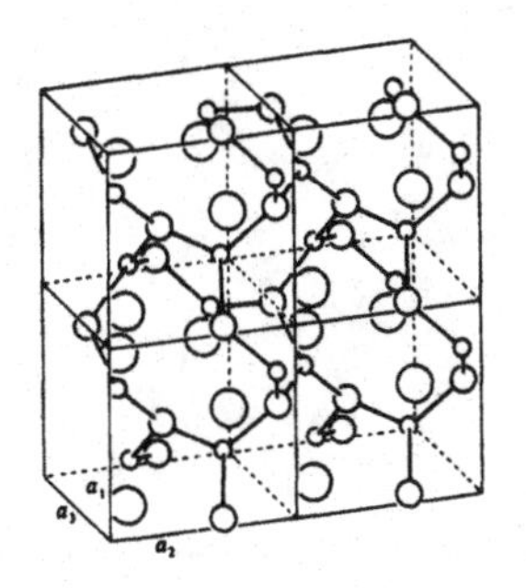

***Interatomic distances* (Å) *for* LaIrSi *up to* 4 Å**

E.s.d.'s are given in parentheses.

La–6La	3·899 (2)	Ir–3La	3·147 (2)	Si–La	3·237 (7)
–3Ir	3·147 (2)	–La	3·202 (2)	–La	3·279 (6)
–Ir	3·202 (2)	–3La	3·794 (2)	–3La	3·616 (6)
–3Ir	3·794 (3)	–3Si	2·315 (7)	–3Ir	2·315 (7)
–Si	3·237 (7)	–6Ir	3·991 (2)	–6Si	3·951 (9)
–3Si	3·279 (6)				
–3Si	3·616 (6)				

Fig. 1. Structure of IrLaSi.

1. Structure Reports, **30**A, 97; **31**A, 63; **38**A, 145; **44**A, 98.
2. Ibid., **11**, 302; **40**A, 216.

IRON NIOBIUM SELENIUM VANADIUM
$FeVNb_2Se_{10}$

CHROMIUM NIOBIUM SELENIUM
$Cr_2Nb_2Se_{10}$

A. BEN SALEM, A. MEERSCHAUT, L. GUEMAS and J. ROUXEL, 1982. Mater. Res. Bull., **17**, 1071-1079.

Monoclinic, $P2_1/m$, a = 9.219, 9.192, b = 3.4688, 3.5086, c = 10.223, 10.409 A, β = 114.16, 115.67°, Z = 2. Mo radiation, R = 0.036, 0.089 for 817, 688 reflexions.

$FeNb_3Se_{10}$-type (1), with Nb in the trigonal prismatic sites and (Fe + V) or Cr in the octahedral sites.

1. Structure Reports, **48**A, 67.

IRON NIOBIUM SILICON
β-$Fe_3Nb_4Si_5$

B. MALAMAN, J. STEINMETZ, G. VENTURINI and B. ROQUES, 1982. J. Less-Common Metals, **87**, 31-43.

Orthorhombic, $P2_1mn$, a = 12.821, b = 4.912, c = 15.521 A, Z = 6. R = 0.052 for 492 reflexions.

The structure (Fig. 1) contains chains of Si, Fe, or disordered Fe/Si atoms, each suurounded by deformed square-, pentagonal-, or hexagonal-antiprisms of Nb, Si, or Fe atoms. An α-form has been described previously (1).

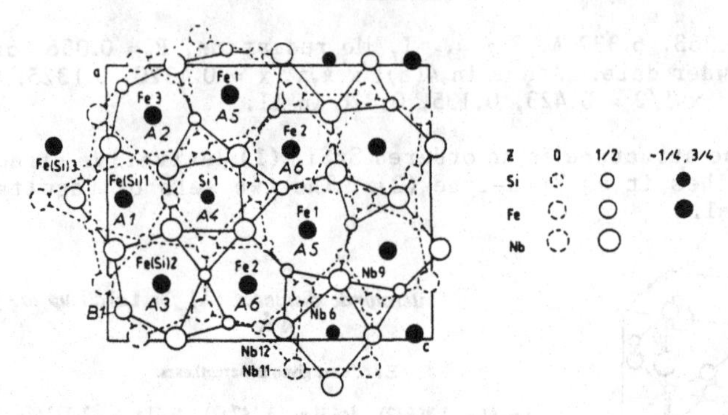

Fig. 1. Structure of β-$Fe_3Nb_4Si_5$.

1. Structure Reports, **46**A, 92.

IRON SELENIUM TITANIUM
$Fe_{0.34}TiSe_2$

J.W. LYDING, M.T. RATAJACK, C.R. KANNEWURF, W.H. GOODMAN, J.A. IBERS and R.E. MARSH, 1982. J. Phys. Chem. Solids, **43**, 599-607.

Trigonal P$\bar{3}$m1, a = 7.148, c = 11.880 A, at 117K, Z = 8. Mo radiation, R = 0.019 for 163 reflexions.

Atomic positions

			x	y	z
Se(1)	in	2(d)	1/3	2/3	0.6272
Se(2)		2(d)	1/3	2/3	0.1159
Se(3)		6(i)	0.1685	-0.1685	-0.1232
Se(4)		6(i)	0.1652	-0.1652	-0.6193
Ti(1)		6(i)	0.4917	-0.4917	0.2501
Ti(2)		2(c)	0	0	0.2496
Fe(1)		1(a)	0	0	0
1.5 Fe(2)		3(f)	1/2	0	1/2

$TiSe_2$ (CdI_2)-type (**1**), with Fe in some octahedral holes between layers.

1. Strukturbericht, **1**, 163.

IRON SULPHUR
FeS

H.E. KING and C.T. PREWITT, 1982. Acta Cryst., B**38**, 1877-1887.

Room-temperature and pressure, hexagonal, P$\bar{6}$2c, a = √3A, c = 2C, where A (∿3.4 A) and C (∿5.9 A) refer to a NiAs-type cell, Z = 12. Mo and Ag radiations, R = 0.039-0.068 for three refinements. Fe in 12(i): 0.3787,0.0553,0.1230; S(1) in 2(a): 0,0,0; S(2) in 4(f): 1/3,2/3,0.0208; S(3) in 6(h): 0.6648, -0.0041,1/4. Troilite-type structure (**1**).

Above 420K and ambient pressure, or 298K and 3.4GPa, orthorhombic, Pnma, a = C, b = A, c = √3A, Z = 4. Mo radiation, R = 0.065-0.149 for three refinements. Fe in 4(c): 0.0035,1/4,0.2352; S in 4(c): 0.2419,1/4,0.5813. MnP-type structure (**2**).

1. Structure Reports, **35**A, 140.
2. Strukturbericht, **3**, 17.

LANTHANON TRANSITION METAL ALLOYS
$Er_{0.5}Gd_{0.5}Ni$, $Er_{0.3}Gd_{0.7}Ni$, $DyNi_{0.35}Pd_{0.65}$, $GdNi_{0.9}Pt_{0.1}$

K. KLEPP and E. PARTHÉ, 1982. J. Less-Common Metals, **85**, 181-194.

Monoclinic, $P2_1/m$, a = 10.647, 18.05, 10.47, 14.52, b = 4.174, 4.205, 4.455, 4.263, c = 5.459, 5.474, 5.501, 5.511 A, β = 97.00, 102.3, 96.73, 100.31°, Z = 6, 10, 6, 8. Ag, Mo radiations, R = 0.066, 0.066, 0.067, 0.090 for 234, 730, 421, 426 reflexions.

Atomic positions

$Gd_{0.5}Er_{0.5}Ni$	x	y	z
R(1)	0.1203	0.25	0.895
R(2)	0.4529	0.25	0.741
R(3)	0.7872	0.25	0.583
Ni(1)	0.024	0.25	0.380
Ni(2)	0.357	0.25	0.211
Ni(3)	0.690	0.25	0.053

R ≡ $Gd_{0.5}Er_{0.5}$

$DyPd_{0.65}Ni_{0.35}$	x	y	z
Dy(1)	0.1207	0.25	0.8865
Dy(2)	0.4547	0.25	0.7441
Dy(3)	0.7883	0.25	0.5943
T(1)	0.0263	0.25	0.358
T(2)	0.3598	0.25	0.227
T(3)	0.6947	0.25	0.079

T ≡ $Pd_{0.65}Ni_{0.35}$

$Gd_{0.7}Er_{0.3}Ni$	x	y	z
R(1)	0.0718	0.25	0.9181
R(2)	0.2721	0.25	0.8264
R(3)	0.4721	0.25	0.7326
R(4)	0.6719	0.25	0.6394
R(5)	0.8720	0.25	0.5434
Ni(1)	0.0150	0.25	0.384
Ni(2)	0.2146	0.25	0.279
Ni(3)	0.4144	0.25	0.184
Ni(4)	0.6143	0.25	0.091
Ni(5)	0.8140	0.25	0.996

R ≡ $Gd_{0.7}Er_{0.3}$

$GdNi_{0.9}Pt_{0.1}$	x	y	z
Gd(1)	0.0900	0.25	0.908
Gd(2)	0.3400	0.25	0.795
Gd(3)	0.5902	0.25	0.677
Gd(4)	0.8401	0.25	0.558
T(1)	0.019	0.25	0.379
T(2)	0.269	0.25	0.256
T(3)	0.519	0.25	0.138
T(4)	0.769	0.25	0.019

T ≡ $Ni_{0.9}Pt_{0.1}$

New examples of FeB-CrB stacking variants, h_2c, h_2c_3, h_2c, h_2c_2, with structures as previously described for related materials (1).

1. Structure Reports, **46**A, 75.

LANTHANON TRANSITION METAL SILICIDES and GERMANIDES
RhScSi, PtScSi, NiYSi, $(Ge,Pt)_2Pr$, RuScGe

E. HOVESTREYDT, N. ENGEL, K. KLEPP, B. CHABOT and E. PARTHÉ, 1982. J. Less-Common Metals, **85**, 247-274.

Silicides, orthorhombic, Pnma, a = 6.474, 6.566, 6.870, b = 4.050, 4.130, 4.155, c = 7.248, 7.275, 7.205 A, Z = 4. Mo radiation, R = 0.052, 0.046, 0.074 for 308, 229, 221 reflexions. Atoms in 4(c): x,1/4,z, x = 0.0094, 0.1568, 0.2857, z = 0.6893, 0.0620, 0.3851 for Sc, Rh, Si in RhScSi, with similar values for the other two compounds. NiTiSi-type structures (**1**).

GePtPr, orthorhombic, Imma, a = 4.438, b = 7.292, c = 7.616 A, Z = 4. Mo radiation, R = 0.088 for 146 reflexions. Pr in 4(e): 0,1/4,0.5417; Ge/Pt in 8(h); 0,0.0449,0.1666. $CeCu_2$-type (**2**).

RuScGe, hexagonal, $P\bar{6}2m$, a = 6.962, c = 3.4683 A, Z = 3. Mo radiation, R = 0.047 for 72 reflexions. Ru in 3(g): 0.2535,0,1/2; Sc in 3(f): 0.599,0,0; Ge(1) in 2(d): 1/3,2/3,1/2; Ge(2) in 1(a): 0,0,0. Fe_2P-type (**3**), as found for AlNiZr (**4**).

1. Structure Reports, **30**, 75.
2. Ibid., **26**, 107.
3. Strukturbericht, **2**, 15.
4. Structure Reports, **32A**, 139.

LANTHANUM PLATINUM SILICON
LaPtSi

K. KLEPP and E. PARTHÉ, 1982. Acta Cryst., B**38**, 1105-1108.

Tetragonal, $I4_1md$, a = 4.2490, c = 14.539 A, Z = 4. Mo radiation, R = 0.052 for 80 reflexions. Atoms in 4(a): 0,0,z, z = 0, 0.5850, 0.419 for La, Pt, Si.

Ordered ternary derivative of the $ThSi_2$-type structure (**1**) (Fig. 1).

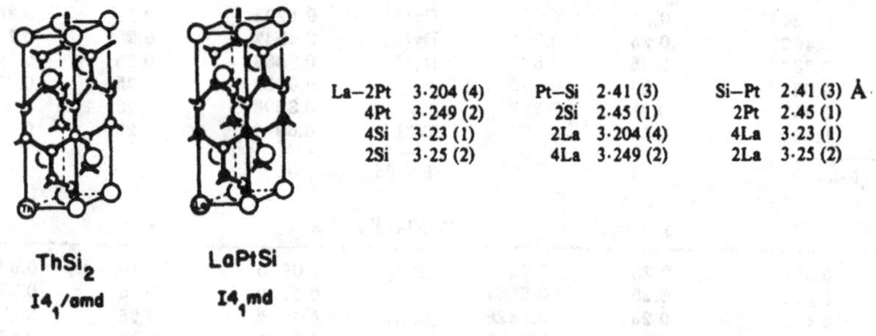

Fig. 1. Structure of LaPtSi.

1. Structure Reports, **9**, 121.

LANTHANUM SELENIUM
$LaSe_2$

I. S. BÉNAZETH, D. CARRÉ and P. LARUELLE, 1982. Acta Cryst., B**38**, 33-37.
II. Idem, 1982. Ibid., B**38**, 37-39.

Monoclinic, $P2_1/a$, a = 8.51, b = 8.58, c (unique axis) = 4.26 A, γ = 90.12°, Z = 4. Mo radiation, R = 0.04 for 678 reflexions and 0.05 for 910 reflexions for crystals twinned on (100) and (201). La: 0.3719,0.2761,0.2803; Se(1): 0.6250,0.3657,0.7417; Se(2): 0.3850,0.0028,0.8267.

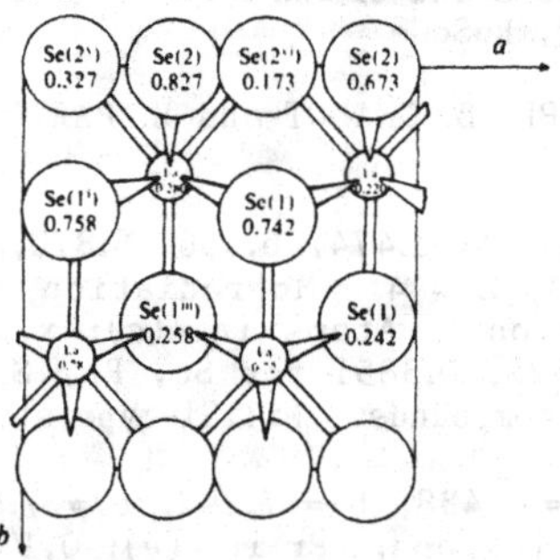

Fig. 1. Structure of $LaSe_2$.

La has tricapped trigonal prismatic 9-coordination (Fig. 1), La-Se = 3.017-3.306(5) A.

LANTHANUM TRANSITION METAL TIN
LaM_xSn_2 (M_x = $Cu_{0.56}$, $Ni_{0.74}$, $Co_{0.52}$, $Fe_{0.34}$), $LaCu_2Sn_2$

W. DÖRRSCHEIDT, G. SAVELSBERG, J. STÖHR and H. SCHÄFER, 1982. J. Less-Common Metals, **83**, 269-278.

LaM_xSn_2, orthorhombic, Cmcm, a = 4.530, 4.526, 4.585, 4.610, b = 18.340, 17.790, 17.414, 17.230, c = 4.460, 4.510, 4.520, 4.490 A, Z = 4. Mo radiation, R = 0.089, 0.055, 0.054, 0.047 for 298, 276, 289, 280 reflexions. Atoms in 4(c):0,y,1/4, y = 0.3941, 0.3928, 0.3952, 0.3974 for La; 0.1776, 0.1853, 0.1893, 0.1967 for M; 0.0465, 0.0484, 0.0531, 0.0562 for Sn(1); 0.7506, 0.7505, 0.7508, 0.7517 for Sn(2). $BaCuSn_2$-type (**1**), for which the structure is also refined: a = 4.790, b = 19.642, c = 4.634 A, y parameters = 0.3917, 0.1769, 0.0423, 0.7500.

$LaCu_2Sn_2$, tetragonal, P4/nmm, a = 4.464, c = 10.509 A, Z = 2. Mo radiation, R = 0.067 for 216 reflexions. La in 2(c): 1/4,1/4,z, z = 0.2390; Cu(1) in 2(a): 3/4,1/4,0; Cu(2) in 2(c): z = 0.6306; Sn(1) in 2(b): 3/4,1/4,1/2; Sn(2) in 2(c): z = 0.8785. $CaBe_2Ge_2$-type (**2**).

1. Structure Reports, **40**A, 33.
2. Ibid., **43**A, 28.

LITHIUM NITROGEN PHOSPHORUS
$LiPN_2$

R. MARCHAND, P. L'HARIDON and Y. LAURENT, 1982. J. Solid State Chem., **43**, 126-130.

Tetragonal, $I\bar{4}2d$, a = 4.567, c = 7.140 A, Z = 4. Powder data. Li in 4(b): 0,0,1/2; P in 4(a): 0,0,0; N in 8(d): 0.146,1/4,1/8.

Isostructural with $CaGeN_2$ (**1**), the structure being derived from that of β-cristobalite. It contains a framework of corner-sharing PN_4 tetrahedra, with Li in holes. P-4N = 1.60(2), Li-4N = 2.17(3) (four further N at 2.99A).

1. Structure Reports, **38**A, 58.

LITHIUM PHOSPHORUS SULPHUR
$Li_4P_2S_6$

R. MERCIER, J.P. MALUGANI, B. FAHYS, J. DOUGLADE and G. ROBERT, 1982. J. Solid State Chem., **43**, 151-162.

Hexagonal, $P6_3/mcm$, a = 6.070, c = 6.577, Z = 1. Mo radiation, R = 0.047 for 90 reflexions. 6 S in 6(g): 0.3237,0,1/4; 2 P in 4(e): 0,0,0.1715; 4 Li in 4(d): 1/3,2/3,0.

ABAB sulphur packing, with four of the six octahedral sites per cell occupied by Li, and the other two by P_2 pairs with 50% occupancy. P-P = 2.26, P-S = 2.03, Li-S = 2.63 A.

LITHIUM PHOSPHORUS SULPHUR (LITHIUM TETRATHIOPHOSPHATE)
Li_3PS_4

R. MERCIER, J.-P. MALUGANI, B. FAHYS, G. ROBERT and J. DOUGLADE, 1982. Acta Cryst., B**38**, 1887-1890.

Orthorhombic, Pnma, a = 13.066, b = 8.015, c = 6.101 A, Z = 4. Mo radiation, R = 0.046 for 255 reflexions.

Tetrahedral PS_4^{3-} anions (mean P-S = 2.050 A), with three types of Li^+ (Fig. 1): Li(1) with tetrahedral coordination (Li-S = 2.41-2.47 A) and partially-occupied Li(2) and Li(3) sites which could result in Li diffusion.

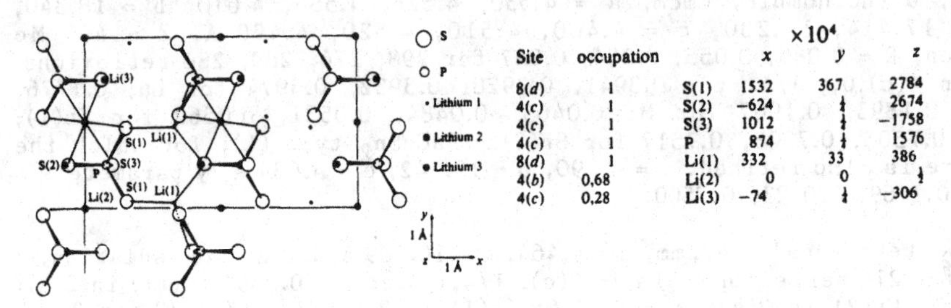

Site	occupation		x	y	z
			×10⁴		
8(d)	1	S(1)	1532	367	2784
4(c)	1	S(2)	−624	¼	2674
4(c)	1	S(3)	1012	¼	−1758
4(c)	1	P	874	¼	1576
8(d)	1	Li(1)	332	33	386
4(b)	0,68	Li(2)	0	0	½
4(c)	0,28	Li(3)	−74	¼	−306

Fig. 1. Structure of Li_3PS_4.

LUTETIUM SULPHUR ZINC
Lu_2ZnS_4

F.H.A. VOLLEBREGT and D.J.W. IJDO, 1982. Acta Cryst., B**38**, 2442-2444.

Orthorhombic, Pnma, a = 13.2180, b = 7.6848, c = 6.2606 A, Z = 4. Neutron powder data.

Atomic positions and interatomic distances (A)

	x	y	z
Zn	0·0955	0·25	0·4196
Lu(1)	0	0	0
Lu(2)	0·2656	0·25	0·9966
S(1)	0·0908	0·25	0·7790
S(2)	0·4235	0·25	0·2451
S(3)	0·1660	0·0082	0·2386

Lu(1)–S(1)*	2·654 (9)	S(1)–S(2)	3·710 (17)
–S(2)*	2·695 (8)		3·957 (16)
–S(3)*	2·655 (11)	*	3·853 (1)
Lu(2)–S(1)	2·682 (16)	S(1)–S(3)*	3·567 (12)
–S(2)	2·603 (15)	*	3·986 (12)
–S(3)*	2·735 (10)	*	3·933 (19)
–S(3)*	2·714 (10)	*	3·786 (19)
Zn–S(1)	2·251 (9)	S(2)–S(3)*	3·878 (18)
–S(2)	2·496 (16)	*	3·706 (18)
–S(3)*	2·368 (10)	*	3·923 (12)
		*	3·858 (12)
		S(3)–S(3)*	3·840 (14)
			3·968 (20)
			3·716 (20)

* Distances marked with an asterisk occur in pairs.

Olivine-type structure (**1**). The Yb, Tm, and Er compounds are isostructural.

1. Strukturbericht, **1**, 352.

MAGNESIUM NITROGEN PHOSPHORUS
Mg_2PN_3

R. MARCHAND and Y. LAURENT, 1982. Mater. Res. Bull., **17**, 399-403.

Orthorhombic, $Cmc2_1$ [not $Pna2_1$, as given in the abstract], a = 9.759, b = 5.635, c = 4.743 A, Z = 4. Powder data. Mg in 8(b): 0.167,0.819,-0.062; P in 4(a): 0,0.366,0; N(1) in 8(b): 0.150,0.753,0.406; N(2) in 4(a): 0,0.324,0.384. Cu_2SiS_3 (Na_2SiO_3) type (**1**), which is an ordered version of the wurtzite structure. P-4N = 1.67, 1.84(6), mean Mg-N = 2.09 A.

1. Structure Reports, **16**, 334; **37**A, 74.

MAGNESIUM STRONTIUM
$Mg_{38}Sr_9$

F. MERLO and M.L. FORNASINI, 1982. Acta Cryst., B**38**, 1797-1798.

Hexagonal, $P6_3/mmc$, a = 10.500, c = 28.251 A, Z = 2. Mo radiation, R = 0.061 for 446 reflexions.

Atomic positions and interatomic distances (A)

		x	y	z
Sr(1)	12(k)	0·1362	0·2724	0·0621
Sr(2)	6(h)	0·5312	0·0624	0·25
Mg(1)	12(k)	0·2337	0·4674	0·5547
Mg(2)	12(k)	0·5037	0·0074	0·6170
Mg(3)	12(k)	0·1674	0·3348	0·6533
Mg(4)	12(k)	0·1743	0·3486	0·1898
Mg(5)	6(h)	0·9011	0·8022	0·25
Mg(6)	6(g)	0·5	0	0
Mg(7)	4(f)	0·3333	0·6667	0·0378
Mg(8)	4(f)	0·3333	0·6667	0·1426
Mg(9)	4(f)	0·3333	0·6667	0·6465
Mg(10)	4(e)	0	0	0·1575

Bond	Distance
Sr(1)-Mg(7)	3·65 (1)
-2Mg(2)	3·66 (1)
-Mg(10)	3·66 (1)
-Mg(4)	3·67 (1)
-2Mg(1)	3·70 (1)
-Mg(1)	3·75 (1)
-2Mg(6)	3·78 (1)
-2Mg(3)	3·81 (1)
-Mg(8)	4·25 (1)
-2Sr(1)	4·29 (1)
-2Sr(1)	4·30 (1)
Sr(2)-4Mg(4)	3·71 (1)
-2Mg(5)	3·71 (2)
-2Mg(2)	3·81 (1)
-2Mg(9)	3·82 (2)
-4Mg(3)	3·90 (1)
-2Sr(2)	4·27 (1)
Mg(1)-Mg(3)	3·04 (1)
-2Mg(6)	3·06 (1)
-2Mg(1)	3·14 (1)
-Mg(9)	3·16 (2)
-Mg(7)	3·18 (2)
-2Mg(2)	3·22 (1)
-2Sr(1)	3·70 (1)
Sr(1)	3·75 (1)
Mg(2)-Mg(8)	3·05 (1)
-Mg(9)	3·21 (1)
-2Mg(1)	3·22 (1)
-2Mg(3)	3·23 (1)
-Mg(6)	3·31 (1)
-2Mg(4)	3·58 (1)
-Mg(7)	3·71 (1)
-2Sr(1)	3·66 (1)
-Sr(2)	3·81 (1)
Mg(3)-Mg(5)	3·00 (1)
-Mg(9)	3·02 (1)
-Mg(1)	3·04 (1)
-Mg(10)	3·05 (1)
-2Mg(2)	3·23 (1)
-2Mg(4)	3·28 (1)
-2Sr(1)	3·81 (1)
-2Sr(2)	3·90 (1)
Mg(4)-Mg(8)	3·18 (1)
-2Mg(5)	3·24 (1)
-2Mg(3)	3·28 (1)
-Mg(10)	3·30 (1)
-Mg(4)	3·40 (1)
-2Mg(2)	3·58 (1)
-Sr(1)	3·67 (1)
-2Sr(2)	3·71 (1)
Mg(5)-2Mg(3)	3·00 (1)
-2Mg(5)	3·12 (2)
-2Mg(10)	3·17 (2)
-4Mg(4)	3·24 (1)
-2Sr(2)	3·71 (2)
Mg(6)-4Mg(1)	3·06 (1)
-2Mg(7)	3·21 (1)
-2Mg(2)	3·31 (1)
-4Sr(1)	3·78 (1)
Mg(7)-Mg(8)	2·96 (3)
-3Mg(1)	3·18 (2)
-3Mg(6)	3·21 (1)
-3Mg(2)	3·71 (1)
-3Sr(1)	3·65 (1)
Mg(8)-Mg(7)	2·96 (3)
-3Mg(2)	3·05 (1)
-3Mg(4)	3·18 (1)
-3Sr(1)	4·25 (1)
Mg(9)-3Mg(3)	3·02 (1)
-3Mg(1)	3·16 (2)
-3Mg(2)	3·21 (1)
-3Sr(2)	3·82 (2)
Mg(10)-3Mg(3)	3·05 (1)
-3Mg(5)	3·17 (2)
-3Mg(4)	3·30 (1)
-3Sr(1)	3·66 (1)

The material was previously described as Mg_4Sr (**1**), but four additional Mg per cell have been found. The atomic coordination of several atoms is increased, and their coordination polyhedra become more regular.

1. Structure Reports, **30**A, 24.

MANGANESE RHODIUM SILICON
GERMANIUM MANGANESE RHODIUM
GERMANIUM MANGANESE PALLADIUM
MnRhSi, GeMnRh, GeMnPd

G. VENTURINI, B. MALAMAN, J. STEINMETZ, A. COURTOIS and B. ROQUES, 1982. Mater. Res. Bull., **17**, 259-267.

MnRhSi, orthorhombic, Pnma, a = 6.200, b = 3.796, c = 7.138 A, Z = 4. Powder data. Mn, Rh, Si in 4(c): x,1/4,z, x = 0.021, 0.152, 0.762, z = 0.174, 0.560, 0.632. TiNiSi (ordered $PbCl_2$ or Co_2P) type (**1**).

GeMnRh, orthorhombic, [Ima2], a = 7.120, b = 11.34, c = 6.559 A, Z = 12. Powder data. Parameters of FeTiSi assumed (**2**), with Mn in the pyramidal and Rh in the tetrahedral sites.

GeMnPd, hexagonal, $P\bar{6}2m$, a = 6.639, c = 3.577 A, Z = 3. R = 0.082 for 206 reflexions. Pd in 3(f): 0.2589,0,0; Mn in 3(g): 0.5914,0,1/2; Ge(1) in 2(c): 1/3,2/3,0; Ge(2) in 1(b): 0,0,1/2. Fe_2P type (**3**).

1. Structure Reports, **24**, 128; **30**A, 75.
2. Ibid., **35**A, 73.
3. Strukturbericht, **2**, 15; Structure Reports, **45**A, 86.

MANGANESE SILICON TITANIUM
$MnTiSi_2$

IRON SILICON TITANIUM
$FeTiSi_2$

J. STEINMETZ, G. VENTURINI, B. ROQUES, N. ENGEL, B. CHABOT and E. PARTHÉ, 1982. Acta Cryst., B**38**, 2103-2108.

Orthorhombic, Pbam, a = 8.703, 8.614, b = 9.541, 9.534, c = 7.765, 7.640 A, Z = 12. Mo radiation, R = 0.037, 0.057 for 769, 615 reflexions.

New structure type, with finite chains of MSi_6 octahedra (M = Mn, Fe) (Fig. 1).

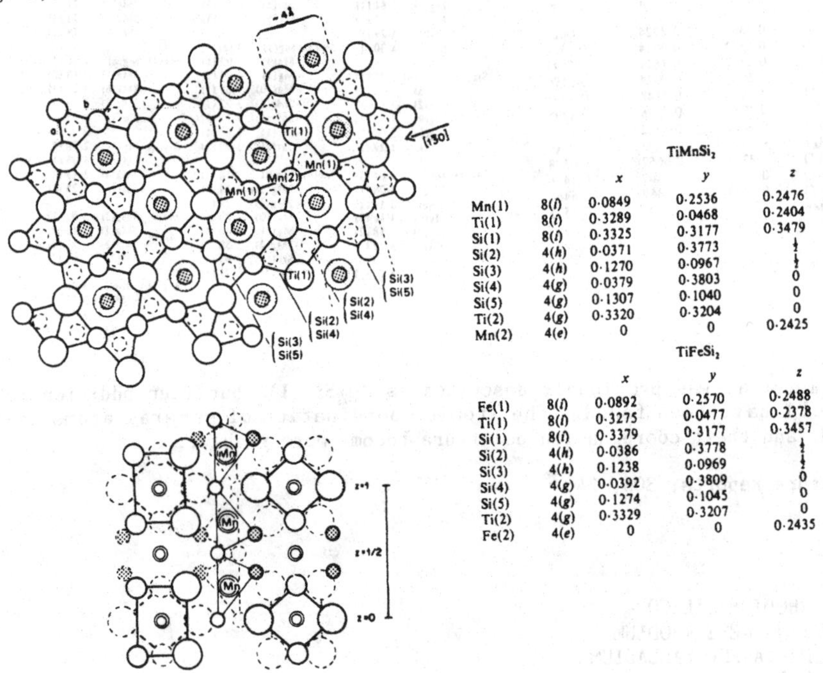

TiMnSi$_2$

		x	y	z
Mn(1)	8(i)	0·0849	0·2536	0·2476
Ti(1)	8(i)	0·3289	0·0468	0·2404
Si(1)	8(i)	0·3325	0·3177	0·3479
Si(2)	4(h)	0·0371	0·3773	½
Si(3)	4(h)	0·1270	0·0967	½
Si(4)	4(g)	0·0379	0·3803	0
Si(5)	4(g)	0·1307	0·1040	0
Ti(2)	4(g)	0·3320	0·3204	0
Mn(2)	4(e)	0	0	0·2425

TiFeSi$_2$

		x	y	z
Fe(1)	8(i)	0·0892	0·2570	0·2488
Ti(1)	8(i)	0·3275	0·0477	0·2378
Si(1)	8(i)	0·3379	0·3177	0·3457
Si(2)	4(h)	0·0386	0·3778	½
Si(3)	4(h)	0·1238	0·0969	½
Si(4)	4(g)	0·0392	0·3809	0
Si(5)	4(g)	0·1274	0·1045	0
Ti(2)	4(g)	0·3329	0·3207	0
Fe(2)	4(e)	0	0	0·2435

Fig. 1. Structure of $MnTiSi_2$ and $FeTiSi_2$.

MANGANESE SILICON ZIRCONIUM
$MnZrSi_2$

G. VENTURINI, J. STEINMETZ and B. ROQUES, 1982. J. Less-Common Metals, **87**, 21-30.

Orthorhombic, Immm, a = 17.324, b = 7.8918, c = 5.1666 A, Z = 12. Cu radiation, R = 0.081 for 260 reflexions.

The structure is related to those of $FeZrSi_2$ and $MnTiSi_2$ (**1**), with octahedral environment for Fe or Mn; these building units are arranged in different ways, but the largest atoms are located in similar holes between the octahedra (Fig. 1). [Note that Fig. 4 in the paper is upside down.]

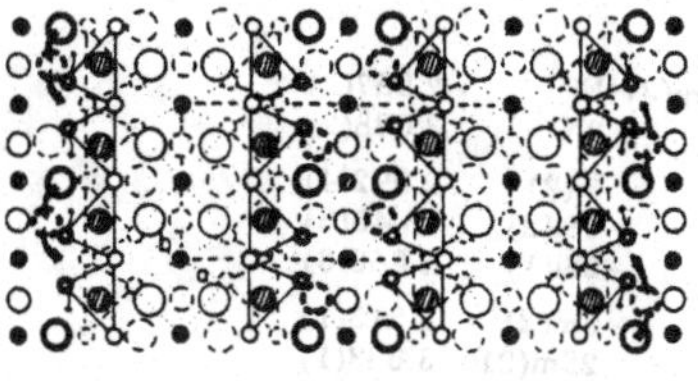

		x	y	z
Si(1)	4i	0	0	0,241
Si(2)	4j	0,5	0	0,255
Si(3)	4e	0,201	0	0
Si(4)	4f	0,790	0,5	0
Si(5)	8n	0,3631	0,353	0
Mn(1)	8k	0,25	0,25	0,25
Mn(2)	4h	0	0,247	0,5
Zr(1)	8n	0,0943	0,2581	0
Zr(2)	4e	0,6409	0	0

Fig. 1. Structure of $MnZrSi_2$.

1. Preceding report.

MANGANESE TITANIUM VANADIUM
$Ti(Mn_{0.6}V_{0.4})_{1.87}$

HYDROGEN MANGANESE TITANIUM VANADIUM
$Ti(Mn_{0.6}V_{0.4})_{1.87}D_{2.36}$

H.W. MAYER, K.M. ALASAFI and O. BERNAUER, 1982. J. Less-Common Metals, **88**, L7-L10.

Hexagonal, $P6_3/mmc$, a = 4.905, 5.229, c = 8.025, 8.581 A, Z = 4. Neutron powder data. Ti in 4(f): 1/3,2/3,z, z = 0.0588, 0.073; Mn/V(1) in 2(a): 0,0,0; Mn/V(2) in 6(h): x,2x,1/4, x = 0.828, 0.829; D distributed in 4(f), 6(h), 12(k), and 24(ℓ) sites. C14-type (**1**).

1. Strukturbericht, **1**, 180.

MERCURY ZINC
$HgZn_3$

M. PUŠELJ, Z. BAN and A. DRAŠNER, 1982. Z. Naturforsch., **37B**, 557-559.

Orthorhombic, $Cmc2_1$, a = 2.708, b = 4.696, c = 5.471 A, Z = 1. Powder data. M = (Hg + 3Zn) in 4(a): 0,y,z, y = 0.357, z = 0.250. Isostructural with β'-Cu_3Ti (**1**).

1. Structure Reports, **15**, 68.

NICKEL SAMARIUM SILICON
$NiSm_3Si_2$

K. KLEPP and E. PARTHÉ, 1982. J. Less-Common Metals, **83**, L33-L35.

Orthorhombic, Pnma, a = 11.505, b = 4.189, c = 11.388 A, Z = 4. R = 0.023 for 375 reflexions.

Atomic positions and interatomic distances (A)

	x	y	z
Sm(1)	0.05775	0.25	0.37559
Ni	0.1286	0.25	0.1332
Sm(2)	0.21373	0.25	0.69768
Si(1)	0.3013	0.25	0.0061
Sm(3)	0.38207	0.25	0.44029
Si(2)	0.4726	0.25	0.6866

Sm(1)-Ni	2.879(2)	Sm(2)-2Ni	2.867(2)	Sm(3)-Ni	2.957(2)
-2Si(2)	3.024(3)	-Si(2)	2.981(4)	-2Ni	3.038(2)
-2Si(1)	3.037(3)	-2Si(1)	3.030(3)	-Si(2)	2.992(4)
-Si(1)	3.243(4)	-Si(2)	3.072(4)	-2Si(2)	3.045(3)
-2Sm(1)	3.766(1)	-Si(1)	3.654(4)	-2Si(1)	3.066(3)
-Sm(3)	3.803(1)	-Sm(3)	3.513(1)	-Sm(2)	3.513(1)
-2Sm(2)	3.852(1)	-2Sm(3)	3.638(1)	-2Sm(2)	3.638(1)
-2Sm(2)	3.925(1)	-2Sm(1)	3.852(1)	-2Sm(3)	3.688(1)
		-2Sm(1)	3.925(1)	-Sm(1)	3.803(1)

Ni-Si(1)	2.459(4)	Si(1)-Ni	2.459(4)	Si(2)-2Ni	2.472(3)
-2Si(2)	2.472(3)	-2Sm(2)	3.030(3)	-Sm(2)	2.981(4)
-2Sm(2)	2.867(2)	-2Sm(1)	3.037(3)	-Sm(3)	2.992(4)
-Sm(1)	2.879(2)	-2Sm(3)	3.066(3)	-2Sm(1)	3.024(3)
-Sm(3)	2.957(2)	-Sm(1)	3.243(4)	-2Sm(3)	3.045(3)
-2Sm(3)	3.038(2)	-Sm(2)	3.654(4)	-Sm(2)	3.072(4)

Isostructural with $NiGd_3Si_2$ (1), as are the Tb and Dy compounds.

1. Structure Reports, 48A, 54.

NICKEL SILICON YTTRIUM

$Y_3(Ni_{0.5}Si_{0.5})_2Si_2$

K. KLEPP and E. PARTHÉ, 1982. Acta Cryst., B38, 2026-2028.

Orthorhombic, Immm, a = 3.9605, b = 4.125, c = 17.63 A, Z = 2. Mo radiation, R = 0.088 for 186 reflexions.

The structure (Fig. 1) is a ternary variant of the Ta_3B_4 type (1), with double layers of face-sharing Y trigonal prisms, centred by Ni and Si in a partially ordered arrangement.

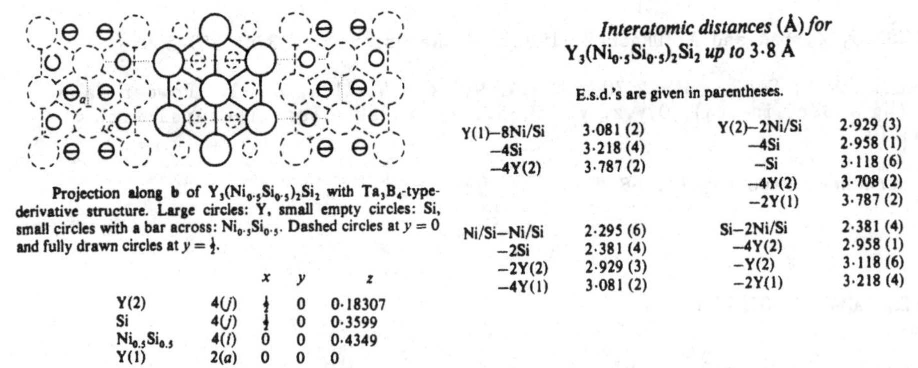

Projection along **b** of $Y_3(Ni_{0.5}Si_{0.5})_2Si_2$ with Ta_3B_4-type-derivative structure. Large circles: Y, small empty circles: Si, small circles with a bar across: $Ni_{0.5}Si_{0.5}$. Dashed circles at $y = 0$ and fully drawn circles at $y = \frac{1}{2}$.

		x	y	z
Y(2)	4(j)	½	0	0·18307
Si	4(j)	½	0	0·3599
$Ni_{0.5}Si_{0.5}$	4(i)	0	0	0·4349
Y(1)	2(a)	0	0	0

Interatomic distances (Å) *for* $Y_3(Ni_{0.5}Si_{0.5})_2Si_2$ *up to* 3·8 Å

E.s.d.'s are given in parentheses.

Y(1)–8Ni/Si	3·081 (2)	Y(2)–2Ni/Si	2·929 (3)
–4Si	3·218 (4)	–4Si	2·958 (1)
–4Y(2)	3·787 (2)	–Si	3·118 (6)
		–4Y(2)	3·708 (2)
		–2Y(1)	3·787 (2)
Ni/Si–Ni/Si	2·295 (6)	Si–2Ni/Si	2·381 (4)
–2Si	2·381 (4)	–4Y(2)	2·958 (1)
–2Y(2)	2·929 (3)	–Y(2)	3·118 (6)
–4Y(1)	3·081 (2)	–2Y(1)	3·218 (4)

Fig. 1. Structure of nickel silicon yttrium.

1. Structure Reports, 12, 32.

NICKEL SULPHUR VANADIUM
NiV_2S_4 (I), NiV_4S_8 (II)

T. MURUGESAN, S. RAMESH, J. GOPALAKRISHNAN and C.N.R. RAO, 1982. J. Solid State Chem., **44**, 119-125.

I, monoclinic, I2/m, a = 5.842, b = 3.279, c = 11.32 A, β = 92.2°, Z = 2. Powder data. Ni in 2(a): 0,0,0; V in 4(i): 0.544,1/2,0.241; S(1) in 4(i): 0.338,0,0.361; S(2) in 4(i): 0.660,0,0.109. V_3S_4-type (**1**).

II, monoclinic, F2/m (unconventional setting of C2/m), a = 11.39, b = 6.63, c = 11.20 A, β = 91.7°, Z = 4. Powder data. Ni in 4(a): 0,0,0; V(1) in 8(g): 1/4,0.312,1/4; V(2), S(1), S(2) in 8(i): x,0,z, x = 0.508, 0.206, 0.195, z = 0.242, 0.153, 0.097; S(3) in 16(j): 0.431,0.246,0.121. V_5S_8-type (**1**).

1. Structure Reports, **44**A, 101.

NICKEL TIN
Ni_3Sn_4

W. JEITSCHKO and B. JABERG, 1982. Acta Cryst., B**38**, 598-600.

Monoclinic, C2/m, a = 12.214, b = 4.060, c = 5.219 A, β = 105.0°, Z = 2. Mo radiation, R = 0.056 for 665 reflexions.

Structure (Fig. 1) essentially as previously described (**1**).

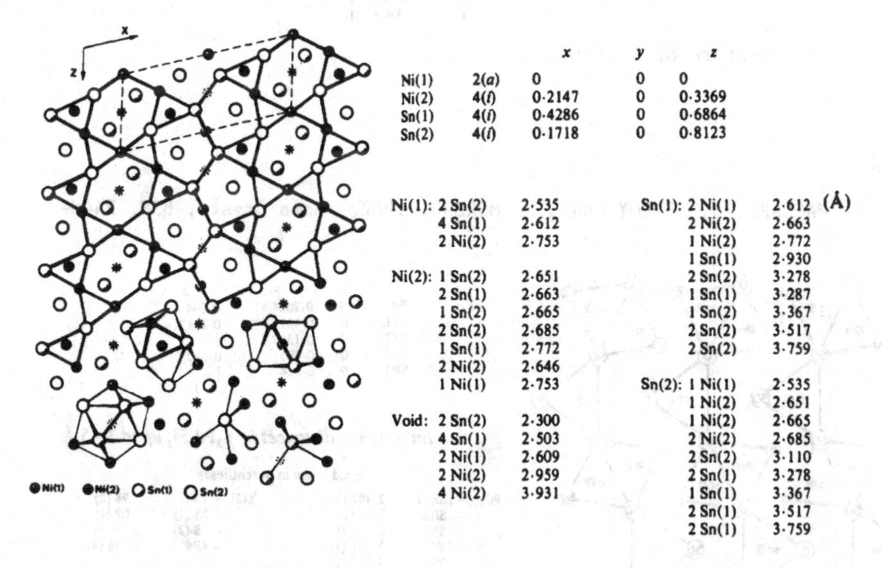

		x	y	z
Ni(1)	2(a)	0	0	0
Ni(2)	4(i)	0·2147	0	0·3369
Sn(1)	4(i)	0·4286	0	0·6864
Sn(2)	4(i)	0·1718	0	0·8123

		(Å)
Ni(1):	2 Sn(2)	2·535
	4 Sn(1)	2·612
	2 Ni(2)	2·753
Ni(2):	1 Sn(2)	2·651
	2 Sn(1)	2·663
	1 Sn(2)	2·665
	2 Sn(2)	2·685
	1 Sn(1)	2·772
	2 Ni(2)	2·646
	1 Ni(1)	2·753
Void:	2 Sn(2)	2·300
	4 Sn(1)	2·503
	2 Ni(1)	2·609
	2 Ni(2)	2·959
	4 Ni(2)	3·931

		(Å)
Sn(1):	2 Ni(1)	2·612
	2 Ni(2)	2·663
	1 Ni(2)	2·772
	1 Sn(1)	2·930
	2 Sn(2)	3·278
	1 Sn(1)	3·287
	1 Sn(2)	3·367
	2 Sn(2)	3·517
	2 Sn(2)	3·759
Sn(2):	1 Ni(1)	2·535
	1 Ni(2)	2·651
	1 Ni(2)	2·665
	2 Ni(2)	2·685
	2 Sn(2)	3·110
	2 Sn(1)	3·278
	1 Sn(1)	3·367
	2 Sn(1)	3·517
	2 Sn(1)	3·759

Fig. 1. Structure of Ni_3Sn_4.

1. Structure Reports, **10**, 77.

PALLADIUM SILICON YTTRIUM

Pd_2YSi

I. J.M. MOREAU, J. LE ROY and D. PACCARD, 1982. Acta Cryst., B38, 2446-2448.

Orthorhombic, Pnma, a = 7.303, b = 6.918, c = 5.489 A, Z = 4. Mo radiation, R = 0.11 for 219 reflexions.

Ordered Fe_3C-type (1) (Fig. 1). Other $LnPd_2Si$ and $LnPt_2Si$ compounds are isostructural.

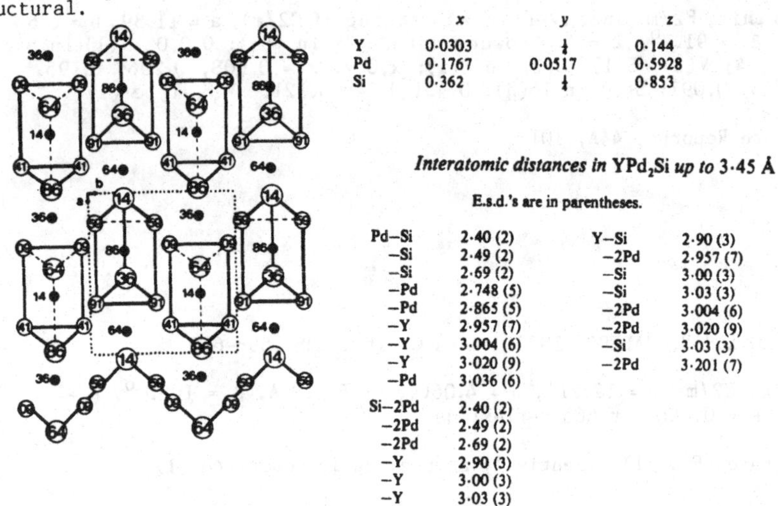

	x	y	z
Y	0·0303	¼	0·144
Pd	0·1767	0·0517	0·5928
Si	0·362	¼	0·853

Interatomic distances in YPd_2Si *up to* 3·45 Å

E.s.d.'s are in parentheses.

Pd–Si	2·40 (2)	Y–Si	2·90 (3)
–Si	2·49 (2)	–2Pd	2·957 (7)
–Si	2·69 (2)	–Si	3·00 (3)
–Pd	2·748 (5)	–Si	3·03 (3)
–Pd	2·865 (5)	–2Pd	3·004 (6)
–Y	2·957 (7)	–2Pd	3·020 (9)
–Y	3·004 (6)	–Si	3·03 (3)
–Y	3·020 (9)	–2Pd	3·201 (7)
–Pd	3·036 (6)		
Si–2Pd	2·40 (2)		
–2Pd	2·49 (2)		
–2Pd	2·69 (2)		
–Y	2·90 (3)		
–Y	3·00 (3)		
–Y	3·03 (3)		

Fig. 1. Structure of Pd_2YSi.

$Pd_2Y_3Si_3$

II. D. PACCARD, J. LE ROY and J.M. MOREAU, 1982, Acta Cryst., B38, 2448-2449.

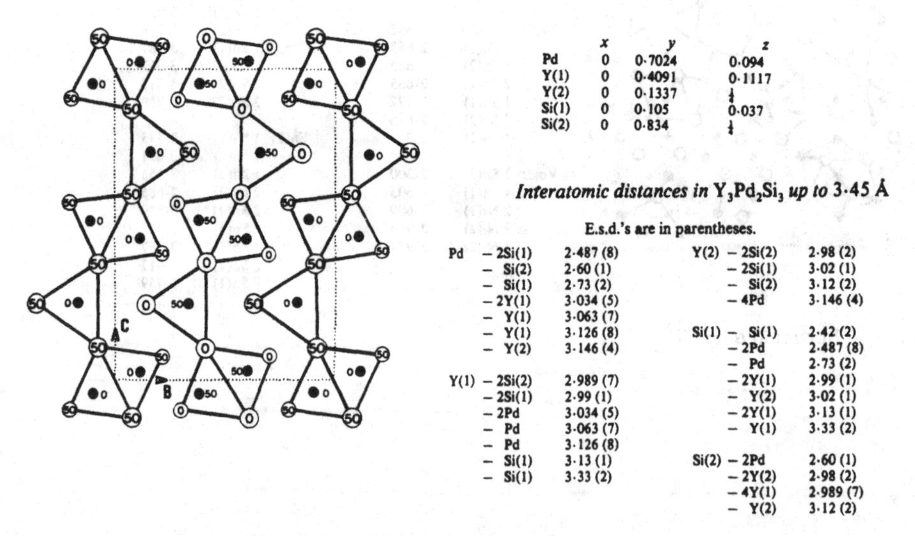

	x	y	z
Pd	0	0·7024	0·094
Y(1)	0	0·4091	0·1117
Y(2)	0	0·1337	¼
Si(1)	0	0·105	0·037
Si(2)	0	0·834	¼

Interatomic distances in $Y_3Pd_2Si_3$ *up to* 3·45 Å

E.s.d.'s are in parentheses.

Pd	– 2Si(1)	2·487 (8)	Y(2)	– 2Si(2)	2·98 (2)
	– Si(2)	2·60 (1)		– 2Si(1)	3·02 (1)
	– Si(1)	2·73 (2)		– Si(2)	3·12 (2)
	– 2Y(1)	3·034 (5)		– 4Pd	3·146 (4)
	– Y(1)	3·063 (7)			
	– Y(1)	3·126 (8)	Si(1)	– Si(1)	2·42 (2)
	– Y(2)	3·146 (4)		– 2Pd	2·487 (8)
				– Pd	2·73 (2)
Y(1)	– 2Si(2)	2·989 (7)		– 2Y(1)	2·99 (1)
	– 2Si(1)	2·99 (1)		– Y(2)	3·02 (1)
	– 2Pd	3·034 (5)		– 2Y(1)	3·13 (1)
	– Pd	3·063 (7)		– Y(1)	3·33 (2)
	– Pd	3·126 (8)			
	– Si(1)	3·13 (1)	Si(2)	– 2Pd	2·60 (1)
	– Si(1)	3·33 (2)		– 2Y(2)	2·98 (2)
				– 4Y(1)	2·989 (7)
				– Y(2)	3·12 (2)

Fig. 2. Structure of $Pd_2Y_3Si_3$

Orthorhombic, Cmcm, a = 4.251, b = 10.406, c = 14.123 A, Z = 4. Mo radiation, R = 0.08 for 182 reflexions.

Isostructural with $Hf_3Ni_2Si_3$ (2), as is $Rh_2Y_3Si_3$. The structure (Fig. 2) contains Si-centred trigonal prisms of Pd and Y atoms.

1. Strukturbericht, **2**, 33.
2. Structure Reports, **43**A, 68.

POTASSIUM SILICON TELLURIUM
$K_4Si_4Te_{10}$

B. EISENMANN and H. SCHÄFER, 1982. Z. anorg. Chem., **491**, 67-72.

Orthorhombic, Pnma, a = 21.258, b = 12.005 c = 10.608 A, Z = 4. Mo radiation, R = 0.077 for 1828 reflexions.

Adamantane-like $Si_4Te_{10}{}^{4-}$ anions of corner-sharing $SiTe_4$ tetrahedra, with anions linked by K^+ ions. Si-Te = 2.43-2.45 (terminal), 2.51-2.55 A (bridging).

POTASSIUM SILVER SELENIUM
$K_2Ag_4Se_3$

W. BRONGER and H. SCHILS, 1982. J. Less-Common Metals, **83**, 287-291.

Monoclinic, C2/m, a = 17.770, b = 4.447, c = 11.856 A, β = 108.4°, Z = 4. Mo radiation, R = 0.072 for 607 reflexions.

Isostructural with $K_2Ag_4S_3$ (1) (Fig. 1). Shortest Ag-Ag distances (2.92-3.33 A) are only a little longer than in Ag metal.

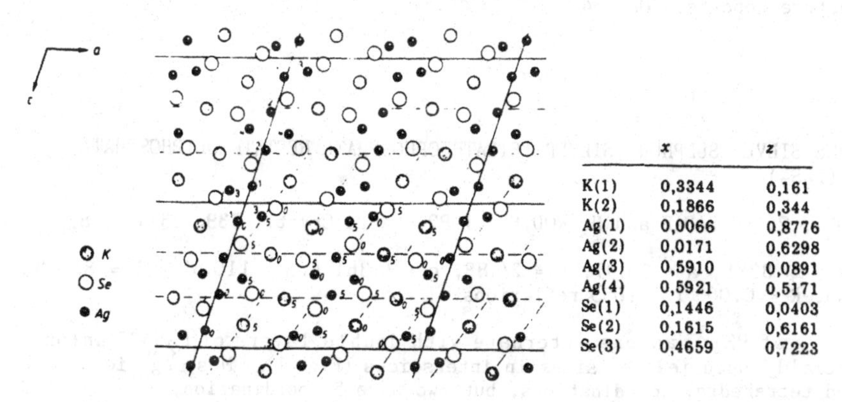

	x	z
K(1)	0,3344	0,161
K(2)	0,1866	0,344
Ag(1)	0,0066	0,8776
Ag(2)	0,0171	0,6298
Ag(3)	0,5910	0,0891
Ag(4)	0,5921	0,5171
Se(1)	0,1446	0,0403
Se(2)	0,1615	0,6161
Se(3)	0,4659	0,7223

Fig. 1. Structure of $K_2Ag_4Se_3$ (y = 0).

1. Structure Reports, **42**A, 120.

PHOSPHORUS SILVER SULPHUR (SILVER HEXATHIODIPHOSPHATE)
$Ag_4P_2S_6$

P. TOFFOLI, A. MICHELET, P. KHODADAD and N. RODIER, 1982. Acta Cryst., **B38**, 706-710.

Monoclinic, $P2_1/b$, a = 6.522, c = 19.616, c = 11.797 A, γ = 93.58°, Z = 6. Mo radiation, R = 0.047 for 1833 reflexions.

Layers of staggered $S_3P-PS_3^{4-}$ anions linked by tetrahedrally coordinated Ag^+ ions (Fig. 1). P-P = 2.26, P-S = 2.01-2.04, Ag-S = 2.46-3.11 A. An orthorhombic form is also known (**1**).

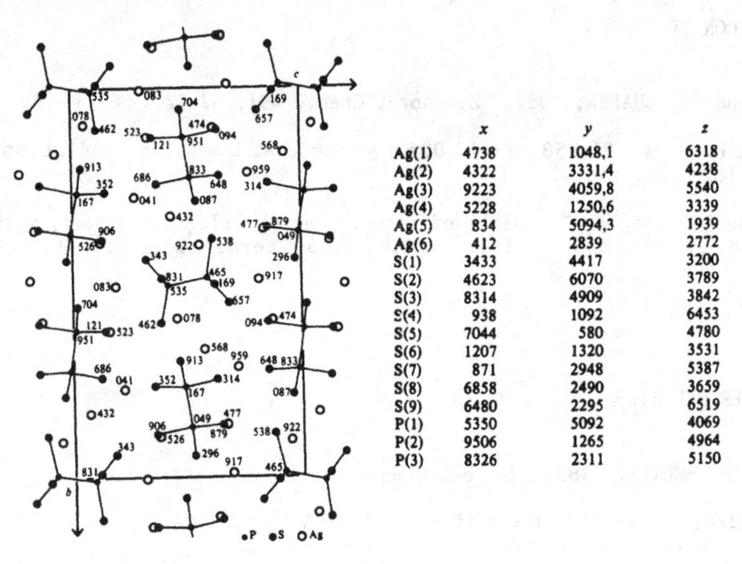

	x	y	z
Ag(1)	4738	1048,1	6318
Ag(2)	4322	3331,4	4238
Ag(3)	9223	4059,8	5540
Ag(4)	5228	1250,6	3339
Ag(5)	834	5094,3	1939
Ag(6)	412	2839	2772
S(1)	3433	4417	3200
S(2)	4623	6070	3789
S(3)	8314	4909	3842
S(4)	938	1092	6453
S(5)	7044	580	4780
S(6)	1207	1320	3531
S(7)	871	2948	5387
S(8)	6858	2490	3659
S(9)	6480	2295	6519
P(1)	5350	5092	4069
P(2)	9506	1265	4964
P(3)	8326	2311	5150

Fig. 1. Structure of $Ag_4P_2S_6$, and atomic positional parameters (x 10^4).

1. Structure Reports, **50**A, 54.

PHOSPHORUS SILVER SULPHUR (SILVER TETRATHIOPHOSPHATE HEPTATHIODIPHOSPHATE)
$Ag_7(PS_4)(P_2S_7)$

P. TOFFOLI, P. KHODADAD and N. RODIER, 1982. Acta Cryst., **B38**, 2374-2378.

Monoclinic, B2/b, a = 23.97, b = 24.88, c = 6.361 A, γ = 110.85°, Z = 8. Mo radiation, R = 0.066 for 1655 reflexions.

Layers of PS_4^{3-} anions alternate with double layers of $P_2S_7^{4-}$ anions, with partially-occupied Ag^+ sites in interstices (Fig. 1). Most Ag ions have distorted tetrahedral coordinations, but two have 5-coordination,

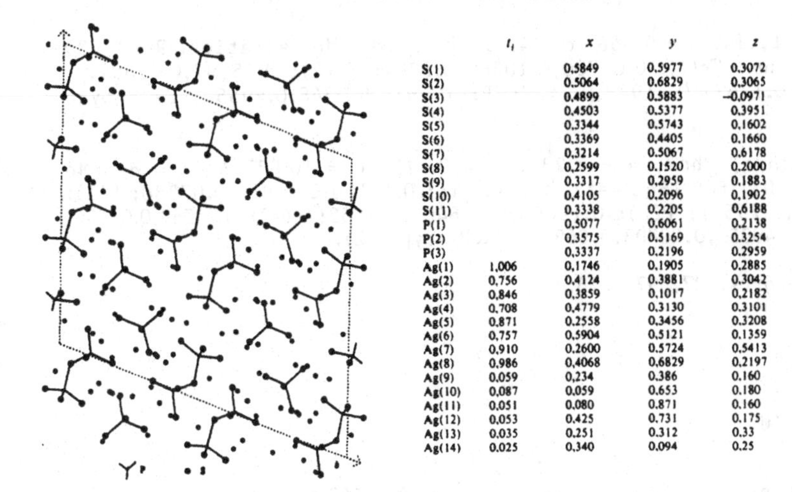

	t_1	x	y	z
S(1)		0,5849	0,5977	0,3072
S(2)		0,5064	0,6829	0,3065
S(3)		0,4899	0,5883	−0,0971
S(4)		0,4503	0,5377	0,3951
S(5)		0,3344	0,5743	0,1602
S(6)		0,3369	0,4405	0,1660
S(7)		0,3214	0,5067	0,6178
S(8)		0,2599	0.1520	0,2000
S(9)		0,3317	0,2959	0,1883
S(10)		0,4105	0,2096	0,1902
S(11)		0,3338	0,2205	0,6188
P(1)		0,5077	0,6061	0,2138
P(2)		0,3575	0,5169	0,3254
P(3)		0,3337	0,2196	0,2959
Ag(1)	1,006	0,1746	0,1905	0,2885
Ag(2)	0,756	0,4124	0,3881	0,3042
Ag(3)	0,846	0,3859	0,1017	0,2182
Ag(4)	0,708	0,4779	0,3130	0,3101
Ag(5)	0,871	0,2558	0,3456	0,3208
Ag(6)	0,757	0,5904	0,5121	0,1359
Ag(7)	0,910	0,2600	0,5724	0,5413
Ag(8)	0,986	0,4068	0,6829	0,2197
Ag(9)	0,059	0,234	0,386	0,160
Ag(10)	0,087	0,059	0,653	0,180
Ag(11)	0,051	0,080	0,871	0,160
Ag(12)	0,053	0,425	0,731	0,175
Ag(13)	0,035	0,251	0,312	0,33
Ag(14)	0,025	0,340	0,094	0,25

Fig. 1. Structure of $Ag_7(PS_4)(P_2S_7)$.

PHOSPHORUS TECHNETIUM
TcP_3

I. R. RÜHL and W. JEITSCHKO, 1982. Acta Cryst., **B38**, 2784-2788.

Orthorhombic, Pnma, a = 15.359, b = 3.092, c = 5.142 A, Z = 4. Mo radiation, R = 0.022 for 996 reflexions.

The structure (Fig. 1) contains TcP_6 octahedra with metal-metal bonds across common edges; P atoms are coordinated tetrahedrally to Tc and P. ReP_3 is isostructural.

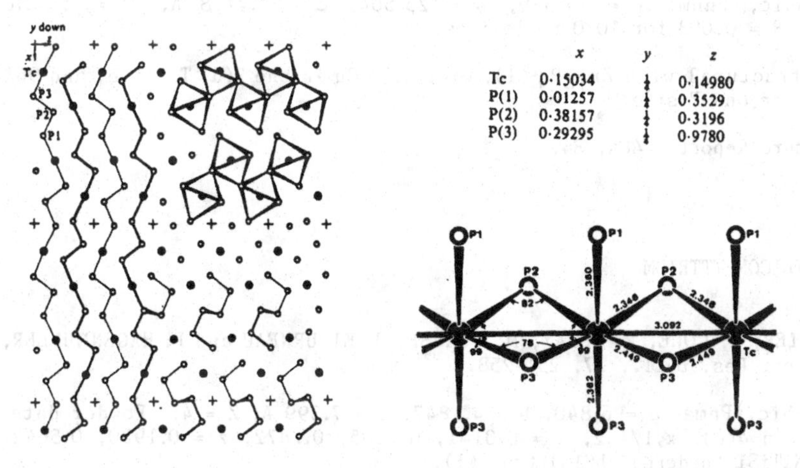

	x	y	z
Tc	0·15034	¼	0·14980
P(1)	0·01257	¼	0·3529
P(2)	0·38157	¼	0·3196
P(3)	0·29295	¼	0·9780

Fig. 1. Structure of TcP_3.

Tc_3P, TcP_4

II. R. RÜHL, W. JEITSCHKO and K. SCHWOCHAU, 1982. J. Solid State Chem., **44**, 134-140.

Tc_3P, tetragonal, $I\bar{4}$, a = 9.568, c = 4.736 A, Z = 8. Mo radiation, R = 0.039 for 1698 reflexions. Tc(1): 0.08700,0.10569,0.2204; Tc(2): 0.35297,0.02591, 0.9865; Tc(3): 0.17672,0.21948,0.7427; P: 0.2916,0.0348,0.4955. Fe_3P-type (**1**).

TcP_4, orthorhombic, Pbca, a = 6.238, b = 9.215, c = 10.837 A, Z = 8. Mo radiation, R = 0.031 for 1848 reflexions. Tc: 0.12136,0.10916,0.07538; P(1): 0.05916,0.28166,0.23571; P(2): 0.19065,0.94640,0.23853; P(3): 0.75589,0.04366, 0.08122; P(4): 0.49952,0.20803,0.06899. ReP_4-type (**2**).

1. Structure Reports, **27**, 97, 275.
2. Ibid., **45**A, 104.

PHOSPHORUS TITANIUM
Ti_7P_4

W. CARRILLO-CABRERA, 1982. Acta Chem. Scand., A**36**, 563-570.

Monoclinic, C2/m, a = 14.474, b = 3.4094, c = 13.616 A, β = 104.597°, Z = 4. Mo radiation, R = 0.056 for 1126 reflexions.

Isostructural with Nb_7P_4 (**1**), with 0.92 Cu in one 4(i) Ti site.

1. Structure Reports, **31**A, 51.

COPPER PHOSPHORUS TITANIUM
$Ti_{13.86}Cu_{0.14}P_9$

W. CARRILLO-CABRERA, 1982. Acta Chem. Scand., A**36**, 571-578.

Orthorhombic, Pnnm, a = 15.509, b = 25.564, c = 3.4718 A, Z = 4. Mo radiation, R = 0.093 for 1070 reflexions.

Isostructural with $Zr_{14}P_9$ (**1**), with 0.56 Cu in the 2(d) Ti site, and 96% occupancy for one P site.

1. Structure Reports, **48**A, 84.

RHODIUM SILICON YTTRIUM
RhYSi

B. CHEVALIER, A. COLE, P. LEJAY, M. VLASSE, J. ETOURNEAU and P. HAGENMULLER, 1982. Mater. Res. Bull., **17**, 251-258.

Orthorhombic, Pnma, a = 6.840, b = 4.1847, c = 7.399 A, Z = 4. Powder data. Rh, Y, Si in 4(c): x,1/4,z, x = 0.0141, 0.1505, 0.7872, z = 0.1920, 0.5642, 0.6070. NiTiSi (ordered $PbCl_2$) type (**1**).

1. Structure Reports, **30**A, 75.

SELENIUM VANADIUM (1T)
VSe_2

J. RIGOULT, C. GUIDI-MOROSINI and A. TOMAS, 1982 Acta Cryst., **B38**, 1557-1559.

Trigonal, $P\bar{3}m1$, a = 3.356, c = 6.104 A, Z = 1. Ag radiation, R = 0.022 for 389 reflexions. 1 V(1) in 1(a): 0,0,0; 0.005 V(2) in 1(b): 0,0,1/2; 2 Se in 2(d): 1/3,2/3,0.25665.

CdI_2-type structure, as previously described (1), with minor occupancy of the normally-vacant octahedral site. V-Se = 2.492, 2.441 A.

1. Structure Reports, **42A**, 146; **45A**, 111.

SILICON SULPHUR
SiS_2

SELENIUM SILICON
$SiSe_2$

J. PETERS and B. KREBS, 1982. Acta Cryst., **B38**, 1270-1272.

Orthorhombic, Ibam, a = 9.583, 9.669, b = 5.614, 5.998, c = 5.547, 5.851 A, at 138, 293K, Z = 4. Mo radiation, R = 0.014, 0.018 for 200, 253 reflexions.

The structures (Fig. 1) are essentially as previously described (1), with chains along **c** of edge-sharing SiX_4 tetrahedra.

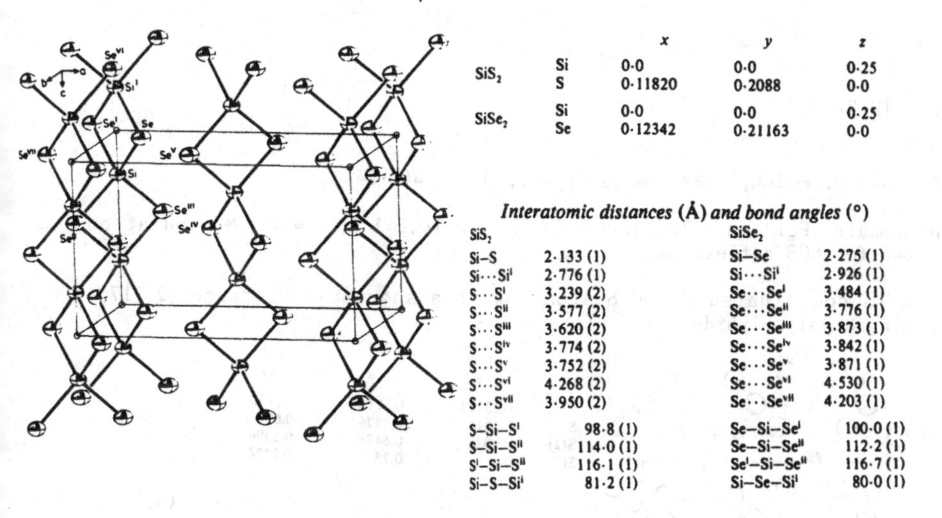

		x	y	z
SiS_2	Si	0·0	0·0	0·25
	S	0·11820	0·2088	0·0
$SiSe_2$	Si	0·0	0·0	0·25
	Se	0·12342	0·21163	0·0

Interatomic distances (Å) *and bond angles* (°)

SiS_2		$SiSe_2$	
Si–S	2·133 (1)	Si–Se	2·275 (1)
Si···Si^{i}	2·776 (1)	Si···Si^{i}	2·926 (1)
S···S^{i}	3·239 (2)	Se···Se^{i}	3·484 (1)
S···S^{ii}	3·577 (2)	Se···Se^{ii}	3·776 (1)
S···S^{iii}	3·620 (2)	Se···Se^{iii}	3·873 (1)
S···S^{iv}	3·774 (2)	Se···Se^{iv}	3·842 (1)
S···S^{v}	3·752 (2)	Se···Se^{v}	3·871 (1)
S···S^{vi}	4·268 (2)	Se···Se^{vi}	4·530 (1)
S···S^{vii}	3·950 (2)	Se···Se^{vii}	4·203 (1)
S–Si–S^{i}	98·8 (1)	Se–Si–Se^{i}	100·0 (1)
S–Si–S^{ii}	114·0 (1)	Se–Si–Se^{ii}	112·2 (1)
S^{i}–Si–S^{ii}	116·1 (1)	Se^{i}–Si–Se^{ii}	116·7 (1)
Si–S–Si^{i}	81·2 (1)	Si–Se–Si^{i}	80·0 (1)

Fig. 1. Structures of SiS_2 and $SiSe_2$.

1. Strukturbericht, **3**, 37, 286; Structure Reports, **17**, 457.

SILICON SULPHUR THALLIUM
SELENIUM SILICON THALLIUM
GERMANIUM SELENIUM THALLIUM
$Tl_4Si_2S_6$, $Tl_4Si_2Se_6$, $Tl_4Ge_2Se_6$

G. EULENBERGER, 1982. Mh. Chem., **113**, 859-867.

Triclinic, $P\bar{1}$, a = 6.699, 6.875, 6.925, b = 6.645, 6.866, 6.934, c = 8.380, 8.731, 8.771 A, α = 90.32, 90.50, 90.55, β = 112.00, 111.69, 111.42, γ = 112.32, 113.70, 114.45°, Z = 1. Mo radiation, R = 0.060, 0.071, 0.098 for 1076, 1009, 964 reflexions.

Isostructural with $Tl_4Ge_2S_6$ (**1**), with $M_2X_6^{4-}$ ions of two edge-sharing MX_4 tetrahedra, linked by 8- and 7-coordinate Tl ions.

1. Structure Reports, **44**A, 66.

SULPHUR THALLIUM TIN
$Tl_2Sn_2S_3$

S. DEL BUCCHIA, J.C. DUMAS, E. PHILIPPOT and M. MAURIN, 1982. Z. anorg. Chem., **487**, 199-206.

Monoclinic, C2/c, a = 13.887, b = 7.742, c = 7.267 A, β = 105.39°, Z = 4. Mo radiation, R = 0.086 for 382 reflexions. Tl in 8(f): 0.6189,0.6168,0.0302; Sn in 8(f): 0.8831,0.6547,0.5271; S(1) in 4(e): 1/2,0.898,3/4; S(2) in 8(f): 0.739,0.916,0.179.

Defect NaCl-type structure. Tl and Sn are each coordinated to 4 S, with stereochemically-active lone-pair electrons. Tl-S = 2.81-3.13, Sn-S = 2.68-3.11 A.

SULPHUR THORIUM
Th_2S_5

H. NOËL and M. POTEL, 1982. Acta Cryst., B**38**, 2444-2445.

Orthorhombic, Pcnb, a = 7.623, b = 7.677, c = 10.141 A, Z = 4. Mo radiation, R = 0.048 for 608 reflexions.

Th is coordinated to 10 S atoms, with a short S-S distance (2.117 A) indicating a polysulphide (Fig. 1).

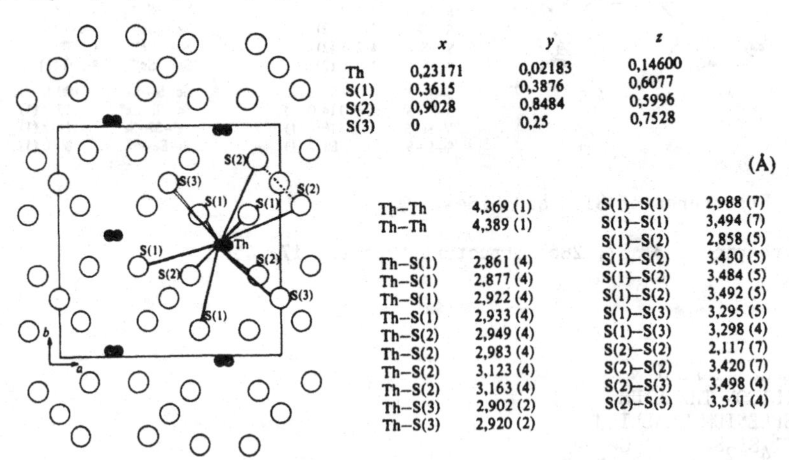

	x	y	z
Th	0,23171	0,02183	0,14600
S(1)	0,3615	0,3876	0,6077
S(2)	0,9028	0,8484	0,5996
S(3)	0	0,25	0,7528

(Å)

Th–Th	4,369 (1)	S(1)–S(1)	2,988 (7)
Th–Th	4,389 (1)	S(1)–S(1)	3,494 (7)
		S(1)–S(2)	2,858 (5)
Th–S(1)	2,861 (4)	S(1)–S(2)	3,430 (5)
Th–S(1)	2,877 (4)	S(1)–S(2)	3,484 (5)
Th–S(1)	2,922 (4)	S(1)–S(2)	3,492 (5)
Th–S(1)	2,933 (4)	S(1)–S(3)	3,295 (5)
Th–S(2)	2,949 (4)	S(1)–S(3)	3,298 (4)
Th–S(2)	2,983 (4)	S(2)–S(2)	2,117 (7)
Th–S(2)	3,123 (4)	S(2)–S(2)	3,420 (7)
Th–S(2)	3,163 (4)	S(2)–S(3)	3,498 (4)
Th–S(3)	2,902 (2)	S(2)–S(3)	3,531 (4)
Th–S(3)	2,920 (2)		

Fig. 1. Structure of Th_2S_5.

SULPHUR TIN
Sn_2S_3

R. KNIEP, D. MOOTZ, U. SEVERIN and H. WUNDERLICH, 1982. Acta Cryst., **B38**, 2022-2023.

Orthorhombic, Pnma, a = 8.878, b = 3.751, c = 14.020 A, Z = 4. Mo radiation, R = 0.04 for 1335 reflexions.

Structure (Fig. 1) as previously described (1).

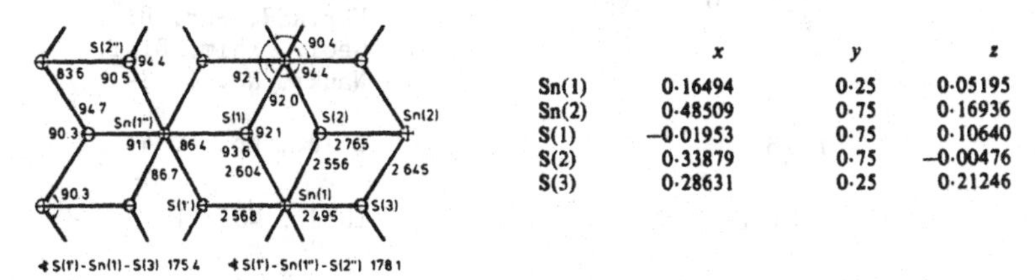

	x	*y*	*z*
Sn(1)	0·16494	0·25	0·05195
Sn(2)	0·48509	0·75	0·16936
S(1)	−0·01953	0·75	0·10640
S(2)	0·33879	0·75	−0·00476
S(3)	0·28631	0·25	0·21246

Fig. 1. Structure of Sn_2S_3 (bond lengths in A, angles in degrees).

1. Structure Reports, **32A**, 128.

TABLE I

Some structural information has also been given for the following materials (listed with abbreviated 1982 references).

Compound	Structure	Reference
$Mg_2NiH_{0.3}$	Mg_2Ni, with an increase in one Mg-Mg distance from 3.06 to 3.73 A, suggesting an H site between these atoms	Acta Chem. Scand., **A36**, 847
K	Anharmonic atom vibrations	Acta Cryst., **A38**, 3
Cd	Anharmonic atom vibrations	Ibid., **A38**, 10
Pyrrhotite, $Fe_{1-x}S$	Modulated structure	Ibid., **A38**, 79
Cr	Aspherical charge distribution	Ibid., **A38**, 103
Li	Anharmonic atom vibrations	Ibid., **A38**, 163
$Au_{22}Mn_6$	One-dimensional antiphase structure	Ibid., **A38**, 269
SiC	45Rb and 66R polytypes	Ibid., **A38**, 477; **B38**, 1703
Fe	Aspherical charge distribution	Ibid., **A38**, 725
AuCu-II	Modulated structure	Ibid., **B38**, 1446
$Ni_{11}Zr_9$	$Pt_{11}Zr_9$	Ibid., **B38**, 2092

$BiCu_9S_6$	Modulated structure	Amer. Min., **67**, 360
$AgGaGeS_4$	Fdd2, a = 12.028, b = 22.918, c = 6.874 A, Z = 2, R = 0.040. Chains of edge-sharing GeS_4 tetrahedra, linked by tetrahedrally-coordinated Ag and Ga	Deposited Doc., VINITI 6319-82, 59-64; Chem. Abs., **100**, 112620
$CoGa_2Zr_6$ GaMnY	β_1-K_2UF_6 FeTiSi	Dopov. Akad. Nauk Ukr.RSR, Ser. B: Geol., Khim. Biol. Nauki, No. 4, 39
$CeNi_2Sb_2$	Al_4B	Ibid., No. 4, 44
CoGaPr CoGaSm CoGaGd CoErGa	TiNiSi	Ibid., No. 11, 57
Fe-Sc-Ga etc.	$ThMn_{12}$ and Th_2Zn_{17} types	Ibid., No. 12, 30
In_2MgS_4	Partially-inverted spinel, x(O) = 0.382	Jpn. J. Appl. Phys., Part 1, **21**, 958
Au-Cd	Antiphase domain structure	J. Appl. Cryst., **15**, 174
Pr (high-pressure)	α-U	J. Appl. Phys., **53**, 9212
Ir_3Yb_5	Rh_3Pu_5 and Mn_5Si_3	J. Less-Common Metals, **83**, L1
PaP_2 Pa_3P_4	anti-Fe_2As Th_3P_4	Ibid., **83**, 169
$NpAs_2$ Np_3As_4	anti-Fe_2As type, P4/nmm, a = 3.930, c = 8.137 A, at 4K, z(Np) = 0.281; z(As) = 0.639 Th_3P_4	Ibid., **83**, 263
Co_2Si_3 Ge_7Re_4	Ru_2Sn_3 Si_7Tc_4	Ibid., **84**, 87
Dy_6Fe_{23}	Th_6Mn_{23}	Ibid., **84**, 93
Np_2Se_5	Th_2S_5	Ibid., **84**, 133
Bi_3Th_5 BiTh Bi_4Th_3 Bi_2Th	Si_3Mn_5 CsCl P_4Th_3 Cu_2Sb	Ibid., **84**, 165
$(Co,Ni)_5Y$	$CaCu_5$	Ibid., **84**, 201
$CaNi_5H_x$	$CaNi_5$ ($CaCu_5$) with doubled **c** axis	Ibid., **84**, 263

$InSm_3$ $InSm_2$ InSm In_5Sm_3 In_3Sm	$AuCu_3$ or Cu $InNi_2$ CsCl or W Orthorhombic, related to Pd_5Pu_3 $AuCu_3$	J. Less-Common Metals, **84**, 281
InLnZn CuLnPb	$CaIn_2$	Ibid., **84**, 301
$Ni_{19}Sm_5$	Polytypes	Ibid., **84**, 317
ReSi $ReSi_2$	FeSi $MoSi_2$	Ibid., **85**, 27
$Al_2(Pd,Si)$	CaF_2	Ibid., **85**, L1
DyPd (low-temp.)	FeB	Ibid., **85**, 181
Ce_3Sn Ce_5Sn_3 Ce_5Sn_4 $CeSn_3$	$AuCu_3$ W_5Si_3 and Mn_5Si_3 Sm_5Ge_4 $AuCu_3$	Ibid., **85**, 195
$CoZr_2$	Ni_2Ti and Al_2Cu	Ibid., **85**, 221
Ca_3Pd Ca_5Pd_2 Ca_3Pd_2 CaPd $CaPd_2$	Fe_3C Mn_5C_2 Er_3Ni_2 CsCl $MgCu_2$	Ibid., **85**, 307
Co_7Y_9	Structure suggested	Ibid., **86**, 29
α-RuP_4 α-OsP_4	CdP_4	Ibid., **86**, 247
$RuTe_2$	Marcasite	Ibid., **86**, L13
CoGe GeRh	FeSi	Ibid., **87**, 53
Al_3Pd_5 Al_3Pd_2 Al_4Pd	Ge_3Rh_5 Al_3Ni_2 Al_4Pt	Ibid., **87**, 117
CuDy Cu_2Dy Cu_5Dy Cu_7Dy	CsCl $CeCu_2$ $CaCu_5$ and $AuBe_5$ Cu_7Tb	Ibid., **87**, 249
As_2Eu_2	Na_2O_2	Ibid., **87**, 327
LnM_2Si_2, Ln = Gd, Dy, Ho, Er M = Ru, Rh, Pd, Ir	$ThCr_2Si_2$	Ibid., **87**, L1
Mg_2NiH_4	Metal positions determined for a monoclinic phase	Ibid., **88**, 63
$TiCr_{1.8}H_{5.3}$	Disordered fluorite	Ibid., **88**, 107

(Cr,Mn)B FeB (low-temp.)	CrB and FeB Structure proposed	J. Solid State Chem., **41**, 195
NbS_3 TaS_3 $TaSe_3$	$NbSe_3$	Ibid., **41**, 315
TiS_2 NbS_2 TaS_2	Phases with intercalated ethylenediamine	Ibid., **45**, 119
Aktashite, $Cu_6Hg_3As_4S_{12}$ Nowackiite, $Cu_6Zn_3As_4S_{12}$	It is pointed out that the two materials are isostructural [already noted in Structure Reports, **32A**, 225; **46A**, 17]	Kristallografija, **27**, 49 [Soviet Physics - Crystallography, **27**, 26]
$(Nb,W)Se_2$ $(Mo,Ta)Se_2$ $(Ta,W)Se_2$	3R-MoS_2	Ibid., **27**, 385 [Ibid., **27**, 232]
V_3Si	Electron-density study	Ibid., **27**, 606 [Ibid., **27**, 367]
Gd	$P6_3/mmc$ and Fm3m	Ibid., **27**, 798 [Ibid., **27**, 480]
FeMoZr	$MgZn_2$ [Strukturbericht, **1**, 180]. $P6_3/mmc$, a = 5.172, c = 8.463 A. Zr in 4(f): z = 0.0625; Mo,Fe in 2(a) and 6(h): x = 0.833	Ibid., **27**, 900 [Ibid., **27**, 540]
Ni_3N	Structure as previously described [Structure Reports, **13**, 140]	Ibid., **27**, 923 [Ibid., **27**, 554]
$Cr_2(Cu,Zn)S_4$	Spinel	Mater. Res. Bull., **17**, 25
AgTl(Se,Te)	AgTlTe [Structure Reports, **46A**, 117]	Ibid., **17**, 533
α-Si_3N_4 β-Si_3N_4	Variation of atomic parameters with fast neutron bombardment	Ibid., **17**, 851
Ln_2MB_6 Ln = lanthanon M = Ru, Os	Y_2ReB_6	Rare Earths Mod. Sci. Technol., **3**, 353
NbS_2	Li intercalated material	Rev. Chim. Minér., **19**, 309
Li_2VSe_2	VSe_2 [Structure Reports, **42A**, 146; **45A**, 111], with Li in two tetrahedral holes	Ibid., **19**, 352

$(Cr,Fe)_{23}C_6$	Non-random site-occupation parameters are given	Scr. Metall., **16**, 453
$MnSi_{0.25}$	$Ni_{12}P_5$	Z. Kristallogr., **158**, 205
Americium, Am	Four phases	Ibid., **158**, 307

PAPERS REFERRED TO LATER YEARS

Preliminary notes have not been reported, since fuller accounts will appear at a later date. The compounds studied, and abbreviated 1982 references, are listed below.

$Cs_7Fe_4Te_8$	Angew. Chem., **94**, 562 [Structure Reports, **50A**, 19]
$Ag_7P_3S_{11}$	C.R. Acad. Sci., Ser II, **294**, 87 [This volume, p. 52]
Cu_3As_4	Z. Kristallogr., **159**, 17
$Cs_2Mn_3S_4$	Ibid., **159**, 26
Wittichenite, Cu_3BiS_3 Skinnerite, Cu_3SbS_3	Ibid., **159**, 92
AuMgSb	Ibid., **159**, 101
$AgTe_3$	Ibid., **159**, 138

STRUCTURE REPORTS

SECTION II

INORGANIC COMPOUNDS

Edited by

J. Trotter

(University of British Columbia)

with the assistance of

S.V. Evans

ARRANGEMENT

To find particular inorganic compounds the subject index or formula index should be used. The general arrangement is: elements, boron hydrides, carbonyls, phosphorus-nitrogen and sulphur-nitrogen compounds, halides, cyanides, oxides, double oxides, hydroxides, sulphides, borates, carbonates, nitrates, phosphates, arsenates, sulphates, perchlorates, iodates, silicates, silicate minerals, electron-diffraction studies. Only complete structure analyses are described; incomplete structural data are given in a Table, and compounds which have been described only in preliminary communications are tabulated.

OXYGEN
α'-O_2

E.M. HÖRL and F. KOHLBECK, 1982. Acta Cryst., B38, 20-23.

Monoclinic, space group not determined, a = 9.225, b = 6.668, c = 3.414 Å, β = 85.05°, at 4.2K, Z = 6. Electron-diffraction patterns. Structure not fully determined.

GADOLINIUM CARBIDE CHLORIDES
$Gd_{10}C_4Cl_{18}$, $Gd_{10}C_4Cl_{17}$

E. WARKENTIN, R. MASSE and A. SIMON, 1982. Z. anorg. Chem., 491, 323-336.

$Gd_{10}C_4Cl_{18}$, monoclinic, $P2_1/c$, a = 9.182, b = 16.120, c = 12.886 Å, β = 119.86°, Z = 2. Mo radiation, R = 0.026 for 3639 reflexions.

$Gd_{10}C_4Cl_{17}$, triclinic, $P\bar{1}$, a = 8.498, b = 9.174, c = 11.462 Å, α = 104.56, β = 95.98, γ = 111.35°, Z = 1. Mo radiation, R = 0.043 for 3560 reflexions.

The structures (Fig. 1) contain $Gd_{10}C_4Cl_{18}$ clusters, which consist of two edge-sharing Gd_6 octahedra, centred by C_2 units (C-C = 1.47 Å) with Cl bridging all available edges; in the Cl_{17} compound the clusters share a Cl atom to form chains.

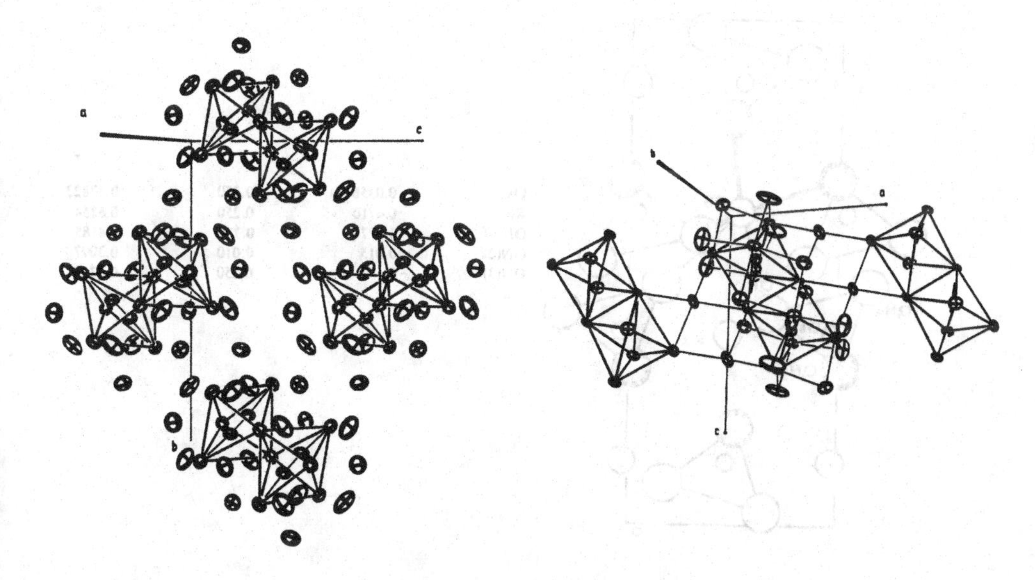

Fig. 1. Structure of $Gd_{10}C_4Cl_{18}$ (left) and $Gd_{10}C_4Cl_{17}$ (right).

POTASSIUM NITRIDOTRIOXOOSMATE(VIII)
$KOsO_3N$

Y. LAURENT, R. PASTUSZAK, P. L'HARIDON and R. MARCHAND, 1982. Acta Cryst., B38, 914-916.

Tetragonal, $I4_1/a$, a = 5.646, c = 13.027 Å, Z = 4. Mo radiation, R = 0.036 for 256 reflexions. Os in 4(b): 0,1/4,5/8; K in 4(a): 0,1/4,1/8; O,N in 16(f): -0.104, 0.479,0.6997 [origin at $\bar{1}$].

Scheelite-type structure (1), with OsX_4 tetrahedra (X = disordered N, O) linked by 8-coordinate K. Os-X = 1.72, K-X = 2.81, 2.88 Å.

1. Strukturbericht, 1, 347; Structure Reports, 29, 331.

RUBIDIUM NITRIDOTRIOXOOSMATE(VIII)
$RbOsO_3N$

P. L'HARIDON, R. PASTUSZAK and Y. LAURENT, 1982. J. Solid State Chem., 43, 29-32.

Orthorhombic, Pnma, a = 5.571, b = 5.794, c = 13.735 Å, Z = 4. Mo radiation, R = 0.046 for 556 reflexions.

The structure (Fig. 1) contains disordered OsO_3N^- tetrahedra linked by 8-coordinate Rb^+ ions. Os-O,N = 1.71-1.72, Rb-O,N = 2.93-3.17(1) Å.

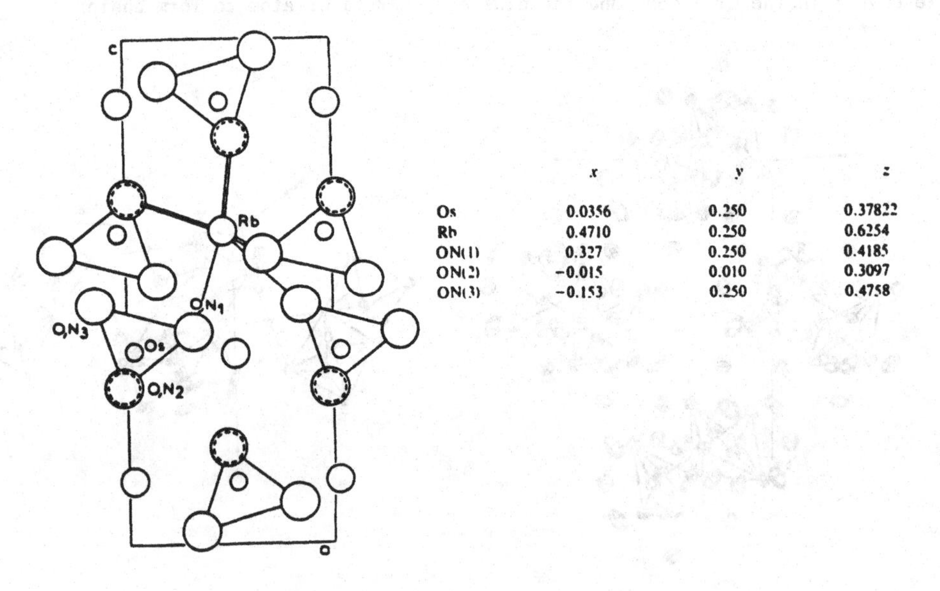

	x	y	z
Os	0.0356	0.250	0.37822
Rb	0.4710	0.250	0.6254
ON(1)	0.327	0.250	0.4185
ON(2)	−0.015	0.010	0.3097
ON(3)	−0.153	0.250	0.4758

Fig. 1. Structure of $RbOsO_3N$.

CAESIUM NITRIDOTRIOXOOSMATE(VIII)
$CsOsO_3N$

R. PASTUSZAK, P. L'HARIDON, R. MARCHAND and Y. LAURENT, 1982. Acta Cryst., B38, 1427-1430.

Orthorhombic, Pnma, a = 8.409, b = 7.242, c = 8.089 Å, Z = 4. Mo radiation, R = 0.043 for 561 reflexions.

The structure (Fig. 1) consists of OsO_3N^- tetrahedra linked by 10-coordinate Cs^+ ions. Os-O = 1.74, Os-N = 1.68, Cs-O,N = 3.09-3.44(1) Å.

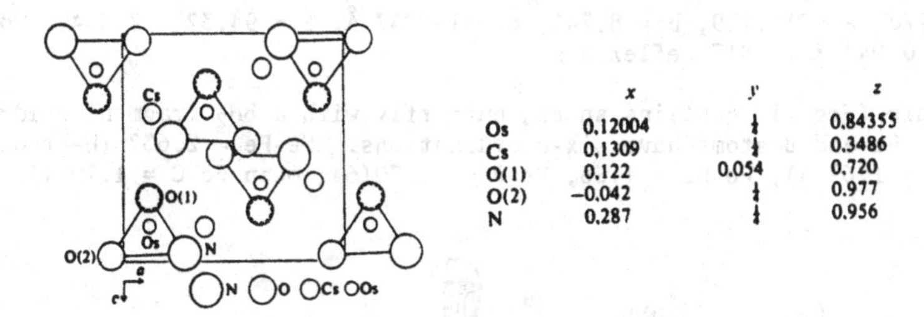

	x	y	z
Os	0.12004	1/4	0.84355
Cs	0.1309	1/4	0.3486
O(1)	0.122	0.054	0.720
O(2)	−0.042	1/4	0.977
N	0.287	1/4	0.956

Fig. 1. Structure of $CsOsO_3N$.

1,10-BIS(DINITROGEN)-closo-DECABORANE(10)
$B_{10}H_8(N_2)_2$

T. WHELAN, P. BRINT, T.R. SPALDING, W.S. McDONALD and D.R. LLOYD, 1982. J. Chem. Soc., Dalton, 2469-2473.

Monoclinic, I2/a, a = 11.411, b = 6.658, c = 13.058 Å, β = 90.76°, Z = 4. Mo radiation, R = 0.043 for 564 reflexions.

The molecule has a closo B_{10} bicapped square antiprism cage with approximate D_{4d} symmetry (Fig. 1). Mean B(1)-B(2) = 1.668, B(2)-B(3) = 1.881, B(2)-B(2') = 1.799(3); B(1)-N(1) = 1.499(2), N(1)-N(2) = 1.091(2) Å.

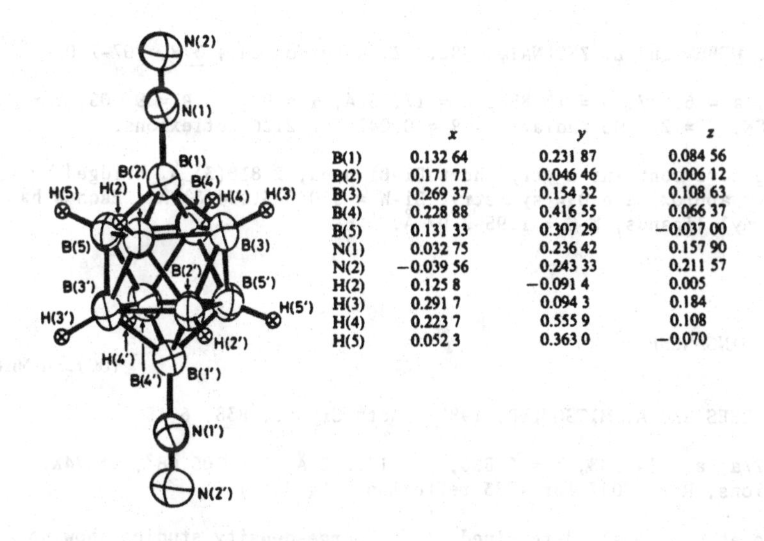

	x	y	z
B(1)	0.132 64	0.231 87	0.084 56
B(2)	0.171 71	0.046 46	0.006 12
B(3)	0.269 37	0.154 32	0.108 63
B(4)	0.228 88	0.416 55	0.066 37
B(5)	0.131 33	0.307 29	−0.037 00
N(1)	0.032 75	0.236 42	0.157 90
N(2)	−0.039 56	0.243 33	0.211 57
H(2)	0.125 8	−0.091 4	0.005
H(3)	0.291 7	0.094 3	0.184
H(4)	0.223 7	0.555 9	0.108
H(5)	0.052 3	0.363 0	−0.070

Fig. 1. Structure of $B_{10}H_8(N_2)_2$.

HYDRIDODODECACARBONYLTETRAFERRABORANE
$HFe_4(BH_2)(CO)_{12}$

K.S. WONG, W.R. SCHEIDT and T.P. FEHLNER, 1982. J. Amer. Chem. Soc., 104, 1111-1113.

Monoclinic, $P2_1/c$, a = 16.429, b = 8.740, c = 13.237 Å, β = 94.32°, Z = 4. Mo radiation, R = 0.044 for 2817 reflexions.

The molecule (Fig. 1) contains an Fe_4 butterfly with a BH_2 fragment bridging the wing tips. Fe and B atoms have six-coordinations. Fe-Fe = 2.637 (H-bridged), 2.667(1), Fe-H = 1.67(4), Fe-HB = 1.56, Fe-B = 1.970(6), mean Fe-C = 1.79(1), C-O = 1.14(6) Å.

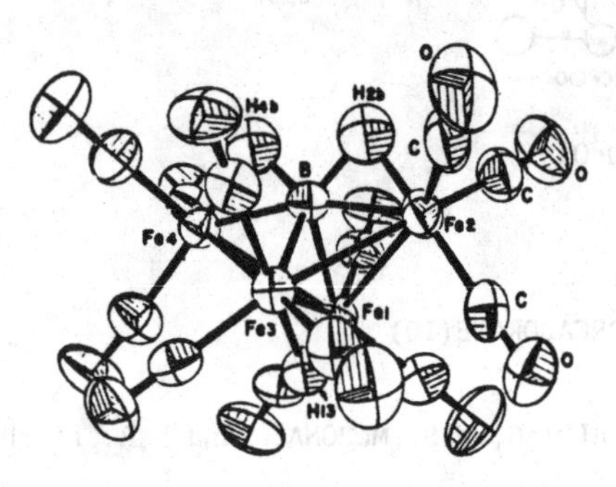

Fig. 1. The $HFe_4(BH_2)(CO)_{12}$ molecule.

TRIS(PENTACARBONYLTUNGSTEN)DIBISMUTH
$Bi_2[W(CO)_5]_3$

G. HUTTNER, U. WEBER and L. ZSOLNAI, 1982. Z. Naturforsch., 37B, 707-710.

Triclinic, $P\bar{1}$, a = 6.927, b = 10.853, c = 17.25 Å, α = 92.52, β = 97.05, γ = 73.30°, at 238K, Z = 2. Mo radiation, R = 0.041 for 2126 reflexions.

The Bi_2W_3 core contains a very short Bi-Bi bond, 2.818(3) Å, bridged by three W atoms to give approximate D_{3h} symmetry, Bi-W = 3.083-3.134(3) Å. Each W has five terminal carbonyl ligands, W-C = 1.95-2.07 Å.

DECACARBONYLDIMANGANESE
$Mn_2(CO)_{10}$

$(OC)_5Mn\text{-}Mn(CO)_5$

M. MARTIN, B. REES and A. MITSCHLER, 1982. Acta Cryst., B38, 6-15.

Monoclinic, I2/a, a = 14.088, b = 6.850, c = 14.242 Å, β = 105.08°, at 74K, Z = 4. Ag, Mo radiations, R = 0.047 for 4583 reflexions (Ag data).

Structure as previously determined (1). Charge-density studies show no significant density on the Mn-Mn bond.

1. Structure Reports, 21, 256; 28, 67.

RHENIUM CARBONYL NITRITE
$Re_3(CO)_{14}NO_2$

RHENIUM CARBONYL CARBOXYLATE
$Re_3(CO)_{14}COOH$

F. OBERDORFER, B. BALBACH and M.L. ZIEGLER, 1982. Z. Naturforsch., 37B, 157-167.

$Re_3(CO)_{14}NO_2$, monoclinic, $P2_1/a$, a = 13.971, b = 17.549, c = 9.230 Å, β = 95.87°, Z = 4. Mo radiation, R = 0.087 for 4238 reflexions.

$Re_3(CO)_{14}COOH$, monoclinic, $P2_1/n$, a = 6.834, b = 17.763, c = 18.444 Å, β = 99.33°, Z = 4. Mo radiation, R = 0.050 for 2727 reflexions.

Both molecules (Fig. 1) contain $Re(CO)_5$ and $(OC)_4Re\text{-}CO\text{-}Re(CO)_4$ groupings, bridged by NO_2 and CO_2H (H not located) groups which are bonded to all three Re atoms; Re atoms have octahedral coordinations.

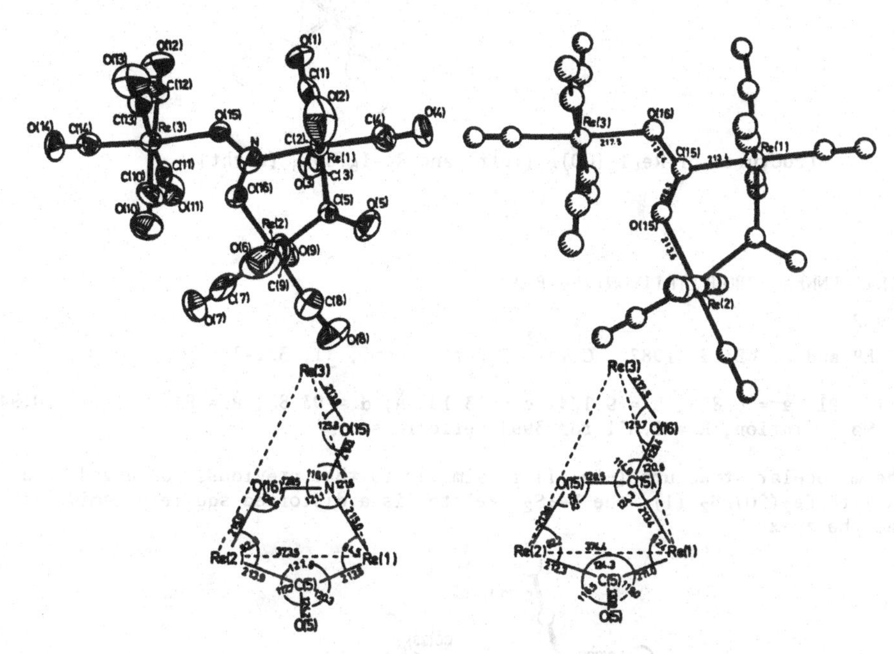

Fig. 1. Structures of $Re_3(CO)_{14}NO_2$ (left) and $Re_3(CO)_{14}COOH$ (right).

RHENIUM CARBONYL IODIDES
$Re_4I_8(CO)_6$, $Re_3I_6(CO)_6$

F. CALDERAZZO, F. MARCHETTI, R. POLI, D. VITALI and P.F. ZANAZZI, 1982. J. Chem. Soc., Dalton, 1665-1670.

$Re_4I_8(CO)_6$, monoclinic, C2/c, a = 22.316, b = 8.679, c = 12.705 Å, β = 96.02°, Z = 4. Mo radiation, R = 0.059 for 1700 reflexions.

$Re_3I_6(CO)_6$, rhombohedral, $R\bar{3}$, a = 6.915, c = 36.939 Å, Z = 3. Mo radiation, R = 0.068 for 490 reflexions.

The Re_4 compound (Fig. 1) contains a non-eclipsed Re_2I_8 complex, with iodine bridging to two terminal $Re(CO)_3$ groups. Re-Re = 2.279(1) (triple metal-metal bond), Re-I = 2.71-2.72 and 2.80-2.81 (bridging), 2.64 (terminal), Re-C = 1.89-1.91, C-O = 1.12-1.15 Å. The Re_3 compound (Fig. 1) has a central ReI_6 core iodine-bridged to two terminal $Re(CO)_3$ groups, with idealized molecular symmetry D_{3d}. Re-I = 2.71, 2.80, Re-C = 1.94, C-O = 1.18 Å.

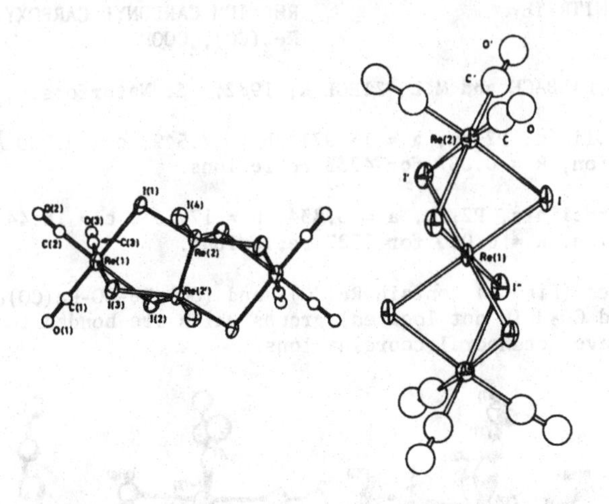

Fig. 1. Structures of $Re_4I_8(CO)_6$ (left) and $Re_3I_6(CO)_6$ (right).

DI-μ_3-THIO-ENNEACARBONYLTRIIRON(2Fe-Fe)
$Fe_3(CO)_9S_2$

P. HÜBENER and E. WEISS, 1982. Cryst. Struct. Comm., 11, 331-336.

Triclinic, $P\bar{1}$, a = 6.805, b = 9.124, c = 13.130 Å, α = 93.83, β = 94.26, γ = 110.94°
Z = 2. Mo radiation, R = 0.051 for 3994 reflexions.

The molecular structure (Fig. 1) is similar to that previously observed in a complex with $Fe_2(CO)_6S_2$ (1); the Fe_3S_2 skeleton is a distorted square pyramid with Fe(2) at the apex.

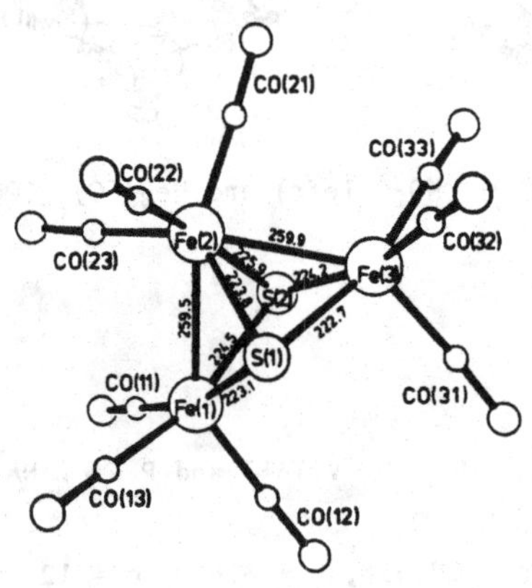

Fig. 1. Structure of $Fe_3(CO)_9S_2$.

1. Structure Reports, 30B, 384.

HEXACARBONYL-μ-CHLORO-(OCTACARBONYLDIIRON)-μ_4-PHOSPHORO-DIIRON
μ_4-ARSENO-HEXACARBONYL-μ-CHLORO-(OCTACARBONYLDIIRON)DIIRON
$(CO)_8Fe_2XFe_2(CO)_6Cl$, X = P, As

G. HUTTNER, G. MOHR, B. PRITZLAFF, J. von SEYERL and L. ZSOLNAI, 1982. Chem. Ber., 115, 2044-2049.

X = P, monoclinic, $P2_1/c$, a = 10.28, b = 18.19, c = 16.08 Å, β = 133.20°, Z = 4. Mo radiation, R = 0.043 for 1750 reflexions.

X = As, monoclinic, $P2_1/c$, a = 9.27, b = 12.91, c = 37.59 Å, β = 91.6°, Z = 8. Mo radiation, R = 0.060 for 3098 reflexions.

Both molecules (Fig. 1) contain $Fe_2(CO)_8$ and $Fe_2(CO)_6(\mu_2\text{-}Cl)$ groups joined by P or As acting as spiro centres.

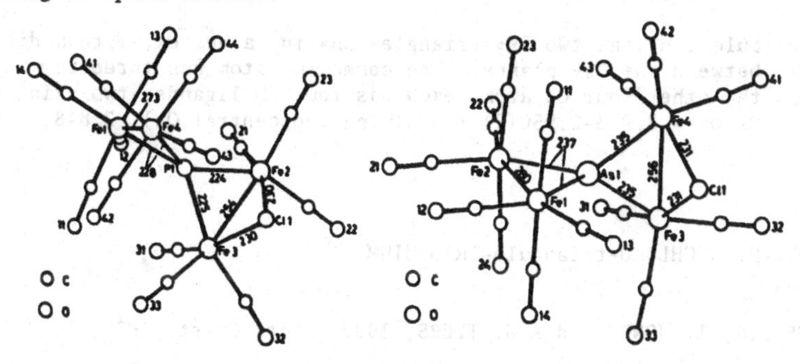

Fig. 1. Structures of $(CO)_8Fe_2XFe_2(CO)_6Cl$, X = P, As; bond lengths in Å x 10^2.

μ-BROMO-μ-HYDRIDO-DECACARBONYLTRIRUTHENIUM(2Ru-Ru)
$Ru_3(CO)_{10}HBr$

N.M. BOAG, C.E. KAMPE, Y.C. LIN and H.D. KAESZ, 1982. Inorg. Chem., 21, 1706-1708.

Monoclinic, $P2_1/c$, a = 8.297, b = 11.549, c = 34.957 Å, β = 93.62°, at -158°C, Z = 8. Mo radiation, R = 0.048 for 4528 reflexions.

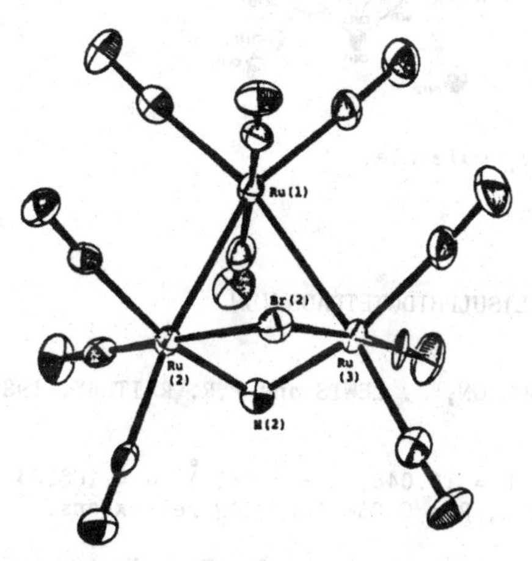

Fig. 1. The $Ru_3(CO)_{10}HBr$ molecule.

The molecule (Fig. 1) contains a Ru_3 triangle, with one Ru-Ru bond (2.819(1) Å) bridged by Br and H; the unbridged bonds are 2.802 and 2.813(1) Å. Ru-Br = 2.560(2), Ru-H = 1.78 Å.

NONADECACARBONYLPENTAOSMIUM
$Os_5(CO)_{19}$

D.H. FARRAR, B.F.G. JOHNSON, J. LEWIS, P.R. RAITHBY and M.J. ROSALES, 1982. J. Chem. Soc., Dalton, 2051-2058.

Triclinic, P$\bar{1}$, a = 8.880, b = 10.244, c = 16.529 Å, α = 99.93, β = 93.44, γ = 110.37°, Z = 2. Mo radiation, R = 0.040 for 2616 reflexions.

The molecule contains two Os_3 triangles sharing a vertex, with a dihedral angle of 21° between the Os_3 planes. The common Os atom has three terminal CO ligands, and the other four Os atoms each has four CO ligands, two axial and two equatorial. Os-Os = 2.913-2.950(2) (involving the central Os), 2.848, 2.853(2) Å.

DECACARBONYL-DI-μ-CHLORO-triangulo-TRIOSMIUM
$Os_3(CO)_{10}Cl_2$

F.W.B. EINSTEIN, T. JONES and K.G. TYERS, 1982. Acta Cryst., B38, 1272-1274.

Orthorhombic, Pbca, a = 25.580, b = 22.832, c = 12.036 Å, D_m = 3.25, Z = 16. Mo radiation, R = 0.041 for 2839 reflexions.

The molecule (Fig. 1) contains an isosceles triangle of Os atoms, Os(2)-Os(3) = 3.233(2) (doubly-bridged, non-bonding), Os(1)-Os(2,3) = 2.852(3) Å (non-bridged, bonding).

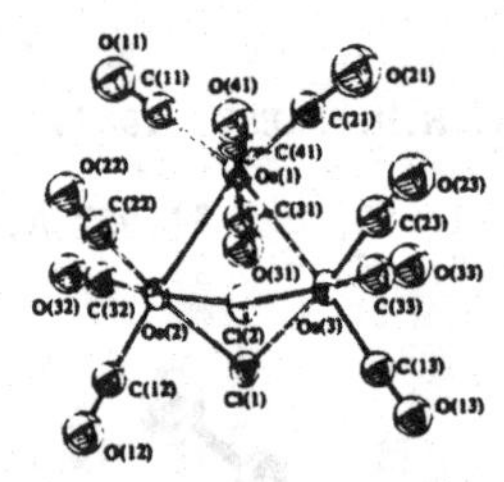

Fig. 1. The $Os_3(CO)_{10}Cl_2$ molecule.

DODECACARBONYL(THIOCARBONYL)SULPHIDOTETRAOSMIUM
$Os_4(CO)_{12}(CS)S$

P.V. BROADHURST, B.F.G. JOHNSON, J. LEWIS and P.R. RAITHBY, 1982. J. Chem. Soc., Dalton, 1641-1644.

Triclinic, P$\bar{1}$, a = 14.834, b = 10.048, c = 7.682 Å, α = 108.84, β = 92.99, γ = 95.64°, Z = 2. Mo radiation, R = 0.058 for 3249 reflexions.

The molecule (Fig. 1) contains a butterfly Os_4 cluster, with a S atom capping three Os atoms; the CS group is on the other side of the framework, with C bonded to all four Os atoms and S to only one Os atom. Os-Os = 2.812-2.859(1), Os-S = 2.419-2.447(5), Os-CS = 2.11-2.21(2), C-S = 1.76(2), mean Os-CO = 2.16(2) Å.

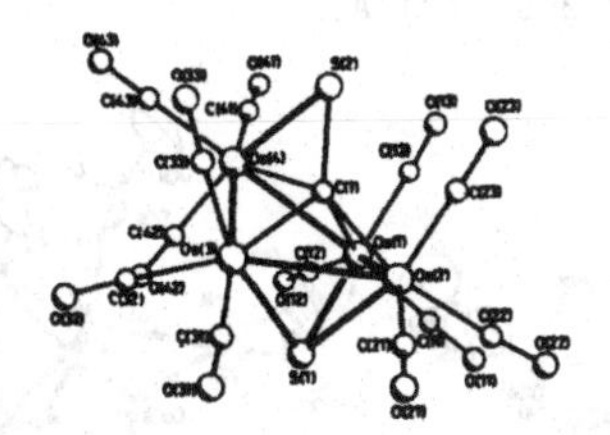

Fig. 1. The $Os_4(CO)_{12}(CS)S$ molecule.

DI-μ-HYDRIDO-TRIDECACARBONYLIRONTRIOSMIUM
$H_2Os_3Fe(CO)_{13}$

M.R. CHURCHILL, C. BUENO, W.L. HSU, J.S. PLOTKIN and S.G. SHORE, 1982. Inorg. Chem., 21, 1958-1963.

Monoclinic, C2/c, a = 31.444, b = 9.700, c = 13.935 Å, β = 110.99°, Z = 8. Mo radiation, R = 0.057 for 2598 reflexions.

The molecule (Fig. 1) contains an Os_3Fe tetrahedron, each Os having three terminal CO ligands, and Fe having four CO, two of which are semi-bridging; two Os-Os bonds are hydrido-bridged. Os-Os = 2.936(1) (hydrido-bridged), 2.847(1), Fe-Os = 2.686(3) (semi-bridged), 2.717(2) Å.

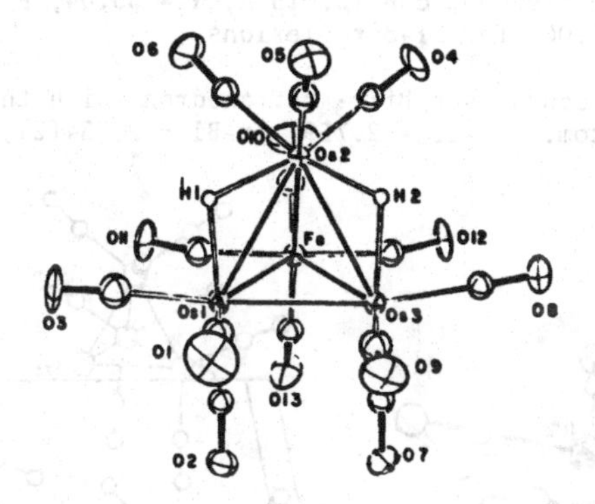

Fig. 1. The $H_2Os_3Fe(CO)_{13}$ molecule.

COBALT CARBONYL SULPHIDES
$[Co_3(CO)_9C](\mu_3\text{-}CS_2)[Co_3(CO)_7S]$ (I) $Co_6(CO)_{16}C_2S_3$
$[Co_3(CO)_8](\mu_5\text{-}CS_2)[Co_3(CO)_7S]$ (II) $Co_6(CO)_{15}CS_3$

G. GERVASIO, R. ROSSETTI, P.L. STANGHELLINI and G. BOR, 1982. Inorg. Chem., 21, 3781-3789.

I. Triclinic, P$\bar{1}$, a = 12.822, b = 13.399, c = 8.851 Å, α = 100.07, β = 85.72, γ = 106.20°, Z = 2. Mo radiation, R = 0.066 for 3224 reflexions.

II. Triclinic, P$\bar{1}$, a = 14.131, b = 10.580, c = 9.768 Å, α = 95.58, β = 83.83, γ = 111.33°, Z = 2. Mo radiation, R = 0.023 for 3730 reflexions.

Both molecules contain two Co_3 subclusters connected by different types of CS_2

bridges (Fig. 1). Co-Co = 2.440-2.533(2) Å.

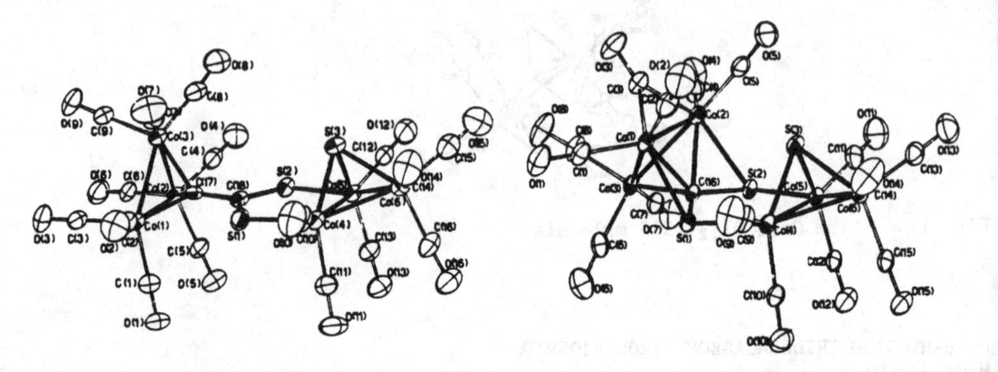

Fig. 1. The $Co_6(CO)_{16}C_2S_3$ (left) and $Co_6(CO)_{15}CS_3$ (right) molecules.

BISMUTH TRIIRIDIUM NONACARBONYL
(μ_3-BISMUTHIO-cyclo-TRIS(TRICARBONYLIRIDIUM)(3Ir-Ir))
$BiIr_3(CO)_9$

W. KRUPPA, D. BLÄSER, R. BOESE and G. SCHMID, 1982. Z. Naturforsch., 37B, 209-213.

Triclinic, $P\bar{1}$, a = 8.100, b = 9.000, c = 12.015 Å, α = 85.04, β = 79.28, γ = 68.85°, Z = 2. Mo radiation, R = 0.060 for 2148 reflexions.

The molecule (Fig. 1) contains a $BiIr_3$ tetrahedron, with three terminal carbony groups bonded to each Ir atom. Ir-Ir = 2.759, Ir-Bi = 2.734(2), mean Ir-C = 1.914, mean C-O = 1.140 Å.

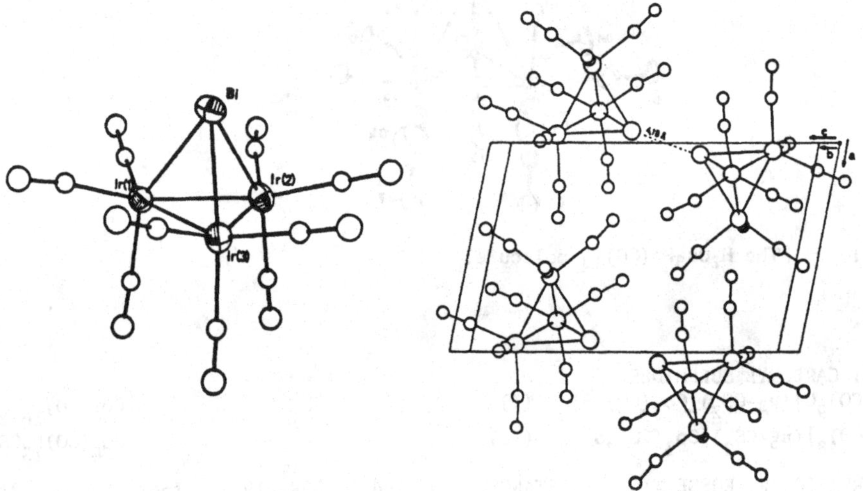

Fig. 1. Structure of $BiIr_3(CO)_9$.

GOLD(I) CARBONYL CHLORIDE
Au(CO)Cl

P.G. JONES, 1982. Z. Naturforsch., 37B, 823-824.

Orthorhombic, Cmcm, a = 4.071, b = 16.422, c = 5.321 Å, Z = 4. Mo radiation, R = 0.037 for 252 reflexions. Atoms in 4(c): 0,y,1/4, y = 0.2363, 0.3739, 0.1190, 0.0515 for Au, Cl, C, O.

The structure contains linear Cl-Au-C-O molecules, Au-Cl = 2.26(1), Au-C = 1.93(2), C-O = 1.11(3) Å; the shortest Au...Au contact is 3.38 Å.

CAESIUM BARIUM AZIDE
$CsBa_2(N_3)_5$

H. KRISCHNER, A.I. SARACOGLU and C. KRATKY, 1982. Z. Kristallogr., 159, 225-229.

Orthorhombic, Pnma, a = 12.928, b = 10.966, c = 8.858 Å, Z = 4. Mo radiation, R = 0.068 for 791 reflexions.

The structure (Fig. 1) contains linear azide anions, linked by 9-coordinate Ba and 7-coordinate Cs ions. N-N = 1.19, Ba-N = 2.82-3.09, Cs-N = 3.08-3.45 Å.

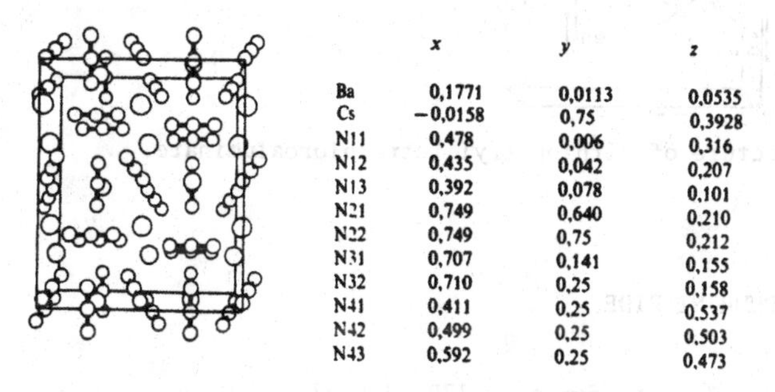

	x	y	z
Ba	0,1771	0,0113	0,0535
Cs	−0,0158	0,75	0,3928
N11	0,478	0,006	0,316
N12	0,435	0,042	0,207
N13	0,392	0,078	0,101
N21	0,749	0,640	0,210
N22	0,749	0,75	0,212
N31	0,707	0,141	0,155
N32	0,710	0,25	0,158
N41	0,411	0,25	0,537
N42	0,499	0,25	0,503
N43	0,592	0,25	0,473

Fig. 1. Structure of caesium barium azide.

AMIDES
$NaCa(NH_2)_3$, $KBa(NH_2)_3$, $RbBa(NH_2)_3$, $RbEu(NH_2)_3$

H. JACOBS, J. KOCKELKORN and J. BIRKENBEUL, 1982. J. Less-Common Metals, 87, 215-224.

Orthorhombic, Fddd, a = 21.745, 24.685, 25.133, 23.975, b = 10.422, 11.727, 11.997, 11.524, c = 7.358, 8.159, 8.407, 8.212 Å, D_m = 1.75, 2.55, 2.84, 3.32, Z = 16. Mo radiation, R = 0.102, 0.047, 0.096, 0.085 for 321, 857, 630, 447 reflexions.

Atomic positions (KBa compound, similar values for others)

			x	y	z
8 Ba(1)	in	8(a)	1/8	1/8	1/8
8 Ba(2)		16(e)	0.2909	1/8	1/8
8 K(1)		8(b)	1/8	1/8	5/8
8 K(2)		16(e)	0.4482	1/8	1/8
16 N(1)		16(f)	1/8	0.368	1/8
32 N(2)		32(h)	0.2925	0.3709	0.1181

Isostructural with related materials (1). $RbSr(NH_2)_3$ is monoclinic, C2/c or Cc.

1. Structure Reports, 44A, 133; 45A, 139.

DITHIONITRYL TETRACHLOROALUMINATE
$NS_2^+AlCl_4^-$

U. THEWALT, K. BERHALTER and P. MÜLLER, 1982. Acta Cryst., B38, 1280-1282.

Orthorhombic, Pnma, a = 10.908, b = 7.291, c = 11.069 Å, Z = 4. Mo radiation, R = 0.073 for 1036 reflexions.

The structure (Fig. 1) contains linear, symmetrical SNS^+ cations, N-S = 1.468(5) Å, and tetrahedral $AlCl_4^-$ anions, Al-Cl = 2.131(2) Å.

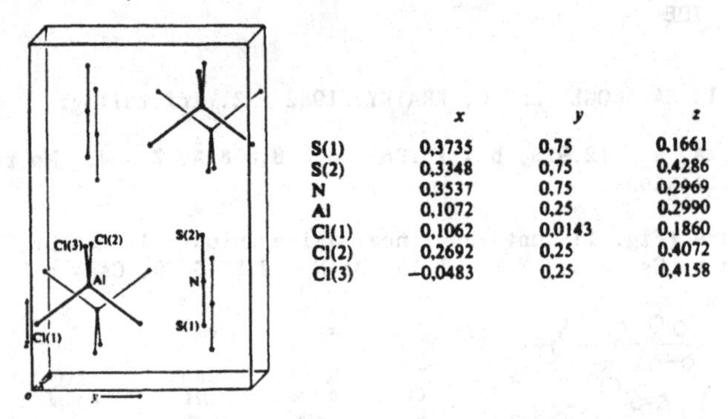

	x	y	z
S(1)	0,3735	0,75	0,1661
S(2)	0,3348	0,75	0,4286
N	0,3537	0,75	0,2969
Al	0,1072	0,25	0,2990
Cl(1)	0,1062	0,0143	0,1860
Cl(2)	0,2692	0,25	0,4072
Cl(3)	−0,0483	0,25	0,4158

Fig. 1. Structure of dithionitryl tetrachloroaluminate.

DIPALLADIUM SULPHUR NITRIDE
$Pd_2(S_3N)_2(S_3N_2)$

U. THEWALT, 1982. Z. Naturforsch., 37B, 276-280.

Rhombohedral, R$\bar{3}$, a = 32.759, c = 6.516 Å, D_m = 2.65, Z = 18. Mo radiation, R = 0.074 for 1803 reflexions.

The molecule (Fig. 1) contains two almost planar PdS_3N five-membered rings, bridged by the SNSNS group, each Pd thus having square-planar coordination to 4 S atoms.

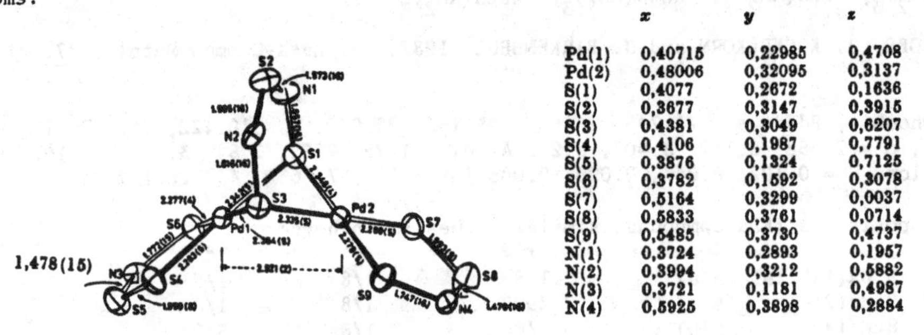

	x	y	z
Pd(1)	0,40715	0,22985	0,4708
Pd(2)	0,48006	0,32095	0,3137
S(1)	0,4077	0,2672	0,1636
S(2)	0,3677	0,3147	0,3915
S(3)	0,4381	0,3049	0,6207
S(4)	0,4106	0,1987	0,7791
S(5)	0,3876	0,1324	0,7125
S(6)	0,3782	0,1592	0,3078
S(7)	0,5164	0,3299	0,0037
S(8)	0,5833	0,3761	0,0714
S(9)	0,5485	0,3730	0,4737
N(1)	0,3724	0,2893	0,1957
N(2)	0,3994	0,3212	0,5882
N(3)	0,3721	0,1181	0,4987
N(4)	0,5925	0,3898	0,2884

Fig. 1. Structure of $Pd_2(S_3N)_2(S_3N_2)$.

PENTAFLUORO-(N-TRITHIADIAZOLYLIDENE)AMINO-CYCLOTRIPHOSPHAZENE
S_3N_2-N-$P_3N_3F_5$

I. RAYMENT, H.M.M. SHEARER and H.W. ROESKY, 1982. J. Chem. Soc., Dalton, 883-885.

Orthorhombic, Pbca, a = 15.915, b = 14.641, c = 6.700 Å, D_m = 2.09, Z = 8. Mo radiation, R = 0.037 for 1528 reflexions.

The molecule (Fig. 1) contains a slightly non-planar six-membered phosphazene ring bonded through a bridging nitrogen atom to a five-membered S_3N_2 ring, in which S(3) is displaced 0.3 Å from the plane of the other four atoms.

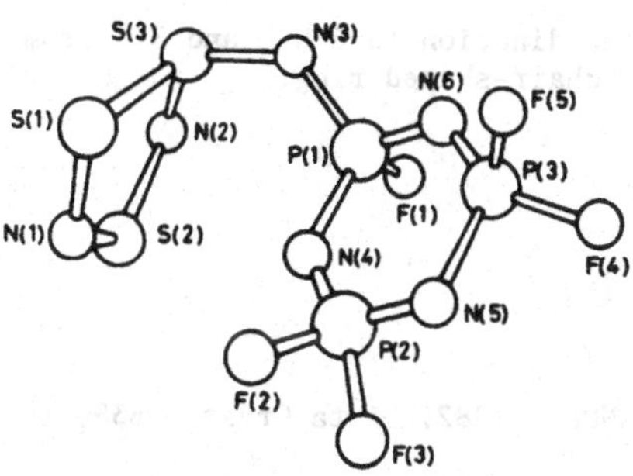

Bond lengths

Bond	Length	Bond	Length
S(1)–S(3)	2.221(2)	P(2)–N(4)	1.555(3)
S(1)–N(1)	1.629(4)	P(2)–N(5)	1.559(4)
S(2)–N(1)	1.540(4)		
S(2)–N(2)	1.572(4)	P(3)–N(5)	1.557(4)
S(3)–N(2)	1.647(4)	P(3)–N(6)	1.554(4)
S(3)–N(3)	1.555(4)		
		P(1)–F(1)	1.540(3)
P(1)–N(3)	1.603(3)	P(2)–F(2)	1.516(3)
P(1)–N(4)	1.579(3)	P(2)–F(3)	1.519(3)
P(1)–N(6)	1.582(4)	P(3)–F(4)	1.518(3)
		P(3)–F(5)	1.525(3)

Bond angles

Angle	Value	Angle	Value
S(3)–S(1)–N(1)	97.8(2)	N(5)–P(2)–F(3)	108.9(2)
N(1)–S(2)–N(2)	109.7(2)	F(2)–P(2)–F(3)	97.3(2)
S(1)–S(3)–N(2)	92.7(1)		
S(1)–S(3)–N(3)	110.8(1)	N(5)–P(3)–N(6)	119.8(2)
N(2)–S(3)–N(3)	112.0(2)	N(5)–P(3)–F(4)	109.3(2)
		N(5)–P(3)–F(5)	108.3(2)
N(3)–P(1)–N(4)	114.4(2)	N(6)–P(3)–F(4)	109.8(2)
N(3)–P(1)–N(6)	108.7(2)	N(6)–P(3)–F(5)	109.3(2)
N(4)–P(1)–N(6)	116.7(2)	F(4)–P(3)–F(5)	98.2(2)
N(3)–P(1)–F(1)	103.9(2)		
N(4)–P(1)–F(1)	106.0(2)	S(1)–N(1)–S(2)	117.6(2)
N(6)–P(1)–F(1)	106.0(2)	S(2)–N(2)–S(3)	119.7(2)
		S(3)–N(3)–P(1)	129.9(2)
N(4)–P(2)–N(5)	119.7(2)	P(3)–N(4)–P(2)	121.7(2)
N(4)–P(2)–F(2)	110.3(2)	P(2)–N(5)–P(3)	120.2(2)
N(4)–P(2)–F(3)	109.9(2)	P(1)–N(6)–P(3)	122.0(2)
N(5)–P(2)–F(3)	108.5(2)		

Fig. 1. The S_3N_2-N-$P_3N_3F_5$ molecule.

5-BROMOSULPHENYL-1,3,2,4-DITHIADIAZOLIUM TRIBROMIDE
$[CS_3N_2Br]^+Br_3^-$

G. WOLMERHÄUSER, C. KRÜGER and Y. TSAY, 1982. Chem. Ber., 115, 1126-1131.

Orthorhombic, $Pna2_1$, a = 7.550, b = 12.466, c = 10.275 Å, Z = 4. Mo radiation, R = 0.048 for 1057 reflexions.

The cation contains an almost planar C-S-N-S-N five-membered ring, with a -S-Br group bonded exocyclically to C; C-S = 1.72, 1.76, C-N = 1.37, S-N = 1.56, 1.67, 1.60, S-Br = 2.21 Å, C-S-Br = 101°. The Br_3^- anion is linear, but asymmetric; Br-Br = 2.42, 2.72 Å, Br-Br-Br = 179°

2,2'-THIOBIS(1,1,3,3-TETRAOXO-1,3,5,2,4,6-TRITHIATRIAZINE)
$(S_3N_3O_4)_2S$

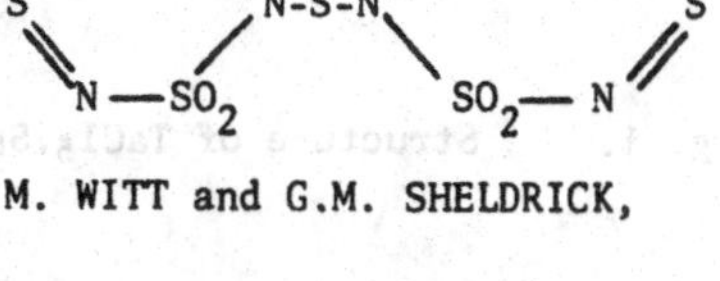

H.W. ROESKY, W. CLEGG, J. SCHIMKOWIAK, M. SCHMIDT, M. WITT and G.M. SHELDRICK, 1982. J. Chem. Soc., Dalton, 2117-2118.

Tetragonal, $P\bar{4}2_1m$, a = 11.838, c = 4.935 Å, Z = 4. Mo radiation, R = 0.027 for 637 reflexions.

The molecule has C_{2v} symmetry, with the tri-coordinate N atoms 0.7 Å out of the planes of the other five ring atoms. S=N = 1.539, S-N = 1.625-1.707, S-O = 1.405(3) Å, angles = 103-126°.

NEODYMIUM TRISULPHIMIDATE HEXAHYDRATE
$Nd(O_6S_3N_3).6H_2O$

M.A. PORAJ-KOŠIC, V.I. SOKOL and M.A. KOP'EVA, 1982. Koord. Khim., 8, 702-708.

Rhombohedral, R32, a = 10.514, c = 19.987 Å, Z = 6. R = 0.053.

Nd has tricapped trigonal prismatic 9-coordination to 6 H_2O and 3 O from the trisulphimidate group; the latter contains a chair-shaped ring.

SULPHUR - TETRASULPHUR TETRAIMIDE
(OCTASULPHUR-CYCLOTETRAAZATHIANE (3:1))
$3S_8.S_4(NH)_4$

M. GASPERIN, R. FREYMANN and H. GARCIA-FERNANDEZ, 1982. Acta Cryst., B38, 1728-1731.

[Duplicate publication of the structure (see 1).]

1. Structure Reports, 48A, 125.

TETRASULPHURTETRANITRIDE-TANTALUM PENTACHLORIDE
$TaCl_5.S_4N_4$

U. THEWALT and G. ALBRECHT, 1982. Z. Naturforsch., 37B, 1098.

Monoclinic, $P2_1/c$, a = 15.926, b = 6.880, c = 22.619 Å, β = 92.33°, Z = 8. Mo radiation, R = 0.082 for 1581 reflexions.

The ring conformation (Fig. 1) differs from that in free S_4N_4, with the four S atoms in a plane and N atoms alternately above and below the plane. Ta is bonded to the ring via N.

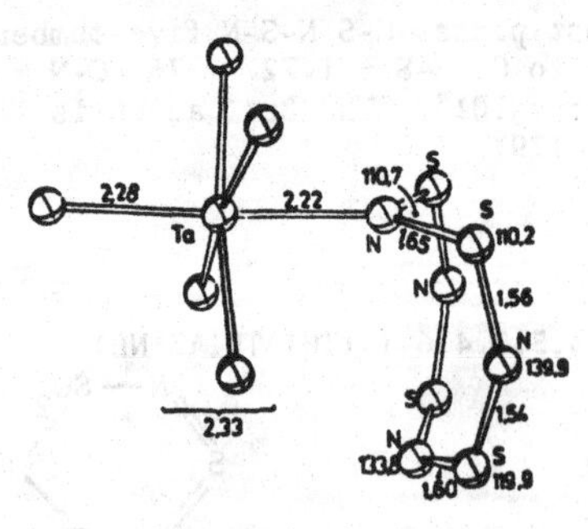

Fig. 1. Structure of $TaCl_5.S_4N_4$.

TETRASULPHURTETRAIMIDE-PENTACARBONYLTUNGSTEN
$(S_4N_4H_4)W(CO)_5$

G. SCHMID, R. GREESE and R. BOESE, 1982. Z. Naturforsch., 37B, 620-626.

Triclinic, P$\bar{1}$, a = 9.085, b = 11.142, c = 17.552 Å, α = 83.54, β = 87.17, γ = 69.80°, Z = 2. Mo radiation, R = 0.048 for 4208 reflexions.

The molecule (Fig. 1) has W bonded to a S atom of the eight-memnered ring. W-S = 2.525(2), W-C = 1.97(1) (trans to S), 2.02-2.04(1), S-N = 1.68-1.71(1), mean C-O = 1.15 Å, N-S-N = 109-110, S-N-S = 120-122°.

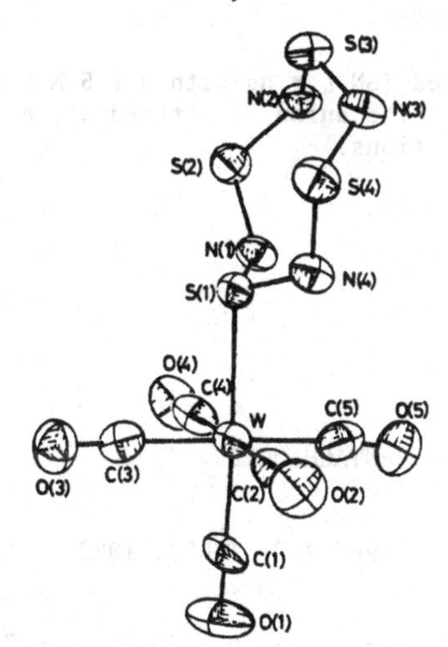

Fig. 1. Structure of $(S_4N_4H_4)W(CO)_5$.

TETRASULPHURTETRANITRIDE-COPPER(I) BROMIDE
$CuBr.S_4N_4$

U. THEWALT and B. MÜLLER, 1982. Z. Naturforsch., 37B, 828-831.

Monoclinic, $P2_1/c$, a = 9.136, b = 11.911, c = 6.935 Å, β = 105.71°, D_m = 2.92, Z = 4. Mo radiation, R = 0.048 for 1728 reflexions.

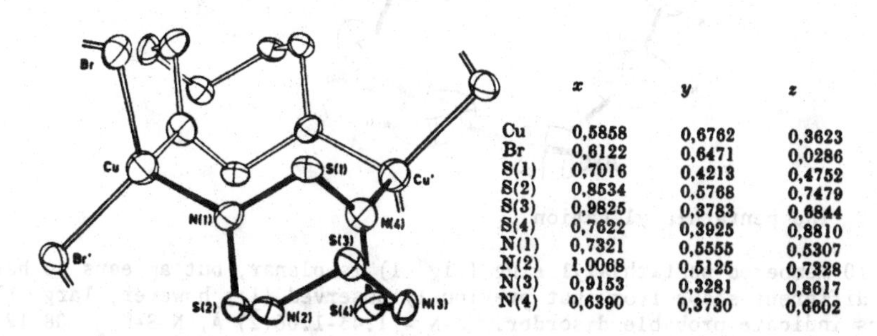

	x	y	z
Cu	0,5858	0,6762	0,3623
Br	0,6122	0,6471	0,0286
S(1)	0,7016	0,4213	0,4752
S(2)	0,8534	0,5768	0,7479
S(3)	0,9825	0,3783	0,6844
S(4)	0,7622	0,3925	0,8810
N(1)	0,7321	0,5555	0,5307
N(2)	1,0068	0,5125	0,7328
N(3)	0,9153	0,3281	0,8617
N(4)	0,6390	0,3730	0,6602

Fig. 1. Structure of $CuBr.S_4N_4$.

The structure (Fig. 1) contains $[CuBr]_\infty$ chains bridged by 1,3-N,N'-bonded S_4N_4 groups which have 8-membered rings with the same conformation and dimensions as in free S_4N_4 (1); Cu has tetrahedral coordination. Cu-Br = 2.381, 2.416(2), Cu-N = 2.091, 2.10$\bar{1}$(5), S-N = 1.626-1.653, S(1,2)...S(3,4) = 2.634, 2.604(3) Å, Br-Cu-Br = 124.6, other angles at Cu = 102.8-111.8, N-S-N = 103.3-105.2, S-N-S = 112.4-113.7(3)°.

1. Strukturbericht, 2, 327; Structure Reports, 16, 278; 28, 54; 44A, 135.

TETRASULPHURPENTANITRIDE HEXAFLUOROARSENATE(V)
$S_4N_5{}^+AsF_6{}^-$

W. ISENBERG and R. MEWS, 1982. Z. Naturforsch., 37B, 1388-1392.

Monoclinic, P2/c, a = 11.927, b = 6.578, c = 12.640 Å, β = 95.55°, Z = 4. Mo radiation, R = 0.057 for 1982 reflexions.

The cation has an eight-membered $(SN)_4$ ring with a 1-5 N bridge, as previously observed (1); S-N = 1.54-1.69(1) Å. The anion is octahedral, As-F = 1.68-1.72(1) Å. There are S...N inter-cation interactions.

1. Structure Reports, 46A, 412.

PENTATHIAZYL HEXACHLOROANTIMONATE(V)
$(S_5N_5{}^+)(SbCl_6{}^-)$

TETRASULPHUR TETRANITRIDE - ANTIMONY PENTACHLORIDE
$S_4N_4.SbCl_5$

R.J. GILLESPIE, J.F. SAWYER, D.R. SLIM and J.D. TYRER, 1982. Inorg. Chem., 21, 1296-1302.

$(S_5N_5{}^+)(SbCl_6{}^-)$, orthorhombic, Fdd2, a = 18.05, b = 56.72, c = 7.380 Å, Z = 16. Mo radiation, R = 0.038 for 1265 reflexions.

$S_4N_4.SbCl_5$, monoclinic, P2/c, a = 6.955, b = 12.400, c = 15.224 Å, β = 107.09°, Z = 4. Mo radiation, R = 0.039 for 2165 reflexions.

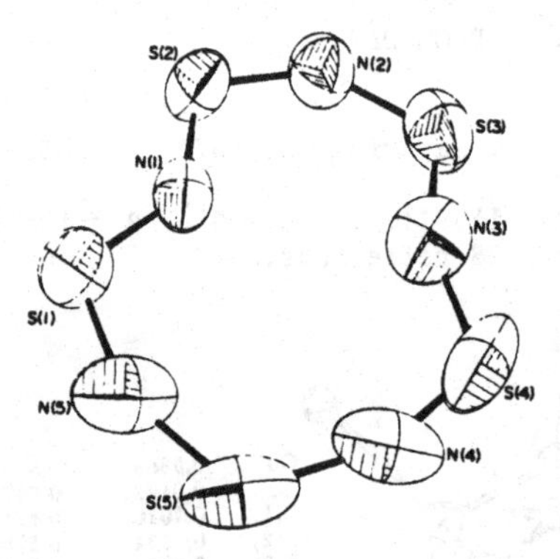

Fig. 1. The pentathiazyl cation.

The 10-membered pentathiazyl ring (Fig. 1) is planar, but appears to have a slightly different shape from that previously observed (1); however, large thermal parameters indicate probable disorder. S-N = 1.43-1.66(2) Å, N-S-N = 108-121, S-N-S = 137-158°. The $SbCl_6{}^-$ anion is octahedral, Sb-Cl = 2.34-2.37 Å.

The $S_4N_4.SbCl_5$ structure is as previously described (2), and is similar to that of other adducts (3), with Sb coordinated to a ring N atom and having octahedral coordination. Sb-Cl = 2.334-2.360(3), Sb-N(1) = 2.134(7), N(1)-S = 1.66, other N-S = 1.53-1.59(1) Å, S-N(1)-S = 123, other S-N-S = 139, N-S-N = 110-121°.

1. Structure Reports, 38A, 193; 42A, 162.
2. Ibid., 24, 295.
3. Ibid., 46A, 147, 148.

HYDROGEN FLUORIDE HYDRATES

$2HF.H_2O$, $4HF.H_2O$

D. MOOTZ and W. POLL, 1982. Z. anorg. Chem., 484, 158-164.

$2HF.H_2O$, monoclinic, $P2_1/c$, a = 3.477, b = 6.024, c = 11.358 Å, β = 96.70°, at -100°C, Z = 4. Mo radiation, R = 0.032 for 1356 reflexions. Previous study in 1.

$4HF.H_2O$, triclinic, $P\bar{1}$, a = 5.574, b = 6.429, c = 6.874 Å, α = 115.79, β = 96.63, γ = 108.79°, at -113°C, Z = 2. Mo radiation, R = 0.049 for 1942 reflexions.

Both structures (Fig. 1) contain H_3O^+, F^-, and HF units, linked by systems of F-H...F and O-H...F hydrogen bonds.

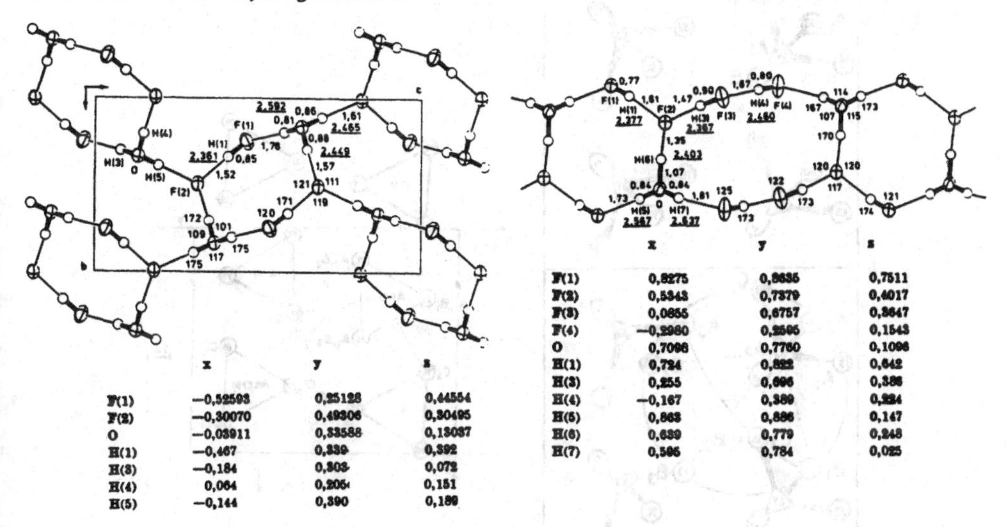

	x	y	z
F(1)	−0,52593	0,25128	0,44554
F(2)	−0,30070	0,49306	0,30495
O	−0,03911	0,33588	0,13037
H(1)	−0,467	0,339	0,392
H(3)	−0,184	0,303	0,072
H(4)	0,064	0,205	0,151
H(5)	−0,144	0,390	0,189

	x	y	z
F(1)	0,8275	0,8835	0,7511
F(2)	0,5343	0,7379	0,4017
F(3)	0,0855	0,6757	0,3647
F(4)	−0,2980	0,2596	0,1543
O	0,7096	0,7760	0,1096
H(1)	0,724	0,882	0,642
H(3)	0,255	0,696	0,386
H(4)	−0,167	0,389	0,224
H(5)	0,863	0,886	0,147
H(6)	0,639	0,779	0,248
H(7)	0,596	0,784	0,025

Fig. 1. Structures of $2HF.H_2O$ (left) and $4HF.H_2O$ (right).

1. Structure Reports, 48A, 131.

ARSENIC(III) FLUORIDE AsF_3 / ARSENIC(III) CHLORIDE $AsCl_3$

J. GALY and R. ENJALBERT, 1982. J. Solid State Chem., 44, 1-23.

Fluoride, orthorhombic, $Pn2_1a$, a = 7.018, b = 7.315, c = 5.205 Å, at 193K, Z = 4. Mo radiation, R = 0.055.

Chloride, orthorhombic, $P2_12_12_1$, a = 9.466, b = 11.335, c = 4.289 Å, at 253K, Z = 4. Mo radiation, R = 0.036. Isostructural with $AsBr_3$ (1).

Atomic positions

Fluoride	x	y	z
As	0.2862	1/4	0.0338
F(1)	0.0736	0.3065	-0.1005 [?]
F(2)	0.3077	0.0490	-0.1260
F(3)	0.1792	0.1493	0.2956
Chloride			
As	0.3010	0.2879	0.4891
Cl(1)	0.2966	0.1236	0.2321
Cl(2)	0.1347	0.3792	0.2380
Cl(3)	0.4782	0.3676	0.2490

Both structures (Fig. 1) contain trigonal pyramidal molecules, with six more-distant F or Cl completing distorted tri-capped trigonal prismatic coordinations. As-F = 1.70-1.72, 2.89-3.63, As-Cl = 2.16-2.17, 3.69-3.97 Å. The paper gives an extensive review of the crystal chemistry of the group VA trihalides.

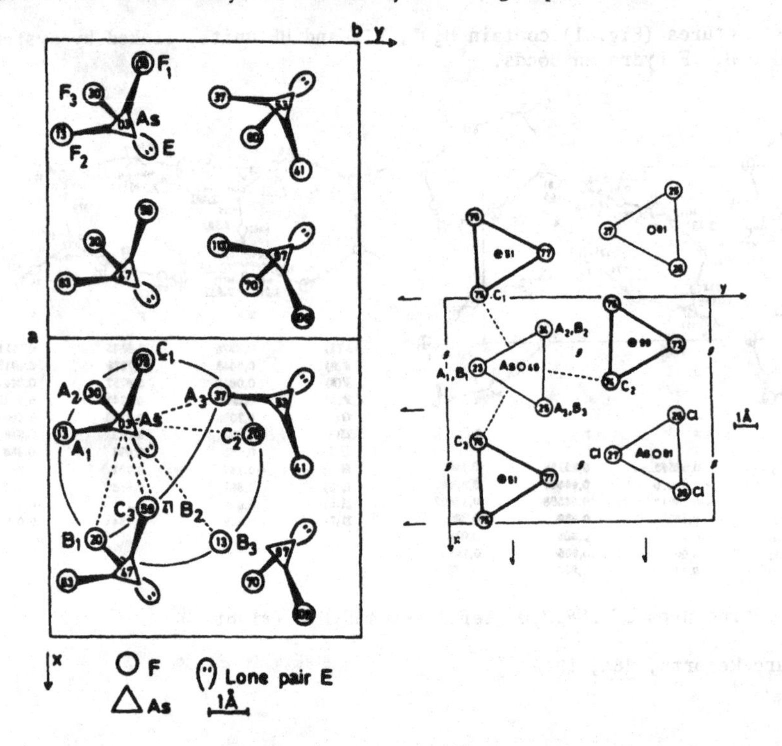

Fig. 1. Structures of AsF_3 (left) and $AsCl_3$ (right).

1. Structure Reports, 30A, 285; 32A, 206.

SULPHUR HEXAFLUORIDE (LOW TEMPERATURE)
SF_6

G. RAYNERD, G.J. TATLOCK and J.A. VENABLES, 1982. Acta Cryst., B38, 1896-1900.

50-94K, trigonal P$\bar{3}$m1, a = 8.01, c = 4.83 Å, D_m = 2.68, Z = 3. S(1) in 1(a): 0,0,0; S(2) in 2(d): 1/3,2/3,0.46; three F in 6(i): x,2x,z [x, z not given]. Electron diffraction patterns.

Isostructural with UCl_6 (1); the structure can be derived from that of high-temperature SF_6 (2) by rotation of one molecule by 60° (Fig. 1). Below 50 K the structure distorts to C-centred monoclinic (or lower) symmetry.

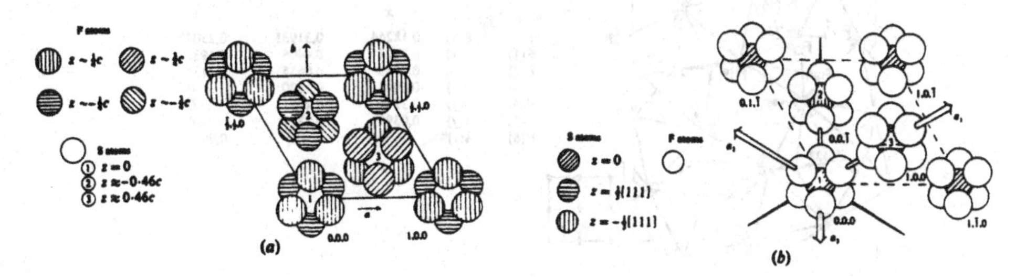

Fig. 1. Structure of (a) trigonal low-temperature, and (b) b.c.c. high-temperature SF_6.

1. Structure Reports, 11, 479; 40A, 150.
2. Ibid., 42A, 165

SCANDIUM FLUORIDE
ScF_3

THALLIUM(III) HEXAFLUOROGALLATE, -INDATE, and -SCANDATE
$TlMF_6$ (M = Ga, In, Sc)

R. LÖSCH, C. HEBECKER and Z. RANFT, 1982. Z. anorg. Chem., 491, 199-202.

ScF_3, rhombohedral, R32, a = 4.021 Å, α = 89.87°, Z = 1 (hexagonal cell has a = 5.680, c = 6.980 Å, Z = 3). Mo radiation, R = 0.049 for 129 reflexions. Sc in 1(a): 0,0,0; F in 3(e): 1/2,0.013,-0.013.

$TlMF_6$, rhombohedral, R$\bar{3}$c, a = 5.268, 5.268, 5.497, c = 14.88, 14.86, 14.47 Å, Z = 6. Powder data. Tl/M in 6(b): 0,0,0; F in 18(e): x,0,1/4, x = -0.395.

The ScF_3 structure is as previously described (1). The $TlMF_6$ compounds have a VF_3-type structure (2), with statistical distribution of the metal ions.

1. Strukturbericht, 7, 12, 96.
2. Structure Reports, 15, 145.

ZIRCONIUM(IV) FLUORIDE (HIGH-TEMPERATURE)
α-ZrF_4

R. PAPIERNIK, D. MERCURIO and B. FRIT, 1982. Acta Cryst., B38, 2347-2353.

Tetragonal, $P4_2/m$, a = 7.896, c = 7.724 Å, D_m = 4.55, Z = 8. Mo radiation, R = 0.030 for 1335 reflexions.

The structure (Fig. 1) contains ZrF_8 dodecahedra sharing corners and edges; Zr-F = 2.082-2.190(2) Å. The β-form (1) is isostructural with UF_4 (2).

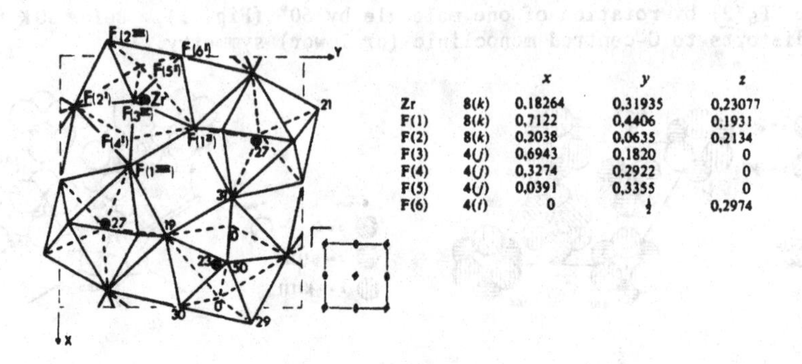

		x	y	z
Zr	8(k)	0,18264	0,31935	0,23077
F(1)	8(k)	0,7122	0,4406	0,1931
F(2)	8(k)	0,2038	0,0635	0,2134
F(3)	4(j)	0,6943	0,1820	0
F(4)	4(j)	0,3274	0,2922	0
F(5)	4(j)	0,0391	0,3355	0
F(6)	4(i)	0	½	0,2974

Fig. 1. Struxture of α-ZrF_4.

1. R.D. BURBANK and F.N. BENSEY, 1956. USAEC Report, K 1280.
2. Structure Reports, 12, 168; 29, 262.

VANADIUM(II) FLUORIDE
VF_2

W.H. BAUR, S. GUGGENHEIM and J.-C. LIN, 1982. Acta Cryst., B38, 351-355.

Tetragonal, $P4_2/mnm$, a = 4.7977, c = 3.2469 Å, Z = 2. Mo radiation, R = 0.035 for 237 reflexions. V in 2(a); F in 4(f): x = 0.3055.

Rutile structure (1), with compressed octahedral coordination for V, V-F = 2.092 (x 4), 2.073(1) (x 2) Å. A computer simulation of a superstructure for $(V,Mg)F_2$ is also described.

1. Strukturbericht, 1, 155; Structure Reports, 45A, 383.

MANGANESE(II) FLUORIDE
MnF_2

I. L.M. ALTE DA VEIGA, L.R. ANDRADE and W. GONSCHOREK, 1982. Z. Kristallogr., 160, 171-178.
II. W. JAUCH, J.R. SCHNEIDER and H. DACHS, 1982. Hahn-Meitner-Inst. Kernforsch. Berlin, HMI-B 370, pp. 129-131.

Tetragonal, $P4_2/mnm$, a = 4.8734, c = 3.3099 Å, Z = 2. Ag radiation, R = 0.02 for 114 reflexions (I), and γ-ray diffraction study (II). Mn in 2(a): 0,0,0; F in 4(f): x,x,0, x = 0.3050, 0.3053.

Rutile-type structure, as previously described (1).

1. Strukturbericht, 1, 158, 192; Structure Reports, 13, 205; 18, 348; 21, 209.

SILVER IODIDE FLUORIDE HYDRATES

$Ag_2IF.H_2O$

I. K. PERSSON and B. HOLMBERG, 1982. J. Solid State Chem., 42, 1-10.

Monoclinic, $P2_1$, a = 4.7206, b = 7.8117, c = 6.3747 Å, β = 93.345°, D_m = 5.35, Z = 2. Mo radiation, R = 0.033 for 744 reflexions, and neutron radiation, R = 0.09 for 262 reflexions.

The structure (Fig. 1) contains AgI_2FO distorted tetrahedra sharing I...I edges and O, F corners to form layers parallel to (001), which are linked by O-H...F hydrogen bonds; the Ag...Ag distance within a pair of tetrahedra is very short, 2.812(2) Å.

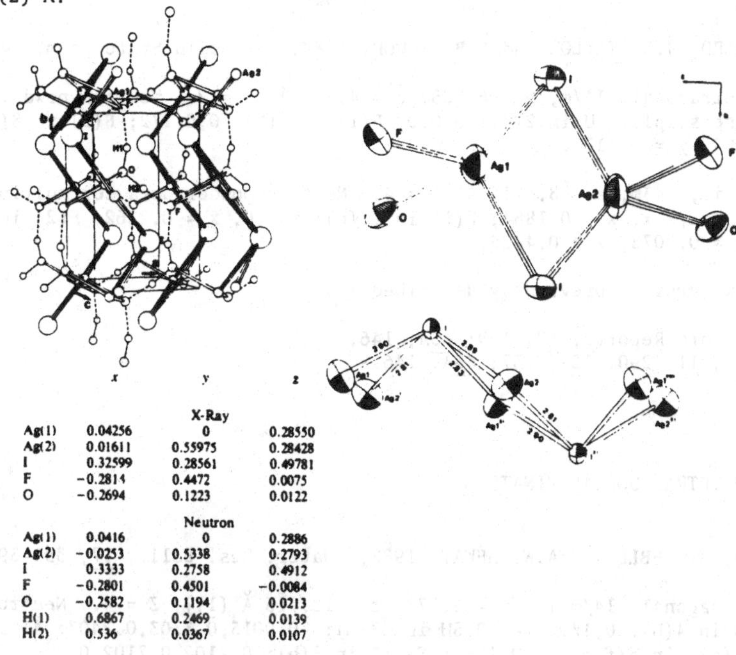

	x	y	z
		X-Ray	
Ag(1)	0.04256	0	0.28550
Ag(2)	0.01611	0.55975	0.28428
I	0.32599	0.28561	0.49781
F	−0.2814	0.4472	0.0075
O	−0.2694	0.1223	0.0122
		Neutron	
Ag(1)	0.0416	0	0.2886
Ag(2)	−0.0253	0.5338	0.2793
I	0.3333	0.2758	0.4912
F	−0.2801	0.4501	−0.0084
O	−0.2582	0.1194	0.0213
H(1)	0.6867	0.2469	0.0139
H(2)	0.536	0.0367	0.0107

Fig. 1. Structure of $Ag_2IF.H_2O$.

$Ag_7I_2F_5.2\cdot5H_2O$

II. K. PERSSON and B. HOLMBERG, 1982. Acta Cryst., B38, 1065-1070.

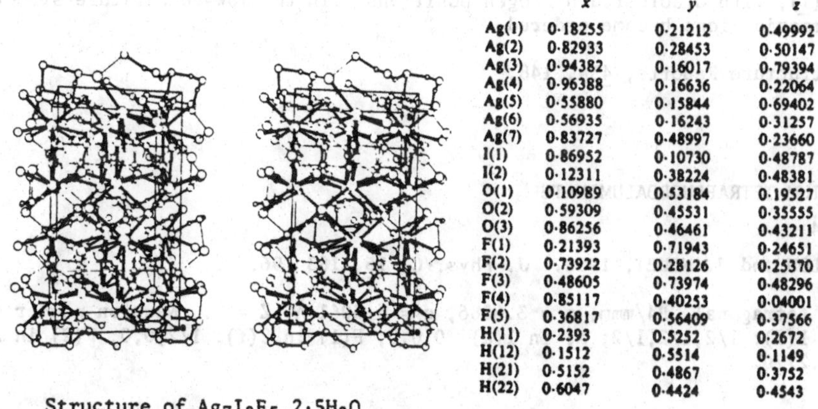

	x	y	z
Ag(1)	0·18255	0·21212	0·49992
Ag(2)	0·82933	0·28453	0·50147
Ag(3)	0·94382	0·16017	0·79394
Ag(4)	0·96388	0·16636	0·22064
Ag(5)	0·55880	0·15844	0·69402
Ag(6)	0·56935	0·16243	0·31257
Ag(7)	0·83727	0·48997	0·23660
I(1)	0·86952	0·10730	0·48787
I(2)	0·12311	0·38224	0·48381
O(1)	0·10989	0·53184	0·19527
O(2)	0·59309	0·45531	0·35555
O(3)	0·86256	0·46461	0·43211
F(1)	0·21393	0·71943	0·24651
F(2)	0·73922	0·28126	0·25370
F(3)	0·48605	0·73974	0·48296
F(4)	0·85117	0·40253	0·04001
F(5)	0·36817	0·56409	0·37566
H(11)	0·2393	0·5252	0·2672
H(12)	0·1512	0·5514	0·1149
H(21)	0·5152	0·4867	0·3752
H(22)	0·6047	0·4424	0·4543

Fig. 2. Structure of $Ag_7I_2F_5.2\cdot5H_2O$.

Monoclinic, $P2_1/n$, a = 7.406, b = 17.830, c = 9.362 Å, β = 91.28°, D_m = 5.9, Z = 4. Mo radiation, R = 0.039 for 2961 reflexions.

The structure (Fig. 2) contains layers of Ag^+, I^-, and F^- ions joined by O-H...F hydrogen bonds (O...F = 2.52-2.59 Å). Ag ions have distorted octahedral coordinations to I, F, and O, and there are two very short Ag...Ag distances, 2.92 and 2.98 Å.

URANIUM(V) FLUORIDE
α-UF_5

URANIUM(IV,V) FLUORIDE
U_2F_9

C.J. HOWARD, J.C. TAYLOR and A.B. WAUGH, 1982. J. Solid State Chem., 45, 396-398.

α-UF_5, tetragonal, I4/m, a = 6.525, c = 4.472 Å, Z = 2. Neutron powder data for multiphase sample. U in 2(a): 0,0,0; F(1) in 2(b): 0,0,1/2; F(2) in 8(h): x,y,0, x = 0.2885, y = 0.1123.

U_2F_9, cubic, $I\bar{4}3m$, a = 8.4716 Å, Z = 4. Neutron powder data for multiphase sample. U in 8(c): x,x,x, x = 0.1884; F(1) in 12(e): x,0,0, x = 0.2262; F(2) in 24(g); x,x,z, x = 0.2078, z = 0.4425.

Structures as previously described (1, 2).

1. Structure Reports, 12, 169; 45A, 146.
2. Ibid., 11, 290; 32A, 157; 45A, 146.

AMMONIUM TETRAFLUOROALUMINATE
NH_4AlF_4

A. BULOU, A. LEBLÉ and A.W. HEWAT, 1982. Mater. Res. Bull., 17, 391-397.

300K, tetragonal, I4/mcm, [a = 5.078, c = 12.715 Å (1)], Z = 4. Neutron powder data. N in 4(b): 0,1/2,1/4; 0.5H in 32(m): -0.0015,0.3303,0.2403; Al in 4(c): 0,0,0; F(ax) in 8(f): 0,0,0.1386; F(eq) in 8(h): 0.2102,0.7102,0.

5K, tetragonal, $P4_2/mbc$, [a ∿ 5.0, c ∿ 12.7 Å], Z = 4. Neutron powder data. N in 4(b): 0,0,1/4; H in 16(i): 0.0017,0.8310,0.2030; Al in 4(c): 0,1/2,0; F(ax) in 8(f): 0,1/2,0.1394; F(eq) in 8(h): 0.2957,0.2980,0.

The room-temperature structure is essentially as previously described in $I\bar{4}c2$ (1), with disordered hydrogen positions. In the low-temperature structure, the ammonium ions become ordered.

1. Structure Reports, 45A, 148.

RUBIDIUM TETRAFLUOROALUMINATE
$RbAlF_4$

A. BULOU and J. NOUET, 1982. J. Phys. C, 15, 183-196.

623K, tetragonal, P4/mmm, a = 3.6586, c = 6.3061 Å, Z = 1. Neutron powder data. Rb in 1(d): 1/2,1/2,1/2; Al in 1(a): 0,0,0; F(1) in 2(f): 1/2,0,0; F(2) in 2(g); 0,0,0.2770.

423 and 293K, tetragonal, P4/mbm, a = 5.1375, 5.1127, c = 6.2912, 6.2815 Å, Z = 2. Neutron powder data. Rb in 2(c): 0,1/2,1/2; Al in 2(a): 0,0,0; F(1) in 4(g): x,1/2-x,0, x = 0.2801, 0.2841; F(2) in 4(e): 0,0,z, z = 0.2771, 0.2776.

200 and 5K, orthorhombic, Pmmn, a = 7.2285, 7.2124, b = 7.2252, 7.2073, c = 6.2624, 6.2396 Å, Z = 4. Neutron powder data. Rb(1) in 2(a): 1/4,1/4,z, z = 0.480, 0.471; Rb(2) in 2(b): 1/4,3/4,z, z = 0.518, 0.533; Al in 4(c): 0,0,0; F(1) in 4(e): 1/4,y, z, y = -0.0304, -0.0356, z = -0.000, -0.021; F(2) in 4(f): x,1/4,z, x = 0.040, 0.0396, z = 0.0330, 0.0350; F(3) in 8(g): x,y,z, x = -0.001, -0.0098, y = -0.0186, -0.0232, z = 0.2786, 0.2786.

The structures all contain AlF_6 octahedra tilted from ideal positions. Previous studies in 1, 2, 3.

1. Strukturbericht, 6, 18, 95.
2. Structure Reports, 45A, 148.
3. Ibid., 46A, 155.

WEBERITE
Na_2MgAlF_7

O. KNOP, T.S. CAMERON and K. JOCHEM, 1982. J. Solid State Chem., 43, 213-221

Orthorhombic, Imm2, a = 7.051, b = 9.968, c = 7.285 Å, Z = 4. Mo radiation, R = 0.019 for 313 reflexions.

Atomic positions

		x	y	z
Na(1)	in 4(c)	0.2492	0	0.4986
Na(2)	4(d)	0	0.2534	0.7503
Mg	4(c)	0.2499	0	0
Al	4(d)	0	0.2499	0.2503
F(1)	2(a)	0	0	0.8844
F(2)	2(b)	0	1/2	0.6119
F(3)	4(d)	0	0.1663	0.4700
F(4)	4(d)	0	0.3367	0.0282
F(5)	8(e)	0.1838	0.1378	0.1805
F(6)	8(e)	0.3156	0.1371	0.8198

Refinements in Imma, Imm2, and $I2_12_12_1$ are indistinguishable, so in the absence of evidence for chirality, Imm2 is accepted as valid (1) as for Na_2NiFeF_7 (2), rather than Imma (3).

1. Structure Reports, 12, 196.
2. Ibid., 44A, 149.
3. Ibid., 44A, 142.

BARIUM ZINC ALUMINUM FLUORIDE
Ba_2ZnAlF_9

T. FLEISCHER and R. HOPPE, 1982. Z. anorg. Chem., 492, 83-94.

Orthorhombic, Pnma, a = 17.841, b = 5.5416, c = 7.3616 Å, D_m = 4.89, Z = 4. Mo radiation, R = 0.076 for 1282 reflexions.

The structure (Fig. 1) contains rings of four MF_6 octahedra (M = Zn/Al) linked to form stairs; Zn/Al-F = 1.79-2.17 Å. Ba ions have 8+4 and 11+1 coordinations, Ba-F = 2.68-2.96, 3.21-3.76 Å. Ba_2ZnGaF_9 has a monoclinically-distorted structure.

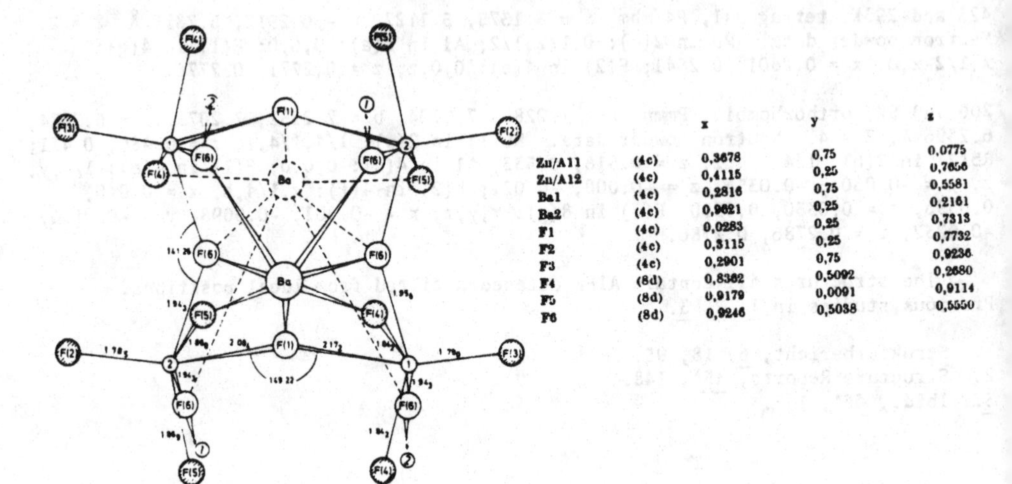

		x	y	z
Zn/Al1	(4c)	0,3678	0,75	0,0775
Zn/Al2	(4c)	0,4115	0,25	0,7656
Ba1	(4c)	0,2816	0,75	0,5581
Ba2	(4c)	0,9621	0,25	0,2161
F1	(4c)	0,0283	0,25	0,7313
F2	(4c)	0,3115	0,25	0,7732
F3	(4c)	0,2901	0,75	0,9236
F4	(8d)	0,8362	0,5092	0,2680
F5	(8d)	0,9179	0,0091	0,9114
F6	(8d)	0,9246	0,5038	0,5550

Fig. 1. Structure of Ba_2ZnAlF_9.

MAGNESIUM HEXAFLUOROARSENATE(V) BIS(SULPHUR DIOXIDE)
$Mg(SO_2)_2(AsF_6)_2$

R. HOPPENHEIT, W. ISENBERG and R. MEWS, 1982. Z. Naturforsch., 37B, 1116-1121.

Monoclinic, $P2_1/n$, a = 10.089, b = 10.397, c = 12.395 Å, β = 91.12°, Z = 4. Mo radiation, R = 0.094 for 1523 reflexions.

The structure contains AsF_6 octahedra linked by $MgF_4(O\text{-}SO)_2$ trans-octahedra. As-F = 1.64-1.78(1), Mg-F = 1.95-1.99(1), Mg-O = 2.08, 2.12(1), S-O = 1.36-1.44(2) Å, Mg-O-S = 146, 152, O-S-O = 115.117°.

CAESIUM TETRAFLUOROANTIMONATE(III)
$CsSbF_4$

V.E. OVČINNIKOV, A.A. UDOVENKO, L.P. SOLOV'EVA, L.M. VOLKOVA and R.L. DAVIDOVIČ, 1982. Koord. Khim., 8, 1539-1541.

Orthorhombic, Pbam, a = 9.566, b = 15.804, c = 6.380 Å, D_m = 4.55, Z = 8. R = 0.065.

The structure contains $(Sb_2F_8{}^{2-})_n$ chains of SbF_5E octahedra (E = lone-pair).

μ-IODO-BIS(4-IODO-CYCLOHEPTASULPHUR) TRIS(HEXAFLUOROANTIMONATE) - BIS(ARSENIC TRIFLUORIDE)
$[(S_7I)_2I](SbF_6)_3.2AsF_3$ (I)

IODO-CYCLOHEPTASULPHUR(1+) TETRASULPHUR(2+) HEXAFLUOROARSENATE
$(S_7I)_4S_4(AsF_6)_6$ (II)

J. PASSMORE, G. SUTHERLAND and P.S. WHITE, 1982. Inorg. Chem., 21, 2717-2723.

I. Triclinic, $P\bar{1}$, a = 9.240, b = 13.321, c = 8.247 Å, α = 91.16, β = 94.22, γ = 111.04°, Z = 1. Mo radiation, R = 0.118 for 1775 reflexions.

II. Tetragonal, P4/n, a = 19.585, c = 8.321 Å, Z = 2. Mo radiation, R = 0.106 for 1307 reflexions.

I contains an $[(S_7I)_2I]^{3+}$ cation (Fig. 1), which consists of two S_7I^+ units linked by an iodine atom via a linear S-I-S bridge, I-S = 2.674(7) Å, and SbF_6^- anions and AsF_3 molecules. II (Fig. 2) contains S_7I^+ cations, square-planar S_4^{2+} cations, and octahedral AsF_6^- anions, with relatively weak anion-cation interactions.

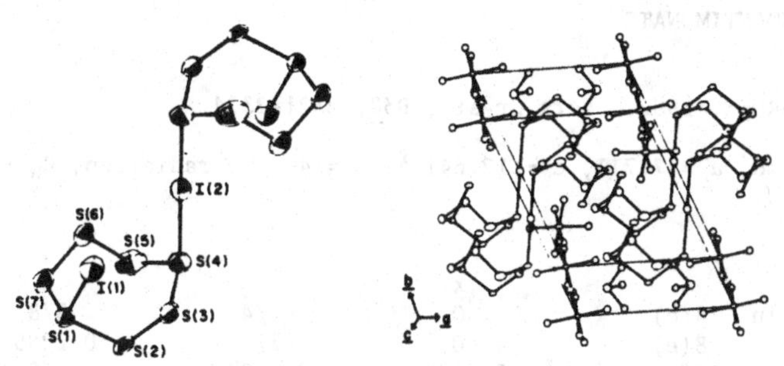

Fig. 1. Structure of $[(S_7I)_2I](SbF_6)_3.2AsF_3$.

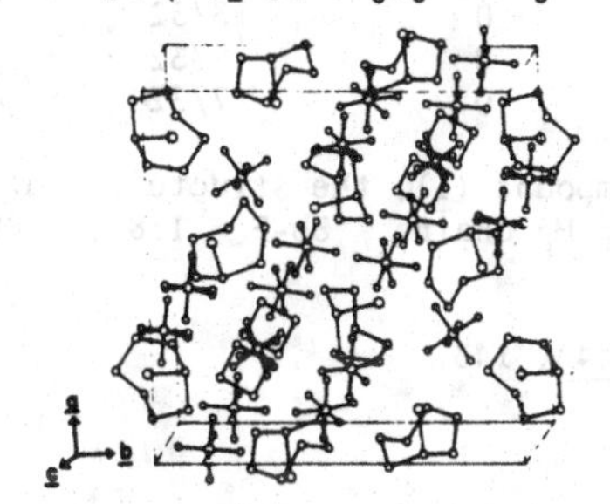

Fig. 2. Structure of $(S_7I)_4S_4(AsF_6)_6$.

TETRASELENIUM(2+) FLUOROANTIMONATE
$(Se_4^{2+})(Sb_2F_4^{2+})(Sb_2F_5^+)(SbF_6^-)_5$ (I)

TETRASELENIUM(2+) TETRACHLOROALUMINATE
$(Se_4^{2+})(AlCl_4^-)_2$ (II)

TETRATELLURIUM(2+) HEXAFLUOROANTIMONATE(V)
$(Te_4^{2+})(SbF_6^-)_2$ (III)

G. CARDINAL, R.J. GILLESPIE, J.F. SAWYER and J.E. VEKRIS, 1982. J. Chem. Soc., Dalton, 765-779.

I. Monoclinic, $P2_1/c$, a = 15.739, b = 13.498, c = 17.040 Å, β = 92.26°, Z = 4. Mo radiation, R = 0.052 for 3634 reklexions.

II. Orthorhombic, Pbam, a = 13.245, b = 13.223, c = 9.266 Å, Z = 4. Mo radiation, R = 0.042 for 912 reflexions.

III. Triclinic, $A\bar{1}$, a = 5.700, b = 16.252, c = 8.076 Å, α = 100.56, β = 102.67, γ = 97.47°, Z = 2. Mo radiation, R = 0.051 for 709 reflexions.

All three structures contain approximately square-planar Se_4^{2+} or Te_4^{2+} cations, mean bond lengths, 2.260(4), 2.286(4), 2.688(3) Å, with large numbers of interionic A...X contacts (A = Se, Te; X = F, Cl). The $Sb_2F_4^{2+}$ and $Sb_2F_5^+$ cations contain two SbF_3 trigonal pyramids sharing an edge and corner, respectively, Sb-F = 1.86-1.98 (terminal), 2.09-2.15 Å (shared); each Sb has five other F neighbours from neighbouring ions at 2.33-3.03 Å. SbF_6^- and $AlCl_4^-$ ions are octahedral and tetrahedral, respectively.

MERCURY HEXAFLUOROANTIMONATE
$Hg_{2.9}SbF_6$

Z. TUN and I.D. BROWN, 1982. Acta Cryst., B38, 2321-2324.

Tetragonal, $I4_1/amd$, a = 7.711, c = 12.641 Å, Z = 4. Mo radiation, R_w = 0.036 for 248 reflexions.

Atomic positions

			x	y	z
Sb	in	4(b)	0	1/4	3/8
F(1)		8(e)	0	1/4	0.2295
F(2)		16(g)	0.6723	0.9223	7/8
0.18 Hg(1)		16(h)	0	1/32	0.0000
0.18 Hg(2)		16(h)	0	3/32	-0.0002
0.18 Hg(3)		16(h)	0	5/32	0.0013
0.18 Hg(4)		16(h)	0	7/32	0.0028

Isostructural with the As compound (1), the structure containing SbF_6^- octahedra and two non-intersecting Hg chains. Sb-F = 1.84, 1.88(1), Hg-Hg = 2.66(2) Å.

1. Structure Reports, 40A, 136; 44A, 143.

AMMONIUM HEXAFLUOROTITANATE(IV)
$(NH_4)_2TiF_6$

Z. TUN and I.D. BROWN, 1982. Acta Cryst., B38, 1792-1794.

Trigonal, $P\bar{3}m1$, a = 5.968, 5.920, c = 4.821, 4.702 Å, at 293, 153K, Z = 1. Mo radiation, R_w = 0.019, 0.018 for 150, 147 reflexions.

Atomic positions (153K)

			x	y	z
Ti	in	1(a)	0	0	0
F		6(i)	0.2993	0.1496	0.2227
N		2(d)	2/3	1/3	-0.3094
H(1)		2(d)	2/3	1/3	-0.120
H(2)		6(i)	0.742	0.484	-0.372

The structure contains a slightly flattened TiF_6^{2-} octahedron, Ti-F = 1.857(1) Å, F-Ti-F = 91.4, 88.6° (at 153K), and an NH_4^+ ion which may be disordered at room temperature, linked by N-H...F hydrogen bonds.

CAESIUM TETRAFLUOROTITANATE(III)
$CsTiF_4$

R. SABATIER, A.-M. VASSON, A. VASSON, P. LETHUILLIER, J.-L. SOUBEYROUX, R. CHEVALIER and J.-C. COUSSEINS, 1982. Mater. Res. Bull., 17, 369-377.

Tetragonal, P4/nmm, a = 7.897, c = 6.505 Å, D_m = 4.19, Z = 4. Mo radiation, R = 0.052 for 379 reflexions.

Atomic positions (origin at centre)

		x	y	z
Cs(1)	in 2(b)	3/4	3/4	1/2
Cs(2)	2(c)	1/4	1/4	0.4483
Ti	4(d)	0	0	0
F(1)	8(i)	1/4	-0.0019	0.0470
F(2)	8(j)	-0.0295	-0.0295	0.2840

Isostructural with $CsFeF_4$ (1), the structure containing layers of corner-sharing TiF_6 distorted octahedra, Ti-F = 2.00 (x 4), 1.88 (x 2) Å , Ti-F-Ti = 162°. The layers are linked by 12-coordinate Cs ions, Cs-F = 3.19, 3.24, 3.28 Å.

1. Structure Reports, 40A, 143.

MANGANESE(II) OCTAFLUOROZIRCONATE(IV) HEXAHYDRATE
$Mn_2ZrF_8.6H_2O$

L.P. OTROŠČENKO, R.L. DAVIDOVIČ, V.I. SIMONOV and N.V. BELOV, 1981. Kristallografija, 26,1191-1194 [Soviet Physics - Crystallography, 26, 675-677].

Monoclinic, B2/b, a = 11.246, b = 13.593, c = 8.127 Å, γ = 119.81°, Z = 4. Mo radiation, R = 0.028 for 2654 reflexions.

Isostructural with the Cd compound (1), the structure (Fig. 1) containing ZrF_8 Thomson cubes linked by $MnF_4(OH_2)_3$ pentagonal bipyramids. Zr-F = 2.07-2.13, Mn-F = 2.17-2.49, Mn-O = 2.12-2.25, O-H...O = 2.79, 2.80, O-H...F = 2.65-2.74 Å.

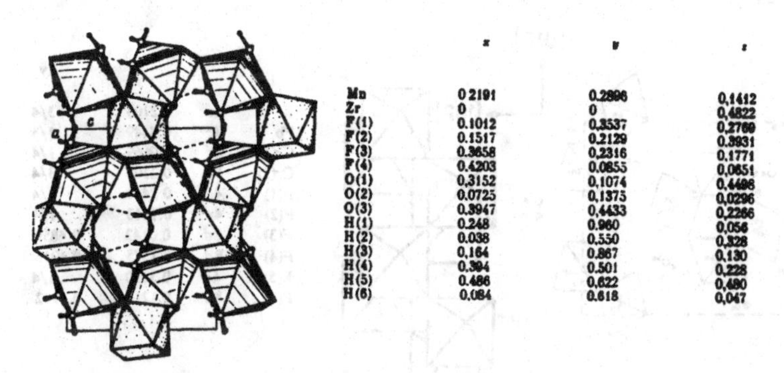

	x	y	z
Mn	0.2191	0.2896	0,1412
Zr	0	0	0,4822
F(1)	0.1012	0,3537	0,2760
F(2)	0.1517	0.2129	0.3931
F(3)	0.3658	0,2316	0.1771
F(4)	0.4203	0.0855	0,0651
O(1)	0,3152	0,1074	0,4496
O(2)	0.0725	0,1375	0,0296
O(3)	0,3947	0,4433	0,2266
H(1)	0.248	0.960	0,056
H(2)	0.038	0,550	0,328
H(3)	0,164	0.867	0,130
H(4)	0.394	0.501	0,228
H(5)	0.486	0,622	0,480
H(6)	0,084	0.618	0,047

Fig. 1. Structure of $Mn_2ZrF_8.6H_2O$.

1. Structure Reports, 48A, 141.

BARIUM PENTAFLUOROCHROMATE(III)
$BaCrF_5$

H. HOLLER, W. KURTZ, D. BABEL and W. KNOP, 1982. Z. Naturforsch., 37B, 54-60.

Orthorhombic, $P2_12_12_1$, a = 13.938, b = 5.711, c = 4.947 Å, D_m = 4.82, Z = 4. Mo radiation, R = 0.048 for 1282 reflexions, and neutron powder data.

Atomic positions

	x	y	z
Ba	0.0923	0.0899	0.0178
Cr	0.1650	0.6026	0.4247
F(1)	0.2667	0.1439	0.7809
F(2)	0.4211	0.1854	0.0871
F(3)	0.0918	0.3481	0.5564
F(4)	0.0807	0.5990	0.1290
F(5)	0.2479	0.3873	0.2326

$BaGaF_5$-type structure (1), with chains of CrF_6 octahedra sharing cis-corners, Cr-F = 1.94 (bridging), 1.88 Å (terminal), Cr-F-Cr = 137°. Ba has 12-coordination, Ba-F = 2.66-3.23 Å.

1. Structure Reports, 45A, 149.

POTASSIUM LEAD(II) ENNEAFLUORODICHROMATE(III)
$KPbCr_2F_9$

M. VLASSE, J.P. CHAMINADE, J.M. DANCE, M. SAUX and P. HAGENMULLER, 1982. J. Solid State Chem., 41, 272-276.

Orthorhombic, Pnma, a = 9.81, b = 5.412, c = 13.93 Å, D_m = 4.68, Z = 4. Mo radiation, R = 0.041 for 1285 reflexions.

Isostructural with $K_2Ta_2O_3F_6$ (1). The structure (Fig. 1) contains double chains along b of corner sharing CrF_6 octahedra, linked by 9-coordinate K^+ and Pb^{2+} ions. Mean Cr-F = 1.88 (terminal), 1.92 Å (bridging); K-F = 2.69-2.99, Pb-F = 2.40-2.87 Å.

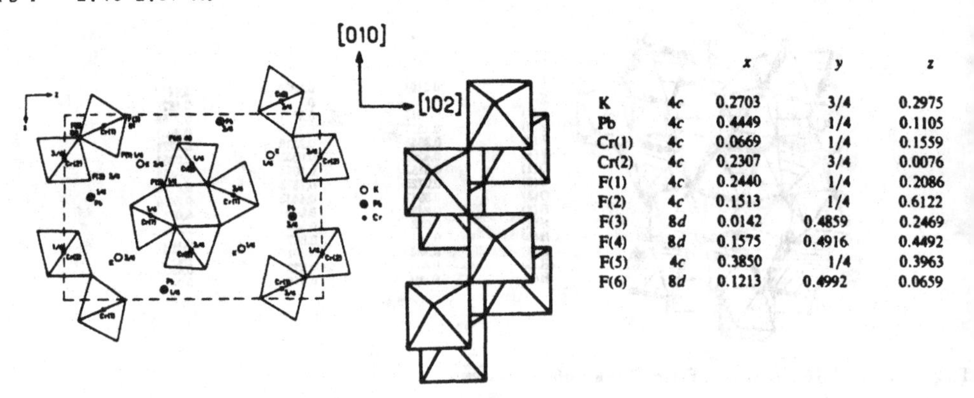

		x	y	z
K	4c	0.2703	3/4	0.2975
Pb	4c	0.4449	1/4	0.1105
Cr(1)	4c	0.0669	1/4	0.1559
Cr(2)	4c	0.2307	3/4	0.0076
F(1)	4c	0.2440	1/4	0.2086
F(2)	4c	0.1513	1/4	0.6122
F(3)	8d	0.0142	0.4859	0.2469
F(4)	8d	0.1575	0.4916	0.4492
F(5)	4c	0.3850	1/4	0.3963
F(6)	8d	0.1213	0.4992	0.0659

Fig. 1. Structure of $KPbCr_2F_9$.

1. Structure Reports, 42A, 203; 43A, 147.

FLUORIDES
K_2MF_4 (M = Mg, Mn, Co, Ni)
$K_3M_2F_7$ (M = Mn, Co, Ni)

D. BABEL and E. HERDTWECK, 1982. Z. anorg. Chem., 487, 75-84.

K_2MF_4
Tetragonal, I4/mmm, a = 3.980, 4.174, 4.073, 4.012 Å, c = 13.179, 13.272, 13.087, 13.076 Å, Z = 2. Mo radiation, R = 0.044, 0.033, 0.038, 0.022 for 291, 170, 499,470 reflexions. K in 4(e): z = 0.3519, 0.3529, 0.3543, 0.3539; M in 2(a); F(1) in 4(e): z = 0.1521, 0.1585, 0.1547, 0.1530; F(2) in 4(c).

$K_3M_2F_7$
Tetragonal, I4/mmm, a = 4.187, 4.074, 4.015, c = 21.586, 21.163, 21.073 Å, Z = 2. Mo radiation, R = 0.035, 0.041, 0.034 for 620, 440, 301 reflexions. K(1) in 4(e): z = 0.3139, 0.3153, 0.3152; K(2) in 2(b); M in 4(e): z = 0.0978, 0.0966, 0.0959; F(1) in 4(e): z = 0.1944, 0.1917, 0.1903; F(2) in 8(g): z = 0.0961, 0.0951, 0.0945; F(3) in 2(a).

K_2NiF_4 (1) and $Sr_3Ti_2O_7$-type (2) structures, as previously reported for the Cu and Zn fluorides (3, 4).

1. Structure Reports, 17, 332; 19, 323.
2. Ibid., 22, 308; 24, 440.
3. Ibid., 48A, 144.
4. Ibid., 46A, 165.

CAESIUM SODIUM HEXAFLUOROMANGANATE(III)
Cs_2NaMnF_6

W. MASSA, 1982. Z. anorg. Chem., 491, 208-216.

α-Phase (high-pressure), cubic, Fm3m, a = 8.762 Å, Z = 4. Powder data. Cs in 8(c): 1/4,1/4,1/4; Na in 4(b): 1/2,0,0; Mn in 4(a): 0,0,0; F in 24(e): 0.215, 0,0. Elpasolite structure (1).

γ-Phase (high-temperature), rhombohedral, R$\bar{3}$m, a = 6.265, c = 30.54 Å, Z = 6. Mo radiation, R = 0.041 for 419 reflexions.

Atomic positions (γ-phase)

			x	y	z
Mn(1)	in	3(a)	0	0	0
Mn(2)		3(b)	0	0	1/2
Na		6(c)	0	0	0.4030
Cs(1)		6(c)	0	0	0.1279
Cs(2)		6(c)	0	0	0.2817
F(1)		18(h)	0.1420	-0.1420	0.4623
F(2)		18(h)	0.1878	-0.1878	0.6306

The high-pressure α-form has an elpasolite structure (1) and the high-temperature γ-form has the 12L-Cs_2NaCrF_6-type structure (2). The normal β-form is a variant of the 12L structure (3).

1. Strukturbericht, 2, 498; Structure Reports, 40A, 137.
2. Structure Reports, 41A, 163; 42A, 179.
3. Ibid., 38A, 204.

SODIUM LITHIUM IRON(III) FLUORIDE
$Na_3Li_3Fe_2F_{12}$

W. MASSA, B. POST and D. BABEL, 1982. Z. Kristallogr., 158, 299-306.

Cubic, Ia3d, a = 12.387 Å, Z = 8. Mo radiation, R_w = 0.018 for 224 reflexions. Fe in 16(a): 0,0,0; Na in 24(c): 1/8,0,1/4; Li in 24(d): 3/8,0,1/4; F in 96(h): -0.02954,0.04737,0.14538.

Garnet structure (1). Fe-6F = 1.929, Li-4F = 1.850, Na-8F = 2.385, 2.549(1) Å.

1. Strukturbericht, 1, 363.

CAESIUM PENTAFLUOROFERRATE(III) MONOHYDRATE
$Cs_2FeF_5.H_2O$

N.V. BELOV, N.I. GOLOVASTIKOV, A.N. IVAŠČENKO, B.Ja. KOTJUŽANSKIJ, O.K. MEL'NIKOV and V.I. FILIPPOV, 1982. Kristallografija, 27, 511-515 [Soviet Physics - Crystallography, 27, 309-312].

Orthorhombic, Cmcm, a = 10.361, b = 8.266, c = 8.401 Å, Z = 4. Mo radiation, R = 0.077 for 722 reflexions.

Atomic positions

			x	y	z
Cs	in	8(e)	0.2080	0	0
Fe		4(c)	0	0.2885	1/4
F(1)		8(g)	0.1840	0.2861	1/4
F(2)		4(c)	0	0.0573	1/4
F(3)		8(f)	0	0.3019	0.0211
O		4(c)	0	0.5447	1/4

The structure (Fig. 1) contains $FeF_5(H_2O)^{2-}$ octahedra, linked by $CsF_{10}O_2$ polyhedra and by hydrogen bonds. Fe-F = 1.91-1.93, Fe-O = 2.12, Cs-F = 2.97-3.44, Cs-O = 3.70 Å.

Fig. 1. Structure of $Cs_2FeF_5.H_2O$.

LITHIUM STRONTIUM HEXAFLUOROFERRATE(III)
$LiSrFeF_6$

T. FLEISCHER and R. HOPPE, 1982. Z. Naturforsch., 37B, 981-987.

Monoclinic, $P2_1/c$, a = 5.2972, b = 8.7989, c = 10.2257 Å, β = 92.212°, D_m = 3.66, Z = 4. Mo radiation, R = 0.092 for 712 reflexions.

The structure (Fig. 1) contains sheets of edge-sharing FeF_6 and LiF_6 octahedra, linked by 8-coordinate Sr ions. Fe-F = 1.90-1.95, Li-F = 1.97-2.18, Sr-F = 2.40-2.95 Å.

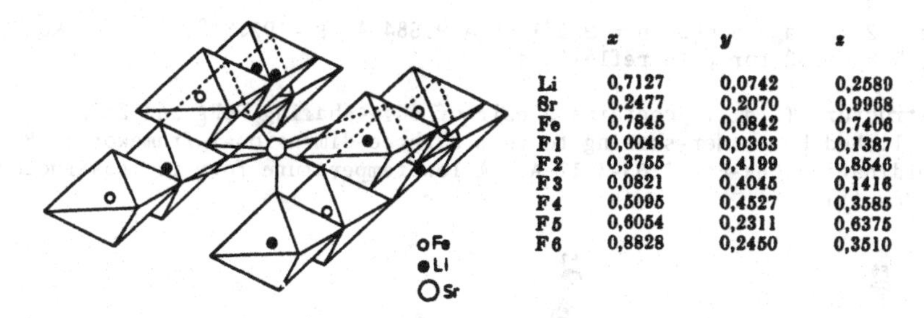

	x	y	z
Li	0,7127	0,0742	0,2580
Sr	0,2477	0,2070	0,9968
Fe	0,7845	0,0842	0,7406
F1	0,0048	0,0363	0,1387
F2	0,3755	0,4199	0,8546
F3	0,0821	0,4045	0,1416
F4	0,5095	0,4527	0,3585
F5	0,6054	0,2311	0,6375
F6	0,8828	0,2450	0,3510

Fig. 1. Structure of $LiSrFeF_6$.

LITHIUM MANGANESE(II) IRON(III) FLUORIDE
$LiMnFeF_6$ (α and β Forms)

G. COURBION, C. JACOBINI and R. DE PAPE, 1982. J. Solid State Chem., 45, 127-134.

Trigonal, P321, a = 8.684, 8.723, c = 4.657, 4.745 Å, D_m = 3.70, 3.59, Z = 3. Mo radiation, R = 0.020 for 1213 reflexions for the α-form; X-ray and neutron powder data for the β-form.

Atomic positions

α-form		x	y	z
Li(1)	in 1(a)	0	0	0
Li(2)	2(d)	1/3	2/3	0.5055
Mn	3(e)	0.3545	0	0
Fe	3(f)	0.6875	0	1/2
F(1)	6(g)	0 5311	0.4235	0.2709
F(2)	6(g)	0.2214	0.4374	0.2769
F(3)	6(g)	0.2215	0.1130	0.2435
β-form				
Fe(1)	1(a)	0	0	0
Fe(2)	2(d)	1/3	2/3	0.506
Mn	3(e)	0.352	0	0
Li	3(f)	0.714	0	1/2
F(1)	6(g)	0.541	0.420	0.268
F(2)	6(g)	0.226	0.462	0.270
F(3)	6(g)	0.221	0.098	0.226

The α-form is isostructural with Na_2SiF_6 (1), with a new type of cation order. The β-form (transition temperature 560°C) is isostructural with $LiMnGaF_6$ (2). Other types of cation ordering are found in related compounds (3).

1. Structure Reports, 29, 264.
2. Ibid., 41A, 155.
3. Ibid., 43A, 127.

BARIUM ZINC IRON(III) FLUORIDE (HIGH-TEMPERATURE)
$BaZnFeF_7$

H. HOLLER and D. BABEL, 1982. Z. anorg. Chem., 491, 137-144.

Monoclinic, $P2_1/c$, a = 5.603, b = 9.971, c = 9.584 Å, β = 92.80°, Z = 4. Mo radiation, R = 0.052 for 2426 reflexions.

The structure (Fig. 1) contains a pair of edge-sharing FeF_6 and ZnF_6 octahedra, linked by corner-sharing to form a three-dimensional framework; Ba has 12-coordination, Ba-F = 2.63-3.17 Å. A low-temperature form is isostructural with $BaMnFeF_7$ (1).

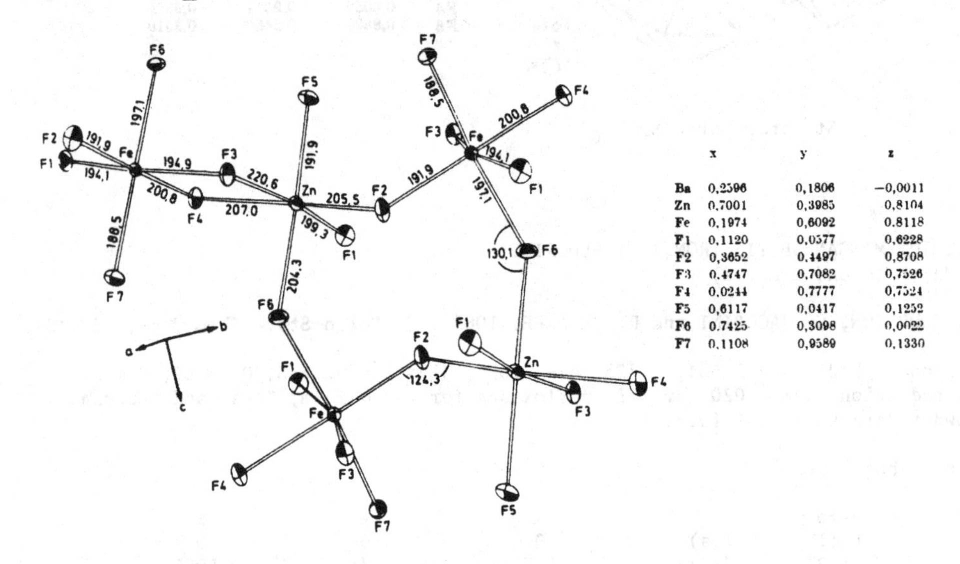

	x	y	z
Ba	0,2596	0,1806	−0,0011
Zn	0,7001	0,3985	0,8104
Fe	0,1974	0,6092	0.8118
F1	0,1120	0,0577	0,6228
F2	0,3652	0,4497	0,8708
F3	0,4747	0,7082	0,7526
F4	0,0244	0,7777	0,7524
F5	0,6117	0,0417	0,1252
F6	0.7425	0,3098	0,0022
F7	0,1108	0,9589	0,1330

Fig. 1. Structure of high-temperature $BaZnFeF_7$.

1. Structure Reports, 48A, 142.

FLUORIDES
A(I)M(II)M(III)F_6

T. FLEISCHER and R. HOPPE, 1982. J. Fluor. Chem., 19, 529-552.

A(I)M(II)M(III) = RbMgCo, RbZnCo, RbNiCo, CsNiCo, CsMgNi, RbMgNi, RbCuFe, cubic, Fd3m, a = 10.185, 10.207, 10.183, 10.271, 10.120, 9.978, 10.216 Å, Z = 8. Powder data. A(I) in 8(b): 3/8,3/8,3/8; M(II)+M(III) in 16(c): 0,0,0; F in 48(f): x,1/8,1/8, x = 0.310, 0.312, 0.310, 0.310, 0.312, 0.316, 0.313. $RbNiCrF_6$-type structure (1).

A(I)M(II)M(III) = RbCuV, RbCuAl, CsCuAl, CsZnAl, orthorhombic, Pnma, a ∿ 7, b ∿ 7, c ∿ 10Å, Z = 4. Powder data, atomic positional parameters not determined. Isostructural with $CsAgFeF_6$ (2), with an ordered arrangement of M(II) in 4(c) and M(III) in 4(a).

A(I)M(II)M(III) = CsNiNi, RbNiNi, orthorhombic, Imma, a = 7.122, 6.946, b = 7.350, 7.333, c = 10.025, 9.768 Å, Z = 4. Powder data.

Atomic positions

			A = Cs			A = Rb		
			x	y	z	x	y	z
A	in	4(e)	1/2	1/4	0.1535	1/2	1/4	0.1596
Ni(1)		4(d)	1/4	1/4	3/4	1/4	1/4	3/4
Ni(2)		4(a)	0	0	0	0	0	0
F(1)		4(e)	0	1/4	0.6627	0	1/4	0.6424
F(2)		4(e)	1/2	1/4	0.4604	1/2	1/4	0.4604
F(3)		16(j)	0.1800	0.9605	0.1320	0.1914	0.9636	0.1288

The $A(I)Ni_2F_6$ structure is very similar to that of $CsAgM(III)F_6$, but in a higher symmetry space group [compare the positional parameters with those in 2].

1. Structure Reports, 38A, 205.
2. Ibid., 48A, 136.

RUTILE-TYPE FLUORIDES
$LiM(II)M(III)F_6$

T. FLEISCHER and R. HOPPE, 1982. Z. Naturforsch., 37B, 988-994.

$LiCaCoF_6$, $LiCdCoF_6$, $LiCaNiF_6$, $LiSrNiF_6$, trigonal, P$\bar{3}$1c, a = 5.1024, 5.0860, 5.0600, 5.1241, c = 9.783, 9.518, 9.745, 10.341 Å, Z = 2. Powder data. Li in 2(c): 1/3,2/3,1/4; M(II) in 2(b): 0,0,0; M(III) in 2(d): 2/3,1/3,1/4; F in 12(i): x,y,z, x = 0.3696, 0.3667, 0.3696, 0.3755, y = 0.0259, 0.0229, 0.0259,0.0319, z = 0.1417, 0.1406, 0.1417, 0.1438. $LiCaAlF_6$-type structures (1).

$LiZnCoF_6$, $LiMgCoF_6$, $LiNiCoF_6$, tetragonal, $P4_2/mnm$, a = 4.6238, 4.6039, 4.6107, c = 9.166, 9.051, 9.071 Å, Z = 2. Powder data. Li/M(II) in 4(e): 0,0,0.334; Co in 2(a): 0,0,0; F(1) in 4(f): 0.296,0.296,0; F(2) in 8(j): 0.308,0.308, 0.339. Trirutile structures.

$LiCuCoF_6$, tetragonal, $P4_2/mnm$, a = 4.6095, c = 3.1051 Å, Z = 2/3. Powder data. Rutile structure (2), with statistical cation distribution.

$LiSrCoF_6$, monoclinic, $P2_1/c$, a = 5.2753, b = 8.7593, c = 10.2133 Å, β = 92.149°, D_m = 3.71, Z = 4. Powder data. $LiSrFeF_6$-type structure (3).

$LiBaCoF_6$, monoclinic, $P2_1/c$, a = 5.5056, b = 10.263, c = 8.4379 Å, β = 90.99°, D_m = 4.40, Z = 4. Powder data. $LiBaCrF_6$-type structure (4).

1. Structure Reports, 37A, 179; T. FLEISCHER and R. HOPPE, 1983. Z. anorg. Chem., in press.
2. Strukturbericht, 1, 155.
3. This volume, p. 95.
4. Structure Reports, 40A, 143.

BARIUM COBALT(II) IRON(III) ENNEAFLUORIDE
Ba_2CoFeF_9

A. de KOZAK, M. LEBLANC, M. SAMOUEL, G. FÉREY, and R. de PAPE, 1981. Rev. Chim. Minér., 18, 659-666.

Monoclinic, $P2_1/n$, a = 7.486, b = 17.757, c = 5.687 Å, β = 90.87°, D_m = 4.96, Z = 4. Mo radiation, R = 0.089 for 589 reflexions.

The structure (Fig. 1) contains double chains of MF_6 octahedra with Co(II)/Fe(III) ordering; Co-F = 1.93-2.08, Fe-F = 1.87-2.04(3) Å. Ba ions have 12- and 10-coordinations, Ba-F = 2.64-3.32 Å.

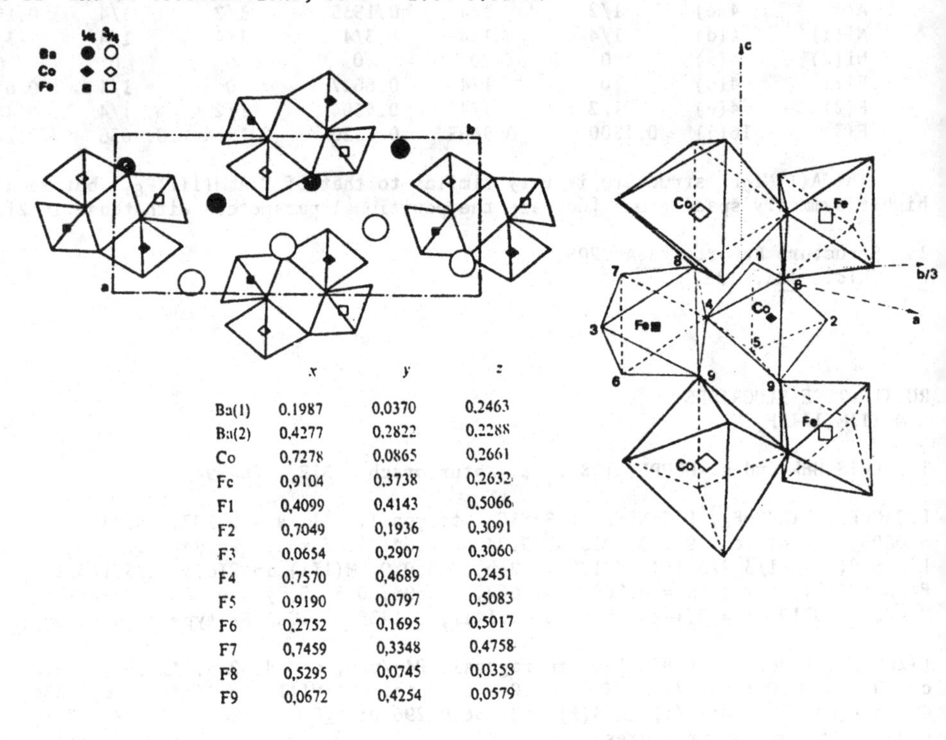

	x	y	z
Ba(1)	0.1987	0.0370	0.2463
Ba(2)	0.4277	0.2822	0.2288
Co	0.7278	0.0865	0.2661
Fe	0.9104	0.3738	0.2632
F1	0.4099	0.4143	0.5066
F2	0.7049	0.1936	0.3091
F3	0.0654	0.2907	0.3060
F4	0.7570	0.4689	0.2451
F5	0.9190	0.0797	0.5083
F6	0.2752	0.1695	0.5017
F7	0.7459	0.3348	0.4758
F8	0.5295	0.0745	0.0358
F9	0.0672	0.4254	0.0579

Fig. 1. Structure of Ba_2CoFeF_9.

FLUOROPALLADATES

I. B.G. MÜLLER, 1982. Z. anorg. Chem., 491, 245-252.

$CsPd_2F_5$
Orthorhombic, Imma, a = 6.533, b = 7.862, c = 10.79 Å, Z = 4. Mo radiation, R = 0.082 for 471 reflexions.

Atomic positions

			x	y	z
Cs	in	4(e)	0	1/4	0.3845
Pd(1)		4(a)	0	0	0
Pd(2)		4(d)	1/4	1/4	3/4
F(1)		4(e)	0	1/4	0.0828
F(2)		16(j)	0.2535	0.4261	0.6228

The structure (Fig. 1) is similar to that of $CsNi_2F_6$ (which has an additional F atom in 4(e): z = 0.66 (1)). Pd(1) has octahedral coordination, Pd-F = 2.16(1) Å, and Pd(2) has square-planar coordination, Pd-F = 1.95(1) Å; Cs has 11 F neighbours at 3.04-3.36 Å. $CsMPdF_5$ (M = Mg, Zn, Ni, Co) are isostructural.

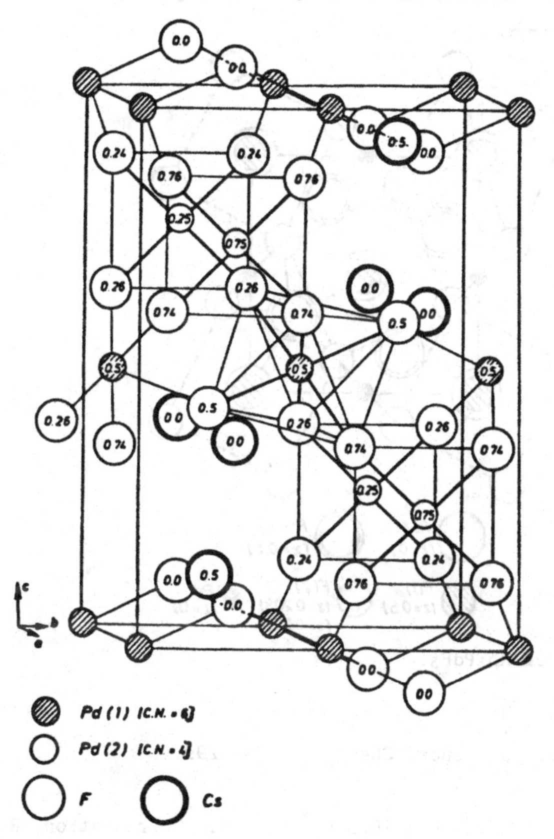

Fig. 1. Structure of $CsPd_2F_5$.

Rb_3PdF_5 and Rb_2CsPdF_5

Tetragonal, P4/mbm, a = 7.462, 7.579, c = 6.457, 6.455 Å, Z = 2. Mo radiation, R = 0.058, 0.070 for 269, 297 reflexions.

Atomic positions

Rb_3PdF_5			x	y	z
Rb(1)	in	2(a)	0	0	0
Rb(2)		4(h)	0.1771	0.6771	1/2
Pd		2(d)	0	1/2	0
F(1)		2(b)	0	0	1/2
	or	4(e)	0	0	0.4573
F(2)		8(k)	0.3697	0.8697	0.2086

Atomic positions

Rb_2CsPdF_5			x	y	z
Cs	in	2(a)	0	0	0
Rb		4(h)	0.1774	0.6774	1/2
Pd		2(d)	1/2	0	0
F(1)		2(b)	0	0	1/2
F(2)		8(k)	0.3725	0.8725	0.2132

The structure (Fig. 2) contains square-planar $PdF(2)_4^{2-}$ and F^- anions linked by Rb or Cs ions with 10- and 8-coordinations. Pd-F = 1.93, 1.94(1), Rb(1) or Cs-F = 3.22-3.29, Rb(2)-F = 2.75-2.99 Å. Cs_3PdF_5 is isostructural.

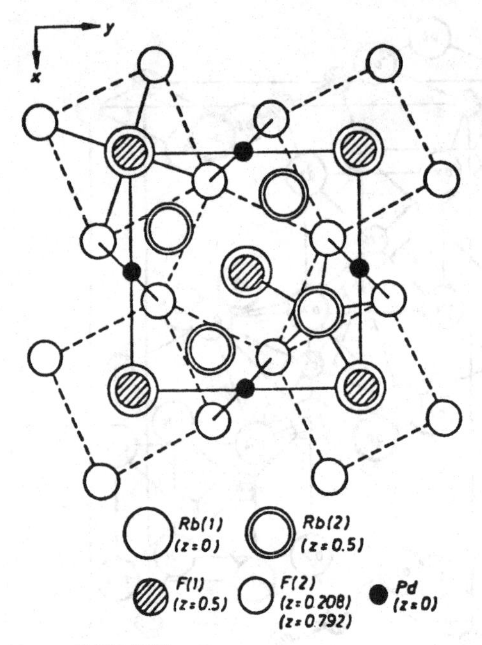

Fig. 2. Structure of Rb_3PdF_5.

$CaPdF_4$, $CdPdF_4$, $HgPdF_4$, PdF_2

II. B.G. MÜLLER, 1982. J. Fluor. Chem., 20, 291-299.

$CaPdF_4$
Tetragonal, I4/mcm, a = 5.521, c = 10.570 Å, Z = 4. Mo radiation, R = 0.083 for 283 reflexions. Ca in 4(a): 0,0,1/4; Pd in 4(d): 0,1/2,0; F in 16(ℓ): x,1/2+x,z, x = 0.1774, z = 0.1354. $KBrF_4$-type structure (2).

$CaPdF_4$, $HgPdF_4$, PdF_2 (high-pressure)
Cubic Pa3, a = 5.403, 5.43, 5.327 Å, Z = 2, 2, 4. Mo radiation, R = 0.047 for 165 reflexions for $CaPdF_4$. Ca/Pd, Hg/Pd, or Pd in 4(a): 0,0,0; F in 8(c): x,x,x, x = 0.335 for $CaPdF_4$. CaF_2-variant structure.

1. This volume, p. 97.
2. Structure Reports, 20, 222; 21, 275; 34A, 202.

CAESIUM HEXAFLUOROPLATINATE(IV)
Cs_2PtF_6

RUBIDIUM HEXAFLUOROPLATINATE(IV)
Rb_2PtF_6

L. GROSSE and R. HOPPE, 1982. Naturwissenschaften, 69, 447.

[Trigonal, P$\bar{3}$m1], a = 6.208, 5.963, c = 5.015, 4.804 Å, Z = 1. Mo, Ag radiations, R = 0.086, 0.067 for 195, 246 reflexions. Pt in 1(a): 0,0,0; Cs or Rb in 2(d): 1/3,2/3,z, z = 0.6989, 0.7036; F in 6(i): x,$\bar{x}$,z, x = 0.1498, 0.1573, z = 0.2175, 0.2225.

Isostructural with K_2PtF_6 (1), Pt-F = 1.945 Å.

1. Structure Reports, 15, 149; 46A, 164.

CAESIUM TETRAFLUOROCUPRATE(III)
$CsCuF_4$

T. FLEISCHER and R. HOPPE, 1982. Z. anorg. Chem., 492, 76-82.

Tetragonal, I4/mcm, a = 5.8488, c = 12.043 Å, Z = 4. Powder data. Cs in 4(a): 0,0,1/4; Cu(III) in 4(d): 0,1/2,0; F in 16(ℓ): 0.148,0.648,0.102.

$KBrF_4$-type structure (1), with square-planar CuF_4^- anions, Cu-F = 1.73 Å

1. Structure Reports, 20, 222; 21, 275; 34A, 202.

TVEITITE
$Ca_{13}Y_6F_{43}$ (or $Ca_{14}Y_5F_{43}$)

D.J.M. BEVAN, J. STRÄHLE and O. GREIS, 1982. J. Solid State Chem., 44, 75-81.

Rhombohedral, R$\bar{3}$, a = 16.692, c = 9.666 Å, Z = 3. Mo radiation, R = 0.122 for 1164 reflexions.

The mineral has an ordered, anion-excess, fluorite related superstructure (Fig. 1), with Y_6F_{37} clusters consisting of an octahedral arrangement of six YF_8 square antiprisms, with a central F capping all six antiprisms; Ca ions have 12-, 8-, and 10-coordinations. Y-F = 2.22-2.34, 2.65, Ca-F = 2.29-2.80(1) Å.

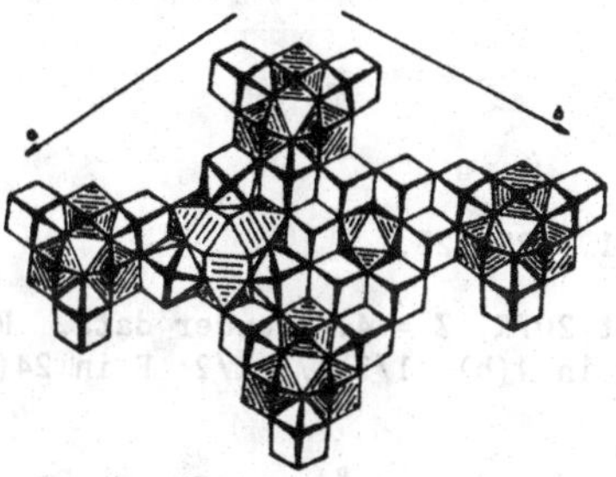

	x	y	z
Y	0.0169	0.1489	0.1786
Ca(1)	0.0	0.0	0.5353
Ca(2)	0.2654	0.2348	0.1557
Ca(3)	0.1749	0.4059	0.1672
F(1)	0.1958	0.0831	0.0040
F(2)	0.1102	0.0975	0.2349
F(3)	0.0190	0.1495	0.4146
F(4)	0.2702	0.1024	0.2573
F(5)	0.1576	0.2686	0.2515
F(6)	0.0324	0.2842	0.0971
F(7)	0.2955	0.3838	0.0744
F(8)	0.0	0.0	0.060

Fig. 1. Structure of tveitite (occupancy = 0.5 for Ca(1) and F(8)).

MAGNESIUM EUROPIUM(II) FLUORIDE
$MgEuF_4$

E. BANKS, R. JENKINS and B. POST, 1982. Rare Earths Mod. Sci. Technol., 3, 329-334.

Orthorhombic, Cmcm, a = 3.935, b = 14.434, c = 5.664 Å, Z = 4. R = 0.102.

Eu has 8-coordination, Eu-F = 2.439-2.925 Å, and Mg has 6-coordination. $MgSmF_4$ and $MgSrF_4$ are isostructural.

POTASSIUM HEPTAFLUORODIHOLMATE
KHo_2F_7

Y. LE FUR, S. ALÉONARD, M.F. GORIUS and M.T. ROUX, 1982. Acta Cryst., B38, 1431-1436.

Monoclinic, Cm, a = 14.287, b = 8.004, c = 11.950 Å, β = 125.33°, Z = 8. Ag radiation, R = 0.033 for 979 reflexions.

The structure (Fig. 1) contains groups of three edge-sharing HoF_8 antiprisms joined by chains of edge-sharing HoF_8 cubes to give Ho_3F_{18} blocks which are linked to form layers. K ions have 10- and 14-coordinations. Ho-F = 2.15-2.47, K-F = 2.54-3.55 Å.

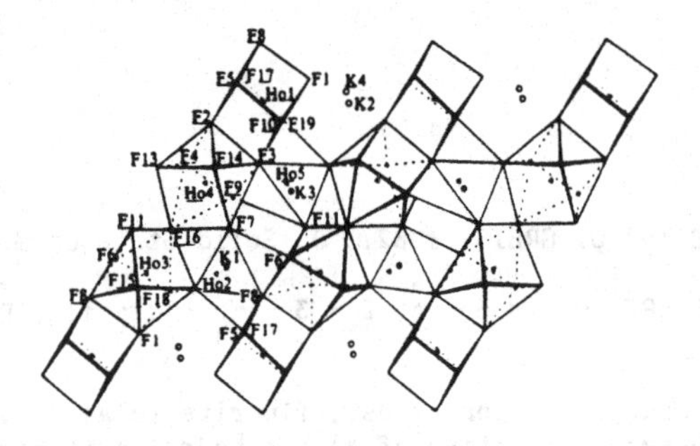

Fig. 1. Structure of KHo_2F_7.

RUBIDIUM SODIUM HOLMIUM FLUORIDE
Rb_2NaHoF_6

J. IHRINGER, 1982. Solid State Comm., 41, 525-527.

Above 172K, cubic, Fm3m, a = 8.8735 Å, at 207K, Z = 4. Powder data. Rb in 8(c): 1/4,1/4,1/4; Na in 4(a): 0,0,0; Ho in 4(b): 1/2,1/2,1/2; F in 24(e): 0.250,0,0.

Below 172K, tetragonal, I4/m, a = 6.2233, c = 8.8957 Å, at 17K, Z = 2. Rb in 4(d): 0,1/2[not 0],1/4; Na in 2(a): 0,0,0; Ho in 2(b): 0,0,1/2; F in 4(e): 0,0,0.258 and 8(h): 0.231,0.279,0.

Above 172K the compound has the elpasolite structure (1, 2). Below the phase change all octahedra are rotated 5° about c and are slightly stretched; Na-F = 2.25, 2.30 Å

1. Strukturbericht, 2, 498; Structure Reports, 40A, 137.
2. Structure Reports, 33A, 186.

POTASSIUM ERBIUM PENTAFLUORIDE
K_2ErF_5

N.V. PODBEREZSKAJA, S.V. BORISOV, V.I. ALEKSEEV, M.N. CEITLIN and K.M. KURBANOV, 1982. Ž. Strukt. Khim., 23, No. 2, 158-160 [J. Struct. Chem., 23, 310-312].

Orthorhombic, $Pc2_1n$, a = 6.592, b = 7.221, c = 10.764 Å, Z = 4. Mo radiation, R = 0.057 for 1474 reflexions.

Atomic positions

	x	y	z
Er	0.06766	-0.00499	0.00366
K(1)	0.046	0.261	0.328
K(2)	0.037	0.748	0.330
F(1)	0.397	-0.019	0.012
F(2)	0.157	0.280	0.064
F(3)	0.154	0.703	0.079
F(4)	0.341	0.011	0.308
F(5)	0.379	0.481	0.330

The structure contains layers of edge-sharing ErF_7 polyhedra linked by edge-sharing to KF_9 polyhedra which are joined to each other by common faces. Er-F = 2.18-2.33, K-F = 2.54-3.38(3) Å.

CAESIUM YTTERBIUM FLUORIDE
$CsYb_3F_{10}$

S. ALÉONARD, M.T. ROUX and B. LAMBERT, 1982. J. Solid State Chem., 42, 80-88.

Monoclinic, Pc, a = 4.2893, b = 6.7437, c = 16.196 Å, β = 90°, Z = 2. Ag radiation, R = 0.031 for 1406 reflexions.

The structure (Fig. 1) contains sheets of edge- and corner-sharing YbF_7 pentagonal bipyramids, with tunnels which contain 8-coordinate Cs ions. Yb-F = 2.12-2.47, Cs-F = 2.89-3.30(1) Å.

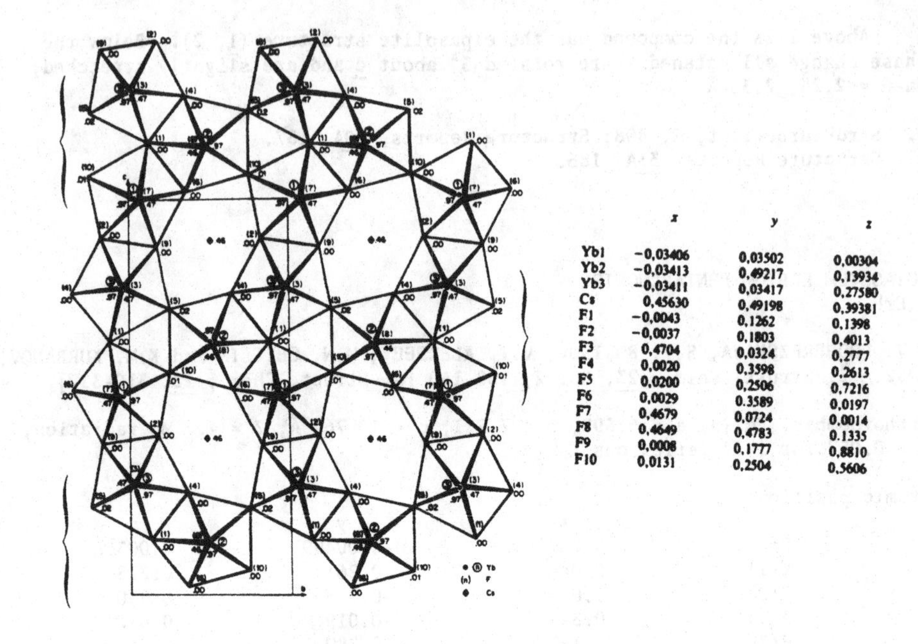

	x	y	z
Yb1	−0,03406	0,03502	0,00304
Yb2	−0,03413	0,49217	0,13934
Yb3	−0,03411	0,03417	0,27580
Cs	0,45630	0,49198	0,39381
F1	−0,0043	0,1262	0,1398
F2	−0,0037	0,1803	0,4013
F3	0,4704	0,0324	0,2777
F4	0,0020	0,3598	0,2613
F5	0,0200	0,2506	0,7216
F6	0,0029	0.3589	0,0197
F7	0,4679	0,0724	0,0014
F8	0,4649	0,4783	0.1335
F9	0,0008	0,1777	0,8810
F10	0,0131	0,2504	0,5606

Fig. 1. Structure of $CsYb_3F_{10}$.

RUBIDIUM LUTETIUM FLUORIDE
β-$RbLu_3F_{10}$

A. ARBUS, M.T. FOURNIER, J.C. COUSSEINS, A. VÉDRINE and R. CHEVALIER, 1982. Acta Cryst., B38, 75-79.

Orthorhombic, Acam, a = 16.013, b = 13.182, c = 8.435 Å, D_m = 5.92, Z = 8. Mo radiation, R = 0.078 for 1326 reflexions.

The structure (Fig. 1) contains layers of LuF_7 pentagonal bipyramids, with 8- and 10-coordinate Rb in tunnels. An α-form is isostructural with β-KYb_3F_{10} (1).

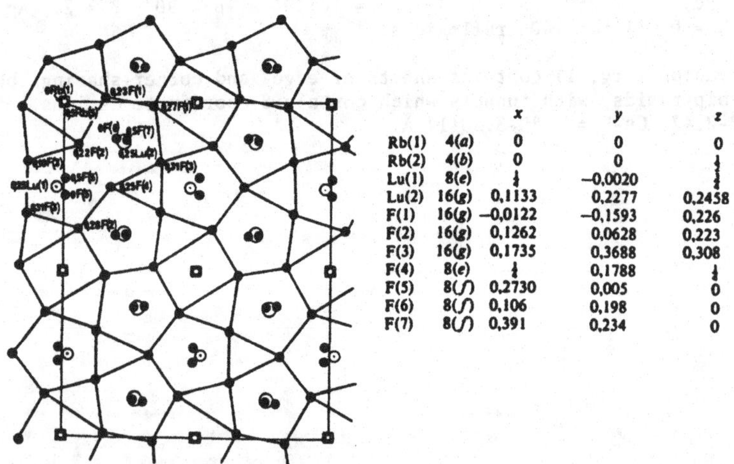

		x	y	z
Rb(1)	4(a)	0	0	0
Rb(2)	4(b)	0	0	$\frac{1}{4}$
Lu(1)	8(e)	$\frac{1}{4}$	−0,0020	$\frac{1}{4}$
Lu(2)	16(g)	0,1133	0,2277	0,2458
F(1)	16(g)	−0,0122	−0,1593	0,226
F(2)	16(g)	0,1262	0,0628	0,223
F(3)	16(g)	0,1735	0,3688	0,308
F(4)	8(e)	$\frac{1}{4}$	0,1788	$\frac{1}{4}$
F(5)	8(f)	0,2730	0,005	0
F(6)	8(f)	0,106	0,198	0
F(7)	8(f)	0,391	0,234	0

Fig. 1. Structure of β-$RbLu_3F_{10}$.

1. Structure Reports, 42A, 182.

SODIUM NONAFLUORODINEPTUNATE
$NaNp_2F_9$

A. COUSSON, H. ABAZLI, A. TABUTEAU, M. PAGÈS and M. GASPERIN, 1982. Acta Cryst., B38, 1801-1803.

Orthorhombic, Pnma, a = 8.617, b = 11.274, c = 6.995 Å, Z = 4. R = 0.030 for 1816 reflexions.

Atomic positions (fractional x 10^4)

	x	y	z
Np	3260	4504	3456
Na	4637	¼	8581
F(1)	−46	5995	759
F(2)	2922	976	269
F(3)	2857	6083	1422
F(4)	5664	801	1984
F(5)	3270	¼	3375

Isostructural with KU_2F_9 (1). Np has 9-coordination (tricapped trigonal prism) and Na has 10 F neighbours; Np-F = 2.26-2.37, Na-F = 2.55-3.17 Å.

1. Structure Reports, 19, 330; 34 A, 199.

SODIUM IRON(II) NEPTUNIUM(IV) FLUORIDE
$NaFeNp_3F_{15}$

A. COUSSON, H. ABAZLI, M. PAGÈS and M. GASPERIN, 1982. Acta Cryst., B38, 2668-2670.

Trigonal, P$\bar{3}$c1, a = 9.802, c = 13.004 Å, Z = 4. Mo radiation, R = 0.042 for 2032 reflexions.

The structure (Fig. 1) contains Np ions with 9-coordination (tricapped trigonal prism), Na with 12-coordination and Fe and disordered Na/Fe with octahedral coordinations. Np-F = 2.20-2.41, Na-F = 2.64-2.69, Fe-F = 2.03, Na/Fe-F = 2.23-2.36(1) Å.

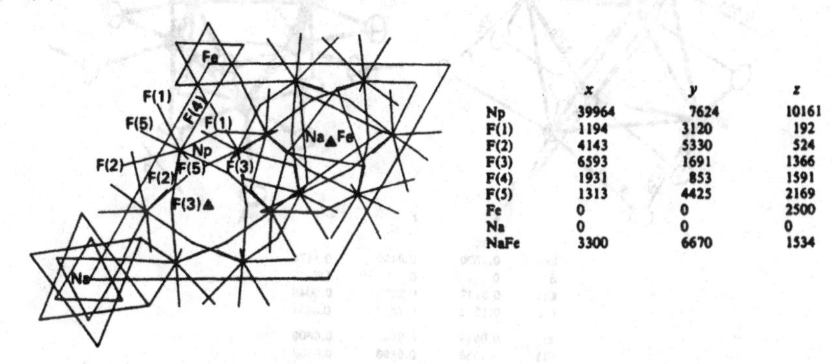

	x	y	z
Np	39964	7624	10161
F(1)	1194	3120	192
F(2)	4143	5330	524
F(3)	6593	1691	1366
F(4)	1931	853	1591
F(5)	1313	4425	2169
Fe	0	0	2500
Na	0	0	0
NaFe	3300	6670	1534

Fig. 1. Structure of $NaFeNp_3F_{15}$, and atomic positional parmeters (x 10^5 for Np, x 10^4 for other atoms).

MAGNESIUM CHLORIDE
β-$MgCl_2$

I.W. BASSI, F. POLATO, M. CALCATERRA and J.C.J. BART, 1982. Z. Kristallogr., 159, 297-302.

Trigonal, $P\bar{3}m1$, a = 3.641, c = 5.927 Å, Z = 1. Powder data. Mg in 1(a): 0,0,0: Cl in 2(d): 1/3,2/3,z, z ∿ 1/4.

CdI_2-type structure (1); Mg-6Cl = 2.51 Å. The more common form has a $CdCl_2$-type structure (2).

1. Strukturbericht, 1, 161.
2. Ibid., 1, 743.

SCANDIUM CHLORIDE
Sc_7Cl_{12}

ZIRCONIUM CHLORIDE
Zr_6Cl_{15}

J.D. CORBETT, K.R. POEPPELMEIER and R.L. DAAKE, 1982. Z. anorg. Chem., 491, 51-59.

Sc_7Cl_{12}, rhombohedral, $R\bar{3}$, a = 12.959, c = 8.825 Å, Z = 3. Mo radiation, R = 0.053 for 483 reflexions.

Zr_6Cl_{15}, cubic, Ia3d, a = 21.141 Å, Z = 16. Mo radiation, R = 0.101 for 404 reflexions.

Sc_7Cl_{12} contains a cubic-close-packing of Sc_6Cl_{12} clusters, with isolated Sc atoms in all the octahedral holes (Fig. 1). Zr_6Cl_{15} is isostructural with Ta_6Cl_{15} (1), with Zr_6Cl_{12} clusters bridged by additional Cl atoms (Fig. 1). Residual density at the centre of each cluster probably results from errors in the data rather than additional O or F atoms. Previous study of both structures in 2.

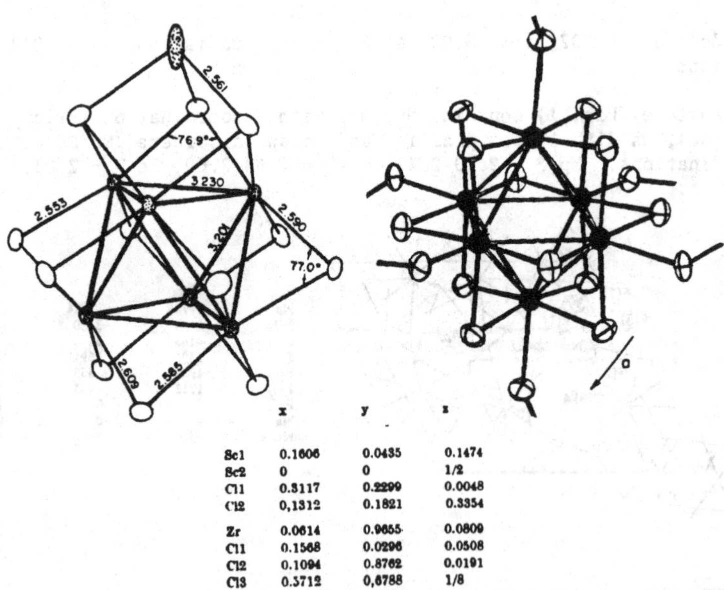

	x	y	z
Sc1	0.1606	0.0435	0.1474
Sc2	0	0	1/2
Cl1	0.3117	0.2299	0.0048
Cl2	0,1312	0.1821	0.3354
Zr	0.0614	0.9655	0.0809
Cl1	0.1568	0.0296	0.0508
Cl2	0.1094	0.8762	0.0191
Cl3	0.3712	0,6788	1/8

Fig. 1. Structures of Sc_7Cl_{12} (left) and Zr_6Cl_{15} (right).

1. Structure Reports, 33A, 203.
2. Ibid., 44A, 153.

ZIRCONIUM CHLORIDE, BROMIDE, and IODIDE

ZrX_3 (X = Cl, Br, I), $Zr_{0.88}I_3$ (0.88 x $ZrI_{3.40}$)

E.M. LARSEN, J.S. WRAZEL and L.G. HOARD, 1982. Inorg. Chem., 21, 2619-2624.

Hexagonal, $P6_3/mcm$, a = 6.383, 6.728, 7.298, 7.239, c = 6.139, 6.299, 6.667, 6.673 Å, Z = 2. Mo radiation, R = 0.030, 0.076, 0.060, 0.057 for 701, 315, 566, 495 reflexions. Zr in 2(b): 0,0,0; X in 6(g): x,0,1/4, x = 0.3167, 0.3211, 0.3255, 0.3246 (for $ZrI_{3.40}$, Zr occupancy = 0.882)

The materials have the β-$TiCl_3$ structure as previously described (1), and contain chains of face-sharing ZrX_6 octahedra. Zr-Zr = 3.07, 3.15, 3.33. 3.34, Zr-X = 2.54, 2.68, 2.90, 2.88 Å.

1. Structure Reports, 29, 249; 31A, 96; 41A, 179.

MOLYBDENUM CHLORIDES

$MoCl_2$ $MoCl_3$ $MoCl_4$

H. SCHÄFER, H.-G. von SCHNERING, J. TILLACK, F. KUHNEN, H. WÖHRLE and H. BAUMANN, 1967. Z. anorg. Chem., 353, 281-310.

[This paper appears not to have been reported previously in Structure Reports.]

$MoCl_2$ (Mo_6Cl_{12})
Orthorhombic, Bbam, a = 11.249, b = 11.280, c = 14.067 Å, Z = 24 (4 x Mo_6Cl_{12}). Mo radiation, R = 0.124 for 748 reflexions (film data).

α-$MoCl_3$
Monoclinic, C2/m, a = 6.092, b = 9.745, c = 7.275 Å, β = 124.6°, D_m = 3.74, Z = 4. Mo radiation, R = 0.068, 0.080 for 41 hk0, 49 0kℓ reflexions (film data).

β-$MoCl_3$
Monoclinic, C2/c, a = 6.115, b = 9.814, c = 11.906 Å, β = 91.0°, Z = 8. No refinement.

$MoCl_4$
Trigonal, P$\bar{3}$1c, a = 6.058, c = 11.674 Å, Z = 3. Mo radiation, R = 0.072, 0.084 for 74 0kℓ, 47 hhℓ reflexions (film data).

Atomic positions

Mo_6Cl_{12}		x	y	z
Mo(1) in	8(d)	0	0	0.1318
Mo(2)	8(f)	0.1502	0.0649	0
Mo(3)	8(f)	0.9343	0.1504	0
Cl(1)	16(g)	0.083	0.202	0.123
Cl(2)	16(g)	0.205	0.919	0.123
Cl(3)	8(d)	0	0	0.301
Cl(4)	8(f)	0.359	0.143	0

Atomic positions

			x	y	z
α-$MoCl_3$					
Mo	in	4(g)	0	0.1417	0
Cl(1)		8(j)	0.502	0.170	0.223
Cl(2)		4(i)	0.008	0	0.274
β-$MoCl_3$					
Mo(1)	in	4(e)	0	0.025	1/4
Mo(2)		4(e)	0	0.309	1/4
Cl(1)		8(f)	-0.185	0.167	0.387
Cl(2)		8(f)	-0.150	0.497	0.363
Cl(3)		8(f)	-0.150	0.837	0.363
$MoCl_4$					
1.5 Mo(1)	in	2(a)	0	0	1/4
1.5 Mo(2)		2(d)	2/3	1/3	1/4
12 Cl		12(i)	-0.327	0.006	0.131

The Mo_6Cl_{12} structure contains Mo_6Cl_8 clusters, with a Mo_6 octahedron and a Cl atom above each face; an additional Cl is bonded to each Mo, four of these Cl being shared with neighbouring clusters to form a two-dimensional arrangement. Mo-Mo = 2.61(1) Å.

α- and β-$MoCl_3$ have layer structures, with cubic and hexagonal close-packed Cl layers, respectively, with Mo in adjacent octahedral holes forming Mo_2 pairs, Mo-Mo = 2.76(1) Å. $MoCl_4$ has a trichloride-like layer structure, with hexagonal close-packed Cl layers, and partially-occupied octahedral metal sites.

MOLYBDENUM CHLORIDE HYDRATE
$(H_3O)_2[(Mo_6Cl_8)Cl_6].6H_2O$

H. OKUMURA, T. TAGA, K. OSAKI and I. TSUJIKAWA, 1982. Bull. Chem. Soc. Japan, 55, 307-308.

Monoclinic, C2/c, a = 17.27, b = 9.17, c = 18.55 Å, β = 98.07°, Z = 4. Cu radiation, R = 0.117 for 865 reflexions (film data).

The structure contains discrete $(Mo_6Cl_8)Cl_6^{2-}$ anions, with a Mo_6 octahedron, eight Cl above the octahedral faces forming a cube, and one additional terminal Cl bonded to each Mo. Mo-Mo = 2.59(1), Mo-Cl = 2.39-2.52(3) Å.

GOLD CHLORIDE
Au_4Cl_8

D.B. DELL'AMICO, F. CALDERAZZO, F. MARCHETTI and S. MERLINO, 1982. J. Chem. Soc., Dalton, 2257-2260.

Triclinic, P$\bar{1}$, a = 7.015, b = 6.830, c = 6.684 Å, α = 94.4, β = 107.5, γ = 88.4°, Z = 1. Mo radiation, R = 0.11 for 989 reflexions (twinned crystal).

The structure (Fig. 1) contains eight-membered Au_4Cl_4 rings with linear coordination at Au(I) and square-planar coordination at Au(III); Au(I)-Cl = 2.30, Au(III)-Cl = 2.32 (ring), 2.25(2) Å (terminal), Cl-Au(I)-Cl = 175, Au(I)-Cl-Au(III) = 103°.

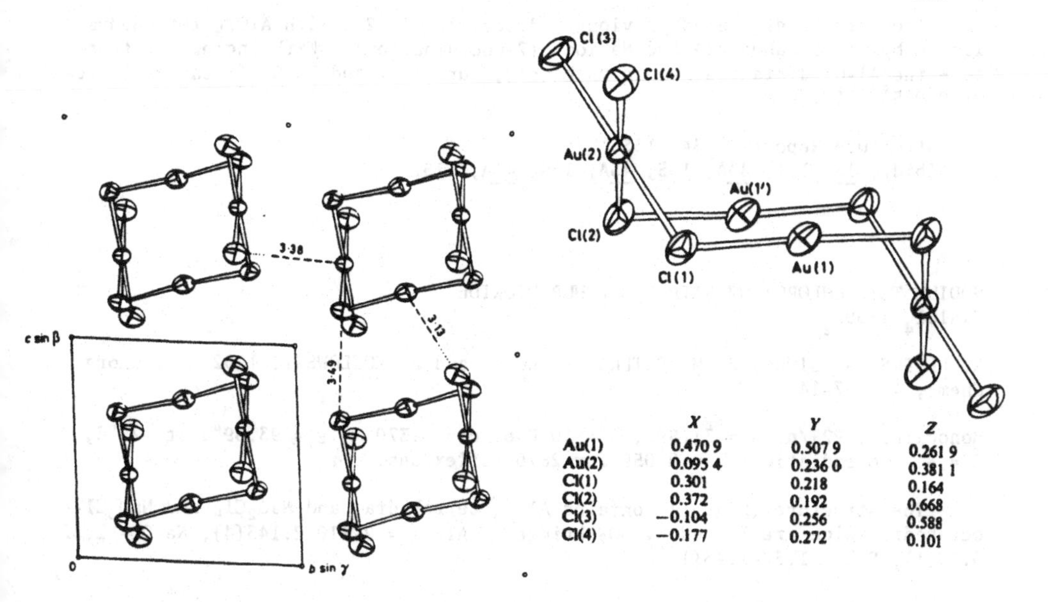

Fig. 1. Structure of Au_4Cl_8.

LITHIUM TETRACHLOROALUMINATE
$LiAlCl_4$

SODIUM TETRACHLOROALUMINATE
$NaAlCl_4$

E. PERENTHALER, H. SCHULZ and A. RABENAU, 1982. Z. anorg. Chem., 491, 259-265.

Lithium salt, monoclinic, $P2_1/c$, a = 7.004, 7.001, 7.023, b = 6.503, 6.526, 6.554, c = 12.996, 13.030, 13.075 Å, β = 93.3, 93.3, 93.3°, at 293, 326, 364 K, Z = 4. Mo radiation, R = 0.018, 0.019, 0.016 for 455, 867, 795 reflexions.

Sodium salt, orthorhombic, $P2_12_12_1$, a = 10.314, 10.382, 10.408, b = 9.887, 9.933, 9.964, c = 6.165, 6.183, 6.196 Å, at 293, 353, 393 K, Z = 4. Mo radiation, R = 0.024, 0.032, 0.032 for 589, 937, 809 reflexions.

Atomic positions (at 293K)

$LiAlCl_4$	x	y	z
Li	0.1577	0.9850	0.3646
Al	0.7059	0.3217	0.8990
Cl(1)	0.6938	0.1832	0.0467
Cl(2)	0.8086	0.6281	0.9286
Cl(3)	0.9255	0.1812	0.8139
Cl(4)	0.4397	0.3140	0.8130

$NaAlCl_4$			
Na	0.1252	0.2132	0.6889
Al	0.0387	0.4858	0.2069
Cl(1)	0.0322	0.4913	0.5524
Cl(2)	0.1481	0.3145	0.1092
Cl(3)	0.3465	0.0226	0.9253
Cl(4)	0.3774	0.3353	0.5735

The structures are as previously described (1, 2), with $AlCl_4$ tetrahedra linked by Li (octahedral) and Na ions (7-coordination). With increasing temperature the Al-Cl distances remain unchanged, but Li-Cl and Na-Cl distances increase by 0.025Å/100K.

1. Structure Reports, 43A, 135.
2. Ibid., 15, 151; 43A, 135; 45A, 169; 46A, 175.

SODIUM TETRACHLOROALUMINATE - SULPHUR DIOXIDE
$NaAlCl_4.1{\cdot}5SO_2$

K. PETERS, A. SIMON, E.-M. PETERS, H. KÜHNL and B. KOSLOWSKI, 1982. Z. anorg. Chem., 492, 7-14.

Monoclinic, $P2_1/n$, a = 31.598, b = 10.046, c = 6.370 Å, β = 93.39°, at -25°C, Z = 8. Mo radiation, R = 0.059 for 2876 reflexions.

The structure (Fig. 1) contains $AlCl_4$ tetrahedra, and NaO_2Cl_4 and NaO_3Cl_3 octahedra which are linked by SO_2 bridges. Al-Cl = 2.110-2.143(4), Na-O = 2.32-3.09(1), S-O = 1.37-1.43(1) Å.

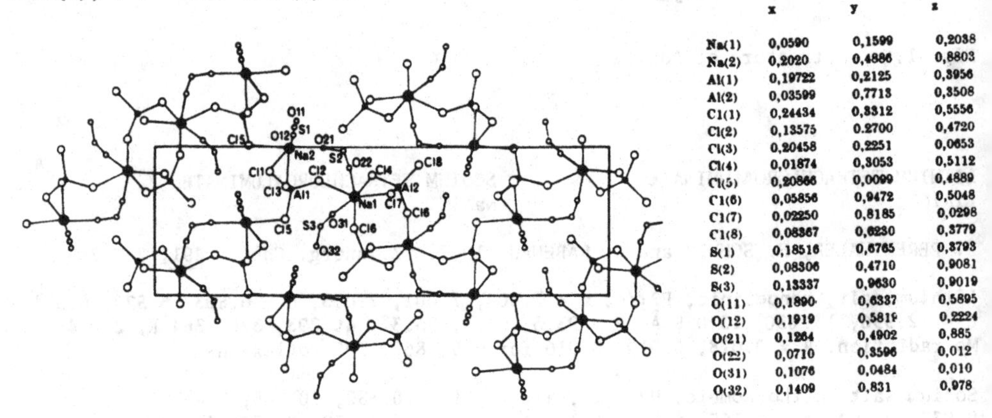

	x	y	z
Na(1)	0,0590	0,1599	0,2038
Na(2)	0,2020	0,4886	0,8803
Al(1)	0,19722	0,2125	0,3956
Al(2)	0,03599	0,7713	0,3508
Cl(1)	0,24434	0,3312	0,5556
Cl(2)	0,13575	0.2700	0,4720
Cl(3)	0,20458	0,2251	0,0653
Cl(4)	0,01874	0,3053	0,5112
Cl(5)	0,20866	0,0099	0,4889
Cl(6)	0,05856	0,9472	0,5048
Cl(7)	0,02250	0,8185	0,0298
Cl(8)	0,08367	0,6230	0,3779
S(1)	0,18516	0,6765	0,3793
S(2)	0,08306	0,4710	0,9081
S(3)	0,13337	0,9630	0,9019
O(11)	0,1890	0,6337	0,5895
O(12)	0,1919	0,5819	0,2224
O(21)	0,1264	0,4902	0,885
O(22)	0,0710	0,3596	0,012
O(31)	0,1076	0,0484	0,010
O(32)	0,1409	0,831	0,978

Fig. 1. Structure of $NaAlCl_4.1{\cdot}5SO_2$.

TITANIUM(II) TETRACHLOROALUMINATE
α-$TiAl_2Cl_8$

A. JUSTNES, E. RYTTER and A.F. ANDRESEN, 1982. Polyhedron, 1, 393-396.

Monoclinic, $P2_1/a$, a = 13.161, b = 7.477, c = 5.981 Å, β = 90.48°, Z = 2. Neutron powder data.

Atomic positions

	x	y	z
Ti	0	0	0
Al	0.119	0.286	0.380
Cl(1)	0.930	0.254	0.253
Cl(2)	0.144	0.014	0.266
Cl(3)	0.180	0.543	0.245
Cl(4)	0.066	0.210	0.747

The structure contains hexagonally close-packed Cl atoms, with Ti octahedrally and Al tetrahedrally coordinated in every second layer, so that chains of one octahedron and two tetrahedra are formed along c. Ti-Cl = 2.35-2.60(6), Al-Cl = 2.17-2.61(12) Å. A high-temperature β-form is isostructural with $CoAl_2Cl_8$ (1).

1. Structure Reports, 27, 459.

COPPER(I) TETRACHLOROALUMINATE
$CuAlCl_4$

K. HILDEBRANDT, P.G. JONES, E. SCHWARZMANN and G.M. SHELDRICK, 1982. Z. Naturforsch., 37B, 1129-1131.

Tetragonal, P$\bar{4}$2c, a = 5.430, c = 10.096 Å, Z = 2. Mo radiation, R = 0.039 for 250 reflexions. Cu in 2(e): 0,0,1/2; Al in 2(d): 0,1/2,1/4; Cl in 8(n): 0.2334, 0.2761,0.3705.

The structure contains distorted cubic-close-packed Cl atoms, with Cu and Al each occupying one eighth of the tetrahedral holes Al-Cl = 2.136, Cu-Cl = 2.359(3) Å, Cl-Al-Cl = 107.2, 110.6, Cl-Cu-Cl = 107.9, 112.7, Cu-Cl-Al = 111.0°.

COPPER(II) TETRACHLOROALUMINATE(III)
$Cu(AlCl_4)_2$

N. KITAJIMA, H. SHIMANOUCHI, Y. ONO and Y. SASADA, 1982. Bull. Chem. Soc. Japan, 55, 2064-2067.

Triclinic, P$\bar{1}$, a = 6.582, b = 7.362, c = 12.265 Å, α = 89.99, β = 85.97, γ = 89.77°, D_m = 2.25, Z = 2. Mo radiation, R = 0.110 for 2594 reflexions.

The structure (Fig. 1) contains two independent centrosymmetric molecules, with square-planar coordination at Cu and tetrahedral coordination at Al atoms. Cu-Cl = 2.286-2.306(4), Al-Cl = 2.193-2.203 (bridging), 2.057-2.106(6) (terminal)Å, Cu-Cl-Al = 88.8-89.3°; longer intermolecular contacts complete tetragonally-elongated octahedral coordination at each Cu atom, Cu...Cl = 2.951 Å. [Compare following report.]

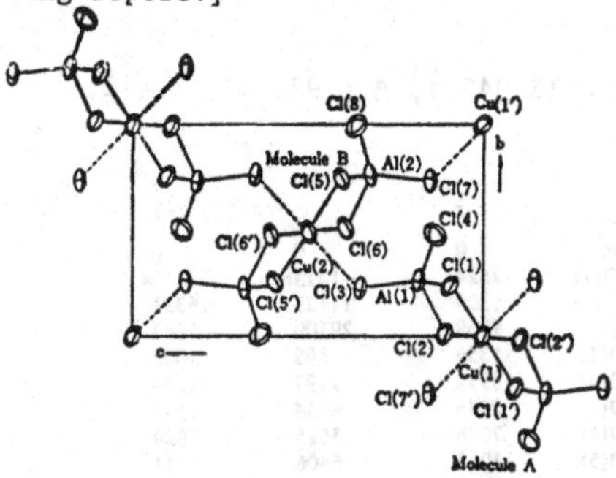

	10^4x	10^4y	10^4z
Cu(1)	0	0	0
Cu(2)	0	5000	5000
Al(1)	−1915	2752	1812
Al(2)	1910	7654	3192
Cl(1)	1109	2442	941
Cl(2)	−2993	130	1082
Cl(3)	−1650	2333	3502
Cl(4)	−3589	4868	1368
Cl(5)	−1123	7441	4053
Cl(6)	3003	5141	3916
Cl(7)	1666	7332	1500
Cl(8)	3555	9871	3637

Fig. 1. Structure of $Cu(AlCl_4)_2$.

COPPER(II) TETRACHLOROGALLATE(III)
$Cu(GaCl_4)_2$

C. VERRIES-PEYLHARD, 1982. C.R. Acad. Sci. Paris, Ser. II, <u>295</u>, 171-174.

Monoclinic, $P2_1/c$, a = 6.543, b = 7.406, c = 12.258 Å, β = 93.9°, Z = 2. Mo radiation, R = 0.065 for 1202 reflexions.

Atomic positions

	x	y	z
Cu	0	0	0
Ga	0.1927	0.2650	0.1852
Cl(1)	0.3001	0.0099	0.1086
Cl(2)	0.1680	0.2287	0.3567
Cl(3)	0.8825	0.2390	0.0966
Cl(4)	0.3591	0.4908	0.1368

The structure contains an isolated $Cu(GaCl_4)_2$ complex, with a central square-planar Cu ion linked by Cu-Cl-Ga bridges to two $GaCl_4$ ions. Cu-Cl = 2.295(3), Ga-Cl = 2.245(2) (bridging), 2.103, 2.135(3) (terminal) Å. Two Cl from neighbouring units complete distorted octahedral coordination at Cu, Cu-Cl = 2.932 Å.

POTASSIUM INDIUM(III) CHLORIDE DIHYDRATE
$K_3InCl_6.2H_2O$

J.-P. WIGNACOURT, G. NOWOGROCKI, G. MAIRESSE and P. BARBIER, 1980. Rev. Inorg. Chem., <u>2</u>, 207-217.

Tetragonal, I4mm, a = 15.723, c = 18.069 Å, D_m = 2.420, Z = 14. R = 0.029.

The structure contains K^+ ions, two $InCl_6{}^{3-}$ ions, an $InCl_5(H_2O)^{2-}$ ion, and a Cl^- ion.

HEXACHLOROSTANNIC ACID DECAHYDRATE
$H_2SnCl_6.10H_2O$

H. HENKE, 1982. Acta Cryst., <u>B38</u>, 920-923.

Monoclinic, $P2_1/n$, a = 8.973, b = 7.248, c = 13.942 Å, β = 92.43°, Z = 2. Mo radiation, R = 0.022 for 2291 reflexions.

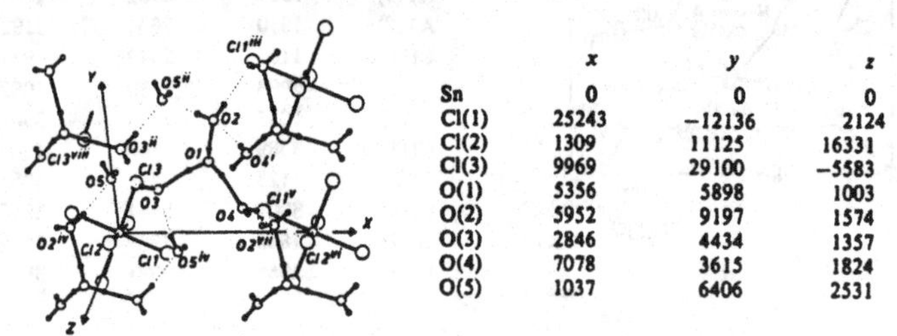

	x	y	z
Sn	0	0	0
Cl(1)	25243	−12136	2124
Cl(2)	1309	11125	16331
Cl(3)	9969	29100	−5583
O(1)	5356	5898	1003
O(2)	5952	9197	1574
O(3)	2846	4434	1357
O(4)	7078	3615	1824
O(5)	1037	6406	2531

Fig. 1. Structure of hexachlorostannic acid decahydrate, and atomic positional parameters (x 10^5 for Cl, x 10^4 for O).

The structure (Fig. 1) contains $SnCl_6^{2-}$ octahedra (Sn-Cl = 2.414, 2.432, 2.436(1) Å), well-defined $H_9O_4^+$ ions (O-H...O = 2.507, 2.557, 2.569(3) Å), and additional water molecules, linked by hydrogen bonds, O-H...O = 2.738-2.845, O-H...Cl = 3.273, 3.275 Å. The formula is therefore $(H_9O_4^+)_2SnCl_6^{2-}.2H_2O$.

CALCIUM ARSENIC CHLORIDE
Ca_3AsCl_3

C. HADENFELDT and H.O. VOLLERT, 1982. Z. anorg. Chem., 491, 113-118.

Cubic, Pm3m, a = 5.760 Å, D_m = 2.62, Z = 1. Mo radiation, R = 0.046 for 119 reflexions. Ca in 3(d): 1/2,0,0; As in 1(a): 0,0,0; Cl in 3(c): 0,1/2,1/2.

Isostructural with Mg_3NF_3 (1); the structure (Fig. 1) contains $CaAs_2Cl_4$ and $AsCa_6$ octahedra and $ClCa_4$ square planes.

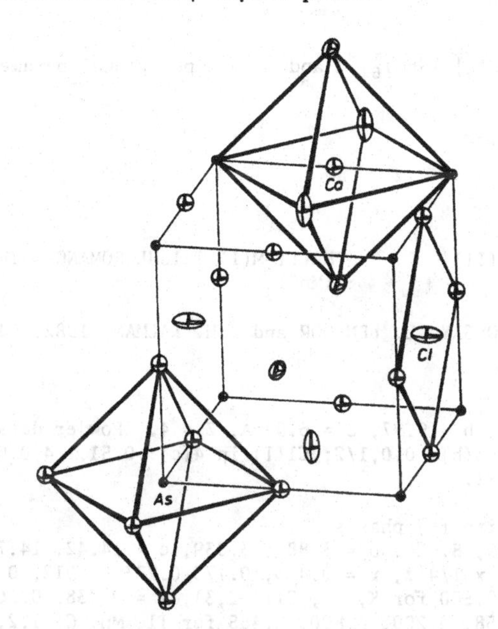

Fig. 1. Structure of Ca_3AsCl_3.

1. Structure Reports, 34A, 189; 35A, 153.

DICHLOROTRIIODINE HEXACHLOROANTIMONATE(V)
I_3SbCl_8 $[I_3Cl_2]^+[SbCl_6^-]$

T. BIRCHALL and R.D. MYERS, 1982. Inorg. Chem., 21, 213-217.

Triclinic, P$\bar{1}$, a = 7.090, b = 11.591, c = 7.126 Å, α = 122.29, β = 98.85, γ = 115.86°, Z = 1. Mo radiation, R = 0.032 for 1545 reflexions. Independent study in 1.

The structure (Fig. 1) contains planar, centrosymmetric $I_3Cl_2^+$ cations and octahedral $SbCl_6^-$ anions linked by strong I...Cl bridges. I-I = 2.906(1), I-Cl = 2.333(1) Å, I-I-I = 180, I-I-Cl = 93°, Sb-Cl = 2.416 (bridging), 2.341, 2.355(2) (terminal), I...Cl = 2.941(1) Å, Sb-Cl...I = 110, Cl-I...Cl = 178°

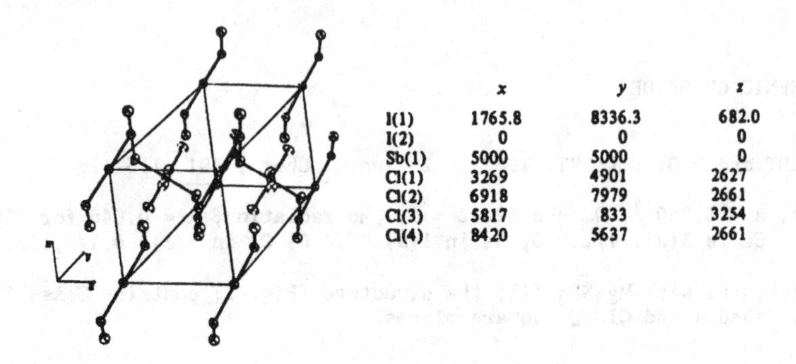

	x	y	z
I(1)	1765.8	8336.3	682.0
I(2)	0	0	0
Sb(1)	5000	5000	0
Cl(1)	3269	4901	2627
Cl(2)	6918	7979	2661
Cl(3)	5817	833	3254
Cl(4)	8420	5637	2661

Fig. 1. Structure of $[I_3Cl_2]^+[SbCl_6]^-$, and atomic positional parameters (x 10^4).

1. Structure Reports, 48A, 153.

POTASSIUM TRICHLOROMANGANATE(II)
$KMnCl_3$

THALLIUM(I) TRICHLOROMANGANATE(II)
$TlMnCl_3$

A. HOROWITZ, M. AMIT, J. MAKOVSKY, L. BEN DOR and Z.H. KALMAN, 1982. J. Solid State Chem., 43, 107-125.

$KMnCl_3$, perovskite (P) phase
Orthorhombic, Pnma, a = 7.08, b = 9.97, c = 6.98 Å, Z = 4. Powder data. K in 4(c): 0.055,1/4,0.991; Mn in 4(b): 0,0,1/2; Cl(1) in 4(c): 0.51,1/4,0.039; Cl(2) in 8(d): 0.30,0.041,0.71.

$KMnCl_3$, $TlMnCl_3$, non-perovskite (K) phases
Orthorhombic, Pnma, a = 8.769, 8.926, b = 3.883, 3.839, c = 14.42, 14.77 Å, Z = 4. Powder data. Atoms in 4(c): x,1/4,z, x = 0.433, 0.175,0.273, 0.012, 0.163, z = 0.828, 0.056, 0.213, 0.902, 0.500 for K, Mn, Cl(1,2,3); x = 0.438, 0.169, 0.277, 0.022, 0.187, z = 0.825, 0.058, 0.209, 0.890, 0.485 for Tl, Mn, Cl(1,2,3).

P-$KMnCl_3$ has perovskite-type structure and the K-phases have a $KCdCl_3$-type structure (1).

1. Strukterbericht, 6, 13; Structure Reports, 11, 434.

YTTRIUM OCTACHLORODITECHNETATE NONAHYDRATE
$YTc_2Cl_8.9H_2O$

F.A. COTTON, A. DAVISON, V.W. DAY, M.F. FREDRICH, C. ORVIG and R. SWANSON, 1982. Inorg. Chem., 21, 1211-1214.

Tetragonal, $P42_12$, a = 11.712, c = 7.661 Å, D_m = 2.31, Z = 4. Mo radiation, R = 0.031 for 621 reflexions.

The structure (Fig. 1) contains $Tc_2Cl_8^{3-}$ anions with approximate D_{4h} symmetry, $Y(H_2O)_8^{3+}$ distorted square antiprisms, and H_2O molecules. Tc-Tc = 2.105(1), Tc-Cl = 2.360, 2.367(2), Y-O = 2.346, 2.373(5) Å.

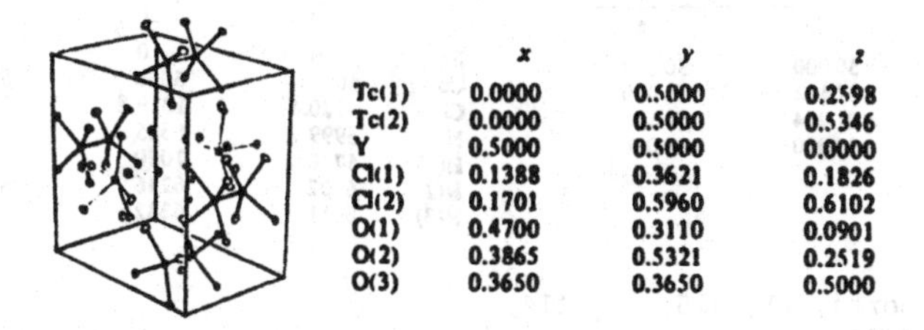

	x	y	z
Tc(1)	0.0000	0.5000	0.2598
Tc(2)	0.0000	0.5000	0.5346
Y	0.5000	0.5000	0.0000
Cl(1)	0.1388	0.3621	0.1826
Cl(2)	0.1701	0.5960	0.6102
O(1)	0.4700	0.3110	0.0901
O(2)	0.3865	0.5321	0.2519
O(3)	0.3650	0.3650	0.5000

Fig. 1. Structure of $YTc_2Cl_8.9H_2O$.

NITROSYL TETRACHLOROFERRATE(III)
$NO[FeCl_4]$

H. PRINZ, K. DEHNICKE and U. MÜLLER, 1982. Z. anorg. Chem., 488, 49-59.

Orthorhombic, Pnna, a = 7.30, b = 9.83, c = 9.65 Å, Z = 4. Mo radiation, R = 0.098 for 331 refleixons.

Atomic positions

	x	y	z
Fe	1/4	0	0.1847
Cl(1)	0.1678	0.1734	0.0566
Cl(2)	0.4878	0.0459	0.3146
N,O	0.8233	0.2295	0.2803

The structure contains tetrahedral $FeCl_4^-$ anions, and disordered NO^+ cations which are surrounded by eight Cl atoms. Fe-Cl = 2.19(1) Å, Cl-Fe-Cl = 106-112°, N-O = 0.71(3) Å.

RINNEITE
NaK_3FeCl_6

HEXAAMMINECOBALT(III) HEXACHLOROFERRATE(III)
$Co(NH_3)_6FeCl_6$

J.K. BEATTIE and C.J. MOORE, 1982. Inorg. Chem., 21, 1292-1295.

NaK_3FeCl_6
Rhombohedral, R$\bar{3}$c, a = 8.3376 Å, α = 92.29°, Z = 2. Mo radiation, R = 0.022 for 450 reflexions. Previous study in 1.

$Co(NH_3)_6FeCl_6$
Cubic, Pa3, a = 11.2506 Å, Z = 4. Mo radiation, R = 0.025 for 1328 reflexions. Isostructural with $Cr(NH_3)_6FeF_6$ (2).

The structures contain octahedral $FeCl_6^{4-}$ and $FeCl_6^{3-}$ ions, respectively, Fe(II)-Cl = 2.5100(5), Fe(III)-Cl = 2.3926(3) Å. Na-6Cl = 2.824, K-8Cl = 3.175-3.337; Co-N = 1.965(1) Å.

Atomic positions

	(x10⁵) for NaK_3FeCl_6		
	x	y	z
Fe	50 000	50 000	50 000
K	12 598	37 402	75 000
Cl	41 962	21 254	54 528
Na	25 000	25 000	25 000

	(x10⁴) for $Co(NH_3)_6FeCl_6$		
	x	y	z
Fe	0	0	0
Co	5000	5000	5000
Cl	-720.4	1319.8	1504.0
N	5999	6315	4430
H(1)	5770	7050	4639
H(2)	6702	6246	4658
H(3)	6011	6382	3675

1. Structure Reports, 11, 415; 18, 414.
2. Ibid., 38A, 205.

HEXAAMMINECOBALT(III) PENTACHLOROCUPRATE(II)
$[Co(NH_3)_6][CuCl_5]$

I. BERNAL, J.D. KORP, E.O. SCHLEMPER and M.S. HUSSAIN, 1982. Polyhedron, 1, 365-369.

Cubic, Fd3c, a = 21.992 Å, Z = 32. Mo and neutron radiations, R = 0.026 and 0.055 for 217 and 334 reflexions.

Atomic positions (neutron data, origin at centre)

			x	y	z
Co	in	32(c)	0	0	0
Cu		32(b)	1/4	1/4	1/4
Cl(ax)		64(e)	0.1899	0.1899	0.1899
Cl(eq)		96(g)	1/4	0.0765	-0.0769
N		192(h)	0.0690	0.0489	-0.0290
H(1)		192(h)	0.1041	0.0474	-0.0037
H(2)		192(h)	0.0585	0.0899	-0.0338
H(3)		192(h)	0.0822	0.0371	-0.0695

Isostructural with the Cr compound (1), the structure containing octahedral cations and trigonal bipyramidal $CuCl_5^{3-}$ anions. Co-N = 1.968(2), N-H = 0.95(1), Cu-Cl = 2.291(1) (axial), 2.392(2) (equatorial) Å.

1. Structure Reports, 26, 326; 33A, 207.

AMMONIUM TETRACHLOROZINCATE
$(NH_4)_2ZnCl_4$

I. H. MATSUNAGA, 1982. J. Phys. Soc. Japan, 51, 864-872.
II. Idem, 1982. Ibid., 51, 873-879.
III. H. MATSUNAGA, K. ITOH and E. NAKAMURA, 1982. Acta Cryst., B38, 898-900.

Phase I, 418K, orthorhombic, Pmcn, a = 7.275, b = 12.745, c = 9.295 Å, Z = 4. Cu radiation, R = 0.037 for 731 reflexions.

Phase II, 333K, orthorhombic, $Pc2_1n$, a = 7.236, b = 12.661, c = 37.116 Å, Z = 16. Cu radiation, R = 0.046 for 1892 reflexions.

Phase III, 298K, monoclinic, Pc, a (unique axis) = 7.213, b = 12.629, c = 37.118 Å, α = 89.99°, Z = 16. Cu radiation, R = 0.050 for 3783 reflexions.

Phase IV, 223K, orthorhombic, $P2_1cn$, a = 7.184, b = 12.568, c = 27.838 Å, Z = 12. Cu radiation, R = 0.038 for 2192 reflexions.

Phase I has a β-K_2SO_4-type structure (1), with large Cl thermal parameters probably corresponding to disorder of the $ZnCl_4$ tetrahedron. Phase II has atomic positional parameters close to those previously reported for the room-temperature phase (III) (2). Phase III is in fact monoclinic (compare 2), with a structure similar to that of phase II, but with reorientation of one $ZnCl_4$ group. Phase IV is isostructural with room-temperature K_2ZnCl_4 (3), the structure (Fig. 1) containing $ZnCl_4$ tetrahedra linked by 8- and 5-coordinate ammonium ions.

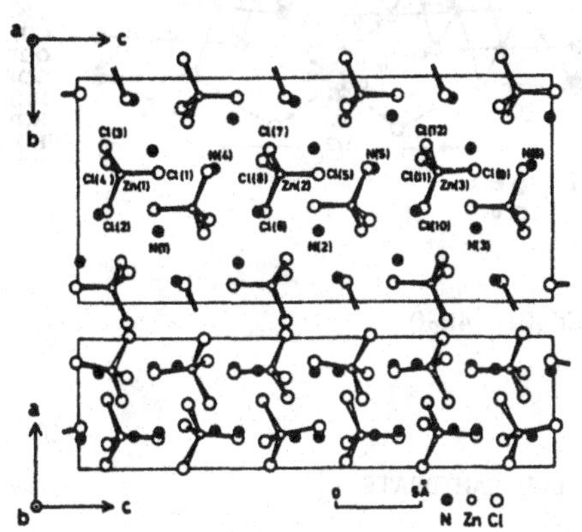

Fig. 1. Structure of ammonium tetrachlorozincate (phase IV).

1. Strukturbericht, 2, 86, 423; Structure Reports, 22, 447; 33A, 367; 38A, 330.
2. Structure Reports, 46A, 187.
3. Ibid., 45A, 174.

RUBIDIUM TETRACHLOROZINCATE (INTERMEDIATE PHASE)
Rb_2ZnCl_4

M. HARADA, 1981. Ann. Rep. Res. React. Inst., Kyoto Univ., 14, 51-58.

The average structure (R = 0.076 for film data at room temperature) is of β-K_2SO_4 type (1). The $ZnCl_4^{2-}$ tetrahedron is significantly distorted.

1. Strukturbericht, 2, 86; Structure Reports, 22, 447; 33A, 367; 38A, 330.

POTASSIUM TRICADMIUM HEPTACHLORIDE TETRAHYDRATE
$KCd_3Cl_7.4H_2O$

M. LEDÉSERT, 1982. Acta Cryst., B38, 1569-1571.

Monoclinic, $P2_1/m$, a = 6.674, b = 16.122, c = 7.037 Å, β = 91.81°, Z = 2. Mo radiation, R = 0.024 for 2689 reflexions.

The structure (Fig. 1) contains chains along a of edge-sharing $CdCl_6$ and $CdCl_5(H_2O)$ octahedra; the chains share octahedral corners to form (001) layers which contain $KCl_5(H_2O)_3$ polyhedra, and the layers are joined by hydrogen bonds. Cd-Cl = 2.56-2.68, Cd-O = 2.32, K-Cl = 3.11-3.54, K-O = 2.71-2.93 Å.

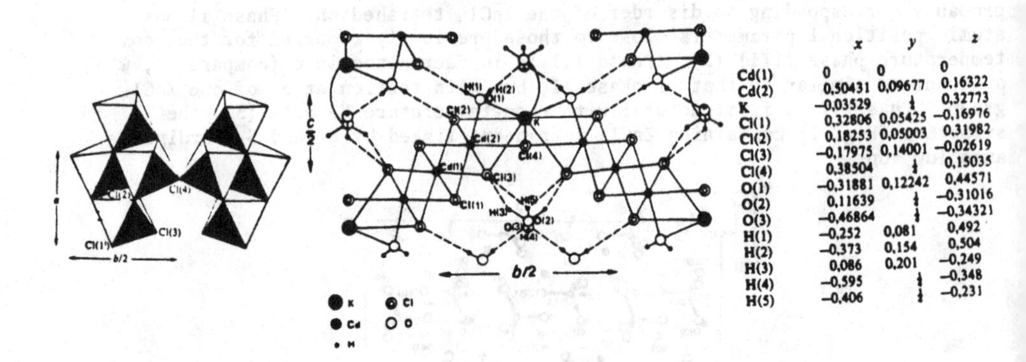

	x	y	z
Cd(1)	0	0	0
Cd(2)	0,50431	0,09677	0,16322
K	−0,03529	¼	0,32773
Cl(1)	0,32806	0,05425	−0,16976
Cl(2)	0,18253	0,05003	0,31982
Cl(3)	−0,17975	0,14001	−0,02619
Cl(4)	0,38504	¼	0,15035
O(1)	−0,31881	0,12242	0,44571
O(2)	0,11639	¼	−0,31016
O(3)	−0,46864	¼	−0,34321
H(1)	−0,252	0,081	0,492
H(2)	−0,373	0,154	0,504
H(3)	0,086	0,201	−0,249
H(4)	−0,595	¼	−0,348
H(5)	−0,406	¼	−0,231

Fig. 1. Structure of $KCd_3Cl_7.4H_2O$.

CADMIUM MAGNESIUM CHLORIDE DODECAHYDRATE
$Cd_2MgCl_6.12H_2O$

M. LEDÉSERT and J.C. MONIER, 1982. Acta Cryst., B38, 237-239.

Orthorhombic, Fdd2, a = 24.587, b = 22.423, c = 7.564 Å, D_m = 2.14, Z = 8. Mo radiation, R = 0.043 for 2322 reflexions.

Isostructural with $Cd_2NiCl_6.12H_2O$ (1). The structure (Fig. 1) contains chains along c of edge-sharing $CdCl_5(OH_2)$ octahedra, $Mg(OH_2)_6$ octahedra, and free water molecules. Cd-Cl = 2.589-2.629(2), Cd-O = 2.364(4), Mg-O = 2.052-2.081(5) Å. The water molecules form O-H...O and O-H...Cl hydrogen bonds.

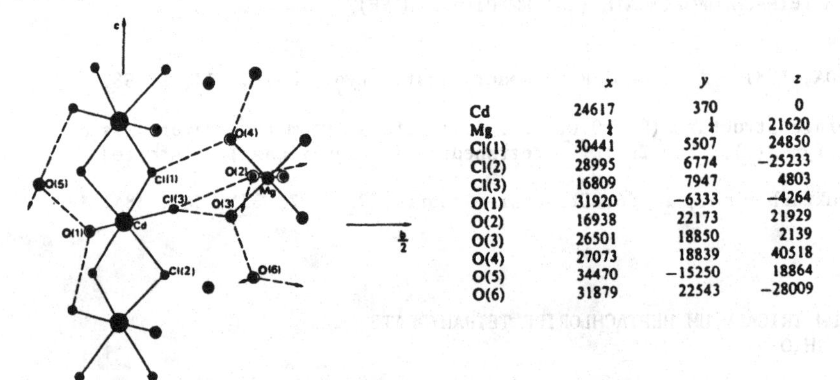

	x	y	z
Cd	24617	370	0
Mg	¼	¼	21620
Cl(1)	30441	5507	24850
Cl(2)	28995	6774	−25233
Cl(3)	16809	7947	4803
O(1)	31920	−6333	−4264
O(2)	16938	22173	21929
O(3)	26501	18850	2139
O(4)	27073	18839	40518
O(5)	34470	−15250	18864
O(6)	31879	22543	−28009

Fig. 1. Structure of $Cd_2MgCl_6.12H_2O$, and atomic positional parameters (x 10^5).

1. Structure Reports, 46A, 185.

CALCIUM CADMIUM CHLORIDE OCTADECAHYDRATE
$Ca_2Cd_3Cl_{10}.18H_2O$ $2[Ca(H_2O)_8].[Cd_3Cl_{10}]_\infty.2H_2O$

H. LELIGNY and J.C. MONIER, 1982. Acta Cryst., B38, 355-358.

Triclinic, $P\bar{1}$, a = 12.792, b = 6.573, c = 11.989 Å, α = 106.54, β = 120.46, γ = 91.97°, D_m = 2.23, Z = 1. Mo radiation, R = 0.034 for 6616 reflexions.

The structure (Fig. 1) contains chains of edge-sharing $CdCl_6$ octahedra, isolated $Ca(H_2O)_8$ Archimedean antiprisms, and additional water molecules, linked by O-H...Cl and O-H...O hydrogen bonds. Cd-Cl = 2.561-2.766(1), Ca-O = 2.406-2.533(5) Å.

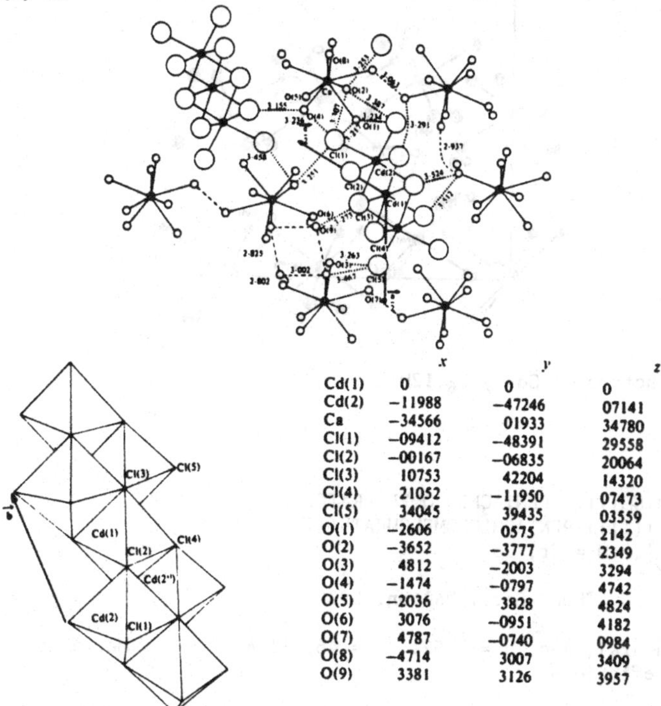

	x	y	z
Cd(1)	0	0	0
Cd(2)	−11988	−47246	07141
Ca	−34566	01933	34780
Cl(1)	−09412	−48391	29558
Cl(2)	−00167	−06835	20064
Cl(3)	10753	42204	14320
Cl(4)	21052	−11950	07473
Cl(5)	34045	39435	03559
O(1)	−2606	0575	2142
O(2)	−3652	−3777	2349
O(3)	4812	−2003	3294
O(4)	−1474	−0797	4742
O(5)	−2036	3828	4824
O(6)	3076	−0951	4182
O(7)	4787	−0740	0984
O(8)	−4714	3007	3409
O(9)	3381	3126	3957

Fig. 1. Structure of $Ca_2Cd_3Cl_{10}.18H_2O$, and atomic positional parameters (x 10^4 for O, x 10^5 for others).

CADMIUM NICKEL CHLORIDE DODECAHYDRATE
$CdNi_2Cl_6.12H_2O$

A. LECLAIRE and M.M. BOREL, 1982. Acta Cryst., B38, 234-236.

Trigonal, P3, a = 9.951, c = 11.239 Å, D_m = 2.23, Z = 2. Mo radiation, R = 0.035 for 3852 reflexions.

The structure (Fig. 1) contains $Ni(OH_2)_6{}^{2+}$ and $CdCl_6{}^{4-}$ octahedra, linked by hydrogen bonds. Ni-O = 2.05-2.08, Cd-Cl = 2.60-2.65 Å.

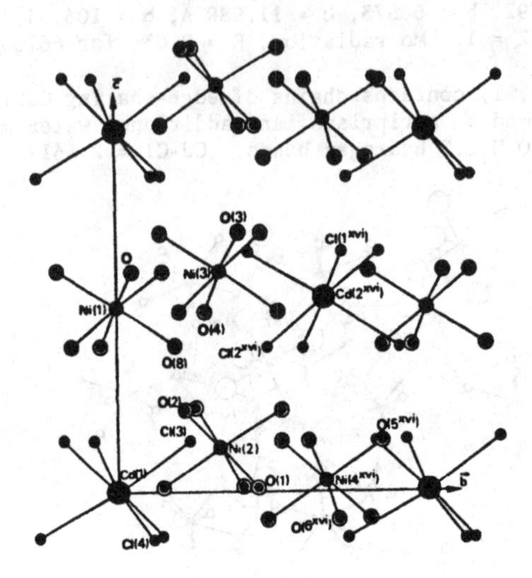

Fig. 1. Structure of $CdNi_2Cl_6.12H_2O$.

HEXAAMMINECHROMIUM(III) PENTACHLOROMERCURATE(II)
HEXAAMMINECOBALT(III) PENTACHLOROMERCURATE(II)
$[M(NH_3)_6][HgCl_5]$, M = Cr, Co

W. CLEGG, 1982. J. Chem. Soc., Dalton, 593-595.

M = Cr, rhombohedral, $R\bar{3}c$, a = 7.521, c = 46.111 Å, Z = 6. Mo radiation, R = 0.041 for 411 reflexions.

M = Co, monoclinic, $P2_1/m$, a = 8.444, b = 7.705, c = 11.357 Å, β = 107.70°, Z = 2. Mo radiation, R = 0.049 for 1591 reflexions. An orthorhombic form is described in 1.

Both structures contain octahedral cations and distorted trigonal-bipyramidal $HgCl_5^{3-}$ anions, which are disordered over two positions in the Cr compound. The equatorial Hg-Cl bonds are short, Hg-Cl = 2.417-2.431 Å, and the axial bonds very long, 2.871-3.038(4) Å. Cr-N = 2.070(8), Co-N = 1.948-1.964(10) Å.

1. Structure Reports, 48A, 162.

POTASSIUM DYSPROSIUM CHLORIDE
KDy_2Cl_7

RUBIDIUM DYSPROSIUM CHLORIDE
$RbDy_2Cl_7$

G. MEYER, 1982. Z. anorg. Chem., 491, 217-224.

Potassium salt, monoclinic, $P2_1/a$, a = 12.739, b = 6.881, c (unique axis) = 12.621 Å, γ = 89.36°, Z = 4. R = 0.095 for 1399 reflexions.

Rubidium salt, orthorhombic, Pnma, a = 12.881, b = 6.935, c = 12.672 Å, Z = 4. R = 0.089 for 816 reflexions.

Both structures (Fig. 1) contain Dy_2Cl_{11} pairs of face-sharing monocapped trigonal prisms, with further edge-sharing producing layers which are linked by K or Rb cations. Dy-Cl = 2.66-2.79(1) Å.

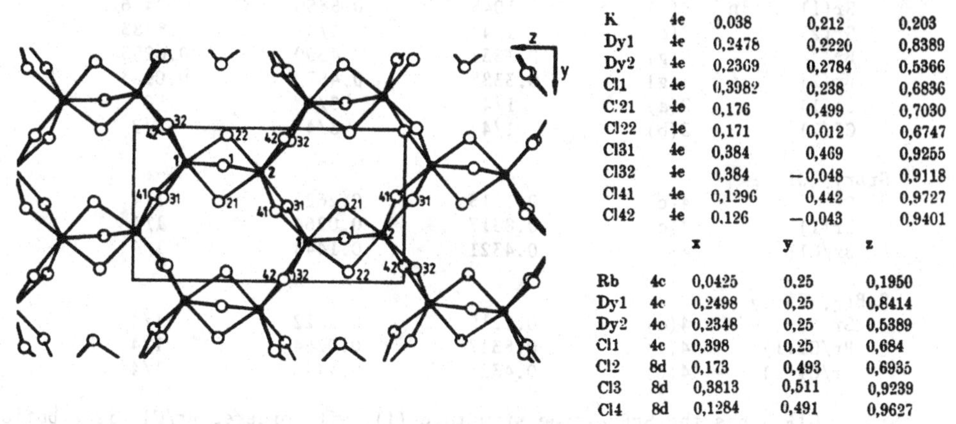

		x	y	z
K	4e	0,038	0,212	0,203
Dy1	4e	0,2478	0,2220	0,8389
Dy2	4e	0,2369	0,2784	0,5366
Cl1	4e	0,3982	0,238	0,6836
Cl21	4e	0,176	0,499	0,7030
Cl22	4e	0,171	0,012	0,6747
Cl31	4e	0,384	0,469	0,9255
Cl32	4e	0,384	−0,048	0,9118
Cl41	4e	0,1296	0,442	0,9727
Cl42	4e	0.126	−0,043	0,9401

		x	y	z
Rb	4c	0,0425	0,25	0,1950
Dy1	4c	0,2498	0,25	0,8414
Dy2	4c	0,2348	0,25	0,5389
Cl1	4c	0,398	0,25	0,684
Cl2	8d	0,173	0,493	0,6935
Cl3	8d	0,3813	0,511	0,9239
Cl4	8d	0,1284	0,491	0,9627

Fig. 1. Dy_2Cl_7 sheets in KDy_2Cl_7 and atomic positional parameters in the potassium and rubidium salts.

YTTERBIUM ERBIUM CHLORIDE

Yb_5ErCl_{13}

H. LÜKE and H.A. EICK, 1982. Inorg. Chem., 21, 965-968.

Monoclinic, C2/c, a = 41.44, b = 6.537, c = 7.004 Å, β = 98.48°, Z = 4. Mo radiation, R = 0.032 for 1012 reflexions.

Atomic positions

	x	y	z
M(1)	0.20421	0.77451	0.62527
Yb(2)	0.12054	0.77154	0.15104
M(3)	0.03981	0.76830	0.49039
Cl(1)	0.2428	0.9819	0.3874
Cl(2)	0.1800	0.5701	0.2844
Cl(3)	0.1623	0.9445	0.9016
Cl(4)	0.1052	0.6038	0.4868
Cl(5)	0.0800	0.9366	0.8194
Cl(6)	0.0341	0.6194	0.1219
Cl(7)	0	0.9404	0.75

The structure contains folded anion layers, with 7-, 8-, and 8-coordinated metal ions. Er is probably distributed in the M(1) and M(3) sites. Yb_6Cl_{13} is probably isostructural.

STRONTIUM BROMIDE CHLORIDE

$SrBr_{1.6}Cl_{0.4}$ $SrBr_{1.4}Cl_{0.6}$ $SrBr_{0.3}Cl_{1.7}$

S.A. HODOROWICZ and H.A. EICK, 1982. J. Solid State Chem., 43, 271-277.

$SrBr_{1.6}Cl_{0.4}$, tetragonal, P4/n, a = 11.503, c = 7.018 Å, Z = 10. Powder data.

$SrBr_{1.4}Cl_{0.6}$, $SrBr_{0.3}Cl_{1.7}$, orthorhombic, Pbnm, a = 9.308, 9.054, b = 7.964, 7.704, c = 4.670, 4.584 Å, Z = 4. Powder data.

Atomic positions

$SrBr_{1.6}Cl_{0.4}$			x	y	z
Sr(1)	in	8(g)	0.1045	0.5856	0.2476
Sr(2)		2(c)	1/4	1/4	0.8483
Br(1)		8(g)	0.1531	0.4590	0.6253
Br(2)		8(g)	0.3388	0.4572	0.0963
Cl(1)		2(a)	1/4	3/4	0
Cl(2)		2(b)	1/4	3/4	1/2
$SrBr_{1.4}Cl_{0.6}$					
Sr		4(c)	0.1215	0.2622	1/4
Br(1)		4(c)	0.8317	0.0264	1/4
Br/Cl(2)		4(c)	0.4321	0.1448	1/4
$SrBr_{0.3}Cl_{1.7}$					
Sr		4(c)	0.1215	0.2622	1/4
Br/Cl(1)		4(c)	0.8317	0.0264	1/4
Br/Cl(2)		4(c)	0.4321	0.1448	1/4

$SrBr_{1.6}Cl_{0.4}$ has the $SrBr_2$-type structure (1), with ordered Br/Cl distribution. The other two phases have the $PbCl_2$-type structure (2), with partial anion ordering in $SrBr_{1.4}Cl_{0.6}$.

1. Structure Reports, 28, 98; 37A, 204.
2. Strukturbericht, 2, 16, 251; Structure Reports, 26, 324; 28, 80; 34A, 176; 42A, 184.

RUBIDIUM TETRABROMOTHALLATE(III) MONOHYDRATE
$RbTlBr_4.H_2O$

POTASSIUM TETRABROMOTHALLATE(III) DIHYDRATE
$KTlBr_4.2H_2O$

AMMONIUM TETRABROMOTHALLATE(III) DIHYDRATE
$NH_4TlBr_4.2H_2O$

H.W. ROTTER and G. THIELE, 1982. Z. Naturforsch., 37B, 995-1004.

Cubic, $F\bar{4}3c$, a = 18.676, 18.544, 18.659 Å, D_m = 3.85, 3.67, 3.45, Z = 24. Ag and Mo radiations, R = 0.066, 0.102, 0.069 for 282, 147 (-50°C), 201 reflexions.

Atomic positions

$RbTlBr_4.H_2O$				x	y	z
	Tl	in	24(d)	1/4	0	0
	Br		96(h)	0.11433	-0.00196	0.17371
	Rb(1)		8(a)	0	0	0
1/12	Rb(2)		96(h)	0.1566	0.1802	0.2005
1/12	Rb(3)		96(h)	0.1479	0.1875	0.2670
1/4	O		96(h)	0.2684	0.2263	0.3714
$KTlBr_4.2H_2O$						
	Tl		24(d)	1/4	0	0
	Br		96(h)	0.1128	0.0091	0.1723
	K(1)		8(a)	0	0	0
1/2	K(2)		32(e)	0.326	0.326	0.326
1/2	O		96(h)	0.29	0.21	0.34

$NH_4TlBr_4.2H_2O$			x	y	z
	Tl in	24(d)	1/4	0	0
	Br	96(h)	0.1115	0.0057	0.1732
	NH_4(1)	8(a)	0	0	0
1/6	NH_4(2)	96(h)	0.197	0.157	0.314
1/2	O	96(h)	0.135	0.259	0.293

The structures (Fig. 1) contain $TlBr_4$ tetrahedra (compare 1), linked by MBr_{12} icosahedra, with cavities containing additional alkali cations and water molecules in a disordered arrangement. Tl-Br = 2.53-2.57(1) Å.

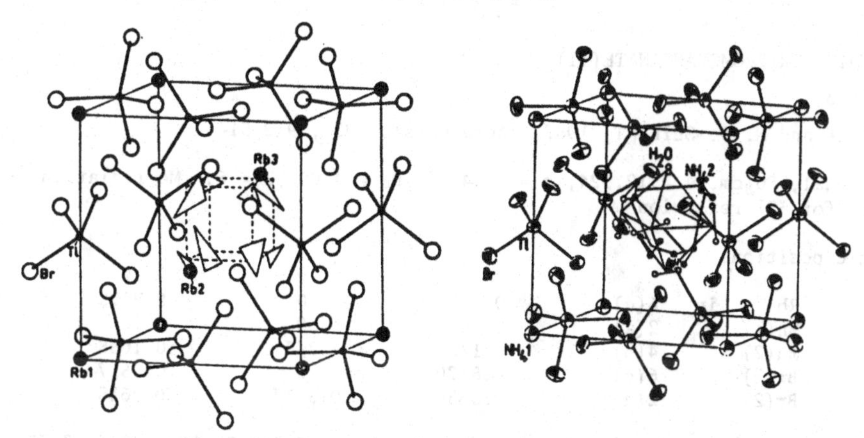

Fig. 1. Structures of $RbTlBr_4.H_2O$ (left) and $NH_4TlBr_4.2H_2O$ (right).

1. Structure Reports, 11, 393.

COPPER TELLURIDE BROMIDE (HIGH-TEMPERATURE)
CuTeBr

R. BACHMANN, K.D. KREUER, A. RABENAU and H. SCHULZ, 1982. Acta Cryst., B38, 2361-2364.

Tetragonal, $I4_1/amd$, a = 16.334, 16.396, c = 4.785, 4.776 Å, at 373, 473K, Z = 16. Mo radiation, R_w = 0.012, 0.012 for 303, 283 reflexions.

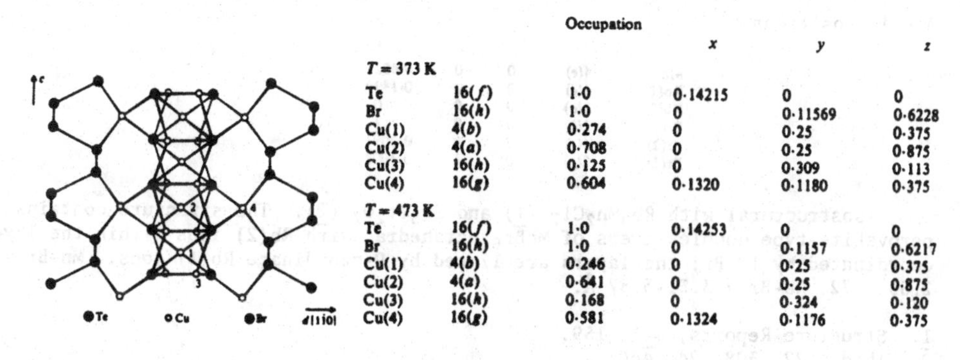

		Occupation	x	y	z
T = 373 K					
Te	16(*f*)	1·0	0·14215	0	0
Br	16(*h*)	1·0	0	0·11569	0·6228
Cu(1)	4(*b*)	0·274	0	0·25	0·375
Cu(2)	4(*a*)	0·708	0	0·25	0·875
Cu(3)	16(*h*)	0·125	0	0·309	0·113
Cu(4)	16(*g*)	0·604	0·1320	0·1180	0·375
T = 473 K					
Te	16(*f*)	1·0	0·14253	0	0
Br	16(*h*)	1·0	0	0·1157	0·6217
Cu(1)	4(*b*)	0·246	0	0·25	0·375
Cu(2)	4(*a*)	0·641	0	0·25	0·875
Cu(3)	16(*h*)	0·168	0	0·324	0·120
Cu(4)	16(*g*)	0·581	0·1324	0·1176	0·375

Fig. 1. Structure of high-temperature CuTeBr.

The high-temperature structure (Fig. 1) contains Te helices and chains of edge-sharing Br_4 tetrahedra along c (1). Cu ions are distributed over four sites, three with tetrahedral coordination to Br, and the fourth coordinated to 2 Br and 2 Te. The compound has a superstructure at room temperature; the iodide has a similar structure (2).

1. Structure Reports, 42A, 216.
2. Ibid., 42A, 217; 45A, 207.

RUBIDIUM TRIBROMOMANGANATE(II)
$RbMnBr_3$

H. FINK and H.-J. SEIFERT, 1982. Acta Cryst., B38, 912-914.

Hexagonal, $P6_3cm$, a = 12.924, c = 6.547 Å, Dm = 4.00, Z = 6. Mo radiation, R = 0.065 for 327 reflexions.

Atomic positions

		x	y	z
Rb	in 6(c)	0.3340	0	0.2985
Mn(1)	2(a)	0	0	0
Mn(2)	4(b)	1/3	2/3	0.1088
Br(1)	6(c)	0.8370	0	0.2517
Br(2)	12(d)	0.3335	0.5037	0.3615

Isostructural with $KNiCl_3$ (1). Mn-6Br = 2.66-2.68, Rb-9Br = 3.61-3.75 Å. Another form ($P6_3/mmc$) has been described previously (2).

1. Structure Reports, 46A, 184.
2. Ibid., 38A, 218; 46A, 195.

RUBIDIUM HEPTABROMODIMANGANATE(II)
$Rb_3Mn_2Br_7$

J. GOODYEAR, E.M. ALI and H.H. SUTHERLAND, 1982. Acta Cryst., B38, 600-602.

Tetragonal, I4/mmm, a = 5.37, c = 27.80 Å, Dm = 3.80, Z = 2. Mo radiation, R = 0.10 for 149 reflexions (films, visual intensities).

Atomic positions

		x	y	z
Mn	4(e)	0	0	0·4021
Rb(1)	4(e)	0	0	0·1899
Rb(2)	2(a)	0	0	0
Br(1)	2(b)	0	0	½
Br(2)	4(e)	0	0	0·3091
Br(3)	8(g)	0	½	0·1002

Isostructural with $Rb_3Mn_2Cl_7$ (1) and $Sr_3Ti_2O_7$ (2). The structure contains perovskite-type double layers of $MnBr_6$ octahedra, with Rb(2) ions within the layers coordinated by 12 Br; the layers are linked by 9-coordinate Rb(1) ions. Mn-Br = 2.59-2.72, Rb-Br = 3.31-3.87 Å.

1. Structure Reports, 44A, 159.
2. Ibid., 22, 308; 24, 440.

POTASSIUM TRIBROMOFERRATE(II)
$KFeBr_3$

E. GUREWITZ and H. SHAKED, 1982. Acta Cryst., B38, 2771-2775.

Orthorhombic, Pnma, a = 9.220, b = 4.026, c = 14.899 Å, Z = 4. Neutron powder data. Atoms in 4(c): x,1/4,z, x = 0.175, 0.468, 0.275, 0.166, 0.015, z = 0.052, 0.822, 0.207, 0.495, 0.902 for Fe, K, Br(1,2,3), at room temperature; parameters given also for 4.2K.

$KCdCl_3$-type structure (1), as previously described (2).

1. Structure Reports, 11, 434.
2. Ibid., 41A, 174.

THALLIUM(I) TRIBROMOFERRATE(II)
$TlFeBr_3$

N. JOUINI, L. GUEN and M. TOURNOUX, 1982. Mater. Res. Bull., 17, 1421-1427.

Hexagonal, $P6_3cm$, a = 12.444, c = 6.23 Å, Z = 6. Mo radiation, R = 0.050 for 229 reflexions.

Atomic positions

			x	y	z
Tl	in	6(c)	0.336	0	1/4
Br(1)		6(c)	0.8390	0	0.323
Br(2)		12(d)	0.1698	0.6660	0.244
Fe(1)		2(a)	0	0	0.052
Fe(2)		4(b)	1/3	2/3	-0.030

The structure is related to that of the 2L ($CsNiCl_3$) hexagonal perovskite type (1), with a = [110] of the 2L phase, and with face-sharing chains of octahedra shifted along c. Fe-6Br = 2.46-2.65, Tl-12Br = 3.55-4.18(1) Å.

1. Structure Reports, 19, 332; 46A, 204.

CAESIUM TETRABROMOCOBALTATE(II) BROMIDE
Cs_3CoBr_5

B.N. FIGGIS and P.A. REYNOLDS, 1981. Aust. J. Chem., 34, 2495-2498.

Tetragonal, I4/mcm, a = 9.46, c = 15.04 Å, at 4.2K, Z = 4. Neutron radiation, R = 0.022 for 416 reflexions.

Atomic positions

			x	y	z
Cs(1)	in	4(a)	0	0	1/4
Cs(2)		8(h)	0.65885	0.15885	0
Co		4(b)	0	1/2	1/4
Br(1)		4(c)	0	0	0
Br(2)		16(ℓ)	0.14468	0.64468	0.15578

Isostructural with the chloride (1). Co-Br = 2.398(2) Å, Br-Co-Br = 107.6, 110.4°, Cs-Br = 3.520-3.895(4) Å.

1. Strukturbericht, 3, 134, 498; Structure Reports, 29, 277; 46A, 184; 48A, 157.

CAESIUM BROMOAURATE
$Cs_{1.5}(AuBr_4)(Br_3)_{0.2}Br_{0.3}$

P. GÜTLICH, B. LEHNIS, K. RÖMHILD and J. STRÄHLE, 1982. Z. Naturforsch., 37B, 550-556.

Cubic, Pm3m, a = 5.475 Å, Z = 2/3. Mo radiation, R = 0.088 for 104 reflexions. Cs in 1(a): 0,0,0; Au,Br in 1(b): 1/2,1/2,1/2; 0.5 Br in 6(f) [not 3(c)]: 0.053, 1/2,1/2.

Disordered perovskite, with statistical distribution of $AuBr_4^-$ and Br_3^- anions (Fig. 1).

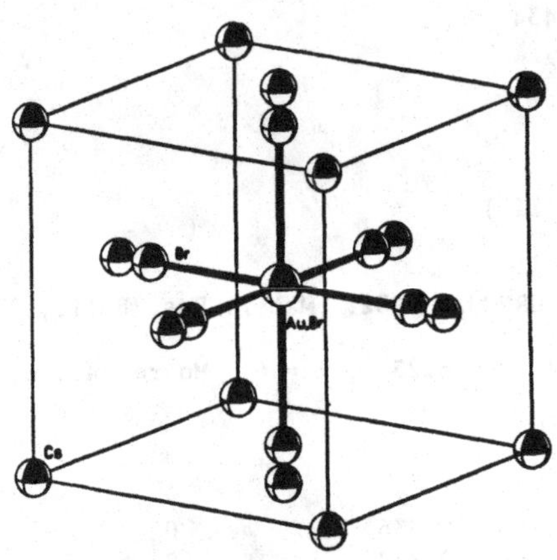

Fig. 1. Structure of caesium bromoaurate.

LITHIUM IODIDE MONOHYDRATE
$LiI.D_2O$

N.H. ANDERSEN, J.K. KJEMS and F.W. POULSEN, 1982. Phys. Scr., 25, 780-784.

α-Phase, room-temperature, cubic, Pm3m, a = 4.2550 Å (supercooled to -150°C), Z = 1. Neutron powder data. Structure as previously described (1).

β-Phase, below -54°C, orthorhombic, $P2_1am$ or Pmam, a = 6.297, b = 5.721, c = 4.352 Å, at -77°C, Z = 2. Neutron powder data. Atomic positional parameters not given explicitly.

1. Structure Reports, 30A, 287.

GALLIUM IODIDES
Ga_2I_4 $Ga^+(GaI_4^-)$
Ga_2I_3 $0.5 \times (Ga^+)_2(Ga_2I_6^{2-})$

G. GERLACH, W. HÖNLE and A. SIMON, 1982. Z. anorg. Chem., 486, 7-21.

Ga_2I_4
Rhombohedral, R3c, a = 25.215, c = 7.839 Å, D_m = 4.36, Z = 18. Mo radiation, R = 0.045 for 1079 reflexions.

Ga_2I_3
Monoclinic, $P2_1/c$, a = 11.294, b = 8.715, c = 13.453 Å, β = 145.6°, D_m = 3.21, Z = 4. Mo radiation, R = 0.056 for 1593 reflexions.

Atomic positions

	x	y	z
		Ga_2I_4	
I(1)	0.0004	0.9034	0
I(2)	-0.0035	0.7381	0.8591
I(3)	0.0963	0.8407	0.2832
I(4)	0.9070	0.7487	0.2968
Ga(3)	0.0001	0.8052	0.1052
Ga(1)	0.0018	0.1574	0.1067
		Ga_2I_3	
I(1)	0.9352	0.1704	0.3558
I(2)	0.5896	0.0079	0.8548
I(3)	0.3102	0.2002	0.3542
Ga(2)	0.1263	0.0970	0.0690
Ga(1)	0.7805	0.3716	0.0477

Ga_2I_4 contains tetrahedral GaI_4^- ions, Ga-I = 2.54-2.61 Å, and Ga^+ ions with 8-coordination, Ga-I = 3.28-3.82(1) Å. Ga_2I_3 contains $Ga_2I_6^{2-}$ ions with a staggered conformation, Ga-Ga = 2.39, Ga-I = 2.59-2.60 Å, and 8-coordinate Ga^+ ions, Ga-I = 3.29-3.78(1) Å.

ZIRCONIUM(II) IODIDE
β-ZrI_2

J.D. CORBETT and D.H. GUTHRIE, 1982. Inorg. Chem., 21, 1747-1751.

Orthorhombic, $Pmn2_1$, a = 3.7442, b = 6.831, c = 14.886 Å, Z = 4. Mo radiation, R = 0.051 for 768 reflexions. Atoms in 2(a): 0,y,z, y = 0.8849, 0.4922, 0.2745, 0.2226, 0.6161, 0.8824, z = 0.5, 0.0066, 0.3945, 0.8567, 0.6510, 0.1135, for Zr(1,2), I(1,2,3,4).

Isostructural with WTe_2 (1), containing sheets with infinite zigzag Zr chains (Zr-Zr = 3.185(3) Å) between puckered iodine layers. α-ZrI_2 has a very similar (monoclinic) structure (2).

1. Structure Reports, 31A, 69.
2. Ibid., 48A, 171.

ZIRCONIUM(II) IODIDE
γ-ZrI_2 1/6 x Zr_6I_{12}

CAESIUM ZIRCONIUM IODIDE
$CsZr_6I_{14}$

D.H. GUTHRIE and J.D. CORBETT, 1982. Inorg. Chem., 21, 3290-3295.

ZrI_2, rhombohedral, $R\bar{3}$, a = 14.502, c = 9.996 Å, Z = 18. Mo radiation, R = 0.11 for 609 reflexions.

$CsZr_6I_{14}$, orthorhombic, Cmca, a = 15.833, b = 14.300, c = 12.951 Å, Z = 4. Mo radiation, R = 0.062 for 1167 reflexions.

Atomic positions

	x	y	z
ZrI_2			
I(1)	0.3115	0.0809	0.0007
I(2)	0.0514	0.1777	0.3246
Zr	0.1430	0.1023	0.1303
$CsZr_6I_{14}$			
I(1)	0.1254	0.0899	0.2501
I(2)	0.1257	0.2573	0.0064
I(3)	1/4	0.3488	1/4
I(4)	0	0.1581	0.7618
I(5)	0.2473	0	0
Cs	0	0	0
Zr(1)	0.3942	0.0643	0.8922
Zr(2)	0	0.3668	0.9021

γ-ZrI_2 (Zr_6I_{12}) contains slightly compressed Zr_6 octahedra, with six edge-bridging I atoms also bridging to neighbouring octahedra; Zr-Zr = 3.19, 3.20, Zr-I = 2.86-3.41 Å. The $CsZr_6I_{14}$ structure is derived from that of Nb_6Cl_{14} and Ta_6I_{14} (1) by addition of Cs to a large interstice; the Zr_6 octahedron is tetragonally distorted. Zr-Zr = 3.29-3.35, Zr-I = 2.86-3.49 Å.

1. Structure Reports, 30A, 263, 264.

ZIRCONIUM(IV) IODIDE
α-ZrI_4

S.I. TROJANOV, 1980. Deposited Doc., VINITI 3150-80, 8 pp. [Chemical Abstracts, 96, 13964z].

Orthorhombic, $Pca2_1$, a = 7.076, b = 8.379, c = 13.640 Å, Z = 4. R = 0.075.

The structure contains edge-sharing ZrI_6 octahedra [as in the monoclinic form (1)].

1. Structure Reports, 45A, 182.

NIOBIUM IODIDE
Nb_6I_{11}

NIOBIUM IODIDE HYDRIDE
HNb_6I_{11}

H. IMOTO and A. SIMON, 1982. Inorg. Chem., 21, 308-319.

Nb_6I_{11} (room-temperature)
Orthorhombic, Pccn, a = 11.32, b = 15.31, c = 13.55 Å, at 298K, Z = 4. Mo radiation, R = 0.051 for 1938 reflexions.

Nb_6I_{11} (low-temperature)
Orthorhombic, $P2_1cn$, a = 11.29, 11.32, b = 15.31, 15.32, c = 13.43, 13.50 Å, at 110, 258K, Z = 4. Mo radiation, R = 0.043, 0.046 for 2959, 3212 reflexions.

HNb_6I_{11} (high-temperature)
Orthorhombic, Pccn, a = 11.35, b = 15.50, c = 13.53 Å, at 347K, Z = 4. Mo radiation, R = 0.067 for 2837 reflexions.

HNb_6I_{11} (low-temperature)
Orthorhombic, $P2_1cn$, a = 11.30, b = 15.47, c = 13.43 Å, at 216K, Z = 4. Mo radiatio
R = 0.052 for 3453 reflexions.

The materials exhibit λ-type phase transitions at 274K (Nb_6I_{11}) and 324K (HNb_6I_{11}). The high-temperature structures are essentially as previously reported (1), with Nb_6 octahedra, 8 iodines above the faces, with these Nb_6I_8 units bridged by three further iodine atoms. In the low-temperature phases the structures are distorted by twists around the pseudo-threefold axes of the clusters. Hydrogen atoms (not located) cause slight increases in Nb-Nb distances.

1. Structure Reports, 32A, 210.

MOLYBDENUM DIIODIDE

MoI_2 (Mo_6I_{12})

Z.G. ALIEV, L.A. KLINKOVA, I.V. DUBROVIN and L.O. ATOVMJAN, 1981. Ž. Neorg. Khim., 26, 1964-1967 [Russ. J. Inorg. Chem., 26, 1060-1062].

Orthorhombic, Abam, a = 12.562, b = 12.562, c = 15.803 Å, D_m = 5.55, Z = 24 (4 x Mo_6I_{12}). Mo radiation, R = 0.076 for 939 reflexions.

Atomic positions

	x	y	z
Mo(1)	0.0653	0.1357	0
Mo(2)	-0.1356	0.0654	0
Mo(3)	0	0	0.1193
I(1)	-0.0733	0.2086	0.1237
I(2)	0.2085	0.0736	0.1233
I(3)	-0.3492	0.1505	0
I(4)	0	0	0.2968

Isostructural with $MoCl_2$ (1), the structure containing a Mo_6 octahedron, surrounded by a cube of 8 I atoms and a larger octahedron of 6 I atoms; four of the latter I atoms are shared to form a layer structure. Mo-Mo = 2.672, Mo-I = 2.772, 2.803, 2.891 Å.

1. This volume, p. 107.

TETRAAMMINEPALLADIUM(II) OCTAIODIDE

$Pd(NH_3)_4I_8$

K.-F. TEBBE and B. FRECKMANN, 1982. Z. Naturforsch., 37B, 542-549.

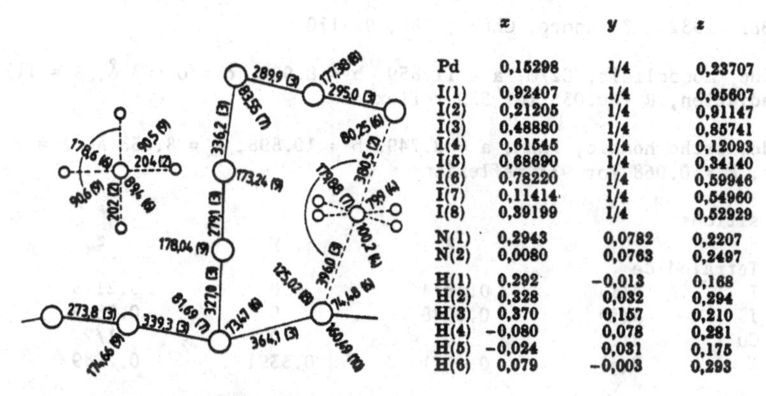

	x	y	z
Pd	0,15298	1/4	0,23707
I(1)	0,92407	1/4	0,95607
I(2)	0,21205	1/4	0,91147
I(3)	0,48880	1/4	0,85741
I(4)	0,61545	1/4	0,12093
I(5)	0,68690	1/4	0,34140
I(6)	0,78220	1/4	0,59946
I(7)	0,11414	1/4	0,54960
I(8)	0,39199	1/4	0,52929
N(1)	0,2943	0,0782	0,2207
N(2)	0,0080	0,0763	0,2497
H(1)	0,292	-0,013	0,168
H(2)	0,328	0,032	0,294
H(3)	0,370	0,157	0,210
H(4)	-0,080	0,078	0,281
H(5)	-0,024	0,031	0,175
H(6)	0,079	-0,003	0,293

Fig. 1. Structure of $Pd(NH_3)_4I_8$; bond lengths (Å x 10^2) and angles (degrees).

Monoclinic, $P2_1/m$, a = 9.590, b = 8.354, c = 13.011 Å, β = 103.37°, Z = 2. Mo radiation, R = 0.057 for 1357 reflexions.

The structure (Fig. 1) contains square-planar cations and planar Z-shaped I_8^{2-} anions, in which the interatomic distances indicate an $I_3^- -I_2-I^- -I_2$ structure; the anions are linked by a longer I...I contact of 3.641 Å.

COPPER(I) IODIDE (12R POLYTYPE)
CuI

R. BATCHELOR and T. BIRCHALL, 1982. Acta Cryst., B38, 1260-1263.

Rhombohedral, $R\bar{3}m$, a = 4.265, c = 41.96 Å, Z = 12. Mo radiation, R = 0.033 for 180 reflexions. Atoms in 6(c): 0,0,z, z = 0.20824, 0.37509, 0.3129, 0.1461, 0.2703, 0.4375 for I(1,2), 0.5 Cu(1,2,3,4).

The structure (Fig. 1) contains close-packed I layers with Cu atoms disordered in tetrahedral holes. Cu-I = 2.604-2.620(6) Å.

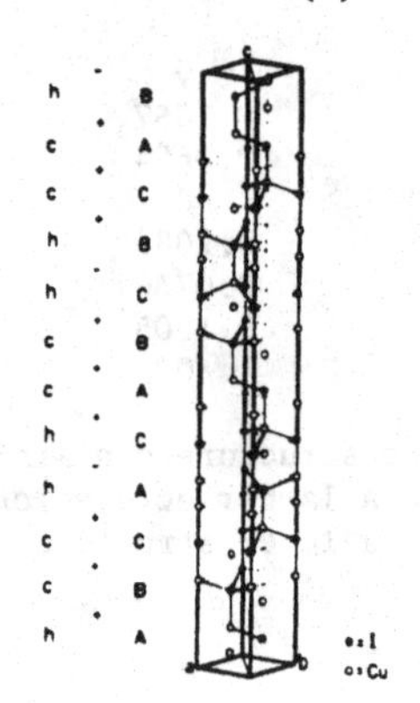

Fig. 1. Structure of 12R-CuI.

TETRAAMMINECOPPER(II) TETRAIODIDE
$Cu(NH_3)_4I_4$

TETRAAMMINECOPPER(II) HEXAIODIDE
$[Cu(NH_3)_4I_3].I_3$

K.-F. TEBBE, 1982. Z. anorg. Chem., 489, 93-110.

Tetraiodide, monoclinic, C2/m, a = 11.859, b = 8.928, c = 6.568 Å, β = 111.10°, Z = 2. Mo radiation, R = 0.036 for 539 reflexions.

Hexaiodide, orthorhombic, Pnnm, a = 8.749, b = 10.898, c = 8.853 Å, Z = 2. Mo radiation, R = 0.068 for 934 reflexions.

Atomic positions

		x	y	z
Tetraiodide				
I	in 4(i)	0.5874	0	0.2118
I^-	4(i)	0.2048	0	0.2835
Cu	2(c)	0	0	1/2
N	8(j)	0.4071	0.3391	0.2889

		x	y	z
Hexaiodide				
I(1) in	2(a)	0	0	0
I(2)	4(g)	0.2205	0.2024	0
I(3)	2(b)	1/2	1/2	0
I(4)	4(g)	0.3025	0.1166	1/2
Cu	2(d)	1/2	0	0
N	8(h)	0.1076	0.4014	0.3354

In both structures (Fig. 1) square-planar $Cu(NH_3)_4{}^{2+}$ ions are connected into zigzag chains, by linear $I_4{}^{2-}$ and $I_3{}^-$ anions; in the hexaiodide an additional $I_3{}^-$ ion lies between the chains. In $I_4{}^{2-}$, I-I = 2.808 (central), 3.344 (terminal) Å; in $I_3{}^-$, I-I = 2.931, 2.936(2) Å. Cu ions have distorted octahedral geometries, Cu-4N = 2.04, Cu-2I = 3.226, 3.294 Å.

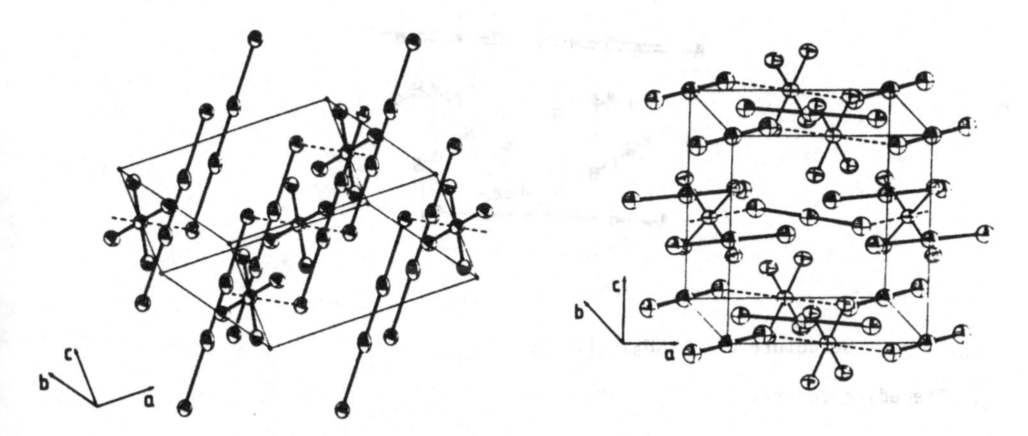

Fig. 1. Structures of $Cu(NH_3)_4I_4$ (left) and $[Cu(NH_3)_4I_3]I_3$ (right).

TETRAAMMINECADMIUM TETRAIODIDE
$Cd(NH_3)_4I_4$

TETRAAMMINECADMIUM HEXAIODIDE
$Cd(NH_3)_4(I.I_2)_2$

K.-F. TEBBE and M. PLEWA, 1982. Z. anorg. Chem., 489, 111-125.

Tetraiodide, monoclinic, C2/m, a = 11.921, b = 9.381, c = 6.601 Å, β = 109.96°, Z = 2. Mo radiation, R = 0.046 for 905 reflexions.

Hexaiodide, tetragonal, $I4_1/amd$, a = 9.497, c = 19.874 Å, Z = 4. Mo radiation, R = 0.050 for 939 reflexions.

Atomic positions

		x	y	z
Tetraiodide				
I in	4(i)	0.4125	0	0.2947
I^-	4(i)	0.7995	0	0.2008
Cd	2(a)	0	0	0
N	8(j)	0.6032	0.3227	0.2415
Hexaiodide				
I in	16(h)	0	0.1055	0.7778
I^-	8(e)	0	1/4	0.2167
Cd	4(b)	0	1/4	3/8
N	16(g)	0.3274	0.5774	7/8

The tetraiodide is isostructural with the corresponding Cu salt (1), containing $Cd(NH_3)_4^{2+}$ square planes, linked by linear I_4^{2-} ions, I-I = 2.793 (central), 3.386 (terminal) Å; Cd-4N = 2.34, Cd-2I = 3.103 Å. The hexaiodide contains square-planar $Cd(NH_3)_4^{2+}$ ions, linked by infinite, nearly linear polyiodide chains (Fig. 1); I-I = 2.793, 3.386(2) Å (alternately, corresponding to $[I^-.I_2]_\infty$). Cd-4N = 2.319(5), Cd-2I = 3.147(2) Å.

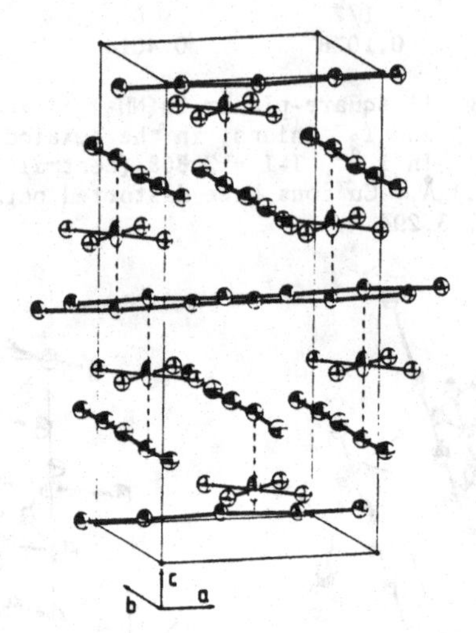

Fig. 1. Structure of $Cd(NH_3)_4(I.I_2)_2$.

1. Preceding report.

CADMIUM IODIDE
CdI_2

I. B. PAŁOSZ, 1982. Acta Cryst., B38, 3001-3009.
II. G.K. CHADHA, 1982. Ibid., B38, 3009-3011.

Structures are given for 19 (I) and 3 (II) new polytypes.

THALLIUM(I) TRIIODOGERMANATE(II)
$TlGeI_3$

N. JOUINI, L. GUEN and M. TOURNOUX, 1982. Ann. Chim., 7, 45-51.

Orthorhombic, Pnma, a = 10.066, b = 4.414, c = 16.359 Å, Z = 4. Mo radiation, R = 0.053 for 720 reflexions.

Atomic positions

	x	y	z
Tl	0.4266	1/4	-0.1724
Ge	0.1636	1/4	0.0599
I(1)	0.1639	1/4	0.4919
I(2)	0.2964	1/4	0.2126
I(3)	0.0218	1/4	-0.1150

Isostructural with NH_4CdCl_3 (1), the structure containing infinite double chains along b of edge-sharing GeI_6 octahedra, linked by 9-coordinate Tl^+ ions (tricapped trigonal prism). Ge-I = 2.83-3.20, Tl-I = 3.59-4.19 Å. Tl_4GeI_6 was also obtained in the TlI-GeI_2 system, and is isostructural with α-Tl_4CrI_6 (2).

1. Strukturbericht, 6, 13, 79; 7, 19, 115; Structure Reports, 44A, 160.
2. Structure Reports, 45A, 185; 46A, 204.

RUBIDIUM HEXAIODOTELLURATE(IV)
Rb_2TeI_6

W. ABRIEL, 1982. Mater. Res. Bull., 17, 1341-1346.

Tetragonal, P4/mnc, a = 8.136, c = 11.810 Å, Z = 2. Mo radiation, R = 0.058 for 528 reflexions. Rb in 4(d): 0,1/2,1/4; Te in 2(a): 0,0,0; I(1) in 4(e): 0,0,0.2480; I(2) in 8(h): 0.2196,0.2856,0.

Tetragonally-distorted K_2PtCl_6-type structure, with the TeI_6 octahedra rotated 7.4° about a fourfold axis from the cubic arrangement. Te-I = 2.947(2), Rb-I = 3.87-4.40 Å. Above 340K, the structure is cubic K_2PtCl_6-type, a = 11.67 Å at 370K.

ARSENIC TELLURIUM IODIDE
As_5Te_7I

R. KNIEP and H.D. RESKI, 1982. Z. Naturforsch., 37B, 151-156.

Monoclinic, Cm, a = 14.601, b = 4.040, c = 13.871 Å, β = 110.62°, Z = 2. Mo radiation, R = 0.098 for 1519 reflexions.

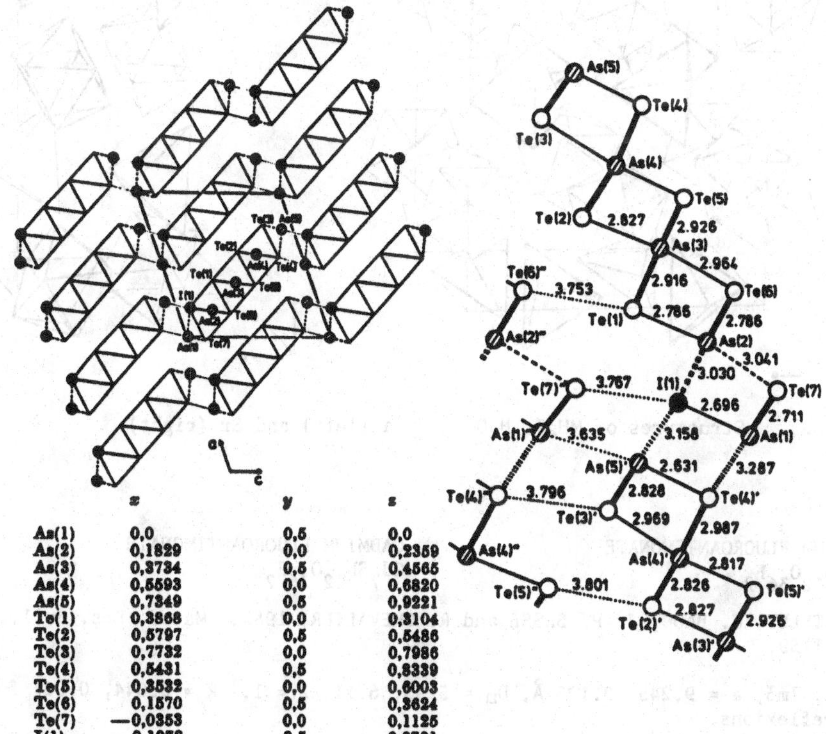

	x	y	z
As(1)	0,0	0,5	0,0
As(2)	0,1829	0,0	0,2359
As(3)	0,3734	0,5	0,4565
As(4)	0,5593	0,0	0,6820
As(5)	0,7349	0,5	0,9221
Te(1)	0,3868	0,0	0,3104
Te(2)	0,5797	0,5	0,5486
Te(3)	0,7732	0,0	0,7986
Te(4)	0,5431	0,5	0,8339
Te(5)	0,3532	0,0	0,6003
Te(6)	0,1570	0,5	0,3624
Te(7)	—0,0353	0,0	0,1125
I(1)	0,1972	0,5	0,0791

Fig. 1. Structure of As_5Te_7I.

The structure (Fig. 1) contains ribbons of condensed octahedral chains sharing edges. Arsenic atoms are at the centres of octahedra and (with trigonal pyramidal coordination) at the borders of the ribbons.

CALCIUM TETRAIODOMERCURATE(II) OCTAHYDRATE
STRONTIUM TETRAIODOMERCURATE(II) OCTAHYDRATE
$MHgI_4.8H_2O$ (M = Ca, Sr)

G. THIELE, K. BRODERSEN and G. PEZZEI, 1982. Z. anorg. Chem., 491, 308-318.

Calcium, tetragonal, $P4_2/mbc$, a = 13.972, c = 18.155 Å, D_m = 3.3, Z = 8. Ag radiation, R = 0.046 for 945 reflexions.

Strontium, orthorhombic, Pcmn, a = 9.616, b = 10.610, c = 17.731 Å, D_m = 3.3, Z = 4. Ag radiation, R = 0.043 for 1361 reflexions.

The structures (Fig. 1) contain tetrahedral HgI_4^{2-} anions and $M(H_2O)_8^{2+}$ cations; the cation polyhedra are dodecahedra in the Ca compound, and antiprisms in the Sr compound. Hg-I = 2.740-2.835 Å, I-Hg-I = 103-131°; Ca-O = 2.49, Sr-O = 2.59-2.64 Å.

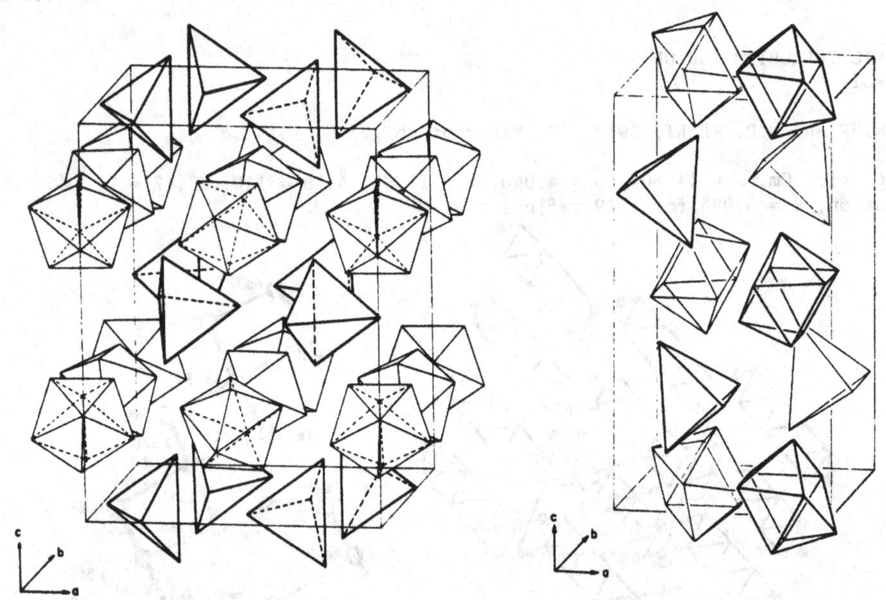

Fig. 1. Structures of $MHgI_4.H_2O$, M = Ca (left) and Sr (right).

CALCIUM FLUOROANTIMONATE
$Ca_7Sb_{12}O_{36}F_2$

CADMIUM FLUOROANTIMONATE
$Cd_7Sb_{12}O_{36}F_2$

H. WATELET, G. BAUD, J.-P. BESSE and R. CHEVALIER, 1982. Mater. Res. Bull., 17, 1155-1159.

Cubic, Im3, a = 9.245, 9.195 Å, D_m = 5.95, 6.11, Z = 1. R = 0.044, 0.067 for 293, 280 reflexions.

Atomic positions

				x	y	z
	Sb	in	12(e)	0.8368, 0.8346	0	1/2
	O(1)		12(d)	0.3602, 0.3646	0	0
	O(2)		24(g)	0	0.3309, 0.3311	0.2823, 0.2926
0.4	M(1)		16(f)	0.1460, 0.1457	x	x
0.05	M(2)		8(c)	1/4	1/4	1/4
	F		2(a)	0	0	0

Isostructural with $KSbO_3.1/6KF$ (1), with statistical distribution of M (Ca, Cd) cations. Sb-O = 1.89-2.04, M-O = 2.45-2.77, M-F = 2.33 Å.

1. Structure Reports, 42A, 200.

POTASSIUM TELLURIUM(IV) OXYFLUORIDE
$KTeOF_3$ (I)

AMMONIUM TELLURIUM(IV) OXYFLUORIDE
$NH_4TeO(OH)F_2$ (II)

TELLURIUM(IV) OXYFLUORIDE
$Te_2O(OH)_2F_4$ (III)

Ju.E. GORBUNOVA, S.A. LINDE, V.I. PAKHOMOV, Ju.V. KOKUNOV and Ju.A. BUSLAEV, Koordin. Khim., 8, 389-395 [Soviet Journal of Coordination Chemistry, 8, 202-203].

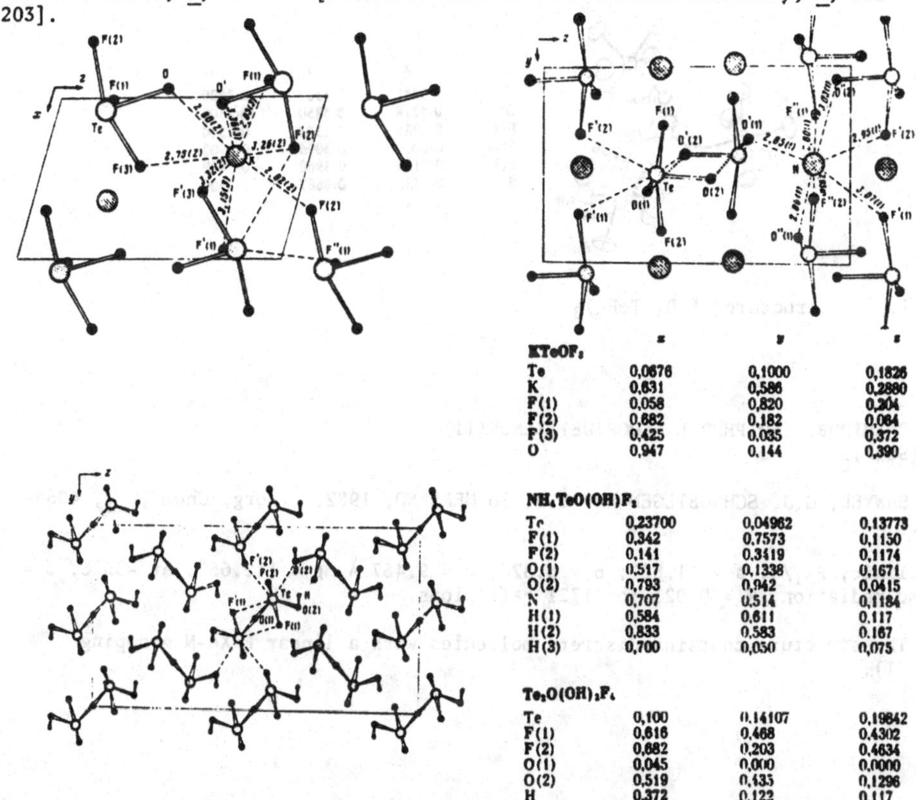

	x	y	z
$KTeOF_3$			
Te	0,0676	0,1000	0,1826
K	0,631	0,586	0,2880
F(1)	0,058	0,820	0,204
F(2)	0,682	0,182	0,064
F(3)	0,425	0,035	0,372
O	0,947	0,144	0,390
$NH_4TeO(OH)F_2$			
Te	0,23700	0,04962	0,13773
F(1)	0,342	0,7573	0,1150
F(2)	0,141	0,3419	0,1174
O(1)	0,517	0,1338	0,1671
O(2)	0,793	0,942	0,0415
N	0,707	0,514	0,1184
H(1)	0,584	0,611	0,117
H(2)	0,833	0,583	0,167
H(3)	0,700	0,050	0,075
$Te_2O(OH)_2F_4$			
Te	0,100	0,14107	0,19842
F(1)	0,616	0,468	0,4302
F(2)	0,682	0,203	0,4634
O(1)	0,045	0,000	0,0000
O(2)	0,519	0,435	0,1296
H	0,372	0,122	0,117

Fig. 1. Structures of $KTeOF_3$ (upper left), $NH_4TeO(OH)F_2$ (upper right), and $Te_2O(OH)_2F_4$ (lower left), and atomic positional parameters.

I. Monoclinic, $P2_1$, a = 4.937, b = 6.746, c = 7.764 Å, β = 106.13°, Z = 2. Mo radiation, R = 0.095 for 641 reflexions.

II. Monoclinic, $P2_1/n$, a = 6.266, b = 6.636, c = 10.709 Å, β = 95.64°, Z = 4. Mo radiation, R = 0.047 for 1474 reflexions.

III. Orthorhombic, F2dd, a = 4.761, b = 11.634, c = 20.931 Å, Z = 8. Mo radiation, R = 0.060 for 467 reflexions.

The structures (Fig. 1) contain distorted trigonal bipyramidal TeF_2X_2E groupings (X = O, OH, F) with the lone pair (E) equatorial; the groupings are discrete in I and II, but form dimers in III. Further linking in the structures is via the cations, hydrogen bonding, and weaker Te...F/O interactions. Te-F/O = 1.95-2.07 (axial), 1.84-1.91(1) Å (equatorial).

TRIS[PENTAFLUOROTELLURATO(VI)]BORON(III)
$B(OTeF_5)_3$

J.F. SAWYER and G.J. SCHROBILGEN, 1982. Acta Cryst., B38, 1561-1563.

Hexagonal, $P6_3/m$, a = 9.218, c = 9.207 Å, Z = 2. Mo radiation, R = 0.031 for 566 reflexions.

The structure (Fig. 1) contains molecules with D_{3h} symmetry.

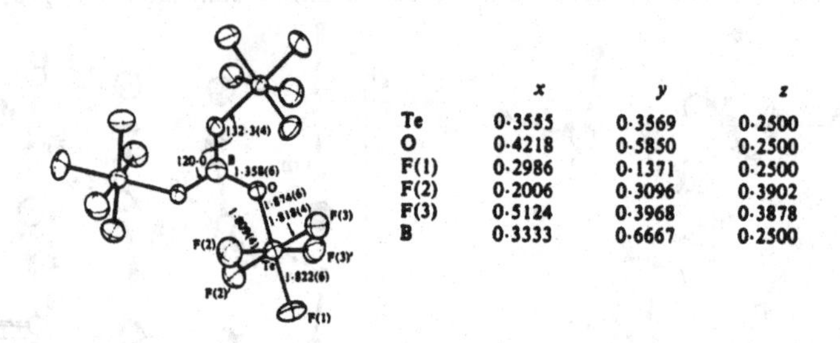

	x	y	z
Te	0·3555	0·3569	0·2500
O	0·4218	0·5850	0·2500
F(1)	0·2986	0·1371	0·2500
F(2)	0·2006	0·3096	0·3902
F(3)	0·5124	0·3968	0·3878
B	0·3333	0·6667	0·2500

Fig. 1. Structure of $B(OTeF_5)_3$.

FLUORO[IMIDOBIS(SULPHURYL FLUORIDE)]XENON(II)
$FXeN(SO_2F)_2$

J.F. SAWYER, G.J. SCHROBILGEN and S.J. SUTHERLAND, 1982. Inorg. Chem., 21, 4064-4072.

Monoclinic, $P2_1/a$, a = 11.827, b = 6.828, c = 9.467 Å, β = 112.65°, at -55°C, Z = 4. Mo radiation, R = 0.023 for 1721 reflexions.

The structure contains discrete molecules with a linear F-Xe-N grouping (Fig. 1).

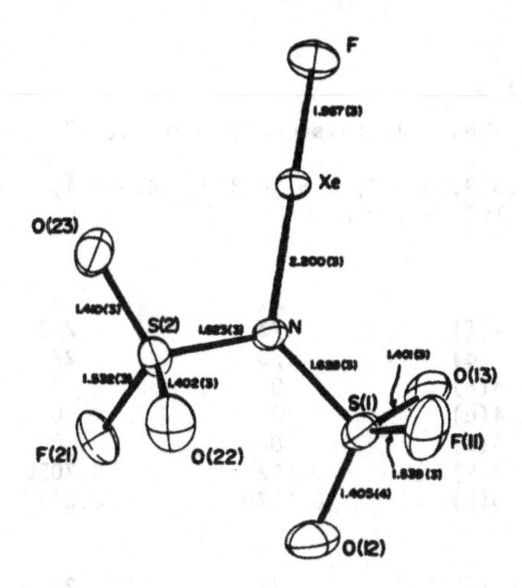

Fig. 1. The $FXeN(SO_2F)_2$ molecule, with bond lengths (Å).

SODIUM TITANIUM MAGNESIUM OXYFLUORIDE
$Na_{0.82}(Ti_{1.20}Mg_{0.80})(O_{3.22}F_{0.78})$

A.M. GOLUBEV, V.N. MOLČANOV, M.Ju. ANTIPIN and V.I. SIMONOV, 1981. Kristallografija, 26, 1254-1258 [Soviet Physics - Crystallography, 26, 713-716].

Orthorhombic, $Pna2_1$, a = 9.238, b = 11.333, c = 2.9792 Å, at 153K, D_m = 3.45, Z = 4. Mo radiation, R = 0.043 for 1194 reflexions.

Atomic positions

	x	y	z
Ti,Mg(1)	0.0176	0.1207	0
Ti,Mg(2)	0.2310	0.3403	0.497
0.64 Na(1)	0.3971	0.1079	0.475
0.18 Na(2)	0.3757	0.0824	0.059
O,F(1)	0.1268	0.1852	0.497
O,F(2)	0.1264	0.4062	0.004
O,F(3)	0.3590	0.2708	0.009
O,F(4)	0.3936	0.4585	0.502

The structure contains edge-sharing $Ti,Mg(O,F)_6$ octahedra, with a conductivity channel containing the Na ions. Ti,Mg-O,F = 1.91-2.09, Na-O = 2.15-2.89, Na(1)... Na(2) = 1.29, 1.78 Å (partial occupancy).

AMMONIUM AQUOTETRAFLUOROOXOVANADATE(IV)
$(NH_4)_2VOF_4(H_2O)$

J. MAZICEK, V.K. TRUNOV and L.A. ASLANOV, 1982. Koordin. Khim., 8, 1550-1551.

Monoclinic, $P2_1/c$, a = 7.026, b = 9.688, c = 10.175 Å, β = 100.50°, Z = 4. R = 0.043.

V has distorted octahedral coordination, with trans O and H_2O.

CAESIUM FLUOROVANADATES
$Cs_3V_2O_4F_5$, $Cs_3V_{1.5}Mo_{0.5}O_3F_6$

R. MATTES and H. FÖRSTER, 1982. J. Less-Common Metals, 87, 237-247.

Hexagonal, $P6_3/mmc$, a = 6.319, 6.359, c = 14.850, 14.999 Å, Z = 2. Mo radiation, R = 0.061, 0.041 for 135, 190 reflexions.

Atomic positions

$Cs_3V_2O_4F_5$		x	y	z
Cs(1) in	4(f)	1/3	2/3	0.4135
Cs(2)	2(d)	1/3	2/3	3/4
0.35 V(1)	4(e)	0	0	0.0894
0.28 V(2)	4(e)	0	0	0.0163
0.40 V(3)	4(e)	0	0	0.1468
F/O(1)	12(k)	0.8525	0.7050	0.0648
F/O(2)	6(h)	0.1370	0.2740	1/4

$Cs_3V_{1.5}Mo_{0.5}O_3F_6$		x	y	z
Cs(1) in	4(f)	1/3	2/3	0.4321
Cs(2)	2(d)	1/3	2/3	3/4
0.73 V + 0.33 Mo	4(e)	0	0	0.1460
F/O(1)	12(k)	0.1419	0.2838	0.5951
F/O(2)	6(h)	0.1315	0.2630	1/4

$Cs_3Fe_2F_9$-type structures (1). $Cs_3V_2O_4F_5$ exhibits a substructure with c' = c/3.

1. Structure Reports, 37A, 187.

CAESIUM AQUOTETRAFLUOROOXOVANADATE(IV) (GREEN FORM)
$Cs_2[VOF_4(H_2O)]$

R. MATTES and H. FOERSTER, 1982. J. Solid State Chem., 45, 154-157.

Orthorhombic, Ccmm, a = 8.231, b = 10.323, c = 8.497 Å, Z = 4. Mo radiation, R = 0.035 for 421 reflexions.

As in the blue form (1) the structure (Fig. 1) contains isolated distorted trans-octahedral anions, linked by 12-coordinated Cs ions and O-H...F hydrogen bonds. V-O = 1.67, $V-OH_2$ = 2.31, V-F = 1.91(1) Å, O-V-F = 95, 98°, Cs-F,O = 2.95-3.61, O-H...F = 2.70(1) Å.

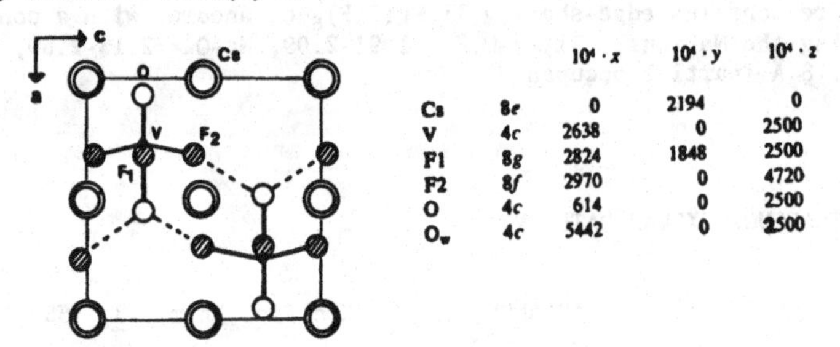

		$10^4 \cdot x$	$10^4 \cdot y$	$10^4 \cdot z$
Cs	8e	0	2194	0
V	4c	2638	0	2500
F1	8g	2824	1848	2500
F2	8f	2970	0	4720
O	4c	614	0	2500
O_w	4c	5442	0	2500

Fig. 1. Structure of the green form of $Cs_2[VOF_4(H_2O)]$.

1. Structure Reports, 45A, 192.

COPPER FLUORONIOBATE
$Cu_{0.6}NbO_{2.6}F_{0.4}$

M. LUNDBERG and P. NDA-LAMBA WA ILUNGA, 1981. Rev. Chim. Minér., 18, 118-124.

Orthorhombic, Pnam, a = 17.694, b = 3.944, c = 3.801 Å, D_m = 4.50, Z = 4. Mo radiation, R = 0.054 for 444 reflexions.

The structure (Fig. 1) contains layers of edge- and corner-sharing NbX_6 octahedra, with an arrangement similar to that of MoO_3; the layers are linked by linearly-coordinated Cu ions. Nb-X = 1.81-2.27, Cu-X = 1.84(3) Å.

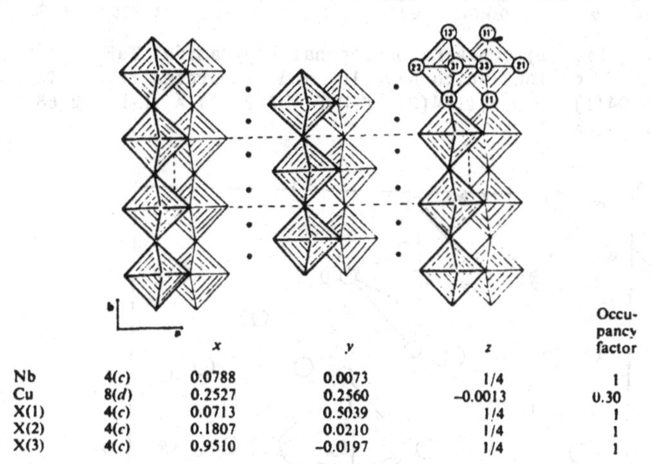

		x	y	z	Occupancy factor
Nb	4(c)	0.0788	0.0073	1/4	1
Cu	8(d)	0.2527	0.2560	−0.0013	0.30
X(1)	4(c)	0.0713	0.5039	1/4	1
X(2)	4(c)	0.1807	0.0210	1/4	1
X(3)	4(c)	0.9510	−0.0197	1/4	1

Fig. 1. Structure of $Cu_{0.6}NbO_{2.6}F_{0.4}$ (X = disordered O and F).

AMMONIUM DIPEROXOTETRAFLUOROTANTALATE(V)
$(NH_4)_3Ta(O_2)_2F_4$

R. SCHMIDT and G. PAUSEWANG, 1982. Z. anorg. Chem., 488, 108-120.

Cubic, Fm3m, a = 9.414 Å, Z = 4. Mo radiation, R_w = 0.032 for 405 reflexions. Ta in 4(a): 0,0,0; N(1) in 4(b): 1/2,0,0; N(2) in 8(c): 1/4,1/4,1/4; 16 F in 24(e): 0.208,0,0; 16 O in 96(j): 0.197,0.088,0.

Disordered elpasolite structure (1). The K salt is monoclinic, with variable stoichiometry cubic phases, and decomposes to a tetragonal phase.

1. Strukturbericht, 2, 498; Structure Reports, 40A, 137.

POTASSIUM PENTAFLUOROPEROXOTANTALATE(V) HYDROGENDIFLUORIDE
$K_3[TaF_5(O_2)][HF_2]$

R. STOMBERG, 1982. Acta Chem. Scand., A36, 423-427.

Orthorhombic, Pnam, a = 6.976, 6.957, b = 13.82, 13.61, c = 9.072, 9.082 Å, at 290, 170K, Z = 4. Mo radiation, R = 0.049, 0.073 for 1203, 1429 reflexions. Previous study in 1.

Atomic positions

		290 K			170 K		
		x	y	z	x	y	z
Ta	4c	0.26481	0.07050	¼	0.26473	0.07003	¼
K(1)	4c	0.2288	0.5983	¼	0.2293	0.5986	¼
K(2)	8d	0.2328	0.3481	0.0234	0.2261	0.3488	0.0217
F(1)	4c	0.5522	0.0780	¼	0.5532	0.0796	¼
F(2)	4c	0.2664	0.2070	¼	0.2653	0.2077	¼
F(3)	8d	0.3181	0.0691	0.0382	0.3208	0.0691	0.0391
F(4)	4c	0.3226	−0.0671	¼	0.3246	−0.0698	¼
F(5)	8d	0.0000	0.2362	0.6264	−0.0042	0.2340	0.6271
O	8d	0.0119	0.0464	0.1734	0.0102	0.0478	0.1706

The structure (Fig. 1) contains pentagonal bipyramidal $TaF_5(O_2)^{2-}$ and linear $F-H-F^-$ anions, and K^+ cations which have 10- and 8-coordinations. Ta-F = 1.87-2.01(1), Ta-O = 1.94(1), O-O = 1.44(2), F-H-F = 2.23(2) Å, K-F = 2.68-2.98(1), K-O = 2.57-3.07(1) Å, at 170K.

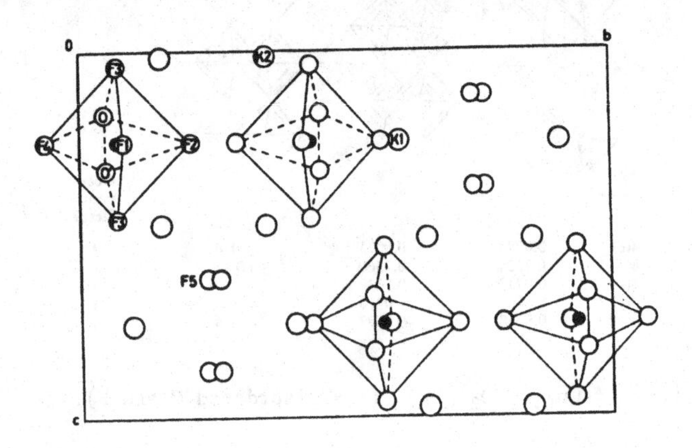

Fig. 1. Structure of $K_3[TaF_5(O_2)][HF_2]$.

1. Structure Reports, 42A, 204.

MOLYBDENUM TELLURIUM OXYFLUORIDE
$MoO(OTeF_5)_4$

OSMIUM TELLURIUM OXYFLUORIDE
$OsO(OTeF_5)_4.F^-.TeF_3^+.2TeF_4$

P. HUPPMANN, H. LABISCHINSKI, D. LENTZ, H. PRITZKOW and K. SEPPELT, 1982. Z. anorg. Chem., 487, 7-25.

Molybdenum compound
Monoclinic, $P2_1/c$, a = 9.18, b = 11.34, c = 20.17 Å, β = 102.9°, Z = 4. Mo radiation, R = 0.058 for 2977 reflexions.

Osmium compound
Triclinic, $P\bar{1}$, a = 15.050, b = 10.317, c = 10.242 Å, α = 72.09, β = 100.97, γ = 101.22°, Z = 2. Mo radiation, R = 0.094 for 5404 reflexions.

The Mo compound contains isolated $O{=}Mo(OTeF_5)_4$ molecules (Fig. 1) with square-pyramidal coordination at Mo. Mo=O = 1.60, Mo-O = 1.83-1.91, O-Te = 1.79-1.82, Te-F = 1.70-1.85(1) Å, O=Mo-O = 101-102, Mo-O-Te = 149-173°; a long intermolecular Mo...F contact of 2.55 Å completes distorted octahedral geometry.

In the Os compound (Fig. 1) octahedral coordination is completed by a F^- ion donated by a chain of fluorine-bridged TeF_4 molecules. Os=O = 1.61, Os-O = 1.87-1.91, O-Te = 1.83-1.95, Te-F = 1.78-1.87(2) Å, O=Os-O = 97-99, Os-O-Te = 134-141°,

Os..F⁻ = 2.05(2) Å; in the TeF_4 chain, Te-F = 1.75-2.01, Te(5)...F⁻ = 2.23, other Te...F = 2.40-2.71 Å.

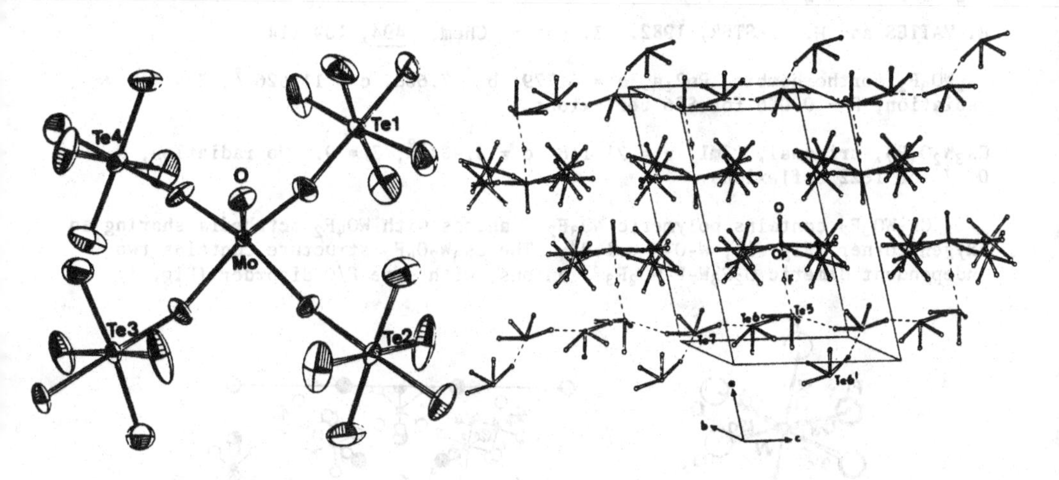

Fig. 1. Structures of the molybdenum (left) and osmium (right) tellurium oxyfluorides.

SODIUM OXYFLUOROTUNGSTATE
$Na_2WO_2F_4$

M. VLASSE, J.-M. MOUTOU, M. CERVERA-MARZAL, J.-P. CHAMINADE and P. HAGENMULLER, 1982. Rev. Chim. Minér., 19, 58-64.

Orthorhombic, Pbcn, a = 5.074, b = 18.253, c = 5.437 Å, D_m = 4.41, Z = 4. Mo radiation, R = 0.052 for 1022 reflexions.

The structure (Fig. 1) contains a three-dimensional arrangement of edge- and corner-sharing WX_6 and NaX_6 octahedra (X = O,F); W-X = 1.75, 1.93, 2.04, Na-X = 2.21-2.45(1) Å.

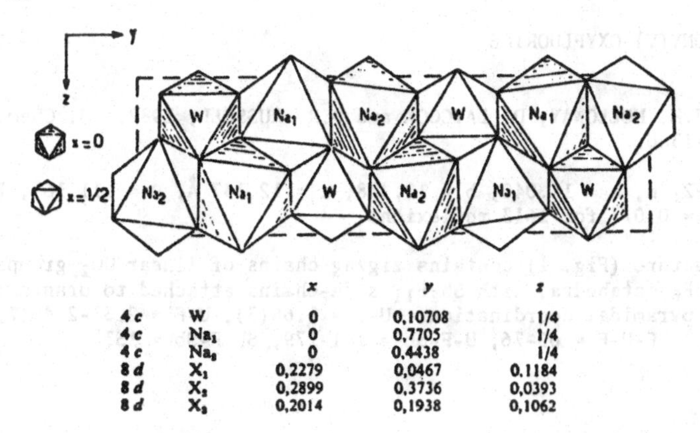

		x	y	z
4 c	W	0	0,10708	1/4
4 c	Na_1	0	0,7705	1/4
4 c	Na_2	0	0,4438	1/4
8 d	X_1	0,2279	0,0467	0,1184
8 d	X_2	0,2899	0,3736	0,0393
8 d	X_3	0,2014	0,1938	0,1062

Fig. 1. Structure of $Na_2WO_2F_4$.

CAESIUM FLUOROTUNGSTATES(VI)
$Cs_2WO_3F_2$, $Cs_3W_2O_4F_7$

R. MATTES and H. FORSTER, 1982. Z. anorg. Chem., 494, 109-114.

$Cs_2WO_3F_2$, orthorhombic, $Pn2_1a$, a = 6.779, b = 7.668, c = 11.626 Å, Z = 4. Mo radiation, R = 0.056 for 556 reflexions.

$Cs_3W_2O_4F_7$, trigonal, $P\bar{3}m1$, a = 21.118, c = 8.434 Å, Z = 9. Mo radiation, R = 0.17 for 1622 reflexions.

$Cs_2WO_3F_2$ contains polymeric $WO_3F_2^{2-}$ anions with WO_4F_2 octahedra sharing an oxygen corner (Fig. 1); W-O-W = 160°. The $Cs_3W_2O_4F_7$ structure contains two independent dimeric O_2F_3W-F-$WO_2F_3^{3-}$ anions, with some F/O disorder (Fig. 1).

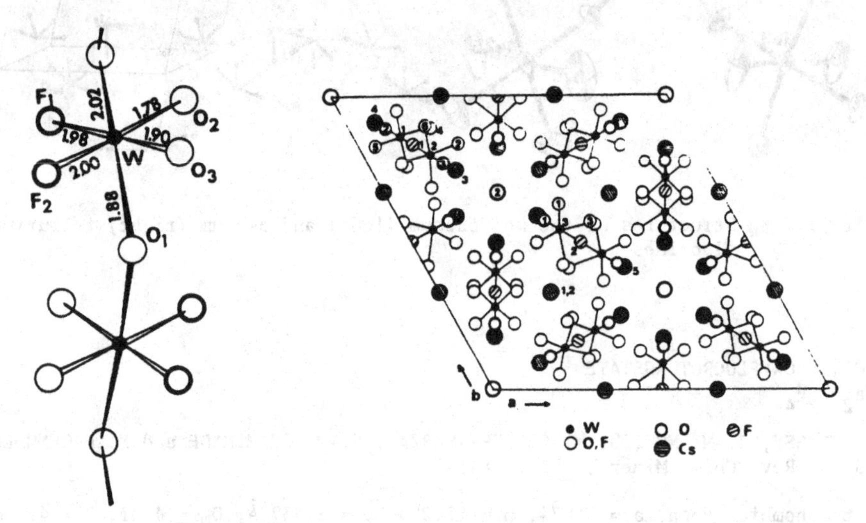

Fig. 1. The polymeric $WO_3F_2^{2-}$ anion (left), and the structure of $Cs_3W_2O_4F_7$ (right).

URANYL ANTIMONY(V) OXYFLUORIDE
$UF_2O_2.3SbF_5$

J. FAWCETT, J.H. HOLLOWAY, D. LAYCOCK and D.R. RUSSELL, 1982. J. Chem. Soc., Dalton, 1355-1360.

Monoclinic, $P2_1/n$, a = 11.040, b = 12.438, c = 12.147 Å, β = 111.16°, Z = 4. Mo radiation, R = 0.077 for 1613 reflexions.

The structure (Fig. 1) contains zigzag chains of linear UO_2 groups fluorine-bridged to SbF_6 octahedra, with Sb_2F_{11} side-chains attached to uranium; U has pentagonal bipyramidal coordination. U-O = 1.68(3), U-F = 2.33-2.45(2), Sb-F = 1.76-2.11(2) Å, F-U-F = 69-76, U-F-Sb = 141-179, Sb-F-Sb = 153°.

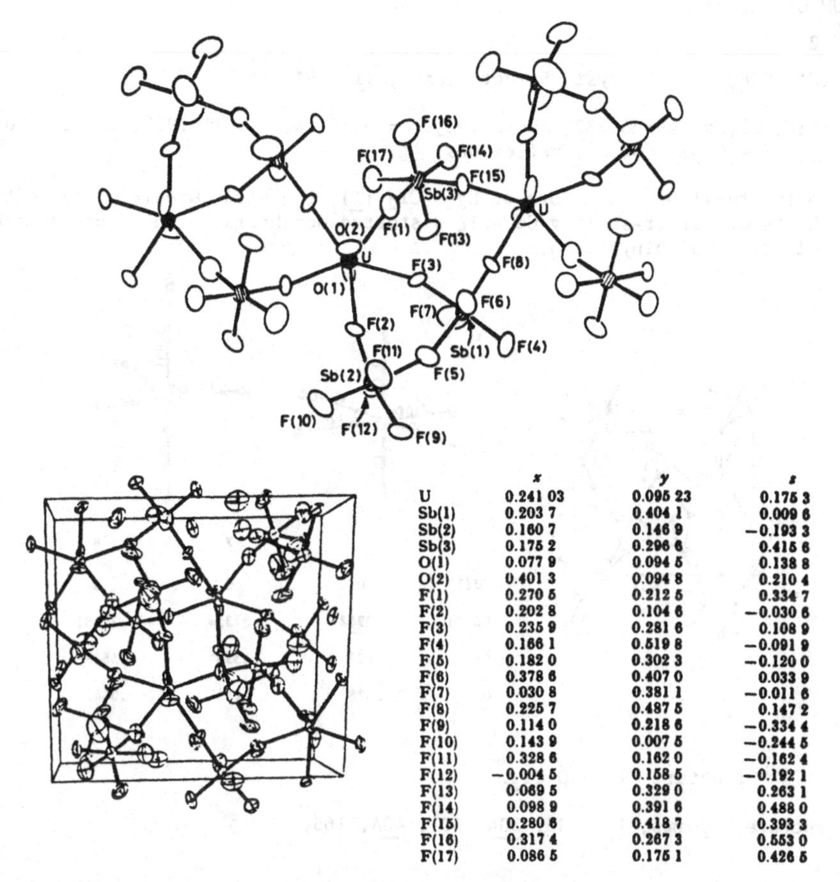

	x	y	z
U	0.241 03	0.095 23	0.175 3
Sb(1)	0.203 7	0.404 1	0.009 6
Sb(2)	0.160 7	0.146 9	−0.193 3
Sb(3)	0.175 2	0.296 6	0.415 6
O(1)	0.077 9	0.094 5	0.138 8
O(2)	0.401 3	0.094 8	0.210 4
F(1)	0.270 5	0.212 5	0.334 7
F(2)	0.202 8	0.104 6	−0.030 6
F(3)	0.235 9	0.281 6	0.108 9
F(4)	0.166 1	0.519 8	−0.091 9
F(5)	0.182 0	0.302 3	−0.120 0
F(6)	0.378 6	0.407 0	0.033 9
F(7)	0.030 8	0.381 1	−0.011 6
F(8)	0.225 7	0.487 5	0.147 2
F(9)	0.114 0	0.218 6	−0.334 4
F(10)	0.143 9	0.007 5	−0.244 5
F(11)	0.328 6	0.162 0	−0.162 4
F(12)	−0.004 5	0.158 5	−0.192 1
F(13)	0.069 5	0.329 0	0.263 1
F(14)	0.098 9	0.391 6	0.488 0
F(15)	0.280 6	0.418 7	0.393 3
F(16)	0.317 4	0.267 3	0.553 0
F(17)	0.086 5	0.175 1	0.426 5

Fig. 1. Structure of $UF_2O_2.3SbF_5$.

cis-DICHLORODIAMMINE-trans-DIHYDROXOPLATINUM(IV)

$PtCl_2(NH_3)_2(OH)_2$

R. FAGGIANI, H.E. HOWARD-LOCK, C.J.L. LOCK, B. LIPPERT and B. ROSENBERG, 1982. Canad. J. Chem., 60, 529-534.

Tetragonal, $P4_2/n$, a = 7.328, c = 11.362 Å, Z = 4. Mo radiation, R = 0.046 for 447 reflexions.

Atomic positions

		x	y	z
Pt	in 4(e)	3/4	1/4	0.0208
Cl	8(g)	0.7501	0.0266	0.1637
N	8(g)	0.750	0.050	-0.0989
O	8(g)	0.477	0.250	0.0256

The crystal contains octahedral molecules, Pt-Cl = 2.306(4), Pt-N = Pt-O = 2.00(1) Å, angles at Pt = 88-94°. Intermolecular distances indicate hydrogen bonding.

CADMIUM OXYCHLORIDE
$Cd_3O_2Cl_2$

C. STÅLHANDSKE, 1982. Cryst. Struct. Comm., 11, 1543-1547.

Monoclinic, $P2_1/c$, a = 6.382, b = 6.692, c = 6.665 Å, β = 115.92°, Z = 2. Mo radiation, R = 0.020 for 631 reflexions.

Isostructural with one form of $Hg_3O_2Cl_2$ (1), the structure (Fig. 1) containing (100) sheets of corner-sharing CdO_2Cl_4 distorted octahedra, with sheets connected by CdO_3Cl_2 trigonal bipyramids.

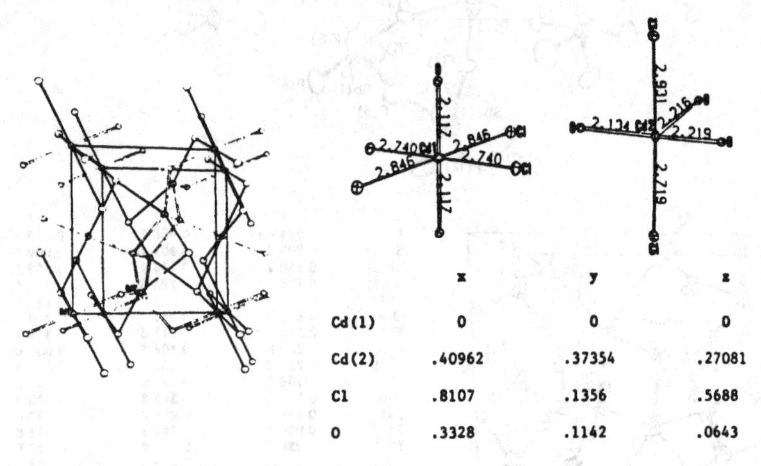

	x	y	z
Cd(1)	0	0	0
Cd(2)	.40962	.37354	.27081
Cl	.8107	.1356	.5688
O	.3328	.1142	.0643

Fig. 1. Structure of $Cd_3O_2Cl_2$.

1. Structure Reports, 19, 343; 30A, 301; 40A, 165.

LANTHANUM CHLOROTUNGSTATES
$La_3WO_6Cl_3$

I. L.H. BRIXNER, H.Y. CHEN and C.M. FORIS, 1982. J. Solid State Chem., 44, 99-107.

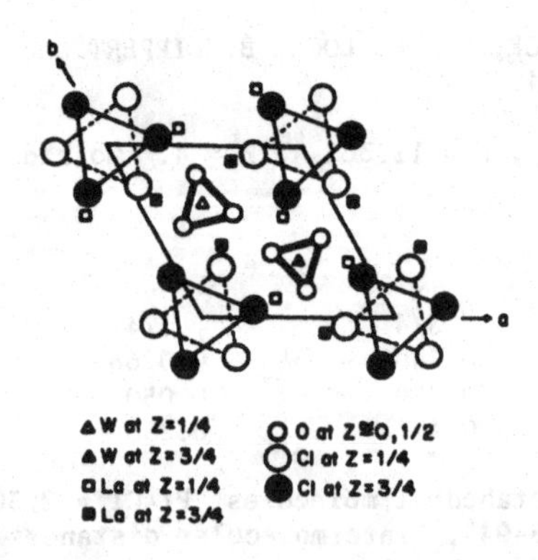

Fig. 1. Structure of lanthanum chlorotungstate, $La_3WO_6Cl_3$.

Hexagonal, $P6_3/m$, a = 9.4048, c = 5.4252 Å, Z = 2. R = 0.028 for 359 reflexions. W in 2(c): 1/3,2/3,1/4; La in 6(h): 0.0906,0.6865,3/4; Cl in 6(h): 0.2402,0.0484, 3/4; O in 12(i): 0.3635,0.8381,0.0216.

The structure (Fig. 1) contains rather unusual WO_6 trigonal prisms, linked by LaO_6Cl_4 polyhedra. W-O = 1.940(2), La-O = 2.436, 2.534, 2.672(2), La-Cl = 2.965, 3.061(1) Å.

$LaWO_4Cl$

II. L.H. BRIXNER, H.Y. CHEN and C.M. FORIS, 1982. J. Solid State Chem., 45, 80-87.

Orthorhombic, $Pbc2_1$, a = 5.899, b = 7.862, c = 19.287 Å, Z = 8 [subcell, Pmcn, b' = b/2]. Mo radiation, R = 0.040 for 2029 reflexions.

The structure (Fig. 2) contains W atoms with (4 + 1) trigonal bipyramidal coordination to oxygen, and LaO_5Cl_2 monocapped trigonal prisms. W-O = 1.72-1.77, 2.16(1), La-O = 2.49-2.55(1), La-Cl = 2.97, 2.98 (two further at 3.43) Å. There is some disorder of W and La sites, and there is a subcell with b' = b/2.

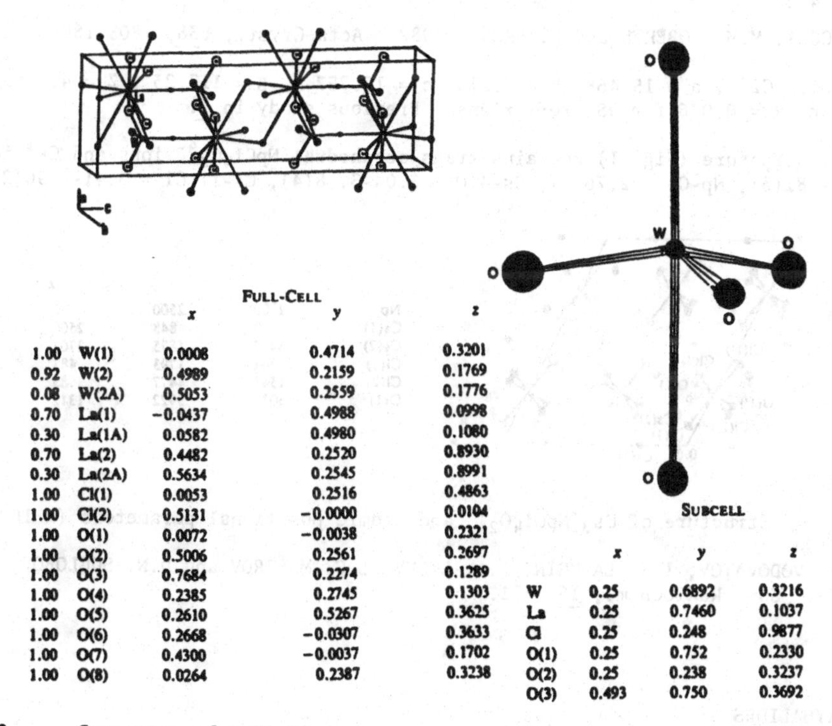

FULL-CELL

		x	y	z
1.00	W(1)	0.0008	0.4714	0.3201
0.92	W(2)	0.4989	0.2159	0.1769
0.08	W(2A)	0.5053	0.2559	0.1776
0.70	La(1)	−0.0437	0.4988	0.0998
0.30	La(1A)	0.0582	0.4980	0.1080
0.70	La(2)	0.4482	0.2520	0.8930
0.30	La(2A)	0.5634	0.2545	0.8991
1.00	Cl(1)	0.0053	0.2516	0.4863
1.00	Cl(2)	0.5131	−0.0000	0.0104
1.00	O(1)	0.0072	−0.0038	0.2321
1.00	O(2)	0.5006	0.2561	0.2697
1.00	O(3)	0.7684	0.2274	0.1289
1.00	O(4)	0.2385	0.2745	0.1303
1.00	O(5)	0.2610	0.5267	0.3625
1.00	O(6)	0.2668	−0.0307	0.3633
1.00	O(7)	0.4300	−0.0037	0.1702
1.00	O(8)	0.0264	0.2387	0.3238

SUBCELL

	x	y	z
W	0.25	0.6892	0.3216
La	0.25	0.7460	0.1037
Cl	0.25	0.248	0.9877
O(1)	0.25	0.752	0.2330
O(2)	0.25	0.238	0.3237
O(3)	0.493	0.750	0.3692

Fig. 2. Structure of $LaWO_4Cl$.

GADOLINIUM TUNGSTATE CHLORIDE
$GdWO_4Cl$

L.H. BRIXNER, H.-Y. CHEN and C.M. FORIS, 1982. Mater. Res. Bull., 17, 1545-1556.

Monoclinic, C2/m, a = 10.324, b = 7.327, c = 6.895 Å, β = 107.2°, D_m = 5.88, Z = 4. Mo radiation, R = 0.040 for 574 reflexions.

Atomic positions

	x	y	z
W	0.6364	0	0.2676
Gd	0.2215	0	0.1139
Cl	-0.0106	0	0.2365
O(1)	0.700	0.186	0.151
O(2)	0.705	0	0.534
O(3)	0.458	0	0.205

The structure contains WO_4 distorted tetrahedra linked by GdO_6Cl_2 distorted square antiprisms. W-O = 1.76-1.80(1), Gd-O = 2.32-2.59(1), Gd-Cl = 2.733, 2.765(3) Å. There is a large void of diameter 5.5 Å centred at 1/2,1/4,1/2. Other $LnMO_4Cl$ compounds (M = Mo, W) are isostructural.

CAESIUM TETRACHLORODIOXONEPTUNATE(V)
$Cs_3[NpCl_4O_2]$

N.W. ALCOCK, M.M. ROBERTS and D. BROWN, 1982. Acta Cryst., B38, 1805-1806.

Monoclinic, C2/c, a = 15.468, b = 7.275, c = 12.757 Å, β = 117.23°, Z = 4. Mo radiation, R = 0.036 for 951 reflexions. Previous study in 1.

The structure (Fig. 1) contains trans-octahedral $NpCl_4O_2{}^{3-}$ ions and Cs^+ ions. Np-O = 1.81(3), Np-Cl = 2.76(2), Cs-4 O = 3.04-3.18(4), Cs-11 Cl = 3.41-3.90(2) Å.

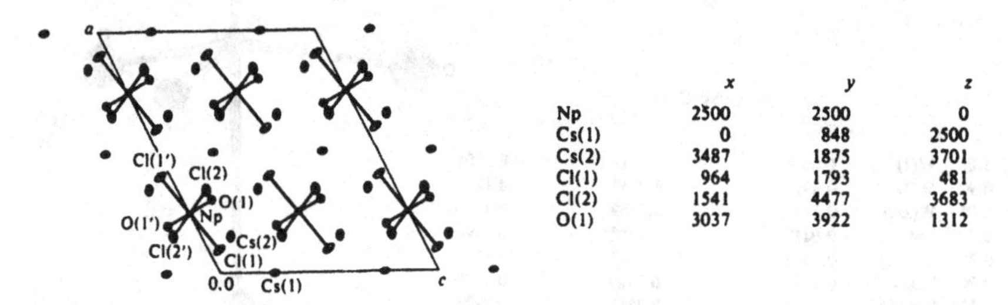

	x	y	z
Np	2500	2500	0
Cs(1)	0	848	2500
Cs(2)	3487	1875	3701
Cl(1)	964	1793	481
Cl(2)	1541	4477	3683
O(1)	3037	3922	1312

Fig. 1. Structure of $Cs_3[NpCl_4O_2]$, and atomic positional parameters (x 10^4).

1. V.A. VODOVATOV, I.N. LADYGIN, A.A. LYČEV, L.G. MAŠIROV and D.N. SUGLOBOV, 1975. Sov. Radiochem., 15, 771.

LEAD OXYHALIDES
$7PbO.MoO_3.PbBr_2$, $7PbO.WO_3.PbBr_2$, $7PbO.PO_{2.5}.PbCl_2$

B. AURIVILLIUS, 1982. Chem. Scripta, 19, 97-107.

Tetragonal, I4/mmm, a = 4.0149, 4.0158, 3.9161, c = 2.2922, 2.2765, 2.2746 Å, Z ∿ 0.9. Mo radiation, R = 0.074, 0.104, 0.107 for 644, 582, 660 reflexions. Pb(1) in 4(e); 0,0,z, z = 0.0854, 0.0849, 0.0838; Pb(2) in 4(e); z = 0.3066, 0.3071, 0.3048; Br in 2(b): 0,0,1/2; O in 8(g); 1/2,0,z, z = 0.1441, 0.1513, 0.1356 (O position half-occupied for P compound)

8 Pb (Mo, W, P), 2 Br, and 8 O were located. The structures contain PbO-like blocks with interstitial oxygen, and Br^- layers.

THALLIUM(I) LEAD(II) OXYBROMIDE
$TlPb_8O_4Br_9$

H.-L. KELLER, 1982. Z. anorg. Chem., 491, 191-198.

Tetragonal, P4/n, a = 12.337, c = 8.214 Å, Z = 2. Mo radiation, R = 0.067 for 770 reflexions.

The structure (Fig. 1) contains $Pb_8O_4{}^{8+}$ units, linked via Br^- and Tl^+ ions. Pb-O = 2.24-2.37, Pb-Br = 2.90-3.86, Tl-8Br = 3.45 Å.

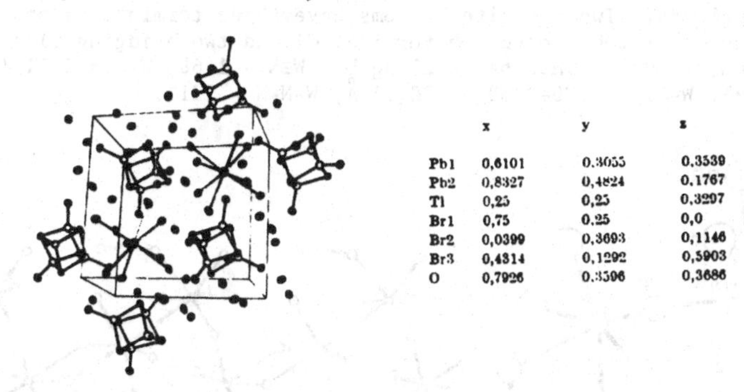

	x	y	z
Pb1	0,6101	0.3055	0.3539
Pb2	0,8327	0,4824	0.1767
Tl	0,25	0,25	0.3297
Br1	0,75	0.25	0,0
Br2	0,0399	0,3693	0,1146
Br3	0,4314	0,1292	0,5903
O	0,7926	0.3596	0,3686

Fig. 1. Structure of $TlPb_8O_4Br_9$.

INDIUM OXIDE NITRIDE FLUORIDE
$In_{32}ON_{17}F_{43}$

N. ABRIAT, J.P. LAVAL, B. FRIT and G. ROULT, 1982. Acta Cryst., B38, 1088-1093.

Cubic, Ia3, a = 10.536 Å, D_m = 6.80, Z = 1. Mo radiation, R = 0.051 for 327 reflexions, and neutron time-of-flight powder data. In(1) in 8(b): 1/4,1/4,1/4; In(2) in 24(d): 0.5346,0,1/4; 0.99 N in 16(c): 0.1342,0.1342,0.1342; 0.85 N,F,O in 48(e): 0.3550,0.0793,0.3598 (neutron data).

This fluorite-related structure contains infinite chains of corner-sharing $In(F,O)_6N_2$ distorted cubes along <111> (Fig. 1).

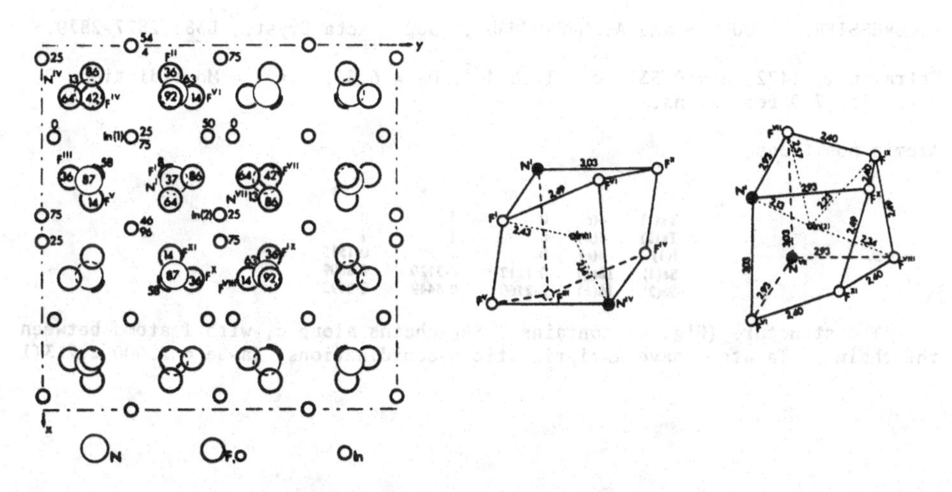

Fig. 1. Structure of $In_{32}ON_{17}F_{43}$, and indium environments.

TUNGSTEN NITRIDE CHLORIDE AZIDE
$[WNCl_3]_4.2HN_3$

I. WALKER, J. STRÄHLE, P. RUSCHKE and K. DEHNICKE, 1982. Z. anorg. Chem., 487, 26-32.

Triclinic, $P\bar{1}$, a = 7.902, b = 8.141, c = 10.161 Å, α = 107.28, β = 84.90, γ = 112.23°, Z = 1. R = 0.08 for 2811 reflexions.

The structure (Fig. 1) contains a square of four W atoms linked by nearly linear W≡N-W bridges. Two opposite W atoms have three terminal Cl and one azide ligand, and the other two W have two terminal Cl and two bridging Cl atoms which link the tetrameric units into bands along b. W≡N = 1.68, W-N = 2.11 (ring), 2.44(3) (azide), W-Cl = 2.26-2.38, 2.80(1) Å, W-N-N = 121°.

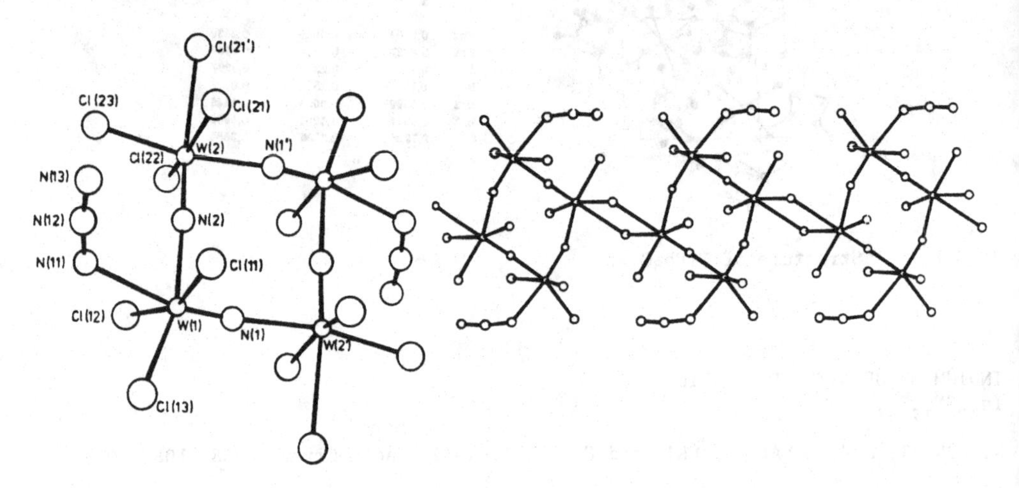

Fig. 1. Structure of $[WNCl_3]_4.2HN_3$.

TANTALUM IODIDE SELENIDE
Ta_2ISe_8

P. GRESSIER, L. GUEMAS and A. MEERSCHAUT, 1982. Acta Cryst., B38, 2877-2879.

Tetragonal, I422, a = 9.531, c = 12.824 Å, D_m = 6.34, Z = 4. Mo radiation, R = 0.057 for 730 reflexions.

Atomic positions

		x	y	z
Ta(1)	4(c)	0	½	0
Ta(2)	4(d)	0	½	¼
I(1)	4(e)	0	0	0·1553
Se(1)	16(k)	0·1212	0·3120	0·8809
Se(2)	16(k)	0·2160	0·5449	0·8693

The structure (Fig. 1) contains $TaSe_4$ chains along c, with I atoms between the chains. Ta atoms have antiprismatic 8-coordinations, Ta-Se = 2.600-2.713(1) Å.

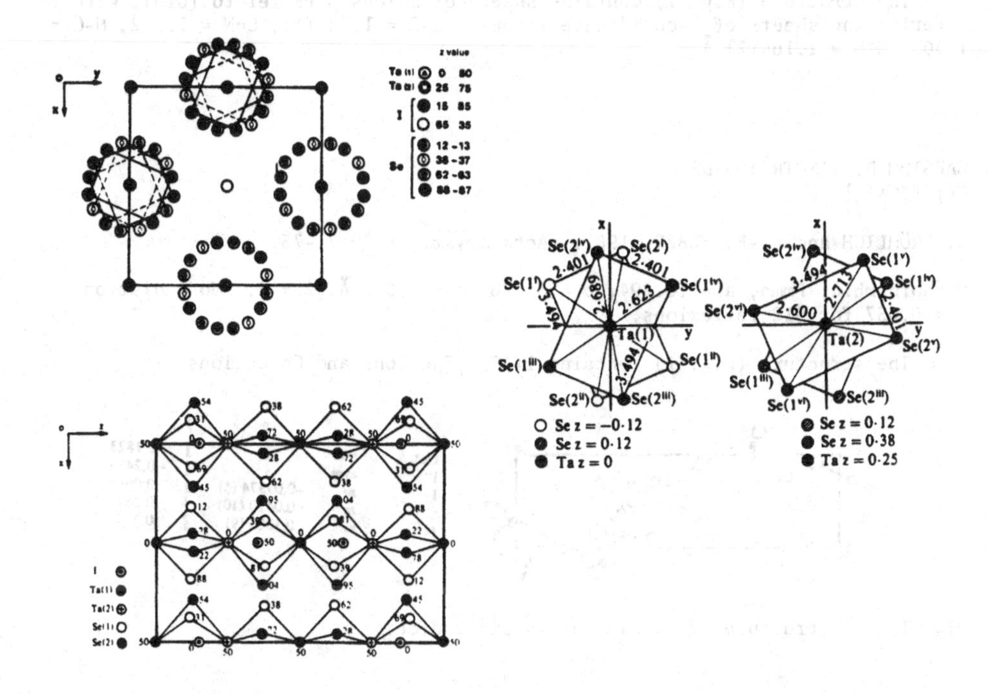

Fig. 1. Structure of Ta_2ISe_8.

POTASSIUM N-CYANODITHIOCARBIMATE MONOHYDRATE
$K_2[S_2C{=}N{-}CN].H_2O$

H. HLAWATSCHEK, M. DRÄGER and G. GATTOW, 1982. Z. anorg. Chem., 491, 145-153.

Orthorhombic, Pnma, a = 10.336, b = 7.862, c = 9.882 Å, D_m = 1.75, Z = 4. Mo radiation, R = 0.029 for 1904 reflexions.

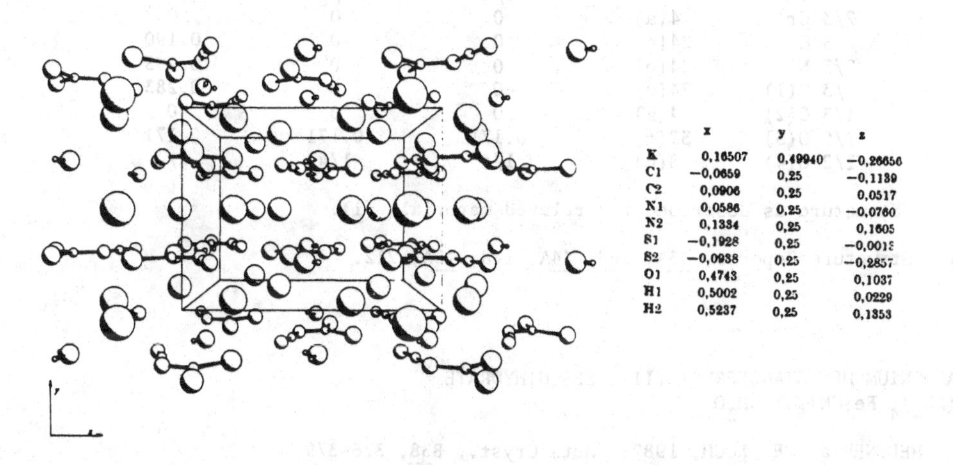

	x	y	z
K	0,16507	0,49940	−0,26656
C1	−0,0659	0,25	−0,1139
C2	0,0906	0,25	0,0517
N1	0,0586	0,25	−0,0760
N2	0,1334	0,25	0,1605
S1	−0,1928	0,25	−0,0013
S2	−0,0938	0,25	−0,2857
O1	0,4743	0,25	0,1037
H1	0,5002	0,25	0,0229
H2	0,5237	0,25	0,1353

Fig. 1. Structure of $K_2[S_2C{=}N{-}CN].H_2O$.

The structure (Fig. 1) contains sheets of anions parallel to (010), with interleaving sheets of 8-coordinate K ions. S-C = 1.724(1), C=N = 1.342, N-C = 1.305, C≡N = 1.164(2) Å

CAESIUM DICYANOTRIIODIDE
$Cs[I(ICN)_2]$

R. FRÖHLICH and K.-F. TEBBE, 1982. Acta Cryst., B38, 71-75.

Orthorhombic, Pmmn, a = 16.494, b = 6.726, c = 4.592 Å, Z = 2. Mo radiation, R = 0.057 for 512 reflexions.

The structure (Fig. 1) contains $I(ICN)_2^-$ anions and Cs cations.

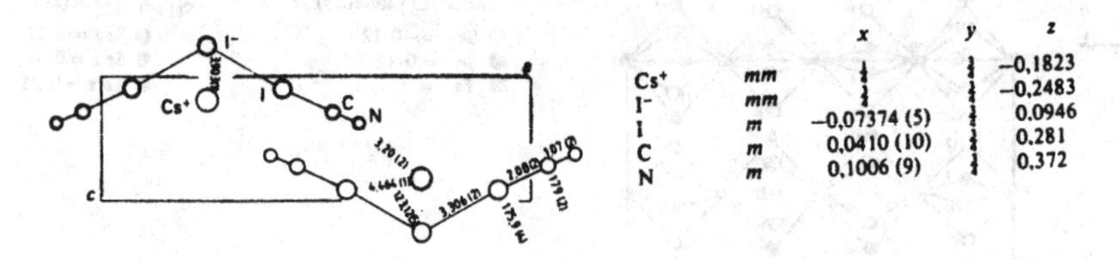

		x	y	z
Cs⁺	mm	[illegible]	[illegible]	−0,1823
I⁻	mm	[illegible]	[illegible]	−0,2483
I	m	−0,07374 (5)	[illegible]	0.0946
C	m	0,0410 (10)	[illegible]	0.281
N	m	0,1006 (9)	[illegible]	0.372

Fig. 1. Structure of caesium dicyanotriiodide.

CADMIUM HEXACYANOCHROMATE(III) TETRADECAHYDRATE
$Cd_3[Cr(CN)_6]_2.14H_2O$

W.O. MILLIGAN, D.F. MULLICA and F.W. HILLS, 1981. J. Inorg. Nucl. Chem., 43, 3119-3124.

Cubic, Fm3m, a = 10.961 Å, D_m = 1.70, Z = 1 1/3. Mo radiation, R = 0.035 for 160 reflexions.

Atomic positions

		x	y	z
Cd	in 4(b)	1/2	1/2	1/2
2/3 Cr	4(a)	0	0	0
2/3 C	24(e)	0	0	0.190
2/3 N	24(e)	0	0	0.295
1/3 O(1)	24(e)	0	0	0.283
1/3 O(2)	4(a)	0	0	0
1/6 O(3)	32(f)	0.171	0.171	0.171
1/2 O(4)	8(c)	1/4	1/4	1/4

Structure as described for related materials (1).

1. Structure Reports, 43A, 164; 44A, 178; 45A, 212.

AMMONIUM HEXACYANOFERRATE(II) SESQUIHYDRATE
$(NH_4)_4 Fe(CN)_6.1\cdot5H_2O$

E. HELLNER and E. KOCH, 1982. Acta Cryst., B38, 376-379.

Cubic, Ia3d, a = 18.261 Å, Z = 16. The structure is as previously described (1); the present paper describes the relationship with the garnet structure.

1. Structure Reports, 44A, 176.

SODIUM ZINC HEXACYANOFERRATE(II) HYDRATE
$Na_2Zn_3[Fe(CN)_6]_2.9H_2O$

E. GARNIER, P. GRAVEREAU and A. HARDY, 1982. Acta Cryst., B38, 1401-1405.

Rhombohedral, R$\bar{3}$c, a = 13.126 Å, α = 56.71°, Z = 2 (hexagonal cell, a = 12.469, c = 32.92 Å, Z = 6). Mo radiation, R = 0.025 for 294 reflexions.

Isostructural with the K salt (1). The structure (Fig. 1) contains $Fe(CN)_6$ octahedra and ZnN_4 tetrahedra linked to form a three-dimensional framework with large cavities in which the Na^+ ions and water molecules are randomly distributed. Fe-C = 1.88, C-N = 1.16, Zn-N = 1.97(1) Å.

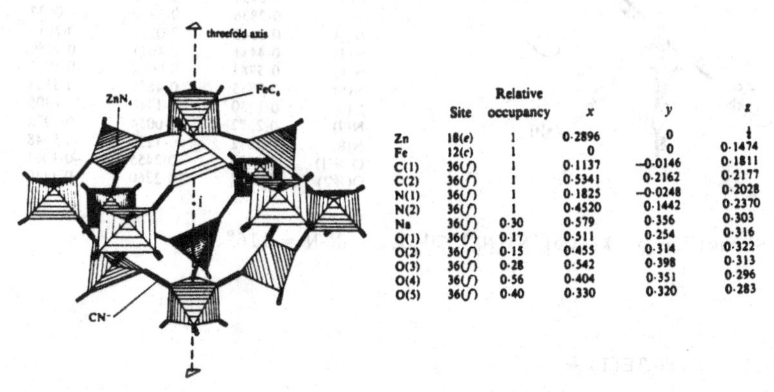

	Site	Relative occupancy	x	y	z
Zn	18(e)	1	0.2896	0	1/4
Fe	12(c)	1	0	0	0.1474
C(1)	36(f)	1	0.1137	−0.0146	0.1811
C(2)	36(f)	1	0.5341	0.2162	0.2177
N(1)	36(f)	1	0.1825	−0.0248	0.2028
N(2)	36(f)	1	0.4520	0.1442	0.2370
Na	36(f)	0.30	0.579	0.356	0.303
O(1)	36(f)	0.17	0.511	0.254	0.316
O(2)	36(f)	0.15	0.455	0.314	0.322
O(3)	36(f)	0.28	0.542	0.398	0.313
O(4)	36(f)	0.56	0.404	0.351	0.296
O(5)	36(f)	0.40	0.330	0.320	0.283

Fig. 1. Structure of sodium zinc hexacyanoferrate(II) hydrate.

1. Structure Reports, 45A, 210.

LANTHANUM HEXACYANOFERRATE(III) TETRAHYDRATE
$LaFe(CN)_6.4H_2O$

F.H. HERBSTEIN and R.E. MARSH, 1982. Acta Cryst., B38, 1051-1055.

Hexagonal, $P6_3/mmc$ (compare $P6_3/m$ in 1), a = 7.541, c = 13.955 Å, Z = 2. Reinterpretation of the data in 1, in the higher symmetry space group.

Atomic positions

			x	y	z
La	in	2(c)	1/3	2/3	1/4
Fe		2(a)	0	0	0
C		12(k)	0.120	0.240	0.079
N		12(k)	0.196	-0.196	0.119
O(1)		4(f)	1/3	2/3	0.904
2/3 O(2)		6(h)	0.525	-0.525	1/4

1. Structure Reports, 46A, 223.

POTASSIUM AZIDOPENTACYANOCOBALTATE(III)
$K_3[Co(CN)_5N_3].2H_2O$

E.E. CASTELLANO, O.E. PIRO, G. PUNTE, J.I. AMALVY, E.L. VARETTI and P.J. AYMONINO, 1982. Acta Cryst., B38, 2239-2242.

Triclinic, P$\bar{1}$, a = 8.514, b = 8.943, c = 9.397 Å, α = 81.47, β = 76.32, γ = 75.95°, D_m = 1.88, Z = 2. Mo radiation, R = 0.035 for 1184 reflexions.

The structure (Fig. 1) contains octahedral anions with a linear, asymmetric azide group, K^+ ions, and water molecules.

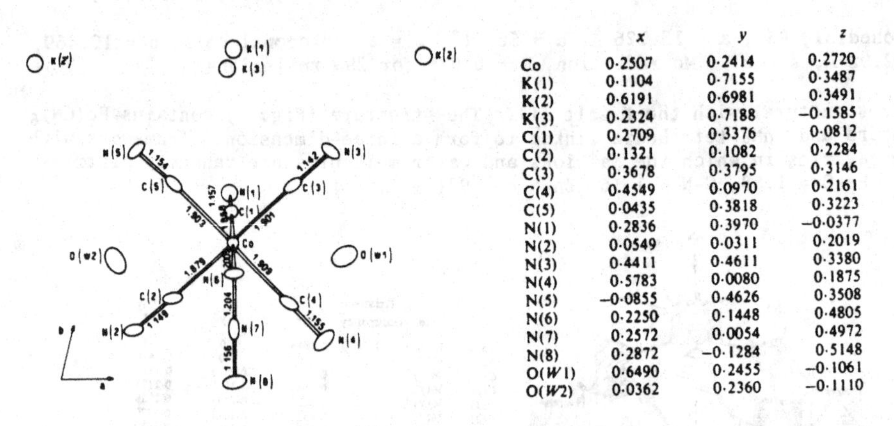

	x	y	z
Co	0·2507	0·2414	0·2720
K(1)	0·1104	0·7155	0·3487
K(2)	0·6191	0·6981	0·3491
K(3)	0·2324	0·7188	−0·1585
C(1)	0·2709	0·3376	0·0812
C(2)	0·1321	0·1082	0·2284
C(3)	0·3678	0·3795	0·3146
C(4)	0·4549	0·0970	0·2161
C(5)	0·0435	0·3818	0·3223
N(1)	0·2836	0·3970	−0·0377
N(2)	0·0549	0·0311	0·2019
N(3)	0·4411	0·4611	0·3380
N(4)	0·5783	0·0080	0·1875
N(5)	−0·0855	0·4626	0·3508
N(6)	0·2250	0·1448	0·4805
N(7)	0·2572	0·0054	0·4972
N(8)	0·2872	−0·1284	0·5148
O(W1)	0·6490	0·2455	−0·1061
O(W2)	0·0362	0·2360	−0·1110

Fig. 1. Structure of $K_3[Co(CN)_5N_3].2H_2O$; Co-N-N = 116°.

COBALT(II) DICYANOAURATE(I)
$Co[Au(CN)_2]_2$

S.C. ABRAHAMS, L.E. ZYONTZ and J.L. BERNSTEIN, 1982. J. Chem. Phys., 76, 5458-5462.

Hexagonal, $P6_422$, a = 8.434, c = 20.695 Å, D_m = 4.32, Z = 6. Mo radiation, R = 0.036 for 717 reflexions.

Atomic positions

	x	y	z
Au	0.31633	0.00058	0.12593
Co	1/2	0	0.3741
N(1)	0.1400	0.7081	0.0172
N(2)	0.4484	0.2895	0.2361
C(1)	0.2180	0.8204	0.0532
C(2)	0.3937	0.1902	0.1951

The structure (Fig. 1) contains slightly non-linear $Au(CN)_2^-$ ions, linked by tetrahedrally coordinated Co^{2+} ions. Mean Au-C = 2.00, C-N = 1.12, Co-N = 1.97(2) Å; C-Au-C = 173, Au-C-N = 171, N-Co-N = 105-113, Co-N-C = 160, 176°. The shortest Au...Au distance is 3.11 Å, compared with 2.88 Å in elemental gold.

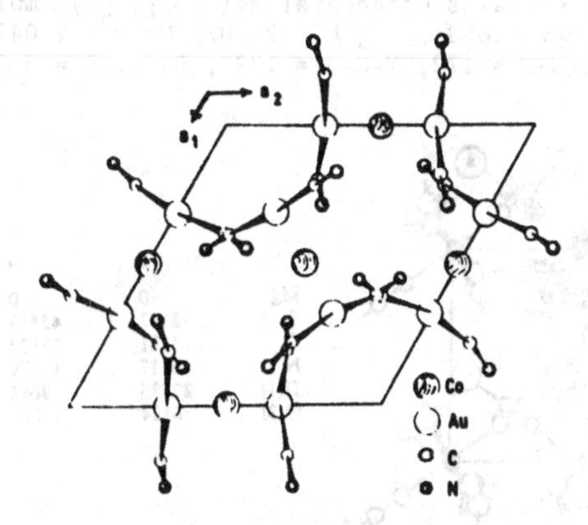

Fig. 1. Structure of $Co[Au(CN)_2]_2$.

POTASSIUM NEODYMIUM HEXACYANOFERRATE(II) TETRAHYDRATE
$KNdFe(CN)_6.4H_2O$

W.O. MILLIGAN, D.F. MULLICA and H.O. PERKINS, 1982. Inorg. Chim. Acta, 60, 35-38.

Hexagonal, $P6_3/m$, a = 7.358, c = 13.780 Å, D_m = 2.40, Z = 2. Mo radiation, R = 0.016 for 858 reflexions.

The structure (Fig. 1) contains octahedral $Fe(CN)_6^{4-}$ anions linked by $NdN_6(H_2O)_3$ polyhedra; the additional water molecule and the K^+ ion occupy holes along a threefold axis. Fe-C = 1.903, Nd-N = 2.518, Nd-O = 2.685, K-O = 2.974(3) Å.

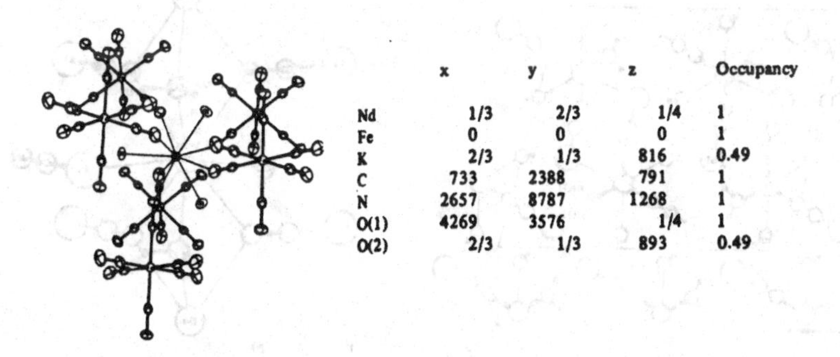

	x	y	z	Occupancy
Nd	1/3	2/3	1/4	1
Fe	0	0	0	1
K	2/3	1/3	816	0.49
C	733	2388	791	1
N	2657	8787	1268	1
O(1)	4269	3576	1/4	1
O(2)	2/3	1/3	893	0.49

Fig. 1. Structure of $KNdFe(CN)_6.4H_2O$, and atomic positional parameters (decimal fractions x 10^4).

MAGNESIUM ISOTHIOCYANATE TETRAHYDRATE
$Mg(NCS)_2.4H_2O$

K. MEREITER and A. PREISINGER, 1982. Acta Cryst., B38, 1263-1265.

Monoclinic, $P2_1/a$, a = 7.488, b = 9.030, c = 7.869 Å, β = 113.63°, Z = 2. Mo radiation, R = 0.029 for 807 reflexions.

The structure (Fig. 1) contains octahedral $Mg(NCS)_2(H_2O)_4$ molecules connected by O-H...S and O-H...O hydrogen bonds. Mg-N = 2.102, Mg-O = 2.047, 2.126, N-C = 1.153, C-S = 1.643(2) Å, Mg-N-C = 172, N-C-S = 179°, O-H...S = 3.28, 3.29, 3.29, O-H...O = 2.96 Å.

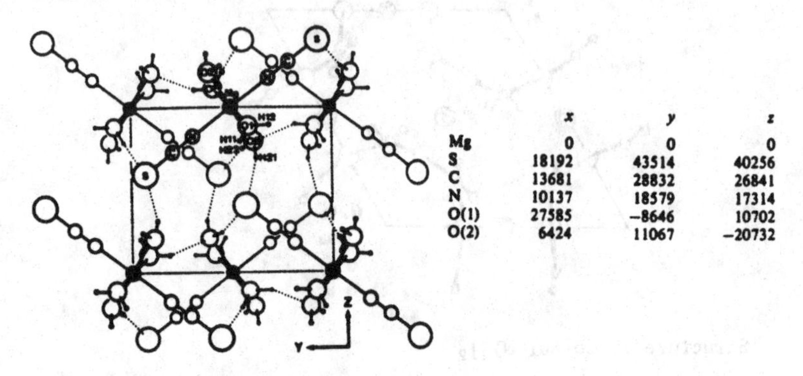

	x	y	z
Mg	0	0	0
S	18192	43514	40256
C	13681	28832	26841
N	10137	18579	17314
O(1)	27585	−8646	10702
O(2)	6424	11067	−20732

Fig. 1. Structure of magnesium isothiocyanate tetrahydrate, and atomic positional parameters (x 10^5).

BARIUM THIOCYANATE TRIHYDRATE
$Ba(SCN)_2.3H_2O$

K. MEREITER and A. PREISINGER, 1982. Acta Cryst., B38, 382-385.

Monoclinic, C2/m, a = 15.981, b = 4.441, c = 13.333 Å, β = 104.65°, Z = 4. Mo radiation, R = 0.032 for 1059 reflexions. The material was previously described as a dihydrate (1).

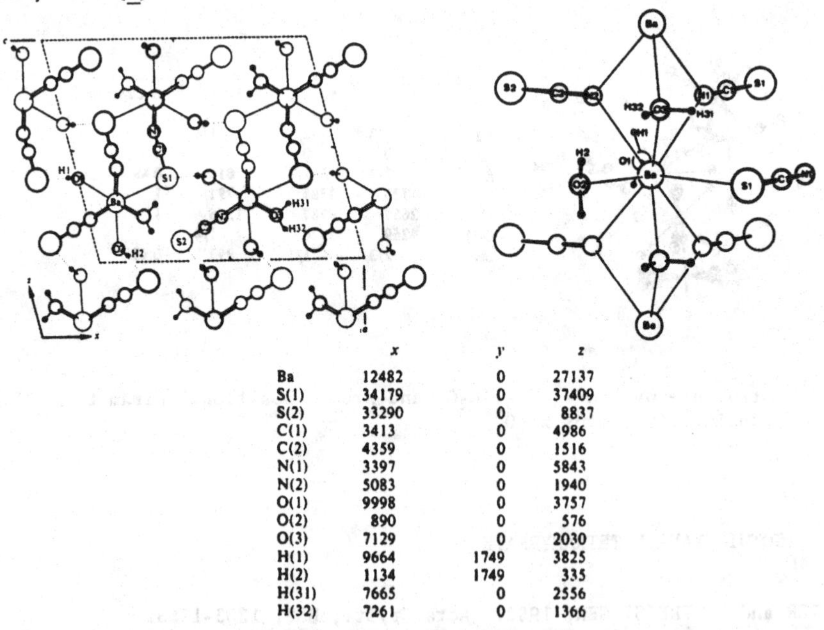

	x	y	z
Ba	12482	0	27137
S(1)	34179	0	37409
S(2)	33290	0	8837
C(1)	3413	0	4986
C(2)	4359	0	1516
N(1)	3397	0	5843
N(2)	5083	0	1940
O(1)	9998	0	3757
O(2)	890	0	576
O(3)	7129	0	2030
H(1)	9664	1749	3825
H(2)	1134	1749	335
H(31)	7665	0	2556
H(32)	7261	0	1366

Fig. 1. Structure of barium thiocyanate trihydrate, and atomic positional parameters (x 10^5 for Ba, S; x 10^4 for N, C, O, H).

The structure (Fig. 1) contains chains of 9-coordinate Ba ions (tricapped trigonal prism), linked by sharing SCN groups and by O-H...S hydrogen bonds. Ba-S = 3.391, Ba-N = 2.90, 2.91, Ba-O = 2.71, 2.76, 2.90(1), S-C = 1.65, 1.66, C-N = 1.15, O-H...S = 3.36-3.77 Å.

1. Structure Reports, 13, 290.

SODIUM TETRA(ISOTHIOCYANATO)COBALTATE(II) OCTAHYDRATE (JULIENITE)
$Na_2[Co(NCS)_4].8H_2O$

K. MEREITER and A. PREISINGER, 1982. Acta Cryst., B38, 1084-1088.

Monoclinic, $P2_1/n$, a = 18.941, b = 19.209, c = 5.460 Å, β = 91.64°, Z = 4. Mo radiation, R = 0.063 for 1022 reflexions. A previous tetragonal description (1) is incorrect.

The structure (Fig. 1) contains $Co(NCS)_4{}^{2-}$ tetrahedra and chains of edge-sharing $Na(OH_2)_6{}^+$ octahedra, with possible weak hydrogen bonding. Co-N = 1.96-1.97, Na-O = 2.36-2.48.

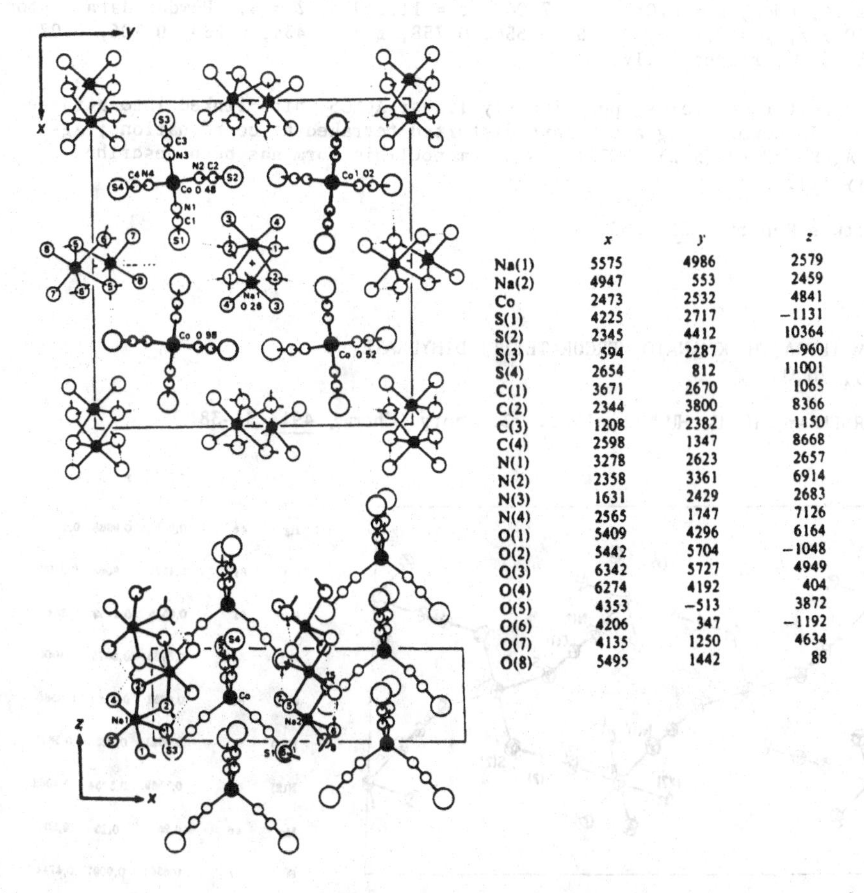

	x	y	z
Na(1)	5575	4986	2579
Na(2)	4947	553	2459
Co	2473	2532	4841
S(1)	4225	2717	−1131
S(2)	2345	4412	10364
S(3)	594	2287	−960
S(4)	2654	812	11001
C(1)	3671	2670	1065
C(2)	2344	3800	8366
C(3)	1208	2382	1150
C(4)	2598	1347	8668
N(1)	3278	2623	2657
N(2)	2358	3361	6914
N(3)	1631	2429	2683
N(4)	2565	1747	7126
O(1)	5409	4296	6164
O(2)	5442	5704	−1048
O(3)	6342	5727	4949
O(4)	6274	4192	404
O(5)	4353	−513	3872
O(6)	4206	347	−1192
O(7)	4135	1250	4634
O(8)	5495	1442	88

Fig. 1. Structure of julienite, and atomic positional parameters (x 10^4).

1. Structure Reports, 17, 462.

COPPER(I) THIOCYANATE (β-FORM, 2H POLYTYPE)
CuNCS

D.L. SMITH and V.I. SAUNDERS, 1982. Acta Cryst., B38, 907-909.

Hexagonal, $P6_3mc$, a = 3.850, c = 10.937 Å, Z = 2. Mo radiation, R = 0.059 for 87 reflexions, for crystals of the 2H polytype with some 3R polytype (1) and disorder and twinning. Atoms in 2(a): 0,0,z, z = 0, 0.1764, 0.2770, 0.4328 for Cu, N, C, S.

The structure is similar to that of the 3R polytype (1). Cu-S = 2.341(2), Cu-N = 1.93(2), S-C = 1.70(3), C-N = 1.10(3) Å.

1. Structure Reports, 48A, 191.

SILVER THIOCYANATE
β-AgSCN

D.L. SMITH, J.E. MASKASKY and L.R. SPAULDING, 1982. J. Appl. Cryst., 15, 488-492.

Orthorhombic, Pmnn, a = 4.083, b = 7.043, c = 11.219 Å, Z = 4. Powder data. Atoms in 4(g): 0,y,z, y = 0.1718, 0.525, 0.656, 0.758, z = 0.1434, 0.239, 0.106, 0.029 for Ag, S, C, N, respectively.

The structure contains approximately linear AgNCS units linked into a three-dimensional framework. Ag and S have distorted tetrahedral coordinations; Ag-N = 2.00(5), Ag-S = 2.64 (x 2), 2.71(2) Å. A monoclinic form has been described previously (1).

1. Structure Reports, 21, 263.

MAGNESIUM TETRA(THIOCYANATO)MERCURATE(II) DIHYDRATE
$MgHg(SCN)_4 \cdot 2H_2O$

K. BRODERSEN and H.-U. HUMMEL, 1982. Z. anorg. Chem., 491, 34-38.

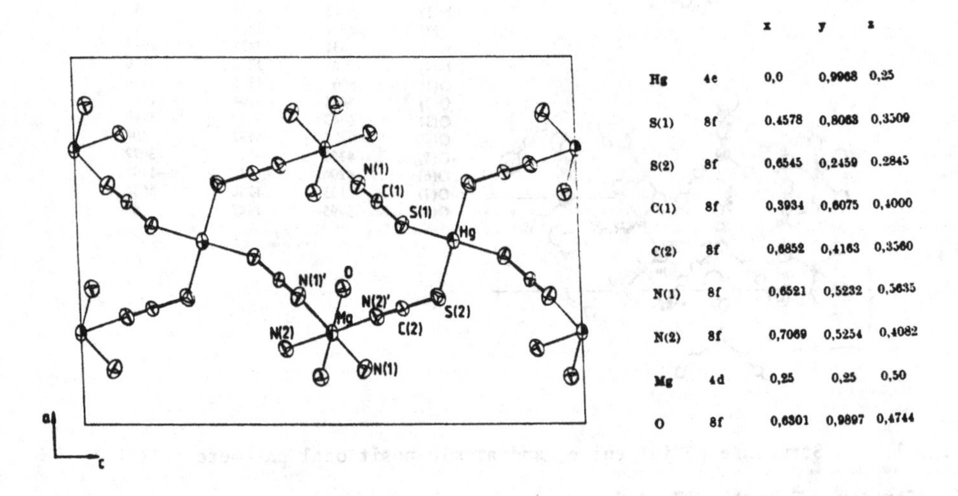

		x	y	z
Hg	4e	0,0	0,9968	0,25
S(1)	8f	0.4578	0,8068	0,3509
S(2)	8f	0,6545	0,2459	0.2845
C(1)	8f	0,3934	0,6075	0,4000
C(2)	8f	0,6852	0,4163	0,3560
N(1)	8f	0,6521	0,5232	0,3635
N(2)	8f	0,7069	0,5254	0,4082
Mg	4d	0,25	0,25	0,50
O	8f	0,6301	0,9897	0,4744

Fig. 1. Structure of $MgHg(SCN)_4 \cdot 2H_2O$.

Monoclinic, C2/c, a = 13.351, b = 5.316, c = 18.670 Å, β = 92.3°, D_m = 2.37, Z = 4. Ag radiation, R = 0.028 for 1102 reflexions.

The structure (Fig. 1) contains tetrahedral $Hg(SCN)_4$ and trans-octahedral $Mg(OH_2)_2(NCS)_4$ groups, linked into layers by thiocyanate bridges. Hg-S = 2.52, 2.58, Mg-N = 2.15, 2.16, Mg-O = 2.09, S-C = 1.66, C-N = 1.16(1) Å, S-Hg-S = 101, 116, Hg-S-C = 96, 97, S-C-N = 177, Mg-N-C = 176, 177°.

CAESIUM DIOXOPENTAKIS(THIOCYANATO)URANATE(VI)
$Cs_3[UO_2(NCS)_5]$

N.W. ALCOCK, M.M. ROBERTS and D. BROWN, 1982. Acta Cryst., B38, 2870-2872.

Orthorhombic, Pnma, a = 13.673, b = 13.283, c = 11.588 Å, Z = 4. Mo radiation, R = 0.062 for 1099 reflexions.

The structure (Fig. 1) is as previously described (1), with Cs^+ cations and distorted pentagonal bipyramidal anions. U-O = 1.77, U-$\overline{N}$ = 2.43, N-C = 1.16, C-S = 1.60(3) Å, O-U-O = 178, U-N-C = 154-173, N-C-S = 177-179°.

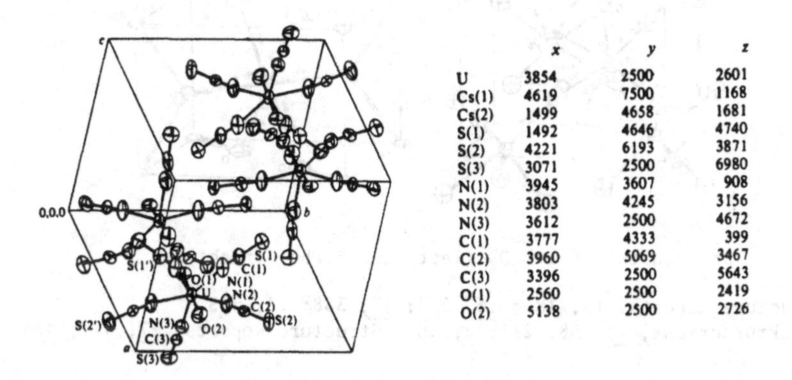

	x	y	z
U	3854	2500	2601
Cs(1)	4619	7500	1168
Cs(2)	1499	4658	1681
S(1)	1492	4646	4740
S(2)	4221	6193	3871
S(3)	3071	2500	6980
N(1)	3945	3607	908
N(2)	3803	4245	3156
N(3)	3612	2500	4672
C(1)	3777	4333	399
C(2)	3960	5069	3467
C(3)	3396	2500	5643
O(1)	2560	2500	2419
O(2)	5138	2500	2726

Fig. 1. Structure of $Cs_3[UO_2(NCS)_5]$, and atomic positional parameters (x 10^4).

1. Structure Reports, 28, 114; 31A, 119.

DINITROGEN TETROXIDE (CUBIC)
N_2O_4

Å. KVICK, R.K. McMULLAN and M.D. NEWTON, 1982. J. Chem. Phys., 76, 3754-3761.

Cubic, Im3, a = 7.694-7.793 Å, at 20-140K, Z = 6. Neutron radiation, R(F^2) = 0.028, 0.034, 0.037 for 192, 195, 197 reflexions, at 20, 60, 100K. N in 12(d): 0,0.3859,0; O in 24(g): 0,0.3260,0.1425, at 20K.

Structure as previously determined (1), with molecular dimensions (extrapolated to 0 K): N-N = 1.756(1), N-O = 1.191(1) Å, O-N-O = 134.5(1)°.

1. Structure Reports, 12, 146.

LEAD DIOXIDE
PbO_2

R.J. HILL, 1982. Mater. Res. Bull., 17, 769-784.

α-PbO_2
Orthorhombic, Pbcn, a = 4.965-4.995, b = 5.947-5.984, c = 5.466-5.485 Å, for three samples, including electrolytically-prepared material, Z = 4. X-ray and neutron powder data. 0.83-0.98 Pb in 4(c): 0,0.1669-0.1779,1/4; O in 8(d): 0.2618-0.2685,0.3960-0.4010,0.4248-0.4252.

β-PbO_2
Tetragonal, $P4_2/mnm$, a = 4.955-4.964, c = 3.382-3.387 Å, for four (including electrolytically-prepared) samples, Z = 2. X-ray and neutron powder data. 0.95-0.99 Pb in 2(a): 0,0,0; O in 4(f): 0.3054-0.3072,x,0.

The structures (Fig. 1) are as previously described (1, 2).

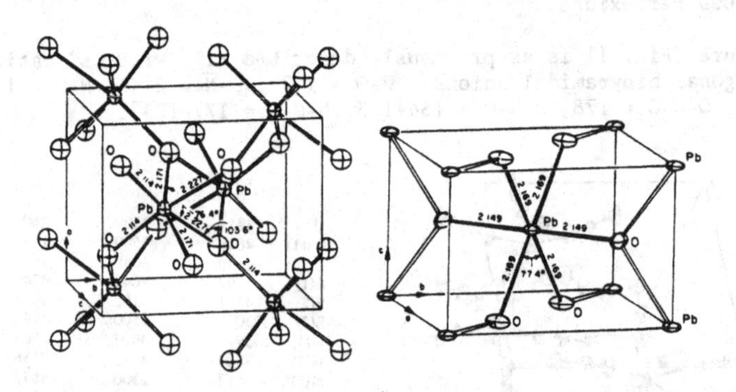

Fig. 1. Structurees of α-PbO_2 (left) and β-PbO_2 (right).

1. Structure Reports, 13, 186; 16, 224; 17, 388; 23, 345.
2. Strukturbericht, 1, 158, 211; 2, 264; Structure Reports, 27, 477; 46A, 229.

MINIUM (RED LEAD)
Pb_3O_4

TIN(IV) DILEAD(II) OXIDE
$SnPb_2O_4$

NICKEL and ZINC ANTIMONATES
MSb_2O_4, M = Ni, Zn

J.R. GAVARRI, 1982. J. Solid State Chem., 43, 12-28.

Tetragonal, $P4_2/mbc$, a $\sim$ 8, c $\sim$ 6 Å, for the four compounds at 5-300K (>170K for Pb_3O_4), Z = 4. Neutron powder data. Pb^{4+}, Sn^{4+}, Ni^{2+}, or Zn^{2+} in 4(d): 0,1/2, 1/4; Pb^{2+} or Sb^{3+} in 8(h): 0.1725,0.1631,0; O(1) in 8(g): 0.6760,0.1760,1/4; O(2) in 8(h): 0.0989,0.6383,0 [parameters for $NiSb_2O_4$ at 300K, similar values for the other compounds and temperatures].

Structures as previously described for Pb_3O_4 (1), $SnPb_2O_4$ (2, 3, where Pbam is used), and MSb_2O_4 (2, 4).

1. Structure Reports, 11, 239; 30A, 313; 41A, 212.
2. Ibid., 9, 174.
3. Ibid., 48A, 205.
4. Ibid., 48A, 209.

TITANIUM OXIDES

Ti_6O_{11}, Ti_7O_{13}, Ti_8O_{15}, Ti_9O_{17} Ti_nO_{2n-1} (n = 6-9)

I. Y. LE PAGE and P. STROBEL, 1982. J. Solid State Chem., 44, 273-281.

Triclinic, $I\bar{1}$, a = 5.552, 5.537, 5.526, 5.524, b = 7.126, 7.132, 7.133, 7.142, c = 32.233, 38.151, 44.059, 50.03 Å, α = 66.94, 66.70, 66.54, 66.41, β = 57.08, 57.12, 57.18, 57.20, γ = 108.51, 108.50, 108.51, 108.53°, Z = 2. Mo radiation, R = 0.039, 0.036, 0.030, 0.025 for 4094, 3007, 3982, 5507 reflexions.

Atomic positions

Ti_6O_{11}

	X	Y	Z
Ti(1,1)	0.01694	0.01249	0.04189
Ti(1,2)	0.04804	0.04599	0.12796
Ti(1,3)	0.08925	0.03991	0.21912
Ti(2,1)	0.01538	0.51222	0.04304
Ti(2,2)	0.04278	0.53903	0.12984
Ti(2,3)	0.09235	0.54489	0.21673
O(1,1)	−0.0195	0.6931	0.07571
O(1,2)	−0.0770	0.6704	0.17792
O(2,0)	0.0336	0.3219	0.01279
O(2,1)	0.0583	0.3383	0.10073
O(2,2)	0.0325	0.3150	0.19623
O(3,1)	0.6036	0.8195	0.05449
O(3,2)	0.6344	0.8401	0.14142
O(3,3)	0.6231	0.8426	0.23277
O(4,0)	0.4055	0.1862	0.03826
O(4,1)	0.4127	0.1885	0.12846
O(4,2)	0.4436	0.1793	0.21666

Ti_7O_{13}

	X	Y	Z
Ti(1,0)	0	0	0
Ti(1,1)	0.02289	0.01798	0.07288
ti(1,2)	0.05373	0.05037	0.14620
Ti(1,3)	0.09013	0.04081	0.22389
Ti(2,0)	0	½	0
Ti(2,1)	0.02534	0.51812	0.07388
Ti(2,2)	0.04851	0.54356	0.14797
Ti(2,3)	0.09374	0.54566	0.22168
O(1,0)	−0.0163	0.6883	0.02573
O(1,1)	−0.0101	0.6993	0.10149
O(1,2)	−0.0736	0.6742	0.18849
O(2,0)	0.0306	0.3227	0.04987
O(2,1)	0.0594	0.3402	0.12404
O(2,2)	0.0336	0.3155	0.20471
O(3,0)	0.6002	0.8098	0.00662
O(3,1)	0.6147	0.8255	0.08371
O(3,2)	0.6411	0.8441	0.15710
O(3,3)	0.6219	0.8418	0.23535
O(4,0)	0.4061	0.1927	0.07037
O(4,1)	0.4158	0.1926	0.14683
O(4,2)	0.4448	0.1816	0.22180

Ti_8O_{15}

	X	Y	Z
Ti(1,1)	0.00490	0.00533	0.032342
Ti(1,2)	0.03156	0.02307	0.095798
Ti(1,3)	0.05726	0.05485	0.159757
Ti(1,4)	0.09216	0.04119	0.227222
Ti(2,1)	0.01200	0.50575	0.032364
Ti(2,2)	0.03377	0.52580	0.096572
Ti(2,3)	0.05330	0.54622	0.161410
Ti(2,4)	0.09541	0.54753	0.225196
O(1,1)	0.9921	0.6946	0.05492
O(1,2)	0.9960	0.7047	0.12063
O(1,3)	0.9271	0.6759	0.19678
O(2,0)	0.0128	0.3120	0.01151
O(2,1)	0.0328	0.3263	0.07661
O(2,2)	0.0627	0.3430	0.14043
O(2,3)	0.0346	0.3147	0.21074
O(3,1)	0.6108	0.8155	0.03858
O(3,2)	0.6231	0.8305	0.10511
O(3,3)	0.6430	0.8459	0.16930
O(3,4)	0.6215	0.8429	0.23719
O(4,0)	0.4016	0.1970	0.02702
O(4,1)	0.4108	0.1977	0.09383
O(4,2)	0.4199	0.1960	0.16018
O(4,3)	0.4458	0.1826	0.22525

Ti_9O_{17}

	X	Y	Z
Ti(1,0)	0	0	0
Ti(1,1)	0.01284	0.01106	0.057343
Ti(1,2)	0.03725	0.02811	0.113554
Ti(1,3)	0.06027	0.05701	0.170154
Ti(1,4)	0.09266	0.04166	0.229844
Ti(2,0)	0	½	0
Ti(2,1)	0.02191	0.51380	0.057078
Ti(2,2)	0.03915	0.52960	0.114287
Ti(2,3)	0.05638	0.54882	0.171596
Ti(2,4)	0.09580	0.54775	0.228114
O(1,0)	0.9956	0.69378	0.01865
O(1,1)	0.9996	0.70035	0.07726
O(1,2)	0.0026	0.70937	0.13528
O(1,3)	0.9292	0.67779	0.20287
O(2,0)	0.0144	0.31481	0.03964
O(2,1)	0.0369	0.33014	0.09674
O(2,2)	0.0655	0.34573	0.15317
O(2,3)	0.0366	0.31665	0.21516
O(3,0)	0.6069	0.80771	0.00511
O(3,1)	0.6180	0.81943	0.06302
O(3,2)	0.6283	0.83343	0.12169
O(3,3)	0.6456	0.84786	0.17851
O(3,4)	0.6226	0.84322	0.23860
O(4,0)	0.4057	0.20318	0.05274
O(4,1)	0.4148	0.20198	0.11185
O(4,2)	0.4225	0.19848	0.17058
O(4,3)	0.4479	0.18366	0.22803

The structures are essentially as previously described (1), with layers of edge- and corner-sharing TiO_6 octahedra parallel to (101), with the layers linked by further corner sharing.

γ-Ti_3O_5

II. S.-H. HONG and S. ÅSBRINK, 1982. Acta Cryst., B38, 2570-2576.

Monoclinic, I2/c, a = 9.9701, b = 5.0747, c = 7.1810 Å, β = 109.865°, Z = 4. Mo radiation, R = 0.019 for 1595 reflexions.

The structure (Fig. 1) can be derived from that of rutile by a crystallographic shear. Two TiO_6 octahedra, one at and one between the shear planes, form two infinite chains, one by edge- and face-sharing, and the other by corner-sharing; chains are connected by sharing edges and corners. Ti-O = 1.881-2.174(1) Å.

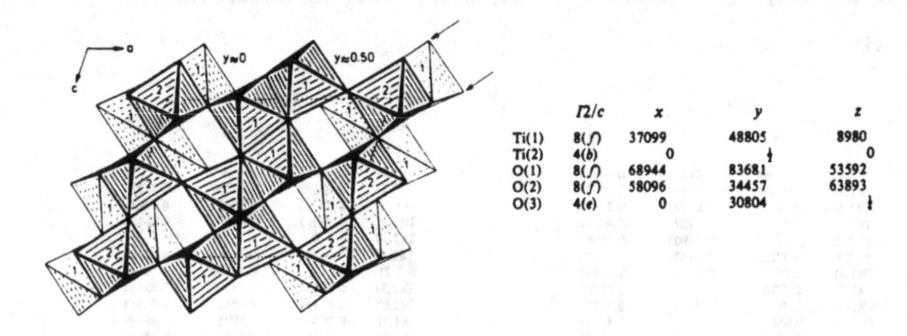

	I2/c	x	y	z
Ti(1)	8(f)	37099	48805	8980
Ti(2)	4(b)	0	½	0
O(1)	8(f)	68944	83681	53592
O(2)	8(f)	58096	34457	63893
O(3)	4(e)	0	30804	¼

Fig. 1. Structure of γ-Ti_3O_5, and atomic positional parameters ($x\ 10^5$).

1. Structure Reports, 24, 311; 37A, 222; 39A, 208.

RUTILE
TiO_2

I. T.M. SABINE and C.J. HOWARD, 1982. Acta Cryst., B38, 701-702.
II. W. GONSCHOREK, 1982. Z. Kristallogr., 160, 187-203.

Tetragonal, $P4_2/mnm$, a = 4.5922, 4.5937, c = 2.9590, 2.9587 Å, Z = 2. Neutron powder data (I); Ag radiation, R = 0.02 for 249 reflexions (II). Ti in 2(a): 0,0,0; O in 4(f): x,x,0, x = 0.3051, 0.3049.

Structure as previously described (1). Charge density analysis indicates covalent character in the bonding (II).

1. Strukturbericht, 1, 155, 204; Structure Reports, 17, 387; 19, 361; 20, 263; 37A, 222; 41A, 402; 45A, 218.

VANADIUM OXIDE (TITANIUM AND CHROMIUM DOPED)
$(V,Ti)_2O_3$, $(V,Cr)_2O_3$

S. CHEN, J.E. HAHN, C.E. RICE and W.R. ROBINSON, 1982. J. Solid State Chem., 44, 192-200.

Rhombohedral, $R\bar{3}c$, a = 4.953, c = 14.006 Å (for V_2O_3; a increases and c decreases with increasing Ti or Cr content), Z = 6. Mo radiation, R = 0.018-0.048 for 98-2554 reflexions. M in 12(c): 0,0,z, z = 0.3463-0.3487; O in 18(e):x,0,1/4, x = 0.3122-0.3073.

α-Corundum structure (1), as previously reported (2).

1. Strukturbericht, 1, 240.
2. Strukturbericht, 1, 242, 264; 2, 310, 314; Structure Reports, 27, 498; 35A, 205; 41A, 216; 46A, 397.

VANADIUM OXIDE (HIGH-TEMPERATURE)
V_3O_5

S.-H. HONG and S. ÅSBRINK, 1982. Acta Cryst., B38, 713-719.

Monoclinic, I2/c, a = 9.846, b = 5.027, c = 7.009 Å, β = 109.54°, at 458K, Z = 4. Mo radiation, R = 0.026 for 1361 reflexions.

As for the low-temperature form (1), the material has a shear structure derived from the rutile structure by crystallographic shear $(1\bar{2}1)$. The structure (Fig. 1) contains two types of octahedral chains; one type contains pairs of face-sharing octahedra linked by further edge sharing, and the other type contains single octahedra sharing corners. The two types of chain are linked together by edge and corner sharing. Bond distances (Fig. 1) suggest that V(1) = 0.5 V(III) + 0.5 V(IV), V(2) = V(III).

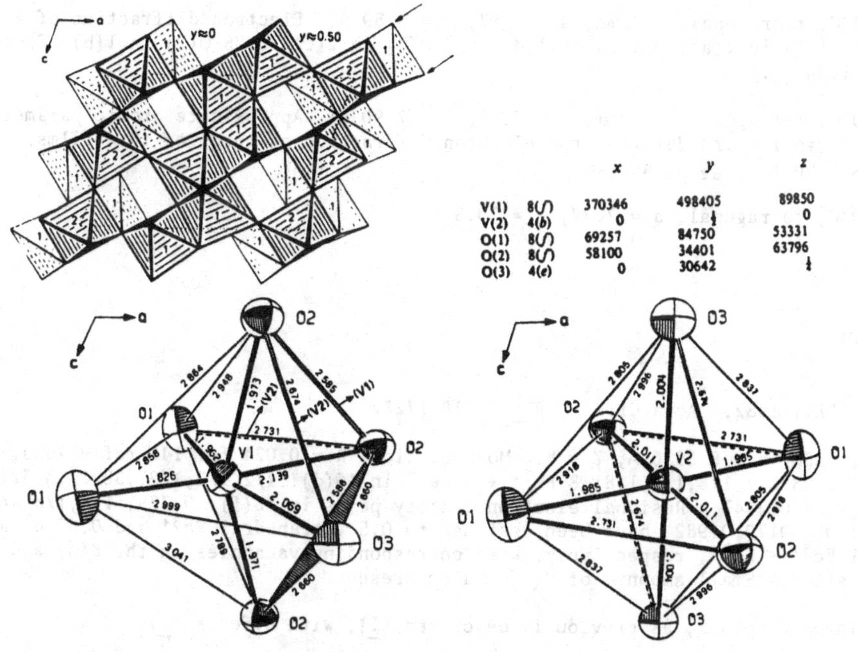

		x	y	z
V(1)	8(f)	370346	498405	89850
V(2)	4(b)	0	½	0
O(1)	8(f)	69257	84750	53331
O(2)	8(f)	58100	34401	63796
O(3)	4(e)	0	30642	¼

Fig. 1. Structure of high-temperature V_3O_5, and atomic positional parameters (x 10^6 for V, x 10^5 for O).

1. Structure Reports, 46A, 231.

NIOBIUM(IV) OXIDE
β-NbO_2

H.-J. SCHWEIZER and R. GRUEHN, 1982. Z. Naturforsch., 37B, 1361-1368.

Tetragonal, $I4_1$, a = 9.693, c = 5.985 Å, Z = 16. R = 0.10 for 859 reflexions.

Atomic positions

	x	y	z
Nb(1)	0.2621	0.5076	0.0299
Nb(2)	0.2455	0.4909	0.4804
O(1)	0.3974	0.1497	0.0077
O(2)	0.1102	0.3642	-0.0089
O(3)	0.6086	0.3609	0.5064
O(4)	0.6500	0.1038	0.2484

Deformed rutile structure, with pairs of edge-sharing NbO_6 octahedra as in α-NbO_2 (1). Nb-O = 1.90-2.15, Nb...Nb = 2.71 Å.

1. Structure Reports, 27, 479; 28, 122; 42A, 236; 43A, 175; 45A, 218.

TANTALUM OXIDES
TaO_x

V.I. KHITROVA and V.V. KLEČKOVSKAJA, 1982. Kristallografija, 27, 736-741 [Soviet Physics - Crystallography, 27, 441-444].

At 600°C, tetragonal, P4/mmm, a = 3.87, c = 3.89 Å. Electron diffraction of thin films. 1 Ta in 1(a); 0.1 Ta in 1(d); 0.4 O(1) in 2(f); 0.25 O(2) in 1(b). Composition $TaO_{0.95}$.

At 750°C, tetragonal, P4/mmm, a = 7.83, c = 7.90 Å. Approximate atomic parameters and occupancies are derived from electron diffraction patterns of thin films. Composition $TaO_{1.62} \sim Ta_3O_5$.

At 850°C, tetragonal, a = 7.72, c = 30.5 Å.

MAGNETITE
Fe_3O_4

M.E. FLEET, 1982. Acta Cryst., B38, 1718-1723.

Cubic, Fd3m, a = 8.3930 Å, Z = 8. Mo radiation, R = 0.024 for 145 reflexions. 8 Fe^{3+} in 8(a): 1/8,1/8,1/8; 8 Fe^{3+} + 8 Fe^{2+} in 16(d): 1/2,1/2,1/2; 32 O in 32(e): x,x,x, x = 0.2547. Residual electron density peaks in 48(f): 0.375, 1/8,1/8 and 96(h): 0,0.0179,0.9821 have been assigned to 0.5 tetrahedral Fe^{3+} and 0.67 octahedral Fe^{3+} + Fe^{2+}, respectively, with corresponding vacancies in the 8(a) and 16(d) sites. Small amounts of Mg are also present.

Inverse spinel, as previously described (1), with defects (2).

1. Strukturberucht, 1, 352; Structure Reports, 22, 300; 33A, 267.
2. Structure Reports, 48A, 195.

MAGNETITE (LOW-TEMPERATURE)
Fe_3O_4

M. IIZUMI, T.F. KOETZLE, G. SHIRANE, S. CHIKAZUMI, M. MATSUI and S. TODO, 1982. Acta Cryst., B38, 2121-2133.

Monoclinic, Cc, a = 11.868, b = 11.851, c = 16.752 Å, β = 90.20°, at 10K (√2a x √2a x 2a of Fd3m spinel cell), Z = 32. Neutron radiation, R = 0.055 for 573 reflexions of an orthorhombic subcell (a/√2 x a/√2 x 2a, Pmca or $Pmc2_1$).

The structure exhibits slight displacements of atoms from the room-temperature cubic spinel positions.

NEODYMIUM OXIDE
Nd_2O_3

NEODYMIUM OXYSULPHIDE
Nd_2O_2S

M. FAUCHER, J. PANNETIER, Y. CHARREIRE and P. CARO, 1982. Acta Cryst., B38, 344-346.

Trigonal, P$\bar{3}$m1, a = 3.827, 3.946, c = 5.991, 6.780 Å, at 4K, Z = 1. Neutron powder data. Nd in 2(d): 1/3,2/3,z, z = 0.2473, 0.2805; O(1) in 2(d): z = 0.6464, 0.6288; O(2) or S in 1(a): 0,0,0.

Structure as previously described (1, 2).

1. Strukturbericht, 1, 244, 261, 745; Structure Reports, 41A, 220.
2. Structure Reports, 12, 174; 39A, 260.

POTASSIUM SODIUM OXIDE
KNaO

RUBIDIUM SODIUM OXIDE
RbNaO

H. SABROWSKY and U. SCHRÖER, 1982. Z. Naturforsch., 37B, 818-819.

Tetragonal, P4/nmm, a = 4.002, 4.093, c = 6.214, 6.531 Å, Z = 2. Mo radiation, R = 0.019 for 159 reflexions for KNaO. Na in 2(a): 0,0,0; K, O in 2(c): 0,1/2,z, z = 0.3638, 0.7938. RbNaO is isostructural.

Anti-PbFCl structure (1).

1. Strukturbericht, 2, 45.

β-ALUMINA (LITHIUM, HYDRATED)
$Li_2Al_{22}O_{34}.1.55H_2O$

J.B. BATES, N.J. DUDNEY, G.M. BROWN, J.C. WANG and R. FRECH, 1982. J. Chem. Phys., 77, 4838-4856.

Hexagonal, $P6_3/mmc$, a = 5.591, c = 22.715 Å, Z = 1. Neutron radiation, R = 0.060 for 674 reflexions.

β-Alumina structure (1), with water oxygen atoms located in the conduction plane and the hydrogen atoms above and below the plane, hydrogen bonded to O(4).

1. Strukturbericht, 5, 72; Structure Reports, 33A, 275; 37A, 226.

INDIUM β-ALUMINA
$In_{2.47}Al_{22}O_{34}$

M.G. PITT and D.J. FRAY, 1982. J. Solid State Chem., 43, 227-236.

Hexagonal, $P6_3/mmc$, a = 5.599, c = 22.901 Å, Z = 1. Mo radiation, R = 0.04 for 446 reflexions.

Atomic positions

			x	y	z
1.18	In(BR) in	2(d)	2/3	1/3	1/4
0.66	In(BR')	6(h)	0.6949	2x	1/4
0.29	In(aBR)	2(b)	0	0	1/4
0.35	In(aBR')	6(h)	0.9324	2x	1/4
12	Al(1)	12(k)	0.8321	2x	0.1046
4	Al(2)	4(f)	1/3	2/3	0.0245
4	Al(3)	4(f)	1/3	2/3	0.1750
2	Al(4)	2(a)	0	0	0
12	O(1)	12(k)	0.1572	2x	0.0491
12	O(2)	12(k)	0.4978	2x	0.1448
4	O(3)	4(f)	2/3	1/3	0.0546
4	O(4)	4(e)	0	0	0.1393
2	O(5)	2(c)	1/3	2/3	1/4

β-Alumina structure (1).

1. Structure Reports, 37A, 226; 42A, 239; 43A, 179.

STRONTIUM ALUMINATE HYDRATE
$Sr_{7.5}Al_{12}O_{24}(OH)_3$ 1.5 x [$5SrO.4Al_2O_3.H_2O$]

L.S. DENT GLASSER, A.P. HENDERSON and R.A. HOWIE, 1982. Acta Cryst., B38, 24-27.

Rhombohedral, $R\bar{3}$, a = 17.91, c = 7.16 Å, Z = 3. Mo radiation, R = 0.064 for 1163 reflexions.

The structure differs from a previous description (1). It contains a framework of AlO_4 tetrahedra sharing all corners (Fig. 1), with disordered Sr and OH in cavities. Al-O = 1.74-1.78(1), Sr-O = 2.44-2.85 Å.

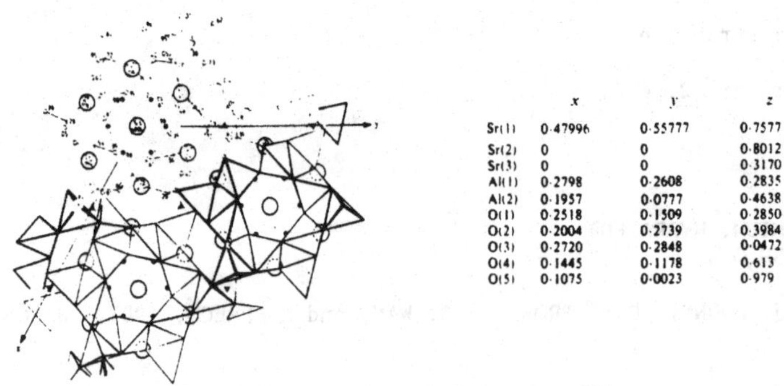

	x	y	z
Sr(1)	0·47996	0·55777	0·7577
Sr(2)	0	0	0·8012
Sr(3)	0	0	0·3170
Al(1)	0·2798	0·2608	0·2835
Al(2)	0·1957	0·0777	0·4638
O(1)	0·2518	0·1509	0·2850
O(2)	0·2004	0·2739	0·3984
O(3)	0·2720	0·2848	0·0472
O(4)	0·1445	0·1178	0·613
O(5)	0·1075	0·0023	0·979

Fig. 1. Structure of strontium aluminate hydrate, and atomic positional parameters; occupancy = 0.5 for Sr(2), 0.25 for Sr(3), [0.5 for O(5)].

1. Structure Reports, 45A, 221.

STRONTIUM TETRAALUMINATE
β-$SrAl_4O_7$

K.-I. MACHIDA, G.-Y. ADACHI, J. SHIOKAWA, M. SHIMADA and M. KOIZUMI, 1982. Acta Cryst., B38, 889-891.

Orthorhombic, Cmma, a = 8.085, b = 11.845, c = 4.407 Å, D_m = 4.80, Z = 4. Mo radiation, R = 0.047 for 324 reflexions; material obtained at high pressure.

The structure (Fig. 1) contains a framework of corner-sharing AlO_6 octahedra and AlO_4 tetrahedra, which is denser than the framework of the α-form (1). Sr ions are located in tunnels along a, and have 10-coordination (compare the 7-coordination in the α-form, 1). Al-O(octahedral) = 1.80 (x 2), 1.97 (x 4), Al-O (tetrahedral) = 1.45-1.54 Å (the tetrahedral distances are very short); Sr-O = 2.50-2.66(1) Å.

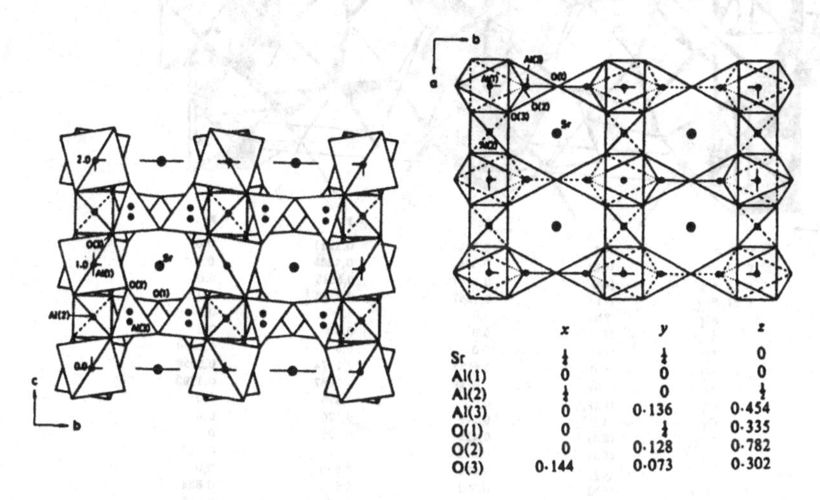

	x	y	z
Sr	¼	¼	0
Al(1)	0	0	0
Al(2)	¼	0	½
Al(3)	0	0·136	0·454
O(1)	0	¼	0·335
O(2)	0	0·128	0·782
O(3)	0·144	0·073	0·302

Fig. 1. Structure of β-$SrAl_4O_7$.

1. Structure Reports, 22, 305; 38A, 247.

LEAD(II) DIALUMINATE
$PbAl_2O_4$

LEAD(II) DIGALLATE
$PbGa_2O_4$

K.-B. PLÖTZ and H. MÜLLER-BUSCHBAUM, 1982. Z. anorg. Chem., 488, 38-44.

Triclinic, P1, a = 5.268, 5.387, b = 8.458, 8.575, c = 5.070, 5.220 Å, α = 90.0, 90.0, β = 118.78, 118.99, γ = 90.0, 90.0°, Z = 2. R = 0.047, 0.074 for 3967, 3715 reflexions.

Stuffed tridymite structures, with AlO_4 and GaO_4 tetrahedra, and Pb ions with [3 (trigonal pyramid) + 2]-coordinations. Al-O = 1.74-1.78, Ga-O = 1.81-1.87, Pb-O = 2.26-2.41, 2.74-2.81 Å.

STRONTIUM LEAD ALUMINATE
$Sr_{1.33}Pb_{0.67}Al_6O_{11}$

K.-B. PLÖTZ and H. MÜLLER-BUSCHBAUM, 1982. Z. anorg. Chem., 491, 253-258.

Orthorhombic, Pnnm, a = 22.129, b = 4.876, c = 8.417 Å, Z = 4. R = 0.085 for 1075 reflexions.

The structure (Fig. 1) contains a framework of interconnected AlO_6 octahedra and AlO_4 tetrahedra, with Sr and Pb statistically occupying cavities in this framework. Al-O = 1.88-2.01 (octahedral), 1.73-1.80 (tetrahedral), Sr(1)-O = 2.57-3.32, Pb(1)-O = 2.39-2.45 (3 distances), 2.68-3.24 (6 distances), Sr/Pb(2)-10 O = 2.56-3.15 Å.

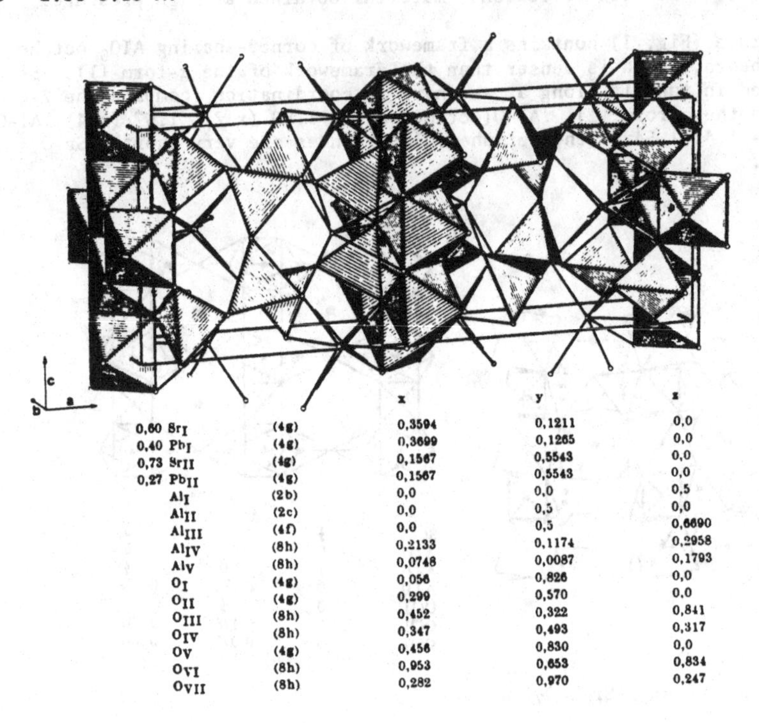

		x	y	z
0,60 Sr_I	(4g)	0,3594	0,1211	0,0
0,40 Pb_I	(4g)	0,3699	0,1265	0,0
0,73 Sr_{II}	(4g)	0,1567	0,5543	0,0
0,27 Pb_{II}	(4g)	0,1567	0,5543	0,0
Al_I	(2b)	0,0	0,0	0,5
Al_{II}	(2c)	0,0	0,5	0,0
Al_{III}	(4f)	0,0	0,5	0,6690
Al_{IV}	(8h)	0,2133	0,1174	0,2958
Al_V	(8h)	0,0748	0,0087	0,1793
O_I	(4g)	0,056	0,826	0,0
O_{II}	(4g)	0,299	0,570	0,0
O_{III}	(8h)	0,452	0,322	0,841
O_{IV}	(8h)	0,347	0,493	0,317
O_V	(4g)	0,456	0,830	0,0
O_{VI}	(8h)	0,953	0,653	0,834
O_{VII}	(8h)	0,282	0,970	0,247

Fig. 1. Structure of strontium lead aluminate.

BARIUM TITANIUM ALUMINATE
$Ba_3TiAl_{10}O_{20}$

M.C. CADÉE, D.J.W. IJDO and G. BLASSE, 1982. J. Solid State Chem., 41, 39-43.

Monoclinic, I2/m, a = 14.888, b = 11.363, c = 4.978 Å, β = 90.82°, Z = 2. Neutron powder data.

Atomic positions

	x	y	z
Ba(1)	0	0	0
Ba(2)	0.2773	0	0.0320
(Ti,Al)	0	0.5	0
(Ti,Al)	0	0.6368	0.5
Al(3)	0.3529	0.3607	0.0116
Al(4)	0.1375	0.2906	0.9768
O(1)	0.4372	0	0.8232
O(2)	0.8952	0	0.4190
O(3)	0.2381	0.3561	0.9359
O(4)	0.4139	0.2441	0.8487
O(5)	0.8608	0.1517	0.8658
O(6)	0.9309	0.3801	0.8216

Isostructural with $Pb_3GeAl_{10}O_{20}$ (1), except for more-regular Ba(2) coordination. The structure contains layers with rings of six corner-sharing AlO_4 tetrahedra, linked by Ba(1) with octahedral coordination, Ba(2) with monocapped tetragonal-prismatic 9-coordination, and edge-sharing $(Ti,Al)O_6$ octahedra. Al-O = 1.69-1.83(1), Ba-O = 2.61-3.12(1), (Ti,Al)-O = 1.87-2.01(2) Å.

1. Structure Reports, 35A, 263.

HÖGBOMITE (8H)

$Al_{14.6}(Fe,Mg,Ti,Zn)_{7.4}O_{30}(OH)_2$ (idealized)

B.M. GATEHOUSE and I.E. GREY, 1982. Amer. Min., 67, 373-380.

Hexagonal, $P6_3mc$, a = 5.734, c = 18.389 Å, Z = 1. Mo radiation, R = 0.031 for 1425 reflexions.

The structure (Fig. 1), which is related to that of nigerite-24R (1), is based on a close-packed anion framework with an 8-layer mixed stacking sequence along c, ABCABACB (cchcccch), with cations ordered into 6 tetrahedral and 16 octahedral sites per unit cell.

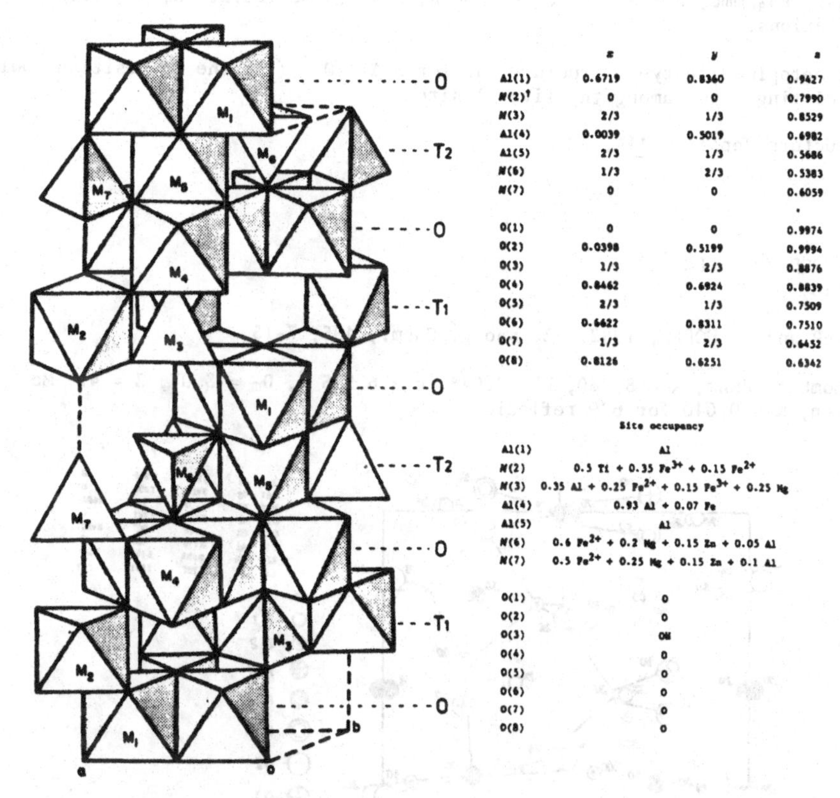

	x	y	z
Al(1)	0.6719	0.8360	0.9427
M(2)†	0	0	0.7990
M(3)	2/3	1/3	0.8529
Al(4)	0.0039	0.5019	0.6982
Al(5)	2/3	1/3	0.5686
M(6)	1/3	2/3	0.5383
M(7)	0	0	0.6059
O(1)	0	0	0.9974
O(2)	0.0398	0.5199	0.9994
O(3)	1/3	2/3	0.8876
O(4)	0.8462	0.6924	0.8839
O(5)	2/3	1/3	0.7509
O(6)	0.6622	0.8311	0.7510
O(7)	1/3	2/3	0.6452
O(8)	0.8126	0.6251	0.6342

Site occupancy

Al(1)	Al
M(2)	0.5 Ti + 0.35 Fe^{3+} + 0.15 Fe^{2+}
M(3)	0.35 Al + 0.25 Fe^{2+} + 0.15 Fe^{3+} + 0.25 Mg
Al(4)	0.93 Al + 0.07 Fe
Al(5)	Al
M(6)	0.6 Fe^{2+} + 0.2 Mg + 0.15 Zn + 0.05 Al
M(7)	0.5 Fe^{2+} + 0.25 Mg + 0.15 Zn + 0.1 Al
O(1)	O
O(2)	O
O(3)	OH
O(4)	O
O(5)	O
O(6)	O
O(7)	O
O(8)	O

Fig. 1. Structure of högbomite-8H.

1. Structure Reports, 45A, 223.

COPPER(I) ALUMINATE
$CuAlO_2$

T. ISHIGURO, N. ISHIZAWA, N. MIZUTANI and M. KATO, 1982. J. Solid State Chem., 41, 132-137.

Rhombohedral, $R\bar{3}m$, a = 2.8584, c = 16.958 Å, at 295K, Z = 3. Mo radiation, R = 0.021-0.036 for 161-189 reflexions, at six temperatures from 295-1200K. Delafossite-type structure, as previously described (1); z(O) = 0.1098. Thermal expansion along a is about three times that along c.

1. Structure Reports, 48A, 197.

LANTHANUM MAGNESIUM ALUMINATE
$LaMgAl_{11}O_{19}$

A. KAHN, A.M. LEJUS, M. MADSAC, J. THÉRY, D. VIVIEN and J.C. BERNIER, 1981. J. Appl. Phys., 52, 6864-6869.

Hexagonal, $P6_3/mmc$, a = 5.588, c = 22.00 Å, Z = 2. Mo radiation, R = 0.039 for 487 reflexions.

Magnetoplumbite-type structure, as for $SrAl_{12}O_{19}$ (1); the Mg position could not be distinguished among the five Al sites.

1. Structure Reports, 41A, 222.

LITHIUM SODIUM GALLATE
$Li_3Na_2GaO_4$

J. KÖHLER and R. HOPPE, 1982. Z. anorg. Chem., 495, 7-15.

Orthorhombic, Pnnm, a = 8.260, b = 7.946, c = 6.515 Å, D_m = 3.07, Z = 4. Mo radiation, R = 0.040 for 679 reflexions.

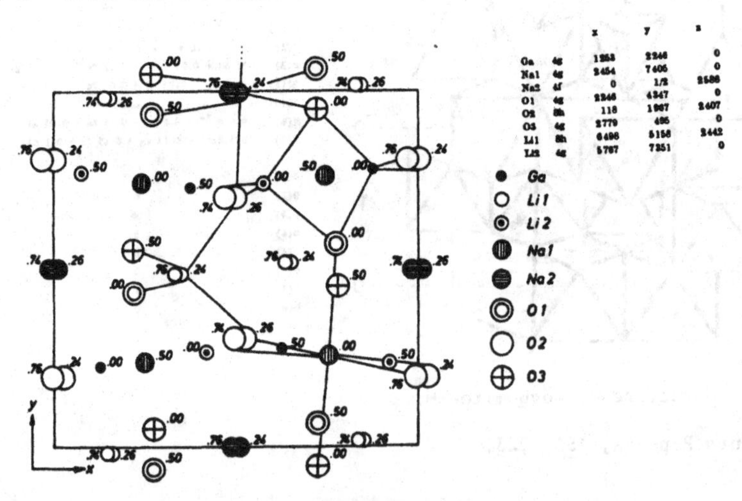

		x	y	z
Ga	4g	1253	2246	0
Na1	4g	2454	7606	0
Na2	4f	0	1/2	2586
O1	4g	2246	4347	0
O2	8h	118	1967	2407
O3	4g	2779	495	0
Li1	8h	6496	5158	2442
Li2	4g	5767	7351	0

Fig. 1. Structure of lithium sodium gallate, and atomic positional parameters (x 10^4).

The structure (Fig. 1) contains GaO_4 tetrahedra linked by 4-coordinate Li and 8- and 5-coordinate Na ions. Ga-O = 1.84-1.88, Li-O = 1.87-2.11, Na-O = 2.41-2.69 Å.

GERMANIUM SODALITE
$Na_8Al_6Ge_6O_{24}(OH)_2$

E.L. BELOKONEVA, L.N. DEM'JANEC, T.G. UVAROVA and N.V. BELOV, 1982. Kristallografija, 27, 995-996 [Soviet Physics - Crystallography, 27, 597-598].

Cubic, $P\bar{4}3n$, a = 9.029 Å, Z = 1. Mo radiation, R = 0.048 for 455 reflexions.

Atomic positions

			x	y	z
6 Ge	in	6(d)	1/4	0	1/2
6 Al		6(c)	1/4	1/2	0
8 Na		8(e)	0.1687	0.1687	0.1687
24 O(1)		24(i)	0.1435	0.4307	0.1448
2 O(2)		8(e)	0.4262	0.4262	0.4262

Sodalite structure (1), but with O(2) distributed over 8(e) (and possibly 2(a)) sites. Ge-O = 1.724(6), Al-O = 1.740(6), Na-O=2.387(5), 2.502(8) Å.

1. Strukturbericht, 2, 150, 562; Structure Reports, 32A, 490.

GERMANATE MICAS
$K(Mg,Mn)_3Ge_3AlO_{10}F_2$

H. TORAYA and F. MARUMO, 1981. Mineral. J., 10, 396-407.

Monoclinic, C2/m, a = 5.435, 5.489, b = 9.413, 9.509, c = 10.458, 10.462 Å, β = 100.03, 100.12°, for two specimens with Mg/Mn ratios of 3 and 1/2, respectively, Z = 2. R = 0.039, 0.050.

Mica structure (1), with random cation distribution in the octahedral sites. Ge,Al-O = 1.75 Å.

1. Structure Reports, 27, 696.

BISMUTH(III) GERMANATE
$Bi_4(GeO_4)_3$

A. DURIF and M.-T. AVERBUCH-POUCHOT, 1982. C.R. Acad. Sci. Paris, Ser II, 295, 555-556.

Cubic, $I\bar{4}3d$, a = 10.497 Å, Z = 4. Ag radiation, R = 0.017 for 169 reflexions. Bi in 16(c): x,x,x, x = 0.08746, Ge in 12(a): 3/8,0,1/4; O in 48(e): 0.06948, 0.12673,0.28873.

Isostructural with the silicate, eulytite (1). Ge-O = 1.740(6) (tetrahedral), Bi-O = 2.161, 2.605(5) (octahedral) Å.

1. Strukturbericht, 2, 122, 522; Structure Reports, 31A, 227.

BARIUM ZINC GERMANATE
$BaZnGeO_4$

K. IIJIMA, F. MARUMO and H. TAKEI, 1982. Acta Cryst., B38, 1112-1116.

Hexagonal, $P6_3$, a = 9.2905, c = 8.728 Å, Z = 6 (subcell, $P6_322$, a = 5.3638, c = 8.728 Å, Z = 2). Cu radiation, R = 0.046 for 306 reflexions.

The material has a stuffed structure derived from the high-tridymite framework (Fig. 1). Tetrahedral bond distances, T(1)-O = 1.76, T(2)-O = 1.90(7) Å, indicate significant Ge/Zn ordering. Ba has 9-coordination, Ba-O = 2.55-3.75(8) Å. Additional reflexions indicate a true cell with quadrupled c axis.

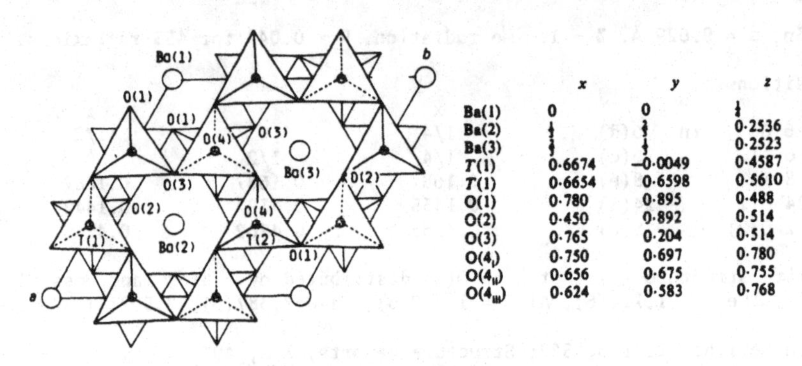

	x	y	z
Ba(1)	0	0	¼
Ba(2)	⅓	⅔	0·2536
Ba(3)	⅔	⅓	0·2523
T(1)	0·6674	−0·0049	0·4587
T(1)	0·6654	0·6598	0·5610
O(1)	0·780	0·895	0·488
O(2)	0·450	0·892	0·514
O(3)	0·765	0·204	0·514
$O(4_I)$	0·750	0·697	0·780
$O(4_{II})$	0·656	0·675	0·755
$O(4_{III})$	0·624	0·583	0·768

Fig. 1. Structure of $BaZnGeO_4$.

YTTRIUM GERMANATE
$Y_3GeO_5(OH,F)_3$

N.I. GOLOVASTIKOV and N.V. BELOV, 1982. Kristallografija, 27, 1087-1089 [Soviet Physics - Crystallography, 27, 650-651].

Monoclinic, $P2_1/m$, a = 5.483, b = 5.886, c = 10.226 Å, γ = 105.14°, Z = 2. Mo radiation, R = 0.066 for 530 reflexions.

The structure (Fig. 1) contains GeO_5 square pyramids linked by 8- and 7-coordinated Y ions. Ge-O = 1.79, 1.82 (x 2), 1.87 (x 2), Y-O = 2.24-2.51 Å.

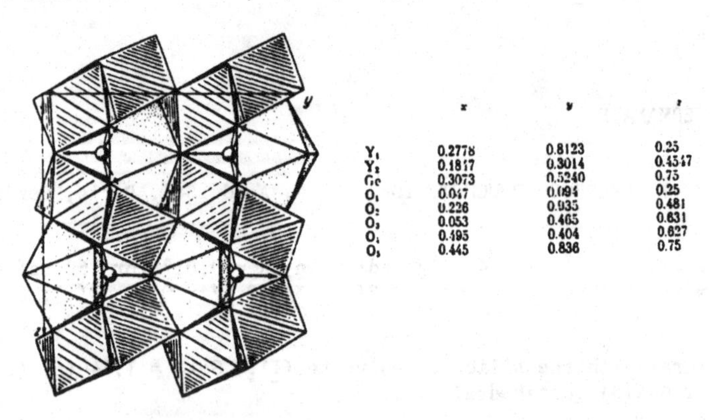

	x	y	z
Y_1	0.2778	0.8123	0.25
Y_2	0.1817	0.3014	0.4547
Ge	0.3073	0.5240	0.75
O_1	0.047	0.094	0.25
O_2	0.226	0.935	0.481
O_3	0.053	0.465	0.631
O_4	0.495	0.404	0.827
O_5	0.445	0.836	0.75

Fig. 1. Structure of yttrium germanate (circles are Ge atoms).

NEODYMIUM DIGERMANATE
$Nd_2Ge_2O_7$

G. VETTER, F. QUEYROUX, P. LABBE and M. GOREAUD, 1982. J. Solid State Chem., 45, 293-302.

Triclinic, P$\bar{1}$, a = 37.609, b = 6.922, c = 6.923 Å, α = 91.46, β = 90.73, γ = 95.15°, Z = 12. Mo radiation, R = 0.061 for 9863 reflexions.

The structure (Fig. 1) contains isolated GeO_4 tetrahedra, Ge_3O_{10} groups, and NdO_7 or NdO_8 polyhedra, arranged in blocks parallel to (100). Ge-O = 1.66-1.87, Nd-O = 2.25-3.13(2) Å.

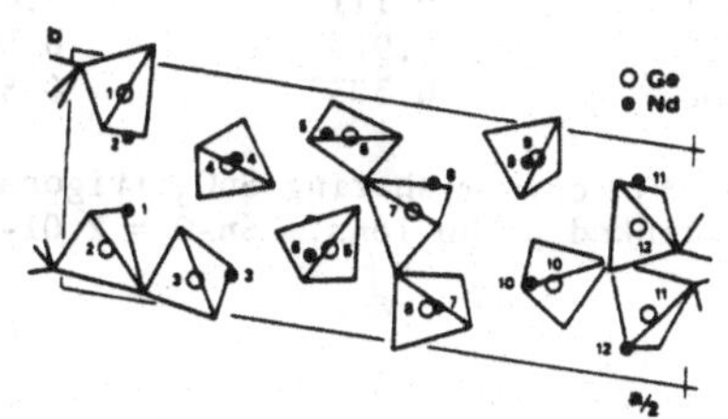

Fig. 1. Structure of $Nd_2Ge_2O_7$.

LITHIUM STANNATE(IV)
Li_2SnO_3

LITHIUM ZIRCONATE(IV)
Li_2ZrO_3

J.L. HODEAU, M. MAREZIO, A. SANTORO and R.S. ROTH, 1982. J. Solid State Chem., 45, 170-179.

Monoclinic, C2/c, a = 5.2889, 5.4218, b = 9.1872, 9.0216, c = 10.0260, 5.4187 Å, β = 100.348, 112.709°, Z = 8, 4. Neutron powder data.

Atomic positions

	Atom		Site	x	y	z
	Li_2SnO_3					
	Sn(1)	in	4(e)	0	0.418	1/4
	Sn(2)		4(e)	0	0.750	1/4
	O(1)		8(f)	0.1387	0.2610	0.1339
	O(2)		8(f)	0.1118	0.5853	0.1340
	O(3)		8(f)	0.1343	0.9078	0.1322
	Li(1)		8(f)	0.231	0.073	-0.0006
0.8	Li(2)		4(d)	1/4	1/4	1/2
0.8	Li(3)		4(e)	0	0.083	1/4
	Li_2ZrO_3					
	Zr		4(e)	0	0.0916	1/4
	O(1)		4(d)	1/4	1/4	1/2
	O(2)		8(f)	0.2721	0.5754	0.4863
0.9	Li(1)		4(e)	0	0.423	1/4
0.9	Li(2)		4(e)	0	0.742	1/4

The structures are essentially as previously described (1, 2), and are NaCl superstructures, with close-packed oxygens and octahedral coordinations for all cations. Sn-O = 2.05-2.08, Zr-O = 2.07-2.12, Li-O = 2.06-2.42 Å.

1. Structure Reports, 18, 445; 35A, 220.
2. Ibid., 41A, 420.

RUBIDIUM STANNATE(II)
Rb_2SnO_2

POTASSIUM STANNATE(II)
K_2SnO_2

R.M. BRAUN and R. HOPPE, 1982. Z. Naturforsch., 37B, 688-694.

Orthorhombic, $P2_12_12_1$, a = 5.761, 5.579, b = 7.493, 7.246, c = 11.167, 10.744 Å, D_m = 4.46, 3.40, Z = 4. R = 0.12 for 800 reflexions for the Rb compound.

Atomic positions

	x	y	z
Rb(1)	0.0112	0.6057	0.4144
Rb(2)	0.0178	0.1415	0.7637
Sn	0.0041	0.1117	0.4424
O(1)	0.0337	0.9161	0.5640
O(2)	0.1638	0.3038	0.5525

The structure contains chains of corner-sharing SnO_3 trigonal pyramids (lone-pair completing a tetrahedron), linked by Rb ions. Sn-O = 2.01-2.10(3) Å, O-Sn-O = 94-107°.

RUBIDIUM DISTANNATE(II)
$Rb_2Sn_2O_3$

R.M. BRAUN and R. HOPPE, 1982. Z. anorg. Chem., 485, 15-22.

Rhombohedral, $R\bar{3}m$, a = 6.086, c = 15.101 Å, D_m = 4.64, Z = 3. Mo radiation, R = 0.053 for 260 reflexions. Rb in 3(a): 0,0,0 and 3(b): 0,0,1/2; Sn in 6(c): 0,0, 0.2351; O in 9(d): 1/2,0,1/2.

Isostructural with the K salt (1). Sn-3 O = 2.04, Rb-6 O = 3.04-3.07 Å.

1. Structure Reports, 48A, 203.

LITHIUM PLUMBATE(IV)
Li_2PbO_3

B. BRAZEL and R. HOPPE, 1982. Z. Naturforsch., 37B, 1369-1374.

Monoclinic, C2/c, a = 5.4452, b = 9.2612, c = 5.4756 Å, β = 111.216°, D_m = 6.89, Z = 4. Mo radiation, R = 0.049 for 533 reflexions.

Atomic positions

			x	y	z
Pb	in	4(e)	0	0.0897	1/4
Li(1)		4(e)	0	0.4268	1/4
Li(2)		4(e)	0	0.7618	1/4
O(1)		8(f)	0.2381	0.0793	0.0135
O(2)		4(d)	1/4	1/4	1/2

The structure is a NaCl variant with alternating Li(2) and Li(1)+Pb layers. All cations have octahedral coordination; Pb-O = 2.14-2.20, Li-O = 2.09-2.37 Å.

POTASSIUM LITHIUM PLUMBATE(IV)

$K_2Li_{14}Pb_3O_{14}$

B. BRAZEL and R. HOPPE, 1982. Z. anorg. Chem., 493, 93-103.

Orthorhombic, Immm, a = 12.7990, b = 7.9446, c = 7.2620 Å, D_m = 4.63, Z = 2. Mo radiation, R = 0.067 for 1022 reflexions.

The structure (Fig. 1) contains a Pb_3O_{14} group of three edge-sharing octahedra, linked by KO_{10} polyhedra and LiO_4 tetrahedra. Pb-O = 2.13-2.39, K-O = 3.02-3.46, Li-O = 1.88-2.18 Å.

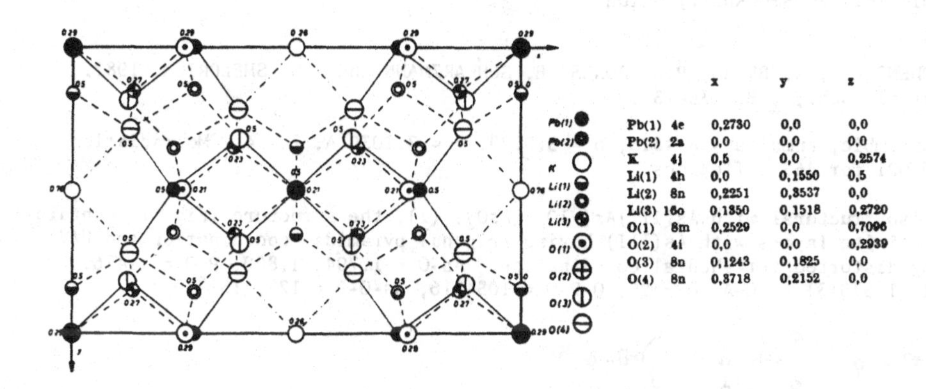

		x	y	z
Pb(1)	4e	0,2730	0,0	0,0
Pb(2)	2a	0,0	0,0	0,0
K	4j	0,5	0,0	0,2574
Li(1)	4h	0,0	0,1550	0,5
Li(2)	8n	0,2251	0,3537	0,0
Li(3)	16o	0,1350	0,1518	0,2720
O(1)	8m	0,2529	0,0	0,7096
O(2)	4i	0,0	0,0	0,2939
O(3)	8n	0,1243	0,1825	0,0
O(4)	8n	0,3718	0,2152	0,0

Fig. 1. Structure of $K_2Li_{14}Pb_3O_{14}$.

LEAD GALLIUM OXIDE

LEAD ALUMINUM OXIDE

$Pb_9Ga_8O_{21}$

I. K.-B. PLÖTZ and H. MÜLLER-BUSCHBAUM, 1982. Z. anorg. Chem., 484, 153-157.

Cubic, Pa3, a = 13.44 Å, Z = 4. R = 0.061 for 993 reflexions.

The structure (Fig. 1) contains rings of six and eight corner-sharing GaO_4 tetrahedra, with six- and eight-coordinate Pb ions. Ga-O = 1.77-1.86, Pb-O = 2.24-3.36 Å.

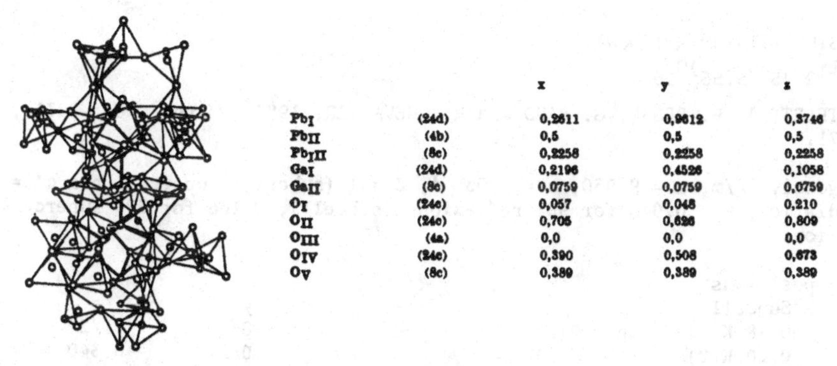

		x	y	z
Pb_I	(24d)	0,2611	0,9612	0,3746
Pb_{II}	(4b)	0,5	0,5	0,5
Pb_{III}	(8c)	0,2258	0,2258	0,2258
Ga_I	(24d)	0,2196	0,4526	0,1058
Ga_{II}	(8c)	0,0759	0,0759	0,0759
O_I	(24e)	0,057	0,048	0,210
O_{II}	(24e)	0,705	0,626	0,809
O_{III}	(4a)	0,0	0,0	0,0
O_{IV}	(24e)	0,390	0,508	0,673
O_V	(8c)	0,389	0,389	0,389

Fig. 1. Structure of $Pb_9Ga_8O_{21}$.

$Pb_8(Pb,Sr)Al_8O_{21}$

II. K.-B. PLÖTZ and H. MÜLLER-BUSCHBAUM, 1982. Z. Naturforsch., 37B, 108-110.

Cubic, Pa3, a = 13.263 Å, Z = 4. R = 0.09 for 644 reflexions.

0.4 Sr substitutes for some of the Pb in the 4(b) site (see I above).

ARSENIC(III) PHOSPHORUS(V) OXIDE
$AsPO_4$

D. BODENSTEIN, A. BREHM, P.G. JONES, E. SCHWARZMANN and G.M. SHELDRICK, 1982. Z. Naturforsch., 37B, 136-137.

Orthorhombic, Pnma, a = 8.4315, b = 5.0229, c = 7.2039 Å, Z = 4. Mo radiation, R = 0.021 for 465 reflexions.

Isostructural with $As(III)As(V)O_4$ (AsO_2) (1), the structure (Fig. 1) containing infinite layers with As(III) having trigonal pyramidal coordination and P(V) having distorted tetrahedral coordination. As-O = 1.794, 1.800, P-O = 1.476, 1.561, 1.570(3) Å, O-As-O = 90, O-P-O = 105-116, As-O-P = 129, 131°.

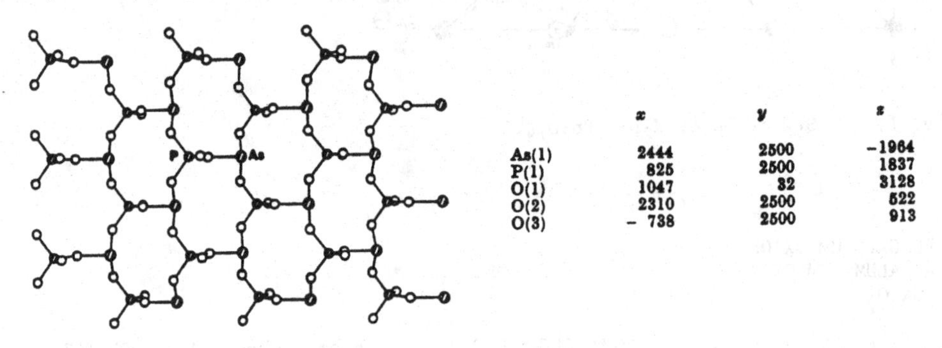

	x	y	z
As(1)	2444	2500	-1964
P(1)	825	2500	1837
O(1)	1047	32	3128
O(2)	2310	2500	522
O(3)	- 738	2500	913

Fig. 1. Structure of $AsPO_4$, and atomic positional parameters (x 10^4).

1. Structure Reports, 46A, 229.

POTASSIUM LITHIUM ANTIMONATE
$K_{1.8}(Li_{2.45}Sb_{5.55})O_{16}$

H. WATELET, J.-P. BESSE, G. BAUD and R. CHEVALIER, 1982. Mater. Res. Bull., 17, 863-871.

Tetragonal, I4/m, a = 9.930, c = 2.936 Å, Z = 1 (subcell; supercell has c' = 3c). Mo radiation, R = 0.048 for 901 reflexions (subcell), 0.106 for 398 supercell reflexions.

Atomic positions

Subcell			x	y	z
0.48 K(1)	in	2(b)	0	0	1/2
0.20 K(2)		4(e)	0	0	0.340
Li,Sb		8(h)	0.165	0.310	0
O(1)		8(h)	0.204	0.154	0
O(2)		8(h)	0.164	0.541	0

Supercell			x	y	z
0.6 K(1)	in	2(b)	0	0	1/2
0.45 K(2)		4(e)	0	0	0.132
0.42 K(3)		4(e)	0	0	0.205
0.18 K(4)		4(e)	0	0	0.459
Li,Sb(1)		8(h)	0.164	0.351	0
Li,Sb(2)		16(i)	0.666	0.849	0.160
O(1)		8(h)	0.203	0.155	0
O(2)		8(h)	0.161	0.543	0
O(3)		16(i)	0.703	0.655	0.168
O(4)		16(i)	0.666	0.041	0.167

The diffraction pattern exhibits diffuse scattering, and heat treatment produces a material with a tripled c axis. The structure is of the hollandite type (1), with some ordering of K ions in the supercell (Fig. 1).

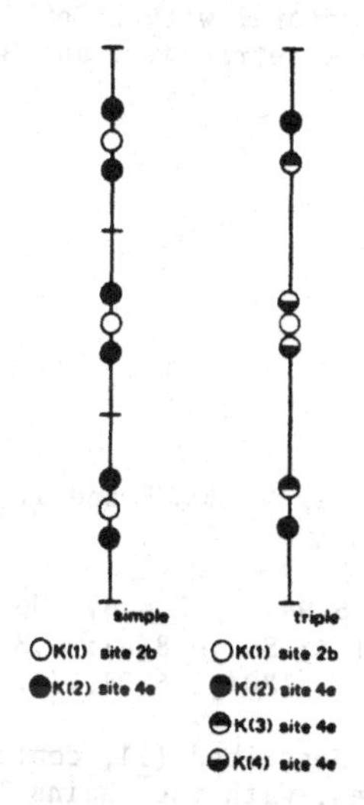

Fig. 1. K ordering in potassium lithium antimonate.

1. Structure Reports, 13, 190.

ANTIMONATES
$Li_2(Cr,M)_3SbO_8$ (M = Al, Ga, Fe)

P. TARTE, R. CAHAY, J. PREUDHOMME, M. HERVIEU, J. CHOISNET and B. RAVEAU, 1982. J. Solid State Chem., 44, 282-289.

$Li_2Cr_2AlSbO_8$, $Li_2CrAl_2SbO_8$, $Li_2Cr_2FeSbO_8$, $Li_2Fe_3SbO_8$, hexagonal, $P6_3mc$, a = 5.796, 5.70, 5.867, 5.923, c = 9.466, 9.39, 9.542, 9.641 Å, Z = 2. Powder data.

High-temperature, $Li_2Cr_2FeSbO_8$, $Li_2Fe_3SbO_8$, cubic, Fd3m, a = 8.398, 8.425 Å, Z = 4. Powder data.

Atomic positions
$P6_3mc$ phases

Li(1)	in 2(b): 1/3,2/3,z	z =	-0.116,	-0.076,	-0.097,	-0.072
Li(2)	2(a): 0,0,z	z =	0.512,	0.555,	0.523,	0.510
Sb	2(b): 1/3,2/3,z	z =	0.491,	0.520,	0.491,	0.480
M	6(c): $x,\bar{x},z$	x =	0.172,	0.166,	0.171,	0.170
		z =	0.215,	0.250,	0.212,	0.204

O(1)	in	2(a): 0,0,z	z =	0.302,	0.338,	0.310,	0.304
O(2)		2(b): 1/3,2/3,z	z =	0.090,	0.165,	0.101,	0.114
O(3)		6(c): x,x̄,z	x =	0.482,	0.490,	0.475,	0.481
			z =	0.346,	0.389,	0.365,	0.344
O(4)		6(c): x,x̄,z	x =	0.165,	0.171,	0.171,	0.180
			z =	0.596,	0.652,	0.598,	0.581

In $Li_2Fe_3SbO_8$, Li(1) = $Li_{0.55}Fe_{0.45}$, Li(2) = $Li_{0.67}Fe_{0.33}$, Sb = $Sb_{0.67}Fe_{0.33}$, M = $Li_{0.78}Fe_{1.89}Sb_{0.33}$

Fd3m phases

Li,Fe in	8(a): 0,0,0		
Li,M,Sb	16(d): 5/8,5/8,5/8		
O	32(e): x,x,x	x =	0.384, 0.382

The hexagonal phases are isostructural with $LiFeSnO_4$ (1), and the cubic phases have the spinel structure (2). Li has tetrahedral and M ions octahedral coordination.

1. Structure Reports, 48A, 240.
2. Strukturbericht, 1, 350.

ZINC DIANTIMONATE(III)
$ZnSb_2O_4$

E. GUTIÉRREZ PUEBLA, E. GUTIÉRREZ RÍOS, A. MONGE and I. RASINES, 1982. Acta Cryst., B38, 2020-2022.

Tetragonal, $P4_2/mbc$, a = 8.527, c = 5.942 Å, Z = 4. Mo radiation, R = 0.015 for 168 reflexions. Zn in 4(d): 0,1/2,1/4; Sb in 8(h): 0.32218,0.33622,0; O(1) in 8(h): 0.0963,0.3600,0; O(2) in 8(g): 0.1804,0.6804,1/4.

The structure is as previously described (1), containing chains of distorted ZnO_6 octahedra sharing opposite edges, with the chains linked by Sb atoms which have trigonal pyramidal coordination. Zn-O = 2.08, 2.18, Sb-O = 1.94, 1.99(1) Å, O-Zn-O = 89-97, 165, 180, O-Sb-O = 93-96°.

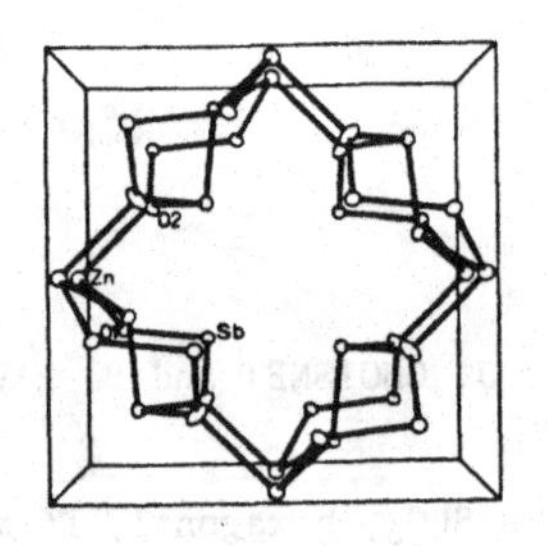

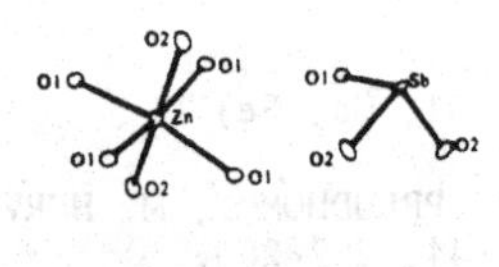

Fig. 1. Structure of $ZnSb_2O_4$.

1. Structure Reports, 9, 174.

CERIUM(III) ANTIMONATE(III)
$CeSbO_3$

P. GARCIA CASADO, A. MENDIOLA and I. RASINES, 1982. Inorg. Chem., 21, 2902-2903.

Cubic, Fd3m, a = 10.8246 Å, Z = 16. Powder data. Atomic positions assumed: Ce in 16(c); Sb in 16(d); 6/7 O(1) in 8(a); 6/7 O(2) in 48(f): x = 0.375 (origin at centre). Defect pyrochlore structure. [See, however, 50A, 346.]

POTASSIUM and SODIUM BISMUTH OXIDES

J. TREHOUX, F. ABRAHAM and D. THOMAS, 1982. Mater. Res. Bull., 17, 1235-1243.

$(K_2O)_x.Bi_2O_3$, cubic, a = 5.543 Å, powder data, fluorite structure. Na compound, a = 5.53 Å.

$(NaBi_{0.5})Bi_2O_6(H_2O)$, pyrochlore, O in 48(f): x = 0.048.

$KBiO_3.xH_2O$, Im3 or Pn3 (see 1), powder data.

1. Structure Reports, 13, 234.

BISMUTH GERMANIUM OXIDE $Bi_4Ge_3O_{12}$ / EULYTITE $Bi_4Si_3O_{12}$

P. FISCHER and F. WALDNER, 1982. Solid State Comm., 44, 657-661.

Cubic, I$\bar{4}$3d, a = 10.495, 10.240 Å, Z = 4. Neutron powder data. Bi in 16(c): x,x,x, x = 0.0876, 0.0860; Ge/Si in 12(a): 3/8,0,1/4; O in 48(e): x,y,z, x = 0.0689, 0.0603, y = 0.1277, 0.1339, z = 0.2875, 0.2863.

Eulytite structures (1), with Ge/SiO_4 tetrahedra linked by Bi^{3+} ions which have 3+3 coordination. Ge-O = 1.736(3), Si-O = 1.619(2), Bi-O = 2.149(4), 2.620(4); 2.125(3), 2.617(3) Å.

1. Strukturbericht, 2, 122, 522; Structure Reports, 31A, 227.

BISMUTH TITANIUM OXIDE
$Bi_{12}TiO_{20}$

S.M. EFENDIEV, T.Z. KULIEVA, V.A. LOMONOV, M.I. CHIRAGOV, M. GRANDOLFO and P. VECCHIA, 1982. Phys. Status Solidi, A, 74, K17-K21.

Cubic, I23, a = 10.188 Å, D_m = 9.10, Z = 2. Mo radiation, R = 0.095 for 750 reflexions.

Atomic positions

			x	y	z
Bi	in	24(f)	0.1689	0.3191	0.0196
Ti		2(a)	0	0	0
O(1)		24(f)	0.1236	0.2361	0.4915
O(2)		8(c)	0.1951	0.1951	0.1951
O(3)		8(c)	0.8975	0.8975	0.8975

Isostructural with $Bi_{12}GeO_{20}$ (1). The structure contains Bi_4O_{20} groups linked by TiO_4 tetrahedra. Bi-O (7-coordination) = 2.16, 2.21, 2.21; 2.51, 2.62; 3.13, 3.37, Ti-O = 1.81 Å.

1. Structure Reports, 32A, 292; 45A, 233.

BARIUM SCANDATE
$BaSc_2O_4$

L.M. KOVBA, L.N. LYKOVA, T.A. KALININA and S.M. ČIŽOV, 1982. Koordin. Khim., 8, 553-556.

Monoclinic, B2/b, a = 9.836, b = 20.578, c = 5.8147 Å, γ = 89°53', Z = 12. R = 0.067.

Sc ions have octahedral and Ba ions cubo-octahedral coordinations.

SODIUM TITANATE
$Na_2Ti_9O_{19}$

Y. BANDO, 1982. Acta Cryst., A38, 211-214.

Monoclinic, C2/m, a = 12.2, b = 3.78, c = 15.6 Å, β = 105°, Z = 2. Electron diffraction and microscopy.

A structure is proposed (Fig. 1), but further work is required to establish the structure fully.

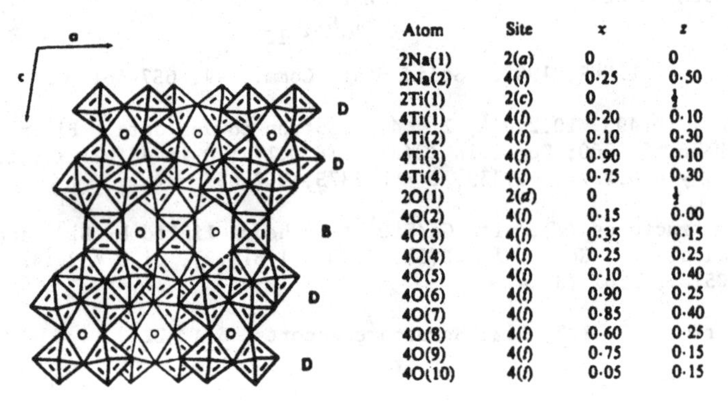

Atom	Site	x	z
2Na(1)	2(a)	0	0
2Na(2)	4(i)	0·25	0·50
2Ti(1)	2(c)	0	½
4Ti(1)	4(i)	0·20	0·10
4Ti(2)	4(i)	0·10	0·30
4Ti(3)	4(i)	0·90	0·10
4Ti(4)	4(i)	0·75	0·30
2O(1)	2(d)	0	½
4O(2)	4(i)	0·15	0·00
4O(3)	4(i)	0·35	0·15
4O(4)	4(i)	0·25	0·25
4O(5)	4(i)	0·10	0·40
4O(6)	4(i)	0·90	0·25
4O(7)	4(i)	0·85	0·40
4O(8)	4(i)	0·60	0·25
4O(9)	4(i)	0·75	0·15
4O(10)	4(i)	0·05	0·15

Fig. 1. Structure of $Na_2Ti_9O_{19}$ (y = 0).

PRIDERITE
$(K,Ba)_{1.6}(Ti,Fe,Mg)_8O_{16}$

W. SINCLAIR and G.M. McLAUGHLIN, 1982. Acta Cryst., B38, 245-246.

Tetragonal, I4/m, a = 10.140, c = 2.965 Å, Z = 1. Mo radiation, R = 0.052 for 987 reflexions.

Atomic positions

			x	y	z
1.2 A	in	2(b)	0	0	1/2
0.4 A(1)		4(e)	0	0	0.327
8 B		8(h)	0.3514	0.1677	0
8 O(1)		8(h)	0.1553	0.2040	0
8 O(2)		8(h)	0.5407	0.1652	0

A = mainly K; A(1) = mainly Ba

Hollandite-type structure, as for related materials (1).

1. Structure Reports, 46A, 243.

BARIUM DITITANATE(III) TETRATITANATE(IV)
$Ba_2Ti_6O_{13}$

W.H. BAUR, E. TILLMANNS and W. HOFMEISTER, 1982. Cryst. Struct. Comm., 11, 2021-2024.

Monoclinic, C2/m, a = 15.004, b = 3.953, c = 9.085 Å, β = 98.01°, Z = 2. R = 0.036 for 1036 reflexions.

Atomic positions

	x	y	z
Ba	0.05153	0	0.77756
Ti(1)	0.62227	0	0.09994
Ti(2)	0.25203	0	0.22287
Ti(3)	0.66877	0	0.44403
O(1)	0.1269	0	0.1121
O(2)	0.7412	0	0.2471
O(3)	0.2003	0	0.4284
O(4)	0.3287	0	0.0824
O(5)	0.3709	0	0.3809
O(6)	0.4280	0	0.6982
O(7)	1/2	0	0

Structure as previously described (1), with full occupancy of Ti sites.

1. Structure Reports, 43A, 350.

BARIUM TRIDECATITANATE
$Ba_4Ti_{13}O_{30}$

E. TILLMANNS, 1982. Cryst. Struct. Comm., 11, 2087-2092.

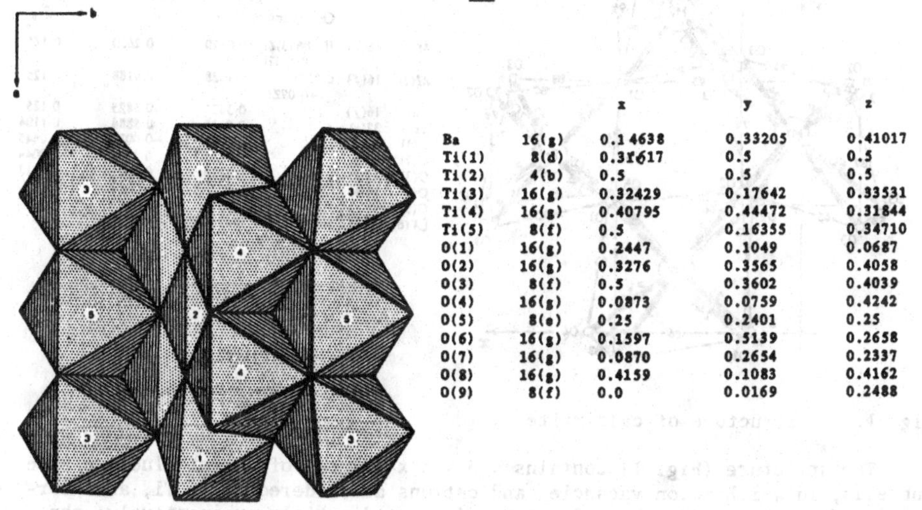

		x	y	z
Ba	16(g)	0.14638	0.33205	0.41017
Ti(1)	8(d)	0.31617	0.5	0.5
Ti(2)	4(b)	0.5	0.5	0.5
Ti(3)	16(g)	0.32429	0.17642	0.33531
Ti(4)	16(g)	0.40795	0.44472	0.31844
Ti(5)	8(f)	0.5	0.16355	0.34710
O(1)	16(g)	0.2447	0.1049	0.0687
O(2)	16(g)	0.3276	0.3565	0.4058
O(3)	8(f)	0.5	0.3602	0.4039
O(4)	16(g)	0.0873	0.0759	0.4242
O(5)	8(e)	0.25	0.2401	0.25
O(6)	16(g)	0.1597	0.5139	0.2658
O(7)	16(g)	0.0870	0.2654	0.2337
O(8)	16(g)	0.4159	0.1083	0.4162
O(9)	8(f)	0.0	0.0169	0.2488

Fig. 1. Structure of barium tridecatitanate.

Orthorhombic, Cmca, a = 17.062, b = 9.862, c = 14.051 Å, D_m = 4.60, Z = 4. Mo radiation, R = 0.03 for 2191 reflexions.

The structure (Fig. 1) contains close packed Ba/O layers with sequence ABCACB or $(hcc)_2$ along c. Ti ions are in octahedral holes, with linear groups of three edge-sharing octahedra connecting blocks of five edge-sharing octahedra by further edge-sharing; four of these units are linked by common corners. Ti-O = 1.79-2.15, Ba-O = 2.72-3.22 Å.

ZIRKELITE (ZIRCONOLITE)
$CaZrTi_2O_7$

W. SINCLAIR and R.E. EGGLETON, 1982. Amer. Min., 67, 615-620.

Monoclinic, C2/c, a = 12.431, b = 7.224, c = 11.483 Å, β = 100.33°, Z = 8. Mo radiation, R = 0.052 for 949 reflexions.

The structure is as previously described (1), except for higher thermal parameters resulting from radiation damage.

1. Structure Reports, 48A, 211.

CALZIRTITE
$Ca_2Zr_5Ti_2O_{16}$

H.J. ROSSELL, 1982. Acta Cryst., B38, 593-595.

Tetragonal, $I4_1/acd$, a = 15.2203, c = 10.1224 Å, at 300K, D_m = 4.98, Z = 8. Powder data. Previous study in 1.

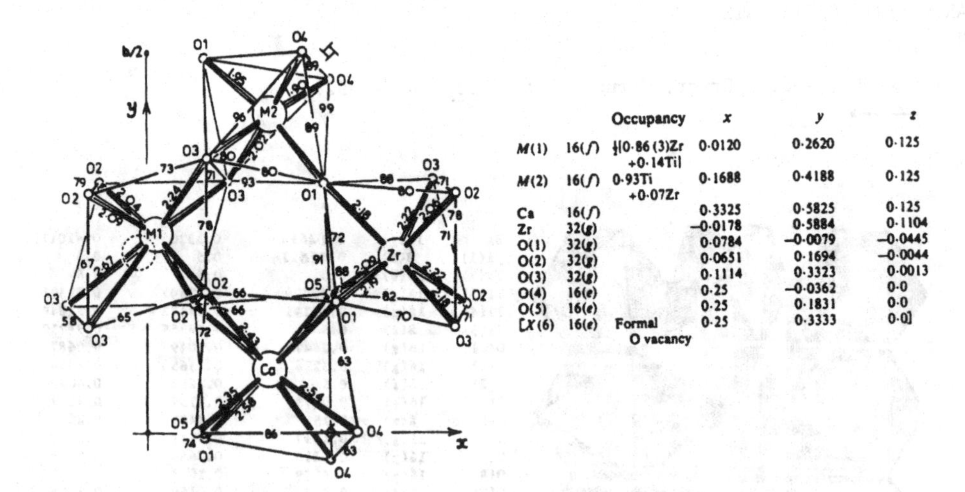

		Occupancy	x	y	z
M(1)	16(f)	½[0·86(3)Zr +0·14Ti]	0·0120	0·2620	0·125
M(2)	16(f)	0·93Ti +0·07Zr	0·1688	0·4188	0·125
Ca	16(f)		0·3325	0·5825	0·125
Zr	32(g)		−0·0178	0·5884	0·1104
O(1)	32(g)		0·0784	−0·0079	−0·0445
O(2)	32(g)		0·0651	0·1694	−0·0044
O(3)	32(g)		0·1114	0·3323	0·0013
O(4)	16(e)		0·25	−0·0362	0·0
O(5)	16(e)		0·25	0·1831	0·0
[X(6)	16(e)	Formal O vacancy	0·25	0·3333	0·0]

Fig. 1. Structure of calzirtite

The structure (Fig. 1) contains a 3 x 3 x 2 array of defect-fluorite type subcells, in which anion vacancies and cations are ordered. Ca, Ti, and four-fifths of the Zr cations have 8-, 6-, and 7-coordinations, respectively; the remaining Zr occupies randomly two sites, M(1), 0.5 Å apart in a distorted cube of oxygen atoms.

1. Structure Reports, 26, 379.

PSEUDOBROOKITES
$FeAlTiO_5$, Fe_2TiO_5

P. TIEDEMANN and H. MÜLLER-BUSCHBAUM, 1982. Z. anorg. Chem., 494, 98-102.

Orthorhombic, Cmcm, a = 3.631, 3.739, b = 9.574, 9.779, c = 9.785, 9.978 Å, Z = 4. R = 0.086, 0.058 for 433, 468 reflexions.

Atomic positions (x = 0)

	$FeAlTiO_5$		Fe_2TiO_5	
	y	z	y	z
M(1)	0.1349	0.5624	0.1360	0.5642
M(2)	0.1867	1/4	0.1890	1/4
O(1)	0.759	1/4	0.766	1/4
O(2)	0.047	0.117	0.048	0.117
O(3)	0.311	0.072	0.311	0.070

Pseudobrookite structures (1), but with disordered metal sites.

1. Strukturbericht, 2, 53, 336; Structure Reports, 21, 279; 22, 312.

LANTHANUM DITITANATE (HIGH-TEMPERATURE)
$La_2Ti_2O_7$

N. ISHIZAWA, F. MARUMO, S. IWAI, M. KIMURA and T. KAWAMURA, 1982. Acta Cryst., B38, 368-372.

Orthorhombic, $Cmc2_1$, a = 3.954, b = 25.952, c = 5.607 Å, at 1173K, Z = 4. Mo radiation, R = 0.066 for 766 reflexions.

Isostructural with $Sr_2Nb_2O_7$ (1), the structure (Fig. 1) containing perovskite-type slabs of TiO_6 octahedra and $\overline{12}$-coordinate La(1) ions, linked by 7-coordinate La(2) ions. Ti-O = 1.76-2.19, La-O = 2.43-2.99 Å. Monoclinic and another orthorhombic forms are also known (2).

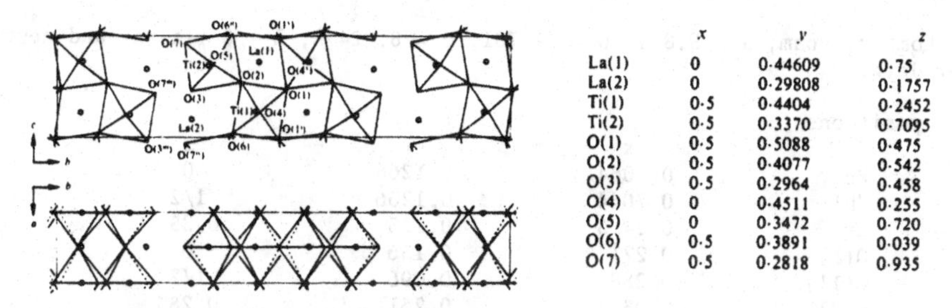

	x	y	z
La(1)	0	0·44609	0·75
La(2)	0	0·29808	0·1757
Ti(1)	0·5	0·4404	0·2452
Ti(2)	0·5	0·3370	0·7095
O(1)	0·5	0·5088	0·475
O(2)	0·5	0·4077	0·542
O(3)	0·5	0·2964	0·458
O(4)	0	0·4511	0·255
O(5)	0	0·3472	0·720
O(6)	0·5	0·3891	0·039
O(7)	0·5	0·2818	0·935

Fig. 1. Structure of high-temperature $La_2Ti_2O_7$.

1. Structure Reports, 41A, 240.
2. Ibid., 41A, 232, 233.

BRANNERITE
UTi_2O_6

J.T. SZYMAŃSKI and J.D. SCOTT, 1982. Canad. Miner., 20, 271-279.

Monoclinic, C2/m, a = 9.8123, b = 3.7697, c = 6.9253 Å, β = 118.957°, D_m = 6.35, Z = 2. Mo radiation, R = 0.022 for 1845 reflexions, for synthetic material.

Isostructural with thorutite, $ThTi_2O_6$ (which was incorrectly called brannerite in 1). The structure (Fig. 1) contains UO_6 and TiO_6 distorted octahedra, U-O = 2.252 (x 2), 2.296 (x 4), Ti-O = 1.854-2.104 (3) Å; U has two further O neighbours at 2.824 Å.

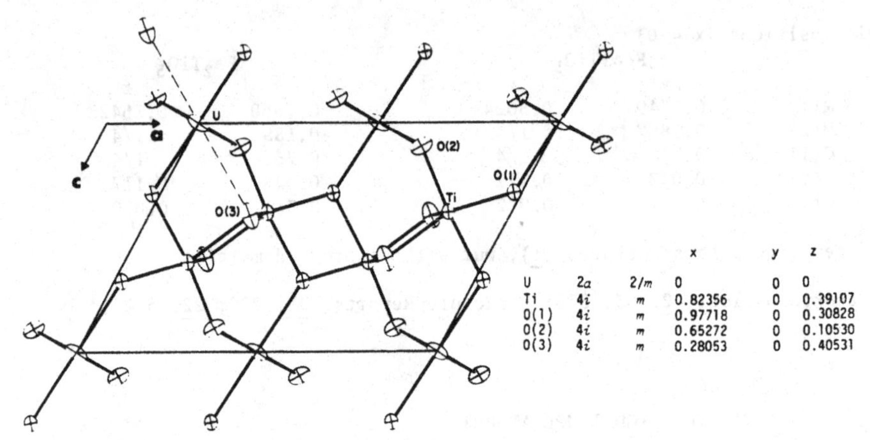

			x	y	z
U	2a	2/m	0	0	0
Ti	4i	m	0.82356	0	0.39107
O(1)	4i	m	0.97718	0	0.30828
O(2)	4i	m	0.65272	0	0.10530
O(3)	4i	m	0.28053	0	0.40531

Fig. 1. Structure of brannerite.

1. Structure Reports, 31A, 139.

LEAD ZIRCONATE (ANTIFERROELECTRIC)
$PbZrO_3$

H. FUJISHITA, Y. SHIOZAKI, N. ACHIWA and E. SAWAGUCHI, 1982. J. Phys. Soc. Japan, 51, 3583-3591.

Orthorhombic, Pbam, a = 5.881, b = 11.781, c = 8.224 Å, Z = 8. X-ray and neutron powder data.

Atomic positions

	x	y	z
Pb	0.7064	0.1266	0
Pb'	0.7064	0.1266	1/2
Zr	0.2423	0.125	0.25
O(1)	0.271	0.156	0
O(1')	0.288	0.096	1/2
O(2)	0.032	0.261	0.283
O(3)	0	1/2	0.206
O(4)	0	0	0.248

The structure is essentially as previously described (1), but the space group is non-polar.

1. Structure Reports, 21, 321.

VANADIUM PYROXENES
$Na_xLi_{2-x}V_2O_6$ (x = 0.15, 0.66, 1.40, 2.00)

R.S. BUBNOVA, S.K. FILATOV, I.V. ROŽDESTVENSKAJA, V.S. GRUNIN and Z.N. ZONN, 1982. Kristallografija, 27, 1094-1097 [Soviet Physics - Crystallography, 27, 654-656].

Monoclinic, C2/c, a = 10.160-10.573, b = 8.471-9.482, c = 5.879-5.888 Å, β = 110.21-108.55°, Z = 4. R = 0.022-0.047 for 718-1062 reflexions.

Diopside structures, as previously described for x = 0, 1, 2 (1). Li ions are concentrated in the more regular M1 octahedra.

1. Structure Reports, 39A, 222; 40A, 188; 45A, 236.

CALCIUM SODIUM VANADATE
$CaNaVO_4$

D.J.W. IJDO, 1982. Acta Cryst., B38, 923-925.

Orthorhombic, Cmcm, a = 5.8726, b = 9.3028, c = 7.1562 Å, Z = 4. Neutron powder data.

Atomic positions

			x	y	z
Na	in	4(c)	0	0.1808	1/4
Ca		4(b)	0	1/2	0
V		4(c)	0	0.8621	1/4
O(1)		8(g)	0.2599	0.4623	1/4
O(2)		8(f)	0	0.2533	0.5597

Isostructural with Na_2CrO_4 (1), as previously reported(2, but with interchange of Na and Ca positions). The structure contains VO_4 tetrahedra linked by chains of edge-sharing CaO_6 octahedra; Na has very-distorted tetrahedral coordination. V-O = 1.69, 1.73, Ca-O = 2.33, 2.38, Na-O = 2.32, 2.47 Å, O-V-O = 110-113, O-Ca-O = 90, 100, O-Na-O = 70, 104, 146°.

1. Strukturbericht, 4, 45, 172; Structure Reports, 18, 474; 48A, 224.
2. Structure Reports, 33A, 310.

BARIUM VANADATE(III)
$BaVO_{2.5}$

D. CHALES de BEAULIEU and H. MÜLLER-BUSCHBAUM, 1982. Mh. Chem., 113, 415-420.

Trigonal, P3m1, a = 5.718, c = 11.613 Å, Z = 5. R = 0.063 for 501 reflexions.

Atomic positions (oxygen occupancies = 5/6)

			x	y	z
Ba(1)	in	1(a)	0	0	0
Ba(2)		1(b)	1/3	2/3	0.7848
Ba(3)		1(b)	1/3	2/3	0.4085
Ba(4)		1(c)	2/3	1/3	0.2191
Ba(5)		1(c)	2/3	1/3	0.5864
V(1)		1(a)	0	0	0.4949
V(2)		1(b)	1/3	2/3	0.0862
V(3)		1(a)	0	0	0.6938
V(4)		1(a)	0	0	0.2753
V(5)		1(c)	2/3	1/3	0.8726

			x	y	z
O(1)	in	3(d)	0.507	0.014	0.011
O(2)		3(d)	0.162	0.324	0.189
O(3)		3(d)	0.151	0.302	0.599
O(4)		3(d)	0.835	0.670	0.799
O(5)		3(d)	0.842	0.684	0.393

The structure contains close-packed Ba/O layers with sequence ccchh, with V in octahedral oxygen holes. V-O = 1.86-2.18, Ba-O (12-coordination) = 2.67-3.06 Å.

BARIUM MAGNESIUM VANADATE
$BaMg_2(VO_4)_2$

Ju.A. VELIKODNIJ, V.K. TRUNOV, V.D. ŽURAVLEV and L.G. MAKAREVIČ, 1982. Kristallografija, 27, 226-228 [Soviet Physics - Crystallography, 27, 138-140].

Tetragonal, $I4_1/acd$, a = 12.424, c = 8.468 Å, Z = 8. Mo radiation, R = 0.022 for 589 reflexions.

Atomic positions (origin at $\bar{1}$)

			x	y	z
Ba	in	8(a)	0	1/4	3/8
V		16(e)	0.0808	0	1/4
Mg		16(f)	0.1686	0.4186	1/8
O(1)		32(g)	0.1585	0.0736	0.3823
O(2)		32(g)	-0.0015	0.0874	0.1540

The structure (Fig. 1) contains spiral chains of edge-sharing MgO_6 octahedra along c, linked by VO_4 tetrahedra into a framework, with cuboctahedral cavities which contain Ba ions. V-4 O = 1.70, 1.74, Mg-6 O = 2.06-2.15, Ba-12 O = 2.75-3.11 Å.

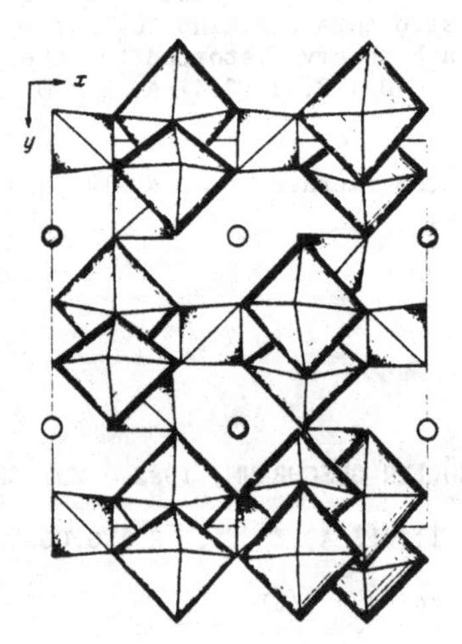

Fig. 1. Structure of barium magnesium vanadate.

SODIUM VANADATOPERIODATE HYDRATE
$Na_6[H_2V_2I_2O_{16}].10H_2O$

R. MATTES and K.-L. RICHTER, 1982. Z. Naturforsch., 37B, 1241-1244.

Triclinic, $P\bar{1}$, a = 11.407, b = 8.865, c = 6.105 Å, α = 82.07, β = 84.24, γ = 70.23°, at -150°C, Z = 1. Mo radiation, R = 0.022 for 1909 reflexions.

The heteropolyanion contains a V_2I_2 ring of VO_5 square pyramids and $IO_5(OH)$ octahedra; one pair of edges is doubly-bridged by O, and the other pair singly bridged. V-O = 1.62, 1.68 (terminal), 1.90-2.09 (bridging) (next nearest at 2.53 Å), I-OH = 1.95, I-O = 1.82, 1.83 (terminal), 1.90-1.92 Å (bridging). Na ions have octahedral coordinations, Na-O = 2.29-2.62 Å, and hydrogen bonding is present.

CAESIUM MANGANESE VANADIUM OXIDE
$Cs_3Mn_3V_4O_{16}$

Y. LE PAGE and P. STROBEL, 1982. Inorg. Chem., 21, 620-623.

Triclinic, $P\bar{1}$, a = 5.1947, b = 7.5017, c = 11.4367 Å, α = 77.70, β = 89.72, γ = 82.62°, Z = 1. Mo radiation, R = 0.052 for 3167 reflexions.

The structure (Fig. 1) contains $(Mn_3O_{10})_n$ ribbons along a of edge-sharing MnO_6 distorted octahedra, linked by V_2O_7 double tetrahedra into ab layers with one Cs ion in a large cavity; two other Cs ions lie between the layers. Mn-O = 1.90-2.22, V-O = 1.61-1.82, Cs-O (12- and 9-coordinations) = 2.94-3.76(1) Å.

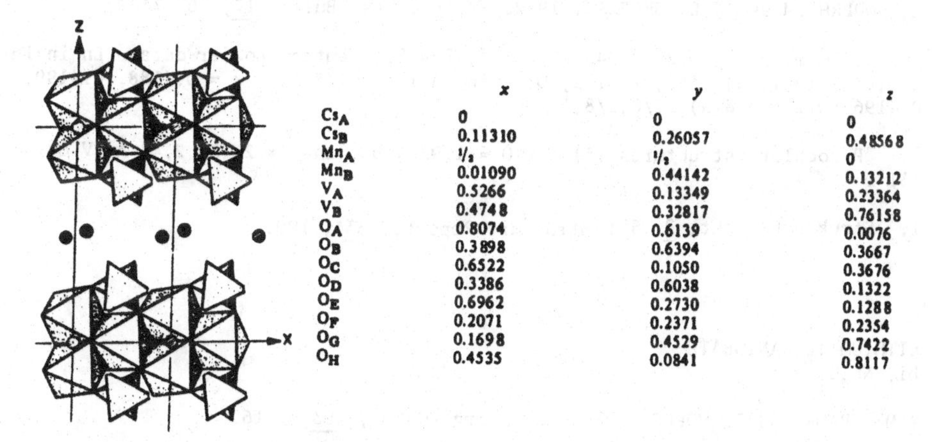

	x	y	z
Cs_A	0	0	0
Cs_B	0.11310	0.26057	0.48568
Mn_A	1/2	1/2	0
Mn_B	0.01090	0.44142	0.13212
V_A	0.5266	0.13349	0.23364
V_B	0.4748	0.32817	0.76158
O_A	0.8074	0.6139	0.0076
O_B	0.3898	0.6394	0.3667
O_C	0.6522	0.1050	0.3676
O_D	0.3386	0.6038	0.1322
O_E	0.6962	0.2730	0.1288
O_F	0.2071	0.2371	0.2354
O_G	0.1698	0.4529	0.7422
O_H	0.4535	0.0841	0.8117

Fig. 1. Structure of $Cs_3Mn_3V_4O_{16}$.

COPPER(II) VANADATE (TRICLINIC FORM)
$Cu_3(VO_4)_2$

J. COING-BOYAT, 1982. Acta Cryst., B38, 1546-1548.

Triclinic, $P\bar{1}$, a = 5.196, b = 5.355, c = 6.505 Å, α = 69.22, β = 88.69, γ = 68.08°, Z = 1. Ag radiation, R = 0.033 for 1602 reflexions.

Isostructural with the phosphate (1), the structure (Fig. 1) containing VO_4

tetrahedra linked by Cu ions with square-planar and trigonal bipyramidal coordinations. V-O = 1.64-1.78, Cu-O = 1.92, 1.97 and 1.91-2.12 Å. A monoclinic form has been described (2).

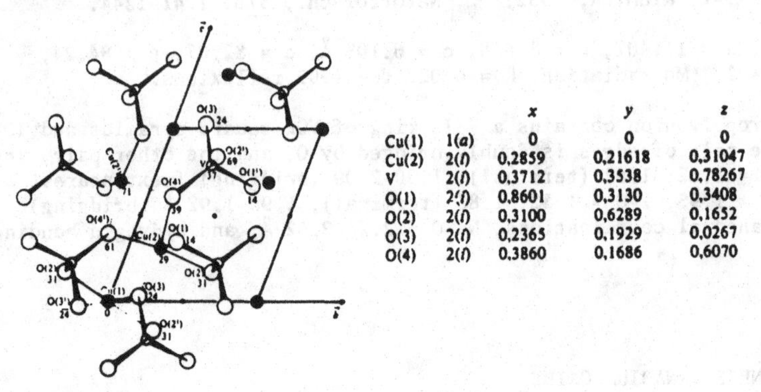

		x	y	z
Cu(1)	1(a)	0	0	0
Cu(2)	2(i)	0,2859	0,21618	0,31047
V	2(i)	0,3712	0,3538	0,78267
O(1)	2(i)	0,8601	0,3130	0,3408
O(2)	2(i)	0,3100	0,6289	0,1652
O(3)	2(i)	0,2365	0,1929	0,0267
O(4)	2(i)	0,3860	0,1686	0,6070

Fig. 1. Structure of triclinic copper(II) vanadate.

1. Structure Reports, 43A, 257.
2. Ibid., 38A, 257.

LANTHANON DIVANADATE(IV)
$Ln_2V_2O_7$ (Ln = Lu, Yb, Tm)

L. SODERHOLM and J.E. GREEDAN, 1982. Mater. Res. Bull., 17, 707-713.

Cubic, Fd3m, a = 9.928, 9.945, 9.973 Å, Z = 8. Neutron powder data. Ln in 16(c): 0,0,0; V in 16(d): 1/2,1/2,1/2; O(1) in 48(f): x,1/8,1/8, x = 0.4188, 0.4199, 0.4196; O(2) in 8(a): 1/8,1/8,1/8.

Pyrochlore structures (1). V-O = 1.93-1.94, Ln-O = 2.15, 2.44 Å, V-O-V = 131°.

1. Strukturbericht, 2, 58; Structure Reports, 33A, 190.

LITHIUM TETRANIOBATE
$Li_{16}Nb_4O_{18}$

R.M. BRAUN and R. HOPPE, 1982. Z. anorg. Chem., 493, 7-16.

Triclinic, P$\bar{1}$, a = 15.210, b = 8.815, c = 5.858 Å, α = 109.8, β = 101.4, γ = 87.0°, Z = 2. Mo radiation, R = 0.057 for 1602 reflexions.

The structure (Fig. 1) is a NaCl variant with anion vacancies and ordered cation distribution. Nb ions have octahedral coordination, and Li ions have 4- to 6-coordinations.

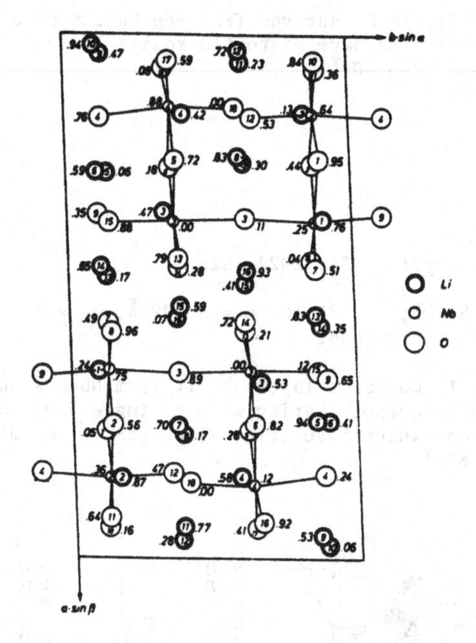

Fig. 1. Structure of $Li_{16}Nb_4O_{18}$.

SODIUM NIOBATE
Na_5NbO_5

J. DARRIET, A. MAAZAZ, J.C. BOULOUX and C. DELMAS, 1982. Z. anorg. Chem., 485, 115-121.

Monoclinic, C2/c, a = 6.24, b = 10.20, c = 10.16 Å, β = 109.3°, D_m = 3.18, Z = 4. Mo radiation, R = 0.03 for 1452 reflexions.

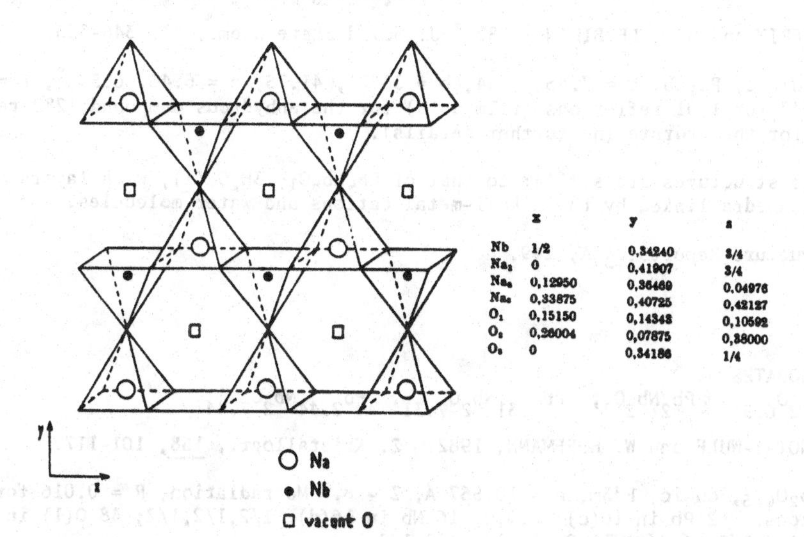

	x	y	z
Nb	1/2	0,34240	3/4
Na_1	0	0,41907	3/4
Na_2	0,12950	0,36469	0,04976
Na_3	0,33875	0,40725	0,42127
O_1	0,15150	0,14348	0,10592
O_2	0,26004	0,07875	0,38000
O_3	0	0,34186	1/4

Fig. 1. Structure of Na_5NbO_5.

The structure (Fig. 1) can be derived from the NaCl-type by ordering cations and oxygen vacancies. Nb and Na have distorted square-pyramidal coordinations. Nb-O = 1.88-1.99, Na-O = 2.27-3.40 Å.

POTASSIUM TRINIOBATE
KNb_3O_8

M. GASPERIN, 1982. Acta Cryst., B38, 2024-2026.

Orthorhombic, Amam, a = 8.903, b = 21.16, c = 3.799 Å, D_m = 4.1, Z = 4. Mo radiation, R = 0.055 for 1781 reflexions.

The structure (Fig. 1) contains layers built from Nb_2O_{10} units of two edge-sharing octahedra linked by corner sharing with a single NbO_6 octahedron. The layers are linked by K ions which have trigonal prismatic coordination. Nb-O = 1.74-2.42, K-O = 2.75-2.85 Å.

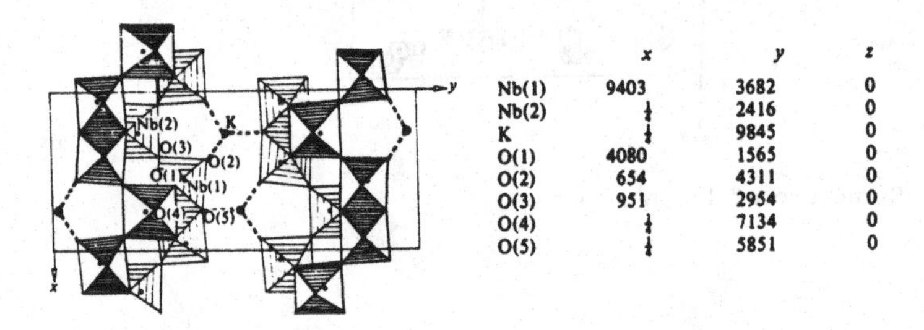

	x	y	z
Nb(1)	9403	3682	0
Nb(2)	1/4	2416	0
K	1/4	9845	0
O(1)	4080	1565	0
O(2)	654	4311	0
O(3)	951	2954	0
O(4)	1/4	7134	0
O(5)	1/4	5851	0

Fig. 1. Structure of KNb_3O_8, and atomic positional parameters (x 10^4).

POTASSIUM HEXANIOBATE
$K_4Nb_6O_{17}$

POTASSIUM RUBIDIUM HEXANIOBATE PENTAHYDRATE
$K_2Rb_2Nb_6O_{17}.5H_2O$

M. GASPERIN and M.T. LE BIHAN, 1982. J. Solid State Chem., 43, 346-353.

Orthorhombic, $P2_1nb$, a = 7.83, 7.84, b = 33.21, 41.75, c = 6.46, 6.50 Å, Z = 4. R = 0.147 for 1401 reflexions (film data) for the anhydrous compound; 289 reflexions for the hydrate (no further details).

The structures are similar to that of $Rb_4Nb_6O_{17}.3H_2O$ (1), with layers of NbO_6 octahedra linked by the alkali-metal cations and water molecules.

1. Structure Reports, 46A, 249.

LEAD NIOBATES
$Pb_{1.5}Nb_2O_{6.5}$, $Pb_2Nb_2O_7$, $Pb_{2.31}Nb_2O_{7.31}$, $Pb_{2.44}Nb_2O_{7.44}$

H. BERNOTAT-WULF and W. HOFFMANN, 1982. Z. Kristallogr., 158, 101-117.

$Pb_{1.5}Nb_2O_{6.5}$, cubic, Fd3m, a = 10.567 Å, Z = 8. Mo radiation, R = 0.016 for 67 reflexions. 12 Pb in 16(c): 0,0,0; 16 Nb in 16(d): 1/2,1/2,1/2; 48 O(1) in 48(f): 0.4373,1/8,1/8; 4 O(2) in 8(a): 1/8,1/8,1/8.

$Pb_2Nb_2O_7$, trigonal, P3m1, a = 7.472, c = 28.351 Å, D_m = 7.32, Z = 8.5. Mo radiation, R = 0.052 for 836 reflexions.

$Pb_{2.31}Nb_2O_{7.31}$, $Pb_{2.44}Nb_2O_{7.44}$, rhombohedral, R3m, a = 7.472, 7.460, c = 66.75, 96.26 Å, D_m = 7.78, 7.79, Z = 19.5, 27. Mo radiation, R = 0.041, 0.034 for 530, 621 reflexions.

$Pb_{1.5}Nb_2O_{6.5}$ has ideal cubic pyrochlore structure (1), with layers of NbO_6 octahedra parallel to (111) in a stacking sequence of hexagonal tungsten bronze type and layers of single octahedra; the Pb site is partially occupied. The other three phases have pyrochlore superstructures, with stacking sequences with 9, 7, and 5 layers (Fig. 1). Pb ions have eight-coordinations and all the structures have oxygen vacancies.

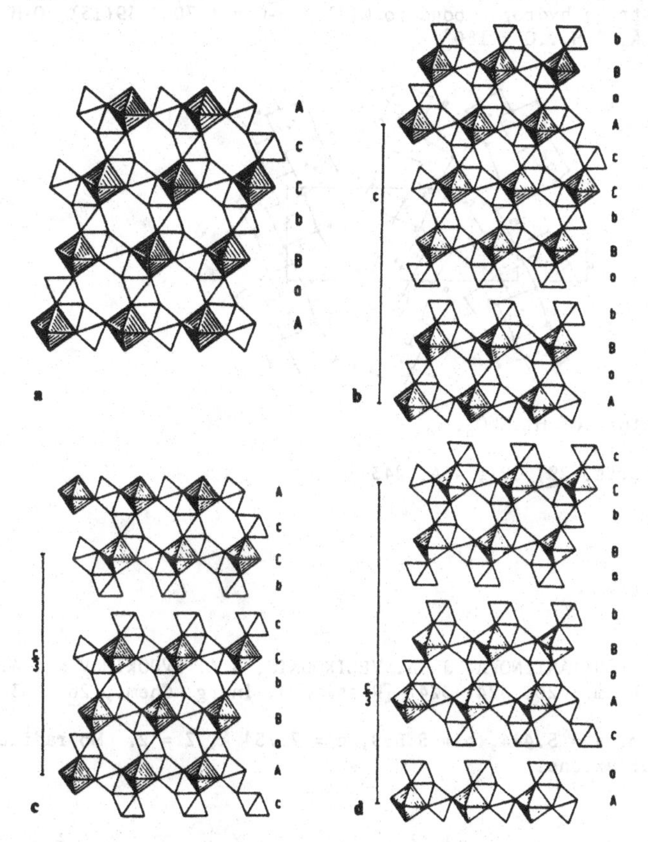

Fig. 1. Structures of (a) $Pb_{1.5}Nb_2O_{6.5}$, (b) $Pb_2Nb_2O_7$, (c) $Pb_{2.31}Nb_2O_{7.31}$, and (d) $Pb_{2.44}Nb_2O_{7.44}$.

1. Structure Reports, 33A, 190.

HYDROGEN TITANATE NIOBATE
$HTiNbO_5$

H. REBBAH, J. PANNETIER and B. RAVEAU, 1982. J. Solid State Chem., 41, 57-62.

Orthorhombic, Pnma, a = 6.534, 6.521, b = 3.777, 3.773, c = 16.675, 16.656 Å, at 300, 10K, Z = 4. Neutron powder data at 300 and 10K, and for deuterium compound at 300K.

Atomic positions

H compound at 300K; similar values at 10K, and for the D compound. Atoms in 4(c): y = 1/4; M(1) = 0.77 Ti + 0.23 Nb, M(2) = 0.23 Ti + 0.77 Nb

	x	z		x	z
H	0.404	0.192	O(2)	0.641	-0.008
M(1)	0.291	0.024	O(3)	0.469	0.131
M(2)	0.794	0.122	O(4)	0.833	0.224
O(1)	0.069	0.077	O(5)	0.231	0.909

The structure (Fig. 1) contains sheets perpendicular to c built up from units of 2 x 2 edge-sharing octahedra, as in $KTiNbO_5$ (1). H is bonded to O(3) and links the layers by a strong hydrogen bond to O(4). M-O = 1.70-2.39(13), O-H = 1.19(12), H...O = 1.35(11) Å, O-H...O = 180°.

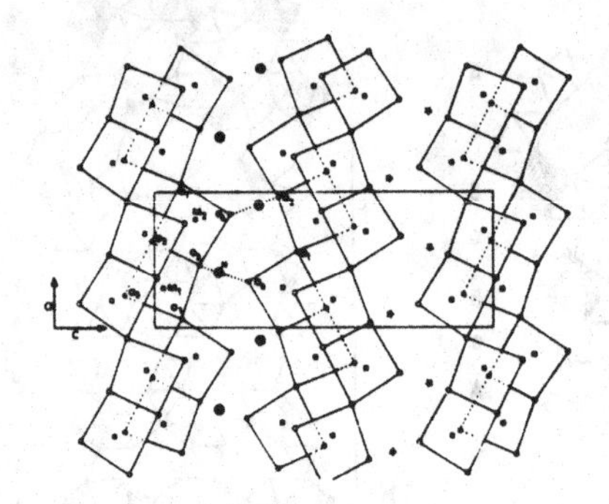

Fig. 1. Structure of $HTiNbO_5$.

1. Structure Reports, 29, 327; 46A, 243.

CALCIUM IRON NIOBATE
Ca_2FeNbO_6

V.K. TRUNOV, L.I. KONSTANTINOVA, Ju.A. VELIKODNIJ, A.A. EVDOKIMOV and A.M. FROLOV, 1981. Ž. Neorg. Khim., 26, 3241-3247 [Russian J. Inorg. Chem., 26, 1738-1741].

Orthorhombic, Pbnm, a = 5.444, b = 5.559, c = 7.754 Å, Z = 2. Mo radiation, R = 0.030 for 1285 reflexions.

Atomic positions

		x	y	z
Nb+Fe in	4(a)	0	0	0
Ca	4(c)	-0.0095	0.4552	1/4
O(1)	8(d)	0.7040	0.2065	0.0434
O(2)	4(c)	0.0844	0.0235	1/4

The structure is related to that of cryolite (1) (compare the related Ca_2LnNbO_6 compounds (2)).

1. Strukturbericht, 6, 29, 120; Structure Reports, 41A, 154.
2. This volume, p. 191.

LANTHANON NIOBATES
$LnNbO_4$ (Ln = La, Sm, Gd, Ho, Yb)

L.N. KINŽIBALO, V.K. TRUNOV, A.A. EVDOKIMOV and V.G. KRONGAUZ, 1982. Kristallografija, 27, 43-48 [Soviet Physics - Crystallography, 27, 22-25].

Monoclinic, I2/a, a = 6.569 [?], 5.422, 5.374, 5.301, 5.243, b = 11.529, 11.178, 11.095, 10.953, 10.841, c = 5.206, 5.121, 5.106, 5.073, 5.046 Å, β = 94.09, 94.69, 94.58, 94.58, 94.50°, Z = 4. Mo radiation, R = 0.042, 0.040, 0.031, 0.034, 0.030 for 789, 817, 596, 492, 831 reflexions.

Atomic positions [for Gd, similar values for others]

		x	y	z
Gd	in 4(e)	1/4	0.62135	0
Nb	4(e)	1/4	0.14534	0
O(1)	8(f)	0.09450	0.45980	0.25430
O(2)	8(f)	-0.00720	0.71720	0.29340

Fergusonite structure (1). Ln-8 O = 2.45-2.54, 2.38-2.45, 2.35-2.45, 2.32-2.41, 2.28-2.38, Nb-6 O = 1.82-1.93 (4 distances), 2.40-2.52 (2 distances) Å.

1. Structure Reports, 23, 388; 46A, 251.

CALCIUM LANTHANON NIOBATES
Ca_2LnNbO_6 (Ln = Nd, Gd, Ho, Er, Lu)

V.K. TRUNOV, L.I. KONSTANTINOVA, Ju.A. VELIKODNIJ, A.A. EVDOKIMOV and A.M. FROLOV, 1981. Ž. Neorg. Khim., 26, 3241-3247 [Russian J. Inorg. Chem., 26, 1738-1741].

Monoclinic, $P2_1/n$, a = 5.615-5.569, b = 5.877-5.767, c = 8.114-8.000 Å, β = 90.20-90.05°, Z = 2. Mo radiation, R = 0.033-0.055 for 1284-1594 reflexions.

Atomic positions (Ln = Nd, similar values for others)

	x	y	z
Nb	0	0	0
M(1)	1/2	1/2	0
M(2)	-0.0148	0.4422	0.2537
O(1)	0.7103	0.1761	0.0525
O(2)	0.6701	0.2163	0.4289
O(3)	0.1162	0.0560	0.2293

Ca/Ln ratios:
M(1) = 2:0, 1.82:0.18, 0.60:1.40, 0.24:1.76, 0.10:1.90
M(2) = 2:2, 2.18:1.82, 3.40:0.60, 3.76:0.24, 3.90:0.10

Cryolite structures (1). There is an abrupt change in cell parameters (accompanied by the appearance of weak superlattice reflexions doubling a and b) between Tb and Dy.

1. Strukturbericht, 6, 29, 120; Structure Reports, 41A, 154.

HOLMIUM SCANDIUM NIOBATE
Ho_2ScNbO_7

SAMARIUM SCANDIUM TANTALATE
Sm_2ScTaO_7

V.S. FILIP'EV, Ja.E. ČERNER, O.A. BUNINA and V.F. SEREGIN, 1982. Kristallografija, 27, 601-602 [Soviet Physics - Crystallography, 27, 364-365].

Cubic, Fd3m, a = 10.332, 10.466 Å, Z = 8. Powder data. Ho or Sm in 16(d): 1/2,1/2,1/2; Sc + Nb or Ta disordered in 16(c): 0,0,0; O(1) in 48(f): x,1/8,1/8, x = 0.199, 0.212; O(2) in 8(b): 3/8,3/8,3/8.

Pyrochlore structures (1), with some additional Ho/Sc disorder.

1. Structure Reports, 33A, 190.

THORIUM NIOBATE

$ThNb_4O_{12}$ — $4 \times Th_{0.25}NbO_3$

M.A. ALARIO-FRANCO, I.E. GREY, J.C. JOUBERT, H. VINCENT and M. LABEAU, 1982. Acta Cryst., A38, 177-186.

Orthorhombic, Immm, a = $3\sqrt{2}\ a_p$ = 16.551, b = $\sqrt{2}b_p$ = 5.517, c = $4c_p$ = 15.716 Å (a_p, b_p, c_p refer to a cubic perovskite cell), Z = 6. Electron and X-ray diffraction patterns.

The material is a cation deficient perovskite with a primary ordering of the Th ions in P4/mmm, a secondary sinusoidal modulation model in Immm (Fig. 1), and, thirdly, a system of octahedral tilts in Pmam. The final model can be described in P2mm.

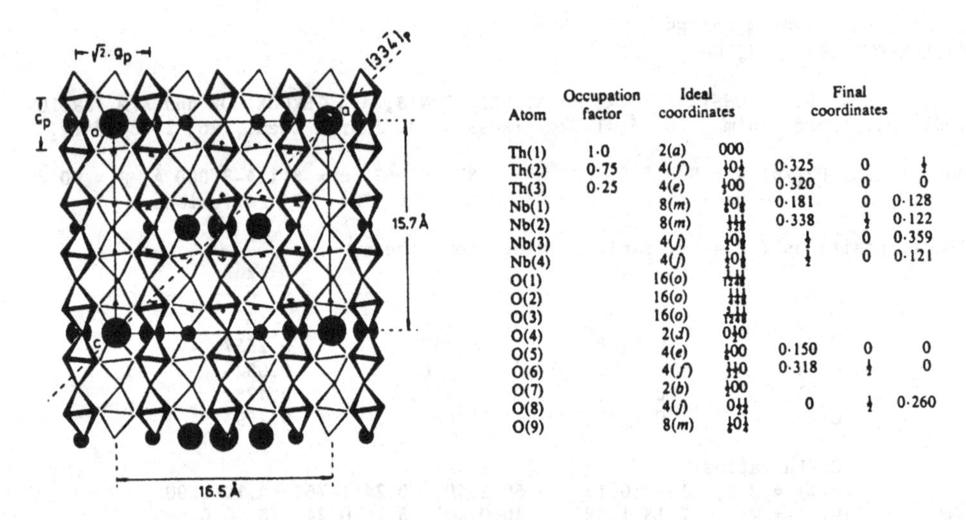

Atom	Occupation factor	Ideal coordinates		Final coordinates		
Th(1)	1·0	2(a)	000			
Th(2)	0·75	4(f)	1/3 0 1/2	0·325	0	1/2
Th(3)	0·25	4(e)	1/3 0 0	0·320	0	0
Nb(1)		8(m)	1/6 0 1/8	0·181	0	0·128
Nb(2)		8(m)	1/3 1/2 1/8	0·338	1/2	0·122
Nb(3)		4(j)	1/2 0 3/8	1/2	0	0·359
Nb(4)		4(j)	1/2 0 1/8	1/2	0	0·121
O(1)		16(o)	1/12 1/4 1/8			
O(2)		16(o)	1/4 1/4 1/8			
O(3)		16(o)	1/12 1/4 1/8			
O(4)		2(d)	0 1/2 0			
O(5)		4(e)	1/6 0 0	0·150	0	0
O(6)		4(f)	1/3 1/2 0	0·318	1/2	0
O(7)		2(b)	1/2 0 0			
O(8)		4(j)	0 1/2 1/4	0	1/2	0·260
O(9)		8(m)	1/6 0 1/4			

Fig. 1. Sinusoidal model for Th ordering in $ThNb_4O_{12}$.

LITHIUM TANTALATES

$LiTaO_3$, $9LiTaO_3.Ta_2O_5$

I. A. SANTORO, R.S. ROTH and M. AUSTIN, 1982. Acta Cryst., B38, 1094-1098.
II. E. PRINCE, 1982. Ibid., B38, 1099-1100.

$LiTaO_3$

Rhombohedral, R3c, a = 5.147, c = 13.766 Å, Z = 6. Neutron powder data. Ta and Li in 6(a): 0,0,z, z = 0, 0.2803; O in 18(b): (0.0492,0.3430,0.0693).

$9LiTaO_3.Ta_2O_5$
Rhombohedral, R3, a = 5.157, c = 13.784 Å, [Z = 0.56]. Neutron powder data.

Atomic positions

				x	y	z
3	Ta(1)	in	3(a)	0	0	0
3	Ta(2)		3(a)	0	0	1/2
0.1	Ta(3)		3(a)	0	0	0.273
0.1	Ta(4)		3(a)	0	0	0.773
9	O(1)		9(b)	0.0498	0.3407	0.0720
9	O(2)		9(b)	0.6593	0.9502	0.5720
2.5	Li(1)		3(a)	0	0	0.2765
2.5	Li(2)		3(a)	0	0	0.7765
0.1	Li(3)		3(a)	0	0	0.62

The $LiTaO_3$ structure is as previously described (1). The non-stoichiometric material has a defect structure (Fig. 1) as previously proposed (2).

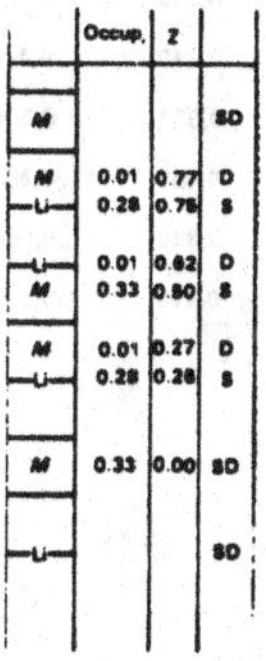

Fig. 1. The defect structure of the non-stoichiometric solid solution, $9LiTaO_3.Ta_2O_5$. S and D indicate atoms present in the stoichiometric or defect structures; approximate occupancies and z parameters are given.

1. Structure Reports, 32A, 312; 39A, 221.
2. K. NASSAU and M.E. LINES, 1970. J. Appl. Phys., 41, 533.

CAESIUM PENTATANTALATE
$Cs_3Ta_5O_{14}$

M. SERAFIN and R. HOPPE, 1982. Z. anorg. Chem., 493, 77-92.

Orthorhombic, Pbam, a = 26.235, b = 7.429, c = 7.388 Å, D_m = 6.95, Z = 4. Mo radiation, R = 0.10 for 1521 reflexions.

The structure contains a complex Ta_5O_{14} framework, with channels occupied by Cs ions. Ta ions have octahedral and trigonal-bipyramidal coordinations, Ta-O = 1.84-2.19 Å; Cs ions have irregular coordinations.

Atomic positions

	x	y	z
Cs 1 4g	0,2914	0,1600	0,0
Cs 2 4h	0,0225	0,2418	0,5
Cs 3 4h	0,3805	0,1576	0,5
Ta 1 4g	0,0547	0,4270	0,0
Ta 2 4g	0,4432	0,4112	0,0
Ta 3 4h	0,2545	0,4005	0,5
Ta 4 8i	0,1519	0,1610	0,2558
O 1 4f	0,0	0,5	0,173
O 2 4g	0,016	0,165	0,0
O 3 4g	0,168	0,130	0,0
O 4 4g	0,418	0,164	0,0
O 5 4h	0,127	0,173	0,5
O 6 4h	0,272	0,140	0,5
O 7 8i	0,202	0,336	0,306
O 8 8i	0,303	0,455	0,683
O 9 8i	0,096	0,336	0,200
O 10 8i	0,409	0,473	0,208

LEAD TANTALATES
$Pb_{1.34}Ta_2O_{6.24}$, $Pb_{14}Ta_{10}O_{39}$

H.G. SCOTT, 1982. J. Solid State Chem., 43, 131-139.

$Pb_{1.3}Ta_2O_{6.3}$, cubic, Fd3m, a = 10.549 Å, Z = 8. Powder data. 16 Ta in 16(c): 0,0,0; 10.7 Pb in 16(d): 1/2,1/2,1/2; 48 O(1) in 48(f): 0.310,1/8,1/8; 1.9 O(2) in 8(b): 3/8,3/8,3/8.

$Pb_{14}Ta_{10}O_{39}$, orthorhombic, B**b, a = 7.513, b = 7.530, c = 32.587 Å, Z = 2. Powder data, structure proposed.

Pyrochlore structures.

COPPER TANTALATE
$Cu_{0.8}Ta_3O_8$

P. NDALAMBA WA ILUNGA, B.-O. MARINDER and M. LUNDBERG, 1981. Chem. Scripta, 18, 217-220.

Orthorhombic, Pmma, a = 16.803, b = 3.891, c = 8.982 Å, D_m = 8.12, Z = 4. Mo radiation, R = 0.049 for 933 reflexions.

$LiNb_6O_{15}F$-type structure (1), but with each Ta atom disordered in two positions out of the mirror plane normal to b. Cu(I) ions partially occupy sites in four-sided tunnels along b, and Cu(II) ions, also with partial occupancy, are in square-planar sites.

1. Structure Reports, 30A, 334.

COPPER(I) TANTALATE(V)
$Cu_5Ta_{11}O_{30}$

L. JAHNBERG, 1982. J. Solid State Chem., 41, 286-292.

Hexagonal, P$\bar{6}$2c, a = 6.2297, c = 32.550 Å, D_m = 8.39, Z = 2. Cu radiation, R = 0.039 for 422 reflexions.

The structure (Fig. 1) contains alternate single and double layers of edge-sharing TaO_7 pentagonal bipyramids, linked by TaO_6 octahedra and linear CuO_2 groups. Ta-O = 1.90-2.43, Cu-O = 1.92, 1.93(1) Å.

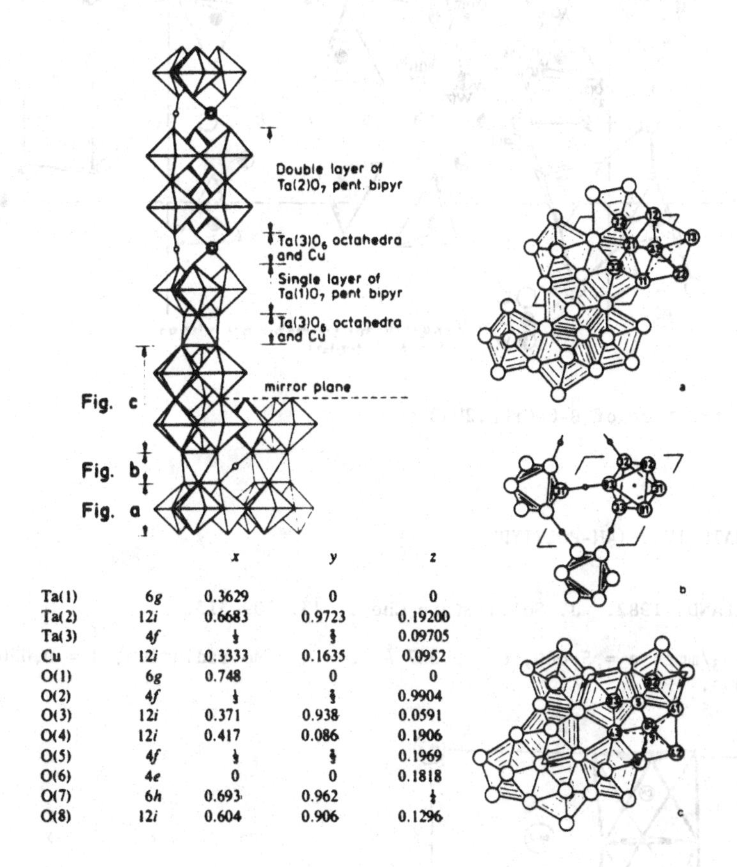

		x	y	z
Ta(1)	6g	0.3629	0	0
Ta(2)	12i	0.6683	0.9723	0.19200
Ta(3)	4f	⅓	⅔	0.09705
Cu	12i	0.3333	0.1635	0.0952
O(1)	6g	0.748	0	0
O(2)	4f	⅓	⅔	0.9904
O(3)	12i	0.371	0.938	0.0591
O(4)	12i	0.417	0.086	0.1906
O(5)	4f	⅓	⅔	0.1969
O(6)	4e	0	0	0.1818
O(7)	6h	0.693	0.962	¼
O(8)	12i	0.604	0.906	0.1296

Fig. 1. Structure of $Cu_5Ta_{11}O_{30}$.

CALCIUM CHROMATE DIHYDRATE
β-$CaCrO_4.2H_2O$

M. BEN AMOR, M. LOUËR and J.-Y. LE MAROUILLE, 1982. C.R. Acad. Sci. Paris, Ser. II, 294, 725-728.

Orthorhombic, Pbcm, a = 5.595, b = 15.715, c = 11.382 Å, Z = 8. Mo radiation, R = 0.026 for 1225 reflexions.

The structure (Fig. 1) contains CrO_4 tetrahedra, linked into sheets parallel to (010) by two independent $CaO_4(OH_2)_4$ distorted antiprisms, with interleaving sheets of water molecules. Cr-O = 1.62-1.65, Ca-O = 2.39-2.55, O-H...O = 2.75-3.00 Å.

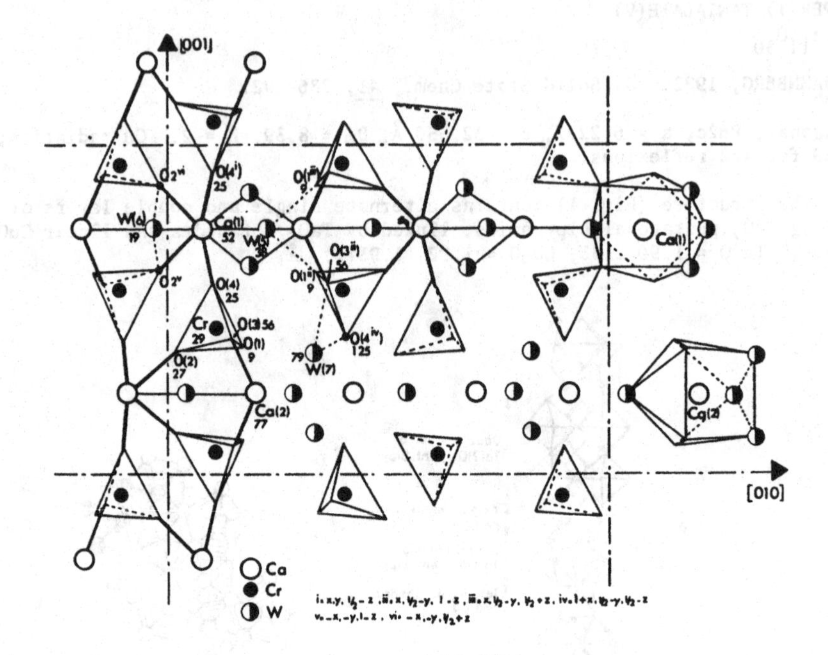

Fig. 1. Structure of β-$CaCrO_4.2H_2O$.

BARIUM CHROMATE(IV) (4H-POLYTYPE)
$BaCrO_3$

B.L. CHAMBERLAND, 1982. J. Solid State Chem., 43, 309-313.

Hexagonal, $P6_3/mmc$, a = 5.660, c = 9.357 Å, Z = 4. Mo radiation, R = 0.030 for 130 reflexions.

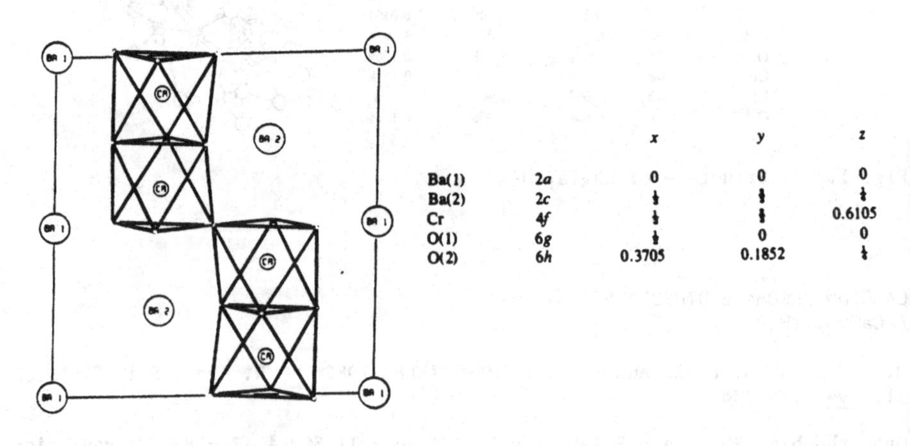

		x	y	z
Ba(1)	2a	0	0	0
Ba(2)	2c	⅓	⅔	¼
Cr	4f	⅓	⅔	0.6105
O(1)	6g	½	0	0
O(2)	6h	0.3705	0.1852	¼

Fig. 1. Structure of 4H-$BaCrO_3$.

Isostructural with β-$BaMnO_3$ (1), $BaRhO_3$ (2), and $SrMnO_3$ (3), the structure (Fig. containing a four-layer stacking sequence of close-packed BaO_3 layers, with Cr in all the octahedral oxygen interstices. Cr-O = 1.934, 1.952 (each x 3), Ba-12 O = 2.830-2.962(4) Å.

1. Structure Reports, 27, 663.
2. Ibid., 48A, 242.
3. Ibid., 48A, 236.

BARIUM CHROMATE(IV) (14H-POLYTYPE)

$BaCrO_3$

B.L. CHAMBERLAND and L. KATZ, 1982. Acta Cryst., B38, 54-57.

Hexagonal, $P6_3/mmc$, a = 5.650, c = 32.467 Å, Z = 14. Mo radiation, R = 0.041 for 346 reflexions.

The structure (Fig. 1) contains a 14-layer stacking sequence of close-packed BaO_3 layers, with Cr in all the O_6 octahedral sites. Cr-O = 1.90-1.99, Ba-O = 2.82-3.06 Å.

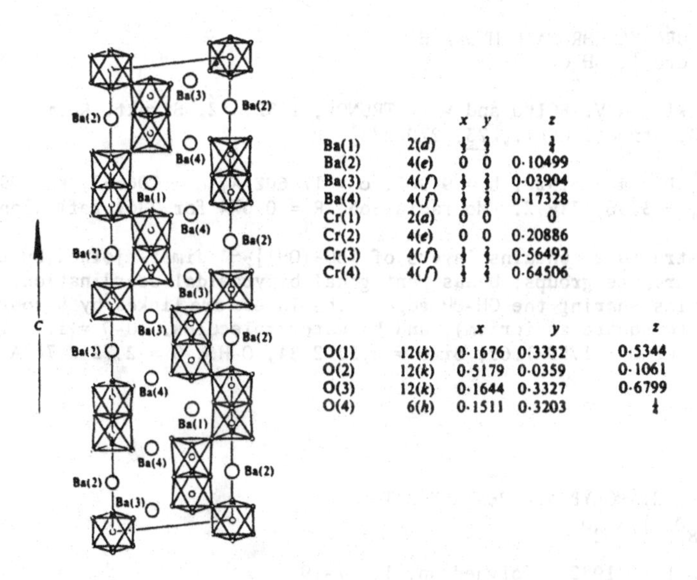

		x	y	z
Ba(1)	2(d)	1/3	2/3	1/4
Ba(2)	4(e)	0	0	0·10499
Ba(3)	4(f)	1/3	2/3	0·03904
Ba(4)	4(f)	1/3	2/3	0·17328
Cr(1)	2(a)	0	0	0
Cr(2)	4(e)	0	0	0·20886
Cr(3)	4(f)	1/3	2/3	0·56492
Cr(4)	4(f)	1/3	2/3	0·64506

		x	y	z
O(1)	12(k)	0·1676	0·3352	0·5344
O(2)	12(k)	0·5179	0·0359	0·1061
O(3)	12(k)	0·1644	0·3327	0·6799
O(4)	6(h)	0·1511	0·3203	1/4

Fig. 1. Structure of $BaCrO_3$ (14H).

POTASSIUM BISMUTH CHROMATE DICHROMATE MONOHYDRATE

$KBi(CrO_4)(Cr_2O_7).H_2O$

A. RIOU, Y. GERAULT and Y. CUDENNEC, 1982. Acta Cryst., B38, 1693-1696.

Triclinic, $P\bar{1}$, a = 6.853, b = 7.280, c = 10.703 Å, α = 80.50, β = 88.70, γ = 85.71°, D_m = 3.79, Z = 2. Mo radiation, R = 0.039 for 2877 reflexions.

The structure (Fig. 1) contains (001) layers which contain Bi_2O_{14} groups of two edge-sharing BiO_8 distorted dodecahedra, and tetrahedral CrO_4^{2-} and $Cr_2O_7^{2-}$ ions. The layers are joined by 8-coordinate K^+ ions. Bi-O = 2.25-2.69, Cr-O = 1.60-1.72 (chromate), 1.59-1.67 (dichromate, terminal), 1.77, 1.79 (dichromate, bridging), K-O = 2.74-3.03, O-H...O = 2.82-2.87 Å, Cr-O-Cr = 118°.

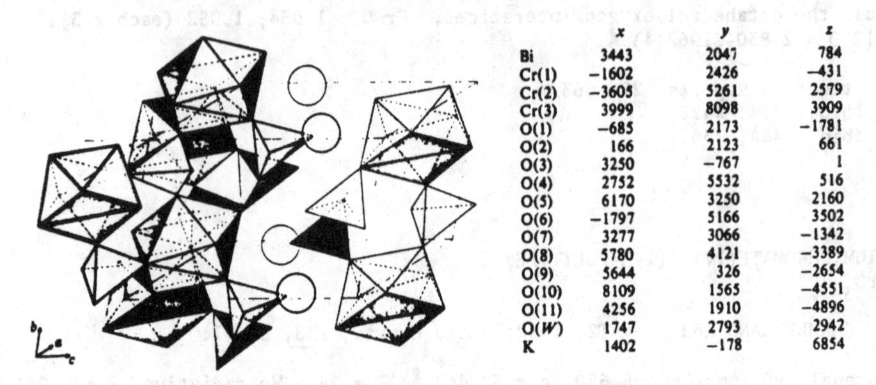

	x	y	z
Bi	3443	2047	784
Cr(1)	−1602	2426	−431
Cr(2)	−3605	5261	2579
Cr(3)	3999	8098	3909
O(1)	−685	2173	−1781
O(2)	166	2123	661
O(3)	3250	−767	1
O(4)	2752	5532	516
O(5)	6170	3250	2160
O(6)	−1797	5166	3502
O(7)	3277	3066	−1342
O(8)	5780	4121	−3389
O(9)	5644	326	−2654
O(10)	8109	1565	−4551
O(11)	4256	1910	−4896
O(W)	1747	2793	2942
K	1402	−178	6854

Fig. 1. Structure of $KBi(CrO_4)(Cr_2O_7).H_2O$, and atomic positional parameters ($\times 10^4$).

STRONTIUM URANYL CHROMATE HYDRATE
$Sr[UO_2(OH)CrO_4]_2.8H_2O$

V.N. SEREŽKIN, N.V. BOIKO and V.K. TRUNOV, 1982. Ž. Strukt. Khim., 23, No. 2, 121-124 [J. Struct. Chem., 23, 270-273].

Triclinic, $P\bar{1}$, a = 8.923, b = 9.965, c = 11.602 Å, α = 106.63, β = 99.09, γ = 97.26°, D_m = 3.56, Z = 2. Mo radiation, R = 0.074 for 2861 reflexions.

The structure contains layers of $[UO_2(OH)]_2^{2+}$ dimers joined by tridentate bridging chromate groups; U has pentagonal bipyramidal coordination, with pairs of bipyramids sharing the OH-OH edge. The layers are linked by 9-coordinate Sr ions (capped square antiprism), and by water molecules. U-O = 1.78-1.80 (uranyl), 2.33-2.44, Cr-O = 1.59-1.69, Sr-O = 2.53-2.84, O-H...O = 2.72-2.78 Å.

AMMONIUM β-OCTAMOLYBDATE PENTAHYDRATE
$(NH_4)_4[Mo_8O_{26}].5H_2O$

T.J.R. WEAKLEY, 1982. Polyhedron, 1, 17-19.

Triclinic, $P\bar{1}$, a = 9.769, b = 9.832, c = 7.848 Å, α = 99.11, β = 101.03, γ = 97.40°, D_m = 3.02, Z = 1. Cu radiation, R = 0.080 for 1568 reflexions (films, densitometer intensities), atomic positional parameters not given . Previous study in 1; the tetrahydrate has also been described (2).

The octamolybdate anion contains eight MoO_6 octahedra sharing edges (Fig. 1), with short terminal Mo-O (1.69-1.75 Å), longer bonds to bicoordinate atoms (1.88-2.00 Å), and long bonds to multiply-shared interior atoms (2.18-2.39 Å). N...O = 2.88-3.09, H_2O...O = 2.97, 2.99 Å.

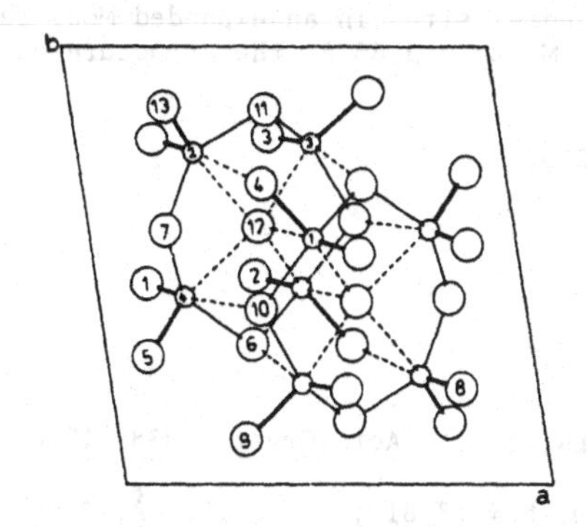

Fig. 1. The $Mo_8O_{26}^{4-}$ anion.

1. Structure Reports, 13, 265.
2. Ibid., 13, 267; 43A, 351.

LITHIUM MOLYBDENUM OXIDE
$LiMoO_2$

MOLYBDENUM(IV) OXIDE
MoO_2

D.E. COX, R.J. CAVA, D.B. McWHAN and D.W. MURPHY, 1982. J. Phys. Chem. Solids, 43, 657-666.

Monoclinic, $P2_1/c$, a = 5.57, 5.611, b = 5.21, 4.856, c = 5.86, 5.629 Å, β = 118.8, 120.95°, Z = 4. Neutron powder data, at 4-440K.

Atomic positions

	$LiMoO_2$			MoO_2		
	x	y	z	x	y	z
Mo	0.2243	-0.0041	0.0063	0.2329	-0.0080	0.0169
O(1)	0.0904	0.2613	0.1832	0.1127	0.2164	0.2339
O(2)	0.4111	0.7349	0.3319	0.3903	0.6966	0.2990
Li	0.260	0.000	0.514			

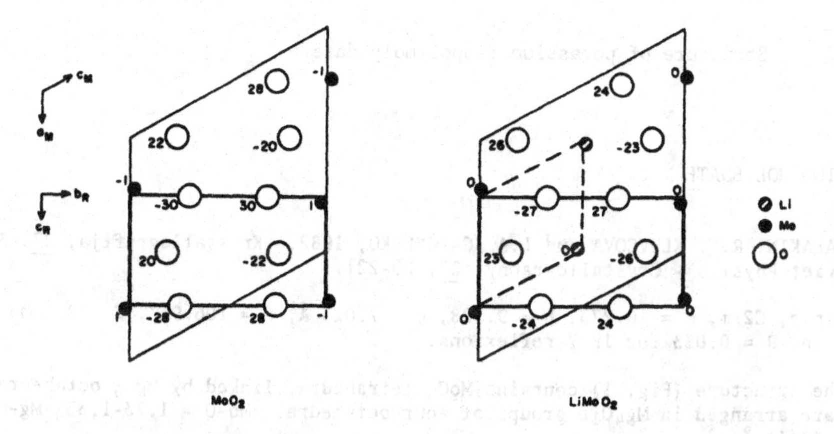

Fig. 1. Structures of MoO_2 and $LiMoO_2$; heavy lines = rutile cell, broken lines = NiAs-type cell.

MoO_2 has a distorted rutile structure (Fig. 1) as previously described (1); Mo has octahedral coordination, mean Mo-O = 2.01 Å, with Mo_2 pairs, Mo-Mo = 2.52 Å. In $LiMoO_2$, Li is in octahedral sites in an expanded MoO_2 lattice (Fig. 1), mean Mo-O = mean Li-O = 2.12, Mo-Mo = 2.46 Å; the structure can also be described as a distorted NiAs-type.

1. Structure Reports, 32A, 259.

POTASSIUM ISOPOLYMOLYBDATE
$K_8[Mo_{36}O_{112}(H_2O)_{16}].36\text{-}40H_2O$

B. KREBS and I. PAULAT-BÖSCHEN, 1982. Acta Cryst., B38, 1710-1718.

Monoclinic, $P2_1/c$, a = 16.550, b = 18.810, c = 27.730 Å, β = 116.82°, D_m = 2.81, Z = 2. Mo radiation, R = 0.057 for 12318 reflexions.

The centrosymmetric isopolymolybdate anion (Fig. 1) contains edge- and corner-sharing MoO_6 octahedra, four MoO_7 polyhedra, and coordinated water molecules, four of which are linked to two Mo. K^+ ions and non-coordinated water molecules are partly disordered.

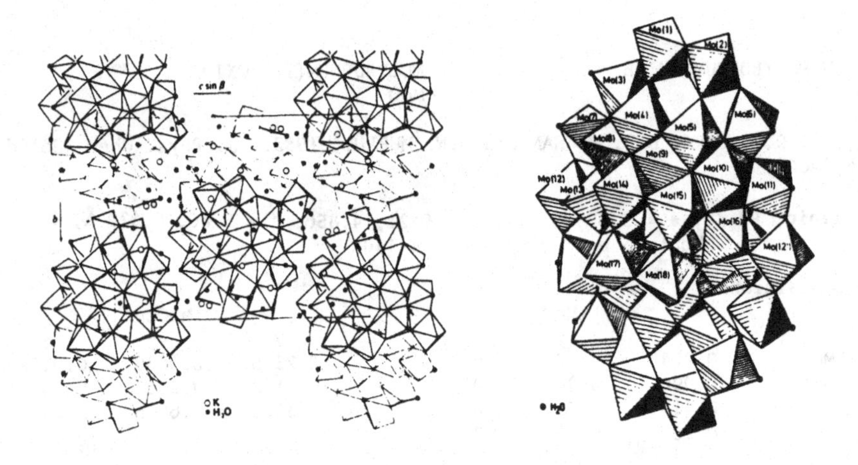

Fig. 1. Structure of potassium isopolymolybdate.

MAGNESIUM MOLYBDATE
$MgMoO_4$

V.V. BAKAKIN, R.F. KLEVCOVA and L.A. GAPONENKO, 1982. Kristallografija, 27, 38-42 [Soviet Physics - Crystallography, 27, 20-22].

Monoclinic, C2/m, a = 10.273, b = 9.288, c = 7.025 Å, β = 106.96°, Z = 8. Mo radiation, R = 0.053 for 1907 reflexions.

The structure (Fig. 1) contains MoO_4 tetrahedra, linked by MgO_6 octahedra which are arranged in Mg_4O_{16} groups of four octahedra. Mo-O = 1.73-1.85, Mg-O = 2.02-2.17(1) Å.

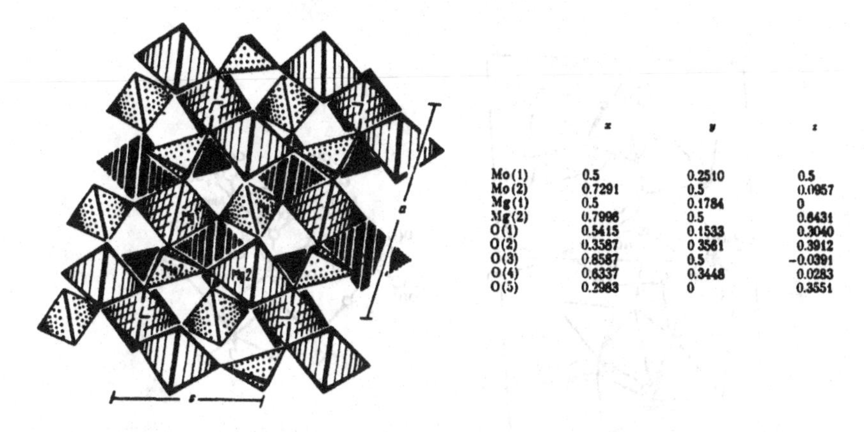

	x	y	z
Mo(1)	0.5	0.2510	0.5
Mo(2)	0.7291	0.5	0.0957
Mg(1)	0.5	0.1784	0
Mg(2)	0.7996	0.5	0.6431
O(1)	0.5415	0.1533	0.3040
O(2)	0.3587	0.3561	0.3912
O(3)	0.8587	0.5	−0.0391
O(4)	0.6337	0.3448	0.0283
O(5)	0.2983	0	0.3551

Fig. 1. Structure of magnesium molybdate.

SODIUM PHOSPHOVANADATOMOLYBDATES

$Na_5[PMo_{10}V_2O_{40}].15H_2O$ (I), $Na_3H_6[PMo_6V_6O_{40}].16H_2O$ (II)

I. V.S. SERGIENKO, L.G. DETUŠEVA, E.N. JURČENKO and M.A. PORAJ-KOŠIC, 1981. Ž. Strukt Khim., 22, No. 6, 37-48 [J. Struct. Chem., 22, 831-839].
II. R.F. KLEVCOVA, E.N. JURČENKO, L.A. GLINSKAJA, L.G. DETUŠEVA and L.I. KUZNECOVA, 1981. Ibid., 22, No. 6, 49-61 [Ibid., 22, 840-850].

I. Triclinic, $P\bar{1}$, a = 11.681, b = 12.373, c = 17.936 Å, α = 82.5, β = 84.5, γ = 65.0°, Z = 2. Mo radiation, R = 0.057 for 4610 reflexions.

II. Monoclinic, $P2_1/m$, a = 13.667, b = 15.260, c = 11.529 Å, β = 106.98°, Z = 2. Mo radiation, R = 0.079 for 2758 reflexions.

Both structures contain Keggin-type heteropolyanions, with a central PO_4 tetrahedron and Mo and V distributed randomly in the twelve MO_6 octahedra. The anions are linked by Na^+ ions and water molecules.

SODIUM TETRAHYDROGENTETRAMOLYBDOTETRAARSENATE(V)(4-) HEXAHYDRATE

$Na_4[H_4As_4Mo_4O_{26}].6H_2O$

Y. TAKEUCHI, A. KOBAYASHI and Y. SASAKI, 1982. Acta Cryst., B38, 242-244.

Triclinic, $P\bar{1}$, a = 10.136, b = 10.697, c = 7.744 Å, α = 111.68, β = 85.06, γ = 118.77°, D_m = 3.18, Z = 1. Mo radiation, R = 0.051 for 3402 reflexions.

The structure (Fig. 1) contains a discrete heteropolyanion, which consists of two pairs of edge-sharing MoO_6 octahedra bridged by corner sharing with four AsO_4 tetrahedra. Mean Mo-O = 1.70, 1.91, 2.00, 2.22, 2.32 Å, for five types of bond; As-O = 1.67-1.69 (bridging), 1.64 (terminal O), 1.71, 1.72(1) Å (terminal OH). Na ions have 5- and 6-coordinations to anion oxygens and water molecules, Na-O = 2.36-2.85 Å, and hydrogen bonds are present, O-H...O = 2.77-3.04 Å.

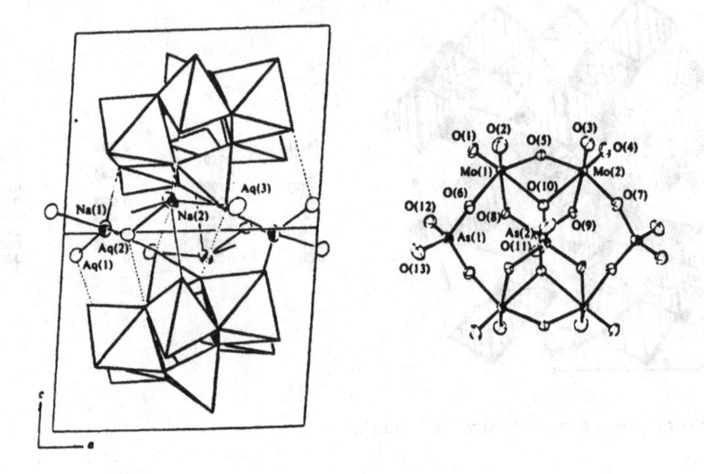

Fig. 1. Structure of $Na_4[H_4As_4Mo_4O_{26}].6H_2O$.

CAESIUM BISMUTH MOLYBDATE
$Cs_5Bi(MoO_4)_4$

V.A. EFREMOV, R.F. KLEVCOVA, B.I. LAZORJAK, L.A. GLINSKAJA, I. MATSIČEK and S.F. SOLODOVNIKOV, 1982. Kristallografija, 27, 461-466 [Soviet Physics - Crystallography, 27, 281-284].

Monoclinic, Cc, a = 8.347, b = 25.338, c = 10.436 Å, β = 110.15°, D_m = 4.76, Z = 4. Mo radiation, R = 0.052 for 2972 reflexions.

The structure (Fig. 1) contains MoO_4 tetrahedra, BiO_6 distorted octahedra, and CsO_{12} polyhedra linked into chains, which are interwoven with columns of CsO_{10} polyhedra. Mo-O = 1.63-1.86(3), Bi-O = 2.14-2.70(3) (suggesting lone-pair influence), Cs-O = 2.81-3.90 Å.

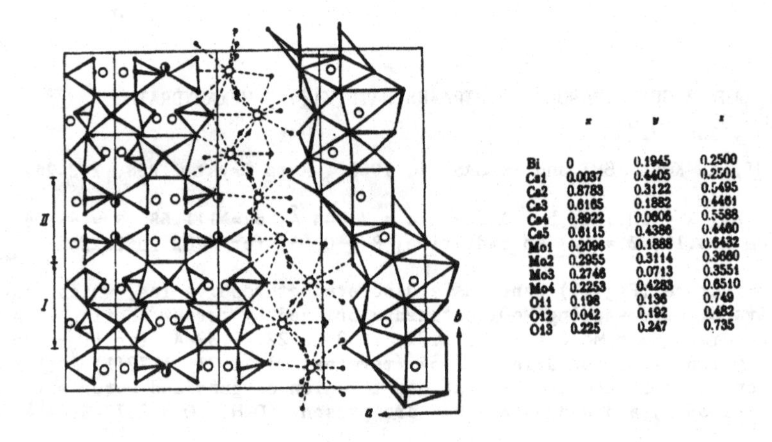

	x	y	z
Bi	0	0.1945	0.2500
Cs1	0.0037	0.4405	0.2501
Cs2	0.8783	0.3122	0.5495
Cs3	0.6165	0.1882	0.4461
Cs4	0.8922	0.0606	0.5588
Cs5	0.6115	0.4386	0.4460
Mo1	0.2096	0.1888	0.6432
Mo2	0.2955	0.3114	0.3660
Mo3	0.2748	0.0713	0.3551
Mo4	0.2253	0.4283	0.6510
O11	0.198	0.136	0.749
O12	0.042	0.192	0.482
O13	0.225	0.247	0.735

Fig. 1. Structure of caesium bismuth molybdate.

AMMONIUM VANADOMOLYBDATE
$(NH_4)_{0.2}V_{0.2}Mo_{0.8}O_3$

L.M. PLJASOVA, S.D. KIPIK and I.P. OLEN'KOVA, 1982. Ž. Strukt Khim., 23, No. 4, 120-124.

Hexagonal, $P6_3/m$, a = 10.53, c = 3.719 Å, D_m = 3.94, Z = 6. Cu radiation, powder data.

Atomic positions

	x	y	z
4.8 Mo + 1.2 V	0.6552	0.5460	1/4
6 O(1)	0.589	0.530	3/4
6 O(2)	0.732	0.744	1/4
6 O(3)	0.476	0.276	1/4
1.2 NH_4	0	0	0

The structure contains edge- and corner-sharing $(Mo/V)O_6$ octahedra. Mo/V-O = 1.65-2.50(2), N...O = 2.88 Å.

IRON(II) MOLYBDENUM(IV) OXIDE
$Fe_2Mo_3O_8$

Y. LE PAGE and P. STROBEL, 1982. Acta Cryst., B38, 1265-1267.

Hexagonal, $P6_3mc$, a = 5.7732, c = 10.0542 Å, at 298K, Z = 2. Mo radiation, R = 0.029 for 966 reflexions.

Atomic positions

	x	y	z
Mo	0·14605	−x	0·25 (fixed)
Fe(1)	1/3	2/3	−0·04810
Fe(2)	1/3	2/3	0·51301
O(1)	0	0	0·3906
O(2)	1/3	2/3	0·1470
O(3)	0·4883	−x	0·3629
O(4)	0·1671	−x	0·6344

Isostructural with $Zn_2Mo_3O_8$ (1). Mo and Fe(2) have distorted octahedral coordinations, and Fe(1) has distorted tetrahedral coordination; the Mo atoms are arranged in Mo_3 clusters. Mo...Mo = 2.530(1), Mo-O = 1.956-2.140, Fe(2)-O = 2.062, 2.163, Fe(1)-O = 1.962, 1.995(5) Å.

1. Structure Reports, 21, 333; 31A, 155.

POTASSIUM NICKEL MOLYBDATE
$K_2Ni_2(MoO_4)_3$

R.F. KLEVCOVA and L.A. GLINSKAJA, 1982. Ž. Strukt. Khim., 23, No. 5, 176-179.

Monoclinic, $P2_1/c$, a = 6.952, b = 8.910, c = 19.733 Å, β = 108.06°, Z = 4. Mo radiation, R = 0.068 for 3368 reflexions.

Isostructural with the Zn compound (1), the structure containing NiO_6 octahedra joined by edge-sharing and by bridging MoO_4 tetrahedra; the chains are linked by 8-coordinate K ions. Ni-O = 2.01-2.17, Mo-O = 1.71-1.83, K-O = 2.65-3.66 Å.

1. Structure Reports, 41A, 252.

SODIUM ZINC MOLYBDATE

$Na_{0.5}Zn_{2.75}(MoO_4)_3$

C. GICQUEL-MAYER and M. MAYER, 1982. Rev. Chim. Minér., 19, 91-98.

Triclinic, P$\bar{1}$, a = 6.983, b = 8.594, c = 10.825 Å, α = 65.87, β = 66.19, γ = 78.17°, Z = 2. Mo radiation, R = 0.037 for 2978 reflexions.

The structure (Fig. 1) contains MoO_4 tetrahedra linked by ZnO_6 octahedra and ZnO_5 trigonal bipyramids and by six-coordinate Na ions. Mo-O = 1.70-1.79, Zn-O = 2.03-2.24, Na-O = 2.30-2.96(1) Å.

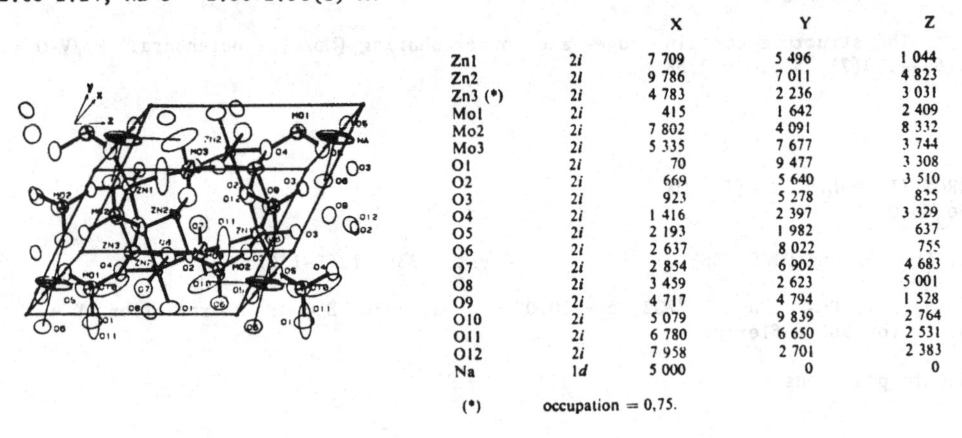

		X	Y	Z
Zn1	2i	7 709	5 496	1 044
Zn2	2i	9 786	7 011	4 823
Zn3 (*)	2i	4 783	2 236	3 031
Mo1	2i	415	1 642	2 409
Mo2	2i	7 802	4 091	8 332
Mo3	2i	5 335	7 677	3 744
O1	2i	70	9 477	3 308
O2	2i	669	5 640	3 510
O3	2i	923	5 278	825
O4	2i	1 416	2 397	3 329
O5	2i	2 193	1 982	637
O6	2i	2 637	8 022	755
O7	2i	2 854	6 902	4 683
O8	2i	3 459	2 623	5 001
O9	2i	4 717	4 794	1 528
O10	2i	5 079	9 839	2 764
O11	2i	6 780	6 650	2 531
O12	2i	7 958	2 701	2 383
Na	1d	5 000	0	0

(*) occupation = 0,75.

Fig. 1. Structure of sodium zinc molybdate, and atomic positional parameters (x 10^4).

CADMIUM YTTRIUM MOLYBDATE

$CdY_4Mo_3O_{16}$

J.B. BOURDET, R. CHEVALIER, J.P. FOURNIER, R. KOHLMULLER and J. OMALY, 1982. Acta Cryst., B38, 2371-2374.

Cubic, Pn3n, a = 10.688 Å, Dm = 4.87, Z = 4. R = 0.04 for 173 (of 1046) reflexions.

The material has a fluorite-type superstructure (Fig. 1), with some Cd/Y disorder. Mo has tetrahedral coordination and Y and Cd,Y have deformed cubic coordinations. Mo-O = 1.75, Y-O = 2.27, 2.50, Cd,Y-O = 2.22, 2.52(3) Å.

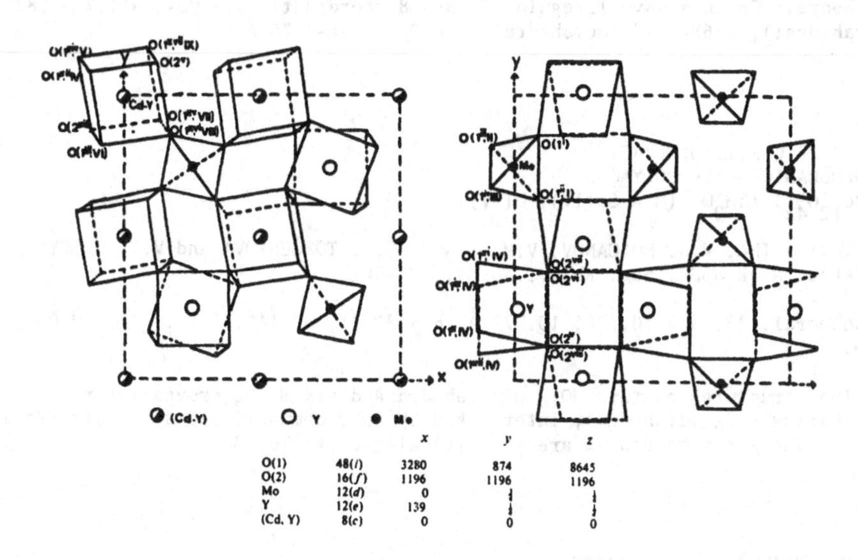

		x	y	z
O(1)	48(*i*)	3280	874	8645
O(2)	16(*f*)	1196	1196	1196
Mo	12(*d*)	0	[illegible]	[illegible]
Y	12(*e*)	139	[illegible]	[illegible]
(Cd, Y)	8(*c*)	0	0	0

Fig. 1. Cation polyhedra at z ∿ 0 and 1/4, and atomic positional parameters (decimal fractions x 10^4) for cadmium yttrium molybdate.

CERIUM(III) MOLYBDATE

$Ce_2Mo_4O_{15}$ $Ce_2(MoO_4)_2(Mo_2O_7)$

G.D. FALLON and B.M. GATEHOUSE, 1982. J. Solid State Chem., 44, 156-161.

Triclinic, P$\bar{1}$, a = 11.903, b = 7.509, c = 7.385 Å, α = 94.33, β = 97.41, γ = 88.56°, Z = 2. Mo radiation, R = 0.035 for 6314 reflexions.

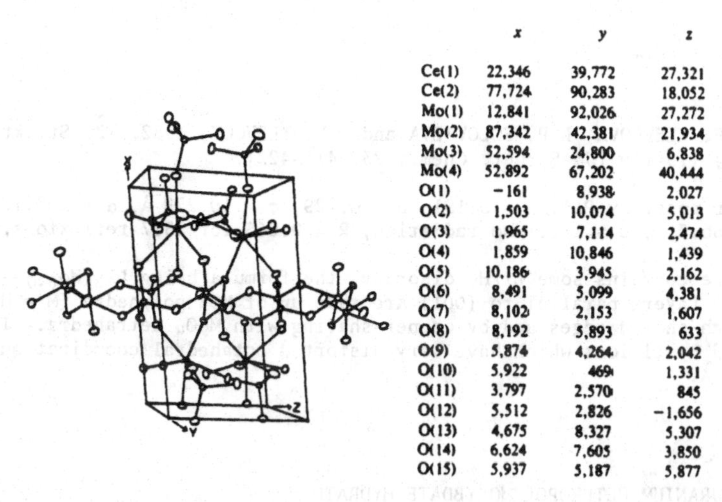

	x	y	z
Ce(1)	22,346	39,772	27,321
Ce(2)	77,724	90,283	18,052
Mo(1)	12,841	92,026	27,272
Mo(2)	87,342	42,381	21,934
Mo(3)	52,594	24,800	6,838
Mo(4)	52,892	67,202	40,444
O(1)	−161	8,938	2,027
O(2)	1,503	10,074	5,013
O(3)	1,965	7,114	2,474
O(4)	1,859	10,846	1,439
O(5)	10,186	3,945	2,162
O(6)	8,491	5,099	4,381
O(7)	8,102	2,153	1,607
O(8)	8,192	5,893	633
O(9)	5,874	4,264	2,042
O(10)	5,922	469	1,331
O(11)	3,797	2,570	845
O(12)	5,512	2,826	−1,656
O(13)	4,675	8,327	5,307
O(14)	6,624	7,605	3,850
O(15)	5,937	5,187	5,877

Fig. 1. Structure of cerium molybdate, and atomic positional parameters (x 10^5 for Ce, Mo; x 10^4 for O).

The structure (Fig. 1) contains isolated MoO_4 tetrahedra and dimolybdate chains which consist of pairs of edge-sharing MoO_6 octahedra bridged by two MoO_4

tetrahedra. Ce ions have irregular 7- and 8-coordinations. Mo-O = 1.73-1.83 (tetrahedral), 1.68-2.42 (octahedral), Ce-O = 2.36-2.75 Å.

MOLYBDOCERIC HETEROPOLYACID
MOLYBDOURANIC HETEROPOLYACID
$H_8[MMo_{12}O_{42}].18H_2O$ (M = Ce(IV), U(IV))

I.V. TAT'JANINA, V.N. MOLČANOV, V.M. IONOV, E.A. TORČENKOVA and V.I. SPICYN, 1982. Izv. Akad. Nauk SSSR, Ser. Khim., No. 4, 803-807.

Rhombohedral, $R\bar{3}$, a = 10.774, 10.777 Å, α = 81.11, 81.14°, Z = 1. R = 0.032, 0.039.

The structures contain MO_{10} eicosahedra and six Mo_2O_9 groups (pairs of face-sharing MoO_6 octahedra), interlinked by four common O atoms to form $MMo_{12}O_{42}$ groups. The water molecules are statistically distributed.

LITHIUM CERIUM(III) MOLYBDATE
$LiCe(MoO_4)_2$

I. A.N. EGOROVA, A.A. MAIER, N.N. NEVSKIJ and M.V. PROVOTOROV, 1982. Izv. Akad. Nauk SSSR, Neorg. Mater., 18, 2036-2038.
II. A.N. EGOROVA, L.E. FYKIN, E.E. RIDER, M.V. PROVOTOROV and A.A. MAIER, 1982. Ibid., 18, 2039-2043.

Tetragonal, $I4_1/a$, a = 5.284, c = 11.563 Å, D_m = 4.79, Z = 2. R = 0.039.

Isostructural with $CaMoO_4$ (1), with MoO_4 tetrahedra linked by Li/Ce ions.

1. Strukturbericht, 1, 349; Structure Reports, 33A, 334.

LITHIUM HOLMIUM MOLYBDATE
$Li_7Ho_3(MoO_4)_8$

E.N. IPATOVA, R.F. KLEVCOVA, L.P. SOLOV'EVA and P.V. KLEVCOV, 1982. Ž. Strukt. Khim., 23, No. 3, 115-119 [J. Struct. Chem., 23, 418-422].

Triclinic, $P\bar{1}$ (or possibly P1), a = 5.191, b = 6.729, c = 10.279 Å, α = 108.10, β = 100.19, γ = 66.31°, Z = 1/2. Mo radiation, R = 0.065 for 2517 reflexions.

The structure contains some Li/Ho disorder, the formula being $Li_3(Ho_{0.75}Li_{0.25})_2(MoO_4)_4$. Layers parallel to (001) are made up of MO_8 polyhedra (M = Ho, Li) joined through shared edges and by corner-sharing with MoO_4 tetrahedra. The layers are linked by Li ions which have very distorted octahedral coordinations, Li-O = 1.94-2.59 Å.

AMMONIUM ERBIUM URANIUM HETEROPOLYMOLYBDATE HYDRATE
$(NH_4)_2[Er_2UMo_{12}O_{42}].22H_2O$

I.V. TAT'JANINA, E.B. FOMIČEVA, V.N. MOLČANOV, V.E. ZAVODNIK, V.K. BEL'SKIJ and E.A. TORČENKOVA, 1982. Kristallografija, 27, 233-238 [Soviet Physics - Crystallography, 27, 142-145].

Triclinic, $I\bar{1}$, a = 13.094, b = 13.195, c = 16.846 Å, α = 89.73, β = 95.15, γ = 93.43°, D_m = 3.23, Z = 2. Mo radiation, R = 0.049 for 6073 reflexions.

The heteropolyanion (Fig. 1) contains six pairs of face-sharing MoO_6 octahedra around a central UO_6 octahedron, with two Er^{3+} ions each coordinated to three molybdate oxygens and five water molecules. The anions are linked by a system of hydrogen bonds via the water molecules and ammonium ions.

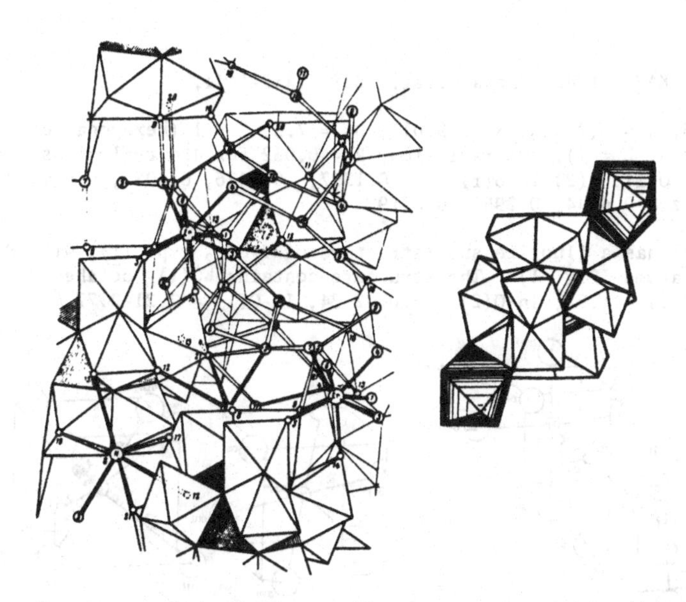

Fig. 1. Structure of the erbium uranium heteropolymolybdate.

POTASSIUM LEAD(II)UNDECATUNGSTOGALLATE(III)(7-) HEXADECAHYDRATE
$K_7GaPbW_{11}O_{39}.16H_2O$

G.F. TOURNÉ, C.M. TOURNÉ and A. SCHOUTEN, 1982. Acta Cryst., B38, 1414-1418.

Orthorhombic, Pnma, a = 13.809, b = 15.087, c = 25.184 Å, D_m = 4.46, Z = 4. Mo radiation, R = 0.06 for 1367 reflexions.

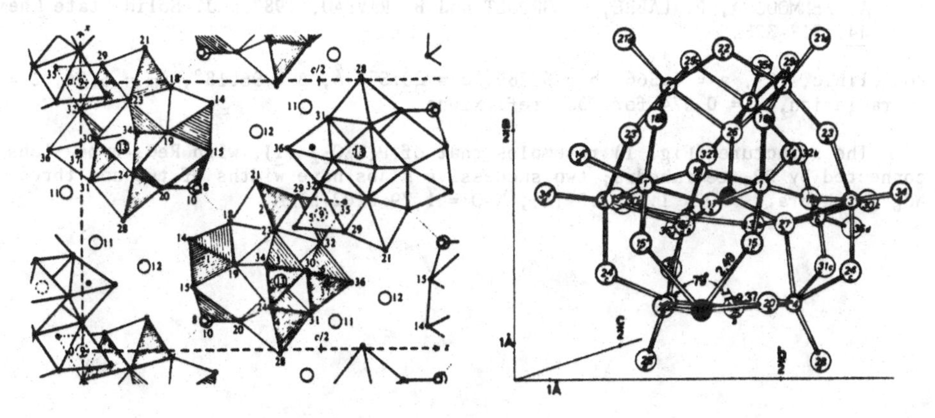

Fig. 1. Structure of $K_7GaPbW_{11}O_{39}.16H_2O$.

The structure (Fig. 1) contains a Keggin-type anion, with Ga in the central tetrahedral site and Pb bonded to 4 oxygen atoms. The anions are linked by K ions and two water molecules, the other water molecules being disordered.

INDIUM TUNGSTATE

In_6WO_{12}

D. MICHEL and A. KAHN, 1982. Acta Cryst., B38, 1437-1441.

Rhombohedral, $R\bar{3}$, a = 6.228 Å, α = 99.01°, D_m = 7.5, Z = 1 (hexagonal cell, a = 9.472, c = 8.939 Å, Z = 3). Mo radiation, R = 0.032 for 484 reflexions. W in 1(a): 0,0,0; In, O(1), O(2) in 6(f): x = 0.1397, -0.0426, 0.4057, y = 0.3214, 0.0974, 0.5838, z = 0.6044, 0.2969, 0.8219.

The material has a fluorite superstructure of Y_6UO_{12} type (1), with O vacancies paired around W atoms (Fig. 1). The structure contains $WO(1)_6$ octahedra and $In_4O(2)$ tetrahedra. W-O(1) = 1.93, In-O(2) = 2.07-2.24, In-O(1) = 2.21-2.77 Å.

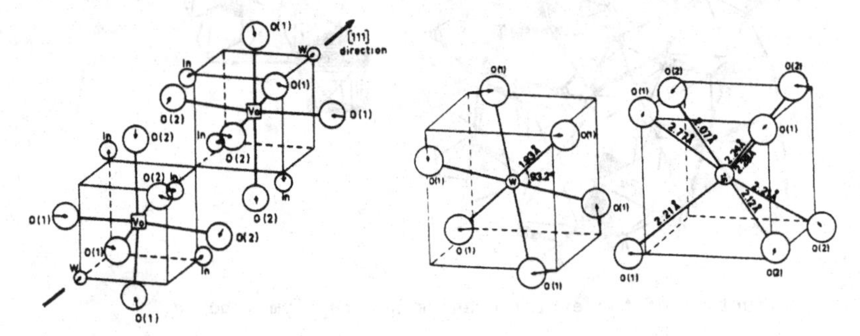

Fig. 1. Structure of In_6WO_{12}.

1. Structure Reports, 31A, 162.

PHOSPHORUS TUNGSTEN OXIDES

$P_4W_{10}O_{38}$

I. A. BENMOUSSA, P. LABBE, D. GROULT and B. RAVEAU, 1982. J. Solid State Chem., 44, 318-325.

Monoclinic, $P2_1$, a = 6.566, b = 5.285, c = 20.573 Å, β = 96.18°, Dm = 6.04, Z = 1. Mo radiation, R = 0.074 for 2339 reflexions.

The structure (Fig. 1) resembles that of $P_4W_8O_{32}$ (1), with ReO_3-type slabs connected by PO_4 tetrahedra; two successive slabs have widths of two and three WO_6 octahedra. P-O = 1.48-1.57(5), W-O = 1.79-2.02(5) Å.

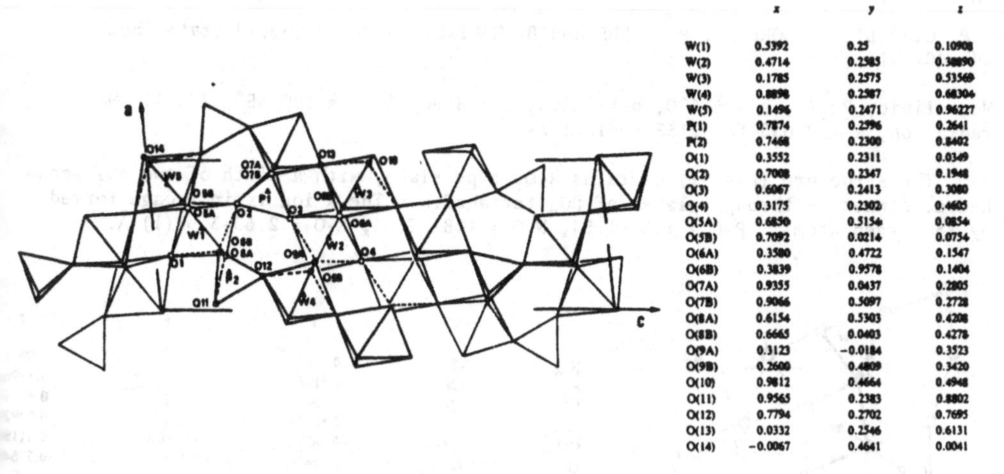

	x	y	z
W(1)	0.5392	0.25	0.10908
W(2)	0.4714	0.2585	0.38890
W(3)	0.1785	0.2575	0.53569
W(4)	0.8898	0.2587	0.68304
W(5)	0.1496	0.2471	0.96227
P(1)	0.7874	0.2596	0.2641
P(2)	0.7468	0.2300	0.8402
O(1)	0.3552	0.2311	0.0349
O(2)	0.7008	0.2347	0.1948
O(3)	0.6067	0.2413	0.3080
O(4)	0.3175	0.2302	0.4605
O(5A)	0.6850	0.5154	0.0854
O(5B)	0.7092	0.0214	0.0754
O(6A)	0.3580	0.4722	0.1547
O(6B)	0.3839	0.9578	0.1404
O(7A)	0.9355	0.0437	0.2805
O(7B)	0.9066	0.5097	0.2728
O(8A)	0.6154	0.5303	0.4208
O(8B)	0.6665	0.0403	0.4278
O(9A)	0.3123	−0.0184	0.3523
O(9B)	0.2600	0.4809	0.3420
O(10)	0.9812	0.4664	0.4948
O(11)	0.9565	0.2383	0.8802
O(12)	0.7794	0.2702	0.7695
O(13)	0.0332	0.2546	0.6131
O(14)	−0.0067	0.4641	0.0041

Fig. 1. Structure of $P_4W_{10}O_{38}$.

$P_8W_{12}O_{52}$

II. B. DOMENGES, M. GOREAUD, P. LABBÉ and B. RAVEAU, 1982. Acta Cryst., B38, 1724-1728.

Orthorhombic, Pnam, a = 11.9866, b = 15.5500, c = 5.3197 Å, Z = 1. Mo radiation, R = 0.036 for 1604 reflexions.

The structure (Fig. 2) contains ReO_3-type blocks of corner-sharing WO_6 octahedra connected by P_2O_7 groups, with pseudo-pentagonal and distorted hexagonal tunnels. W-O = 1.79-2.04, P-O = 1.58 (bridging), 1.40-1.52 Å (terminal), P-O-P = 140°.

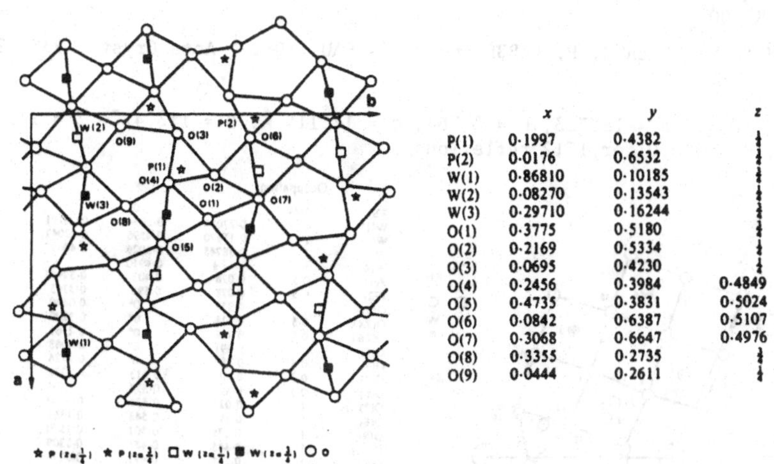

	x	y	z
P(1)	0·1997	0·4382	¼
P(2)	0·0176	0·6532	¼
W(1)	0·86810	0·10185	¼
W(2)	0·08270	0·13543	¼
W(3)	0·29710	0·16244	¼
O(1)	0·3775	0·5180	¼
O(2)	0·2169	0·5334	¼
O(3)	0·0695	0·4230	¼
O(4)	0·2456	0·3984	0·4849
O(5)	0·4735	0·3831	0·5024
O(6)	0·0842	0·6387	0·5107
O(7)	0·3068	0·6647	0·4976
O(8)	0·3355	0·2735	¼
O(9)	0·0444	0·2611	¼

Fig. 2. Structure of $P_8W_{12}O_{52}$.

1. Structure Reports, 48A, 232.

POTASSIUM PHOSPHOTUNGSTATE BRONZE
$K_{0.4}P_2W_4O_{16}$

J.P. GIROULT, M. GOREAUD, P. LABBE and B. RAVEAU, 1982. J. Solid State Chem., 44, 407-414.

Monoclinic, $P2_1/m$, a = 6.670, b = 5.323, c = 8.909 Å, β = 100.55°, Z = 1. Mo radiation, R = 0.033 for 2155 reflexions.

The structure (Fig. 1) contains ReO_3-type slabs, with a width of two WO_6 octahedra, connected through planes of PO_4 tetrahedra. The K ion is in a cage formed by 18 oxygen atoms. P-O = 1.51-1.54, W-O = 1.81-2.03, K-O = 2.63-3.08(1) Å.

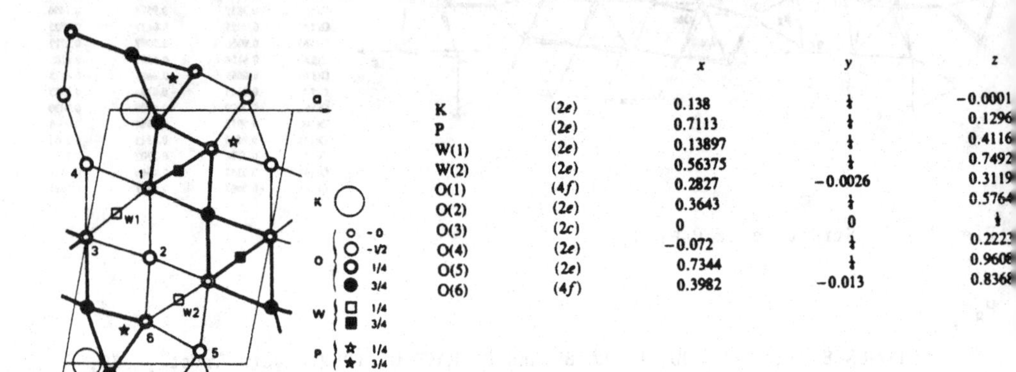

		x	y	z
K	(2e)	0.138	$\frac{1}{4}$	−0.0001
P	(2e)	0.7113	$\frac{1}{4}$	0.1296
W(1)	(2e)	0.13897	$\frac{1}{4}$	0.4116
W(2)	(2e)	0.56375	$\frac{1}{4}$	0.7492
O(1)	(4f)	0.2827	−0.0026	0.3119
O(2)	(2e)	0.3643	$\frac{1}{4}$	0.5764
O(3)	(2c)	0	0	$\frac{1}{2}$
O(4)	(2e)	−0.072	$\frac{1}{4}$	0.2223
O(5)	(2e)	0.7344	$\frac{1}{4}$	0.9608
O(6)	(4f)	0.3982	−0.013	0.8368

Fig. 1. Structure of $K_{0.4}P_2W_4O_{16}$; K occupancy = 0.2.

RUBIDIUM PHOSPHOTUNGSTATE BRONZE
$Rb_{1.74}P_8W_{28}O_{100}$

J.P. GIROULT, M. GOREAUD, P. LABBÉ and B. RAVEAU, 1982. Acta Cryst., B38, 2342-2347.

Monoclinic, P2/c, a = 15.723, b = 3.764, c = 17.118 Å, β = 113.42°, Z = 1/2. Mo radiation, R = 0.030 for 1512 reflexions.

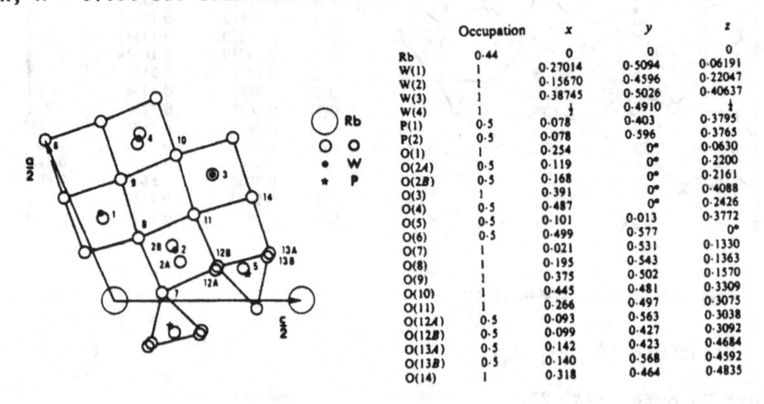

	Occupation	x	y	z
Rb	0·44	0	0	0
W(1)	1	0·27014	0·5094	0·06191
W(2)	1	0·15670	0·4596	0·22047
W(3)	1	0·38745	0·5026	0·40637
W(4)	1	$\frac{1}{2}$	0·4910	$\frac{1}{4}$
P(1)	0·5	0·078	0·403	0·3795
P(2)	0·5	0·078	0·596	0·3765
O(1)	1	0·254	0*	0·0630
O(2A)	0·5	0·119	0*	0·2200
O(2B)	0·5	0·168	0*	0·2161
O(3)	1	0·391	0*	0·4088
O(4)	0·5	0·487	0*	0·2426
O(5)	0·5	0·101	0·013	0·3772
O(6)	0·5	0·499	0·577	0*
O(7)	1	0·021	0·531	0·1330
O(8)	1	0·195	0·543	0·1363
O(9)	1	0·375	0·502	0·1570
O(10)	1	0·445	0·481	0·3309
O(11)	1	0·266	0·497	0·3075
O(12A)	0·5	0·093	0·563	0·3038
O(12B)	0·5	0·099	0·427	0·3092
O(13A)	0·5	0·142	0·423	0·4684
O(13B)	0·5	0·140	0·568	0·4592
O(14)	1	0·318	0·464	0·4835

Fig. 1. Structure of rubidium phosphotungstate bronze [asterisked parameters did not converge].

The structure (Fig. 1) is similar to that of related phases (1), with ReO_3-type W-O slabs connected through P_2O_7 groups. The observed disorder suggests a superstructure, $P2_1/c$, with doubled b axis.

1. Structure Reports, 46A, 276; 48A, 233.

POTASSIUM DIAQUOTRICUPROOCTADECATUNGSTODIARSENATE(III)(12-) UNDECAHYDRATE
$K_{12}[As_2W_{18}O_{66}Cu_3(H_2O)_2].11H_2O$

F. ROBERT, M. LEYRIE and G. HERVÉ, 1982. Acta Cryst., B38, 358-362.

Orthorhombic, Pnma, a = 30.378, b = 14.985, c = 19.198 Å, D_m = 4.10, Z = 4. Mo radiation, R = 0.049 for 4026 reflexions.

The heteropolyanion (Fig. 1) contains two AsW_9O_{33} subunits (1) linked by 3 Cu, one with square-planar and two with square-pyramidal coordinations; As has trigonal-pyramidal coordination.

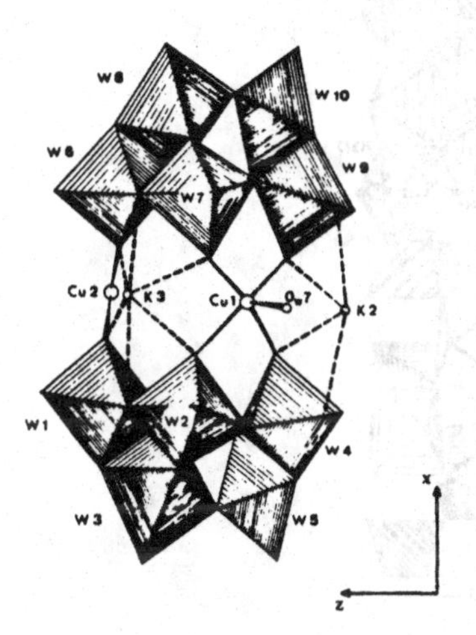

Fig. 1. The $[As_2W_{18}O_{66}Cu_3(H_2O)_2]^{12-}$ anion.

1. Structure Reports, 45A, 261; 46A, 277.

SODIUM ZIRCONIUM TUNGSTATE
$Na_2ZrW_3O_{12}$

R.F. KLEVCOVA, V.V. BAKAKIN, S.F. SOLODOVNIKOV and L.A. GLINSKAJA, 1981. Ž. Strukt. Khim., 22, No. 6, 6-12 [J. Struct. Chem., 22, 807-812].

Tetragonal, $I4_1$, a = 5.040, c = 36.743 Å, Z = 4. Mo radiation, R = 0.063 for 810 reflexions.

The structure (Fig. 1) exhibits a combination of wolframite and scheelite arrangements; the former contains zigzag ribbons of edge-sharing $W(2)O_6$ and $W(3)O_6$ octahedra bordered by ZrO_6 and $Na(2)O_6$ octahedra, and the latter contains $W(1)O_4$ tetrahedra and $Na(1)O_8$ polyhedra. W-O = 1.76-2.31 (octahedra), 1.79 (tetrahedron), Zr-O = 2.05-2.20, Na-O = 2.28-2.39 (octahedron), 2.26-3.05 Å (8-coordination).

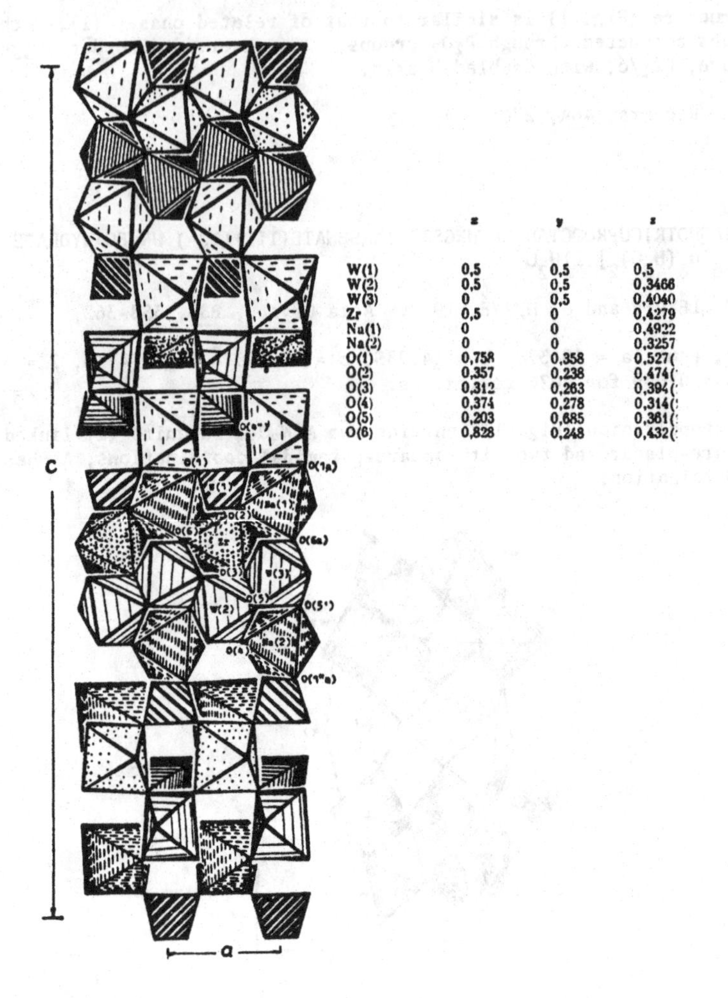

	x	y	z
W(1)	0,5	0,5	0,5
W(2)	0,5	0,5	0,3466
W(3)	0	0,5	0,4040
Zr	0,5	0	0,4279
Na(1)	0	0	0,4922
Na(2)	0	0	0,3257
O(1)	0,758	0,358	0,527(
O(2)	0,357	0,238	0,474(…
O(3)	0,312	0,263	0,394(…
O(4)	0,374	0,278	0,314(…
O(5)	0,203	0,685	0,361(…
O(6)	0,828	0,248	0,432(…

Fig. 1. Structure of sodium zirconium tungstate.

TETRAMETHYLAMMONIUM HETEROPOLYVANADATOTUNGSTOSULPHATE
$[N(CH_3)_4]_6S_2V_2W_{16}O_{62}$ (I)

SODIUM TETRAMETHYLAMMONIUM VANADATOTUNGSTATE HYDRATE
$Na[N(CH_3)_4]_2V\ W_5O_{19}.H_2O$ (II)

A. BOTAR and J. FUCHS, 1982. Z. Naturforsch., 37B, 806-814.

I. Monoclinic, $P2_1/c$, a = 13.361, b = 25.378, c = 28.847 Å, β = 91.37°, Z = 4. Cu radiation, R = 0.080 for 5862 reflexions.

II. Orthorhombic, Pmnm, a = 12.810, b = 10.842, c = 9.703 Å, Z = 2. Mo radiation, R = 0.074 for 1245 reflexions.

I contains a heteropolysulphate anion with structure (Fig. 1) similar to that of $P_2W_{18}O_{62}{}^{6-}$ (1); it contains two $S(V,W)_9O_{34}$ groups sharing six oxygen atoms. II contains an M_6O_{19}-type anion (2), with some V/W disorder (Fig. 1); Na has 5-coordination.

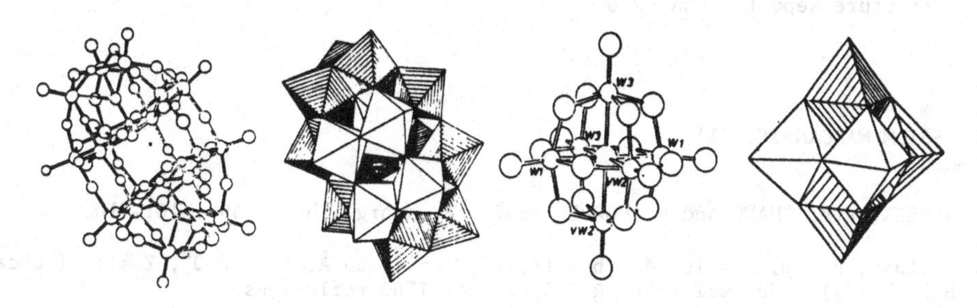

Fig. 1. The $S_2V_2W_{16}O_{62}^{6-}$ (left) and $VW_5O_{19}^{3-}$ (right) anions.

1. Structure Reports, 17, 402,
2. Ibid., 46A, 248.

HYDROGEN TANTALUM TUNGSTATES
$DTaWO_6$

I. F.J. ROTELLA, J.D. JORGENSEN, R.M. BIEFELD and B. MOROSIN, 1982. Acta Cryst., B38, 1697-1703.

Cubic, Fd3m, a = 10.4269, 10.4281 Å, at 12, 300K, Z = 8. Neutron powder data. Ta/W in 16(c): 0,0,0; O in 48(f): x,1/8,1/8, x = 0.3101; 1/6 D in 48(f): x = 0.4082.

Defect pyrochlore structure, with D atoms forming spiral chains in interconnecting open channels along <110>.

$TaWO_{5.5}$, $HTaWO_6$, $H_2Ta_2O_6$, $HTaWO_6.H_2O$

II. D. GROULT, J. PANNETIER and B. RAVEAU, 1982. J. Solid State Chem., 41, 277-285.

Cubic, Fd3m, a = 10.437, 10.444, 10.603, 10.400 Å, at room temperature, Z = 8. Neutron powder data. Ta,W in 16(c); O in 48(f): x = 0.31; 1/6 or 1/3 H in 48(f): x = 0.40; 1/4 O(water) in 32(e): x = 0.355; 1/6 H(water) in 96(g): x = 0.36, z = 0.28. Defect pyrochlore structures.

SODIUM TERBIUM TUNGSTATE
$Na_5Tb(WO_4)_4$

SODIUM LANTHANUM TUNGSTATE
$Na_5La(WO_4)_4$

SODIUM LANTHANUM MOLYBDATE
$Na_5La(MoO_4)_4$

V.A. EFREMOV, V.K. TRUNOV and T.A. BEREZINA, 1982. Kristallografija, 27, 134-139 [Soviet Physics - Crystallography, 27, 77-81].

Tetragonal $I4_1/a$, a = 11.468, 11.630, 11.574, c = 11.385, 11.560, 11.620 Å, Z = 4. R = 0.034, 0.045, 0.026 for 793, 1071, 1089 reflexions.

Isostructural with other $Na_5Ln(MO_4)_4$ compounds (1).

1. Structure Reports, 46A, 270.

POTASSIUM MANGANATE(III)
$KMnO_2$

M. JANSEN, F.M. CHANG and R. HOPPE, 1982. Z. anorg. Chem., 490, 101-110.

Monoclinic, $P2_1/m$, a = 10.142, b = 11.309, c = 6.269 Å, β = 95.0°, Z = 12 (subcell with b' = b/4). Mo radiation, R = 0.089 for 1703 reflexions.

Atomic positions

	x	y	z
K(1)	0.1757	1/4	0.0747
K(2)	0.1173	0.6040	0.0628
K(3)	0.4142	0.4615	0.1871
K(4)	0.4233	3/4	0.1684
Mn(1)	0.2674	0.5013	0.6360
Mn(2)	0.2785	3/4	-0.3593
Mn(3)	0.2710	1/4	-0.3814
Mn(4)	-0.0107	-0.3745	0.5138
O(1)	0.0562	0.4975	0.6935
O(2)	-0.0554	1/4	0.2960
O(3)	0.0540	1/4	0.6886
O(4)	0.2279	0.1201	0.4220
O(5)	0.2616	0.6259	0.4368
O(6)	0.3301	0.6195	0.8333
O(7)	0.3436	0.1265	0.8048

The structure contains edge-sharing MnO_5 square pyramids, linked by six-coordinate K ions. Mn-O = 1.88-1.95 (4 basal distances), 2.20-2.33 (1 apical distance), K-O = 2.62-3.08 Å.

CALCIUM MANGANATES
$CaMnO_{2.5}$, $CaMnO_3$, $Ca_2MnO_{3.5}$

I. K.R. POEPPELMEIER, M.E. LEONOWICZ and J.M. LONGO, 1982. J. Solid State Chem., 44, 89-98.
II. K.R. POEPPELMEIER, M.E. LEONOWICZ, J.C. SCANLON, J.M. LONGO and W.B. YELON, 1982. Ibid., 45, 71-79.

$CaMnO_{2.5}$, orthorhombic, Pbam, a = 5.424, b = 10.230, c = 3.735 Å, Z = 4. X-ray and neutron powder data.

$CaMnO_3$, orthorhombic, Pnma, a = 5.279, b = 7.448, c = 5.264 Å, Z = 4. Mo radiation, R = 0.039 for 1104 reflexions.

$Ca_2MnO_{3.5}$, orthorhombic, Bbam, a = 5.30, b = 10.05, c = 12.24 Å, Z = 8. X-ray powder data.

Atomic positions

$CaMnO_{2.5}$			x	y	z
Ca	in	4(h)	0.2990	0.3610	1/2
Mn		4(g)	0.2826	0.1201	0
O(1)		4(h)	0.2875	0.1027	1/2
O(2)		4(g)	0.0800	0.2824	0
O(3)		2(a)	0	0	0

$CaMnO_3$		x	y	z
Ca	in 4(c)	0.0333	1/4	-0.0057
Mn	4(b)	0	0	1/2
O(1)	4(c)	0.4899	1/4	0.0659
O(2)	8(d)	0.2873	0.0336	-0.2879

$Ca_2MnO_{3.5}$
Atomic parameters given are approximate [and do not correspond to Bbam, but apparently to Pbam, Z = 4].

$CaMnO_3$ is an orthorhombic perovskite; the reduced Mn(III) phases are oxygen deficient and contain MnO_5 square pyramids (Fig. 1).

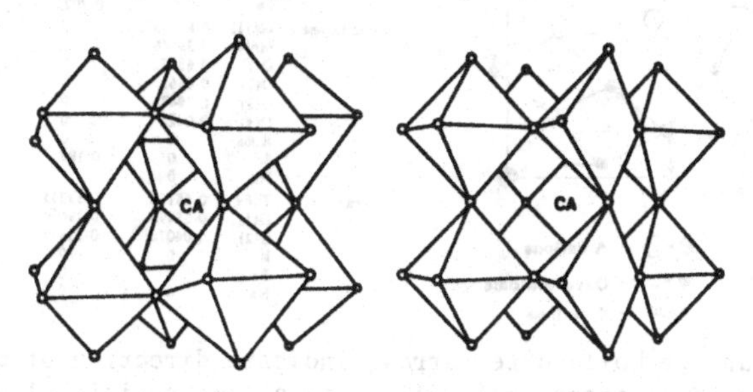

Fig. 1. Structures of $CaMnO_3$ (left) and $CaMnO_{2.5}$ (right).

$Ca_2Mn_3O_8$

III. G.B. ANSELL, M.A. MODRICK, J.M. LONGO, K.R. POEPPELMEIER and H.S. HOROWITZ, 1982. Acta Cryst., B38, 1795-1797.

Monoclinic, C2/c, a = 11.014, b = 5.851, c = 4.942 Å, β = 109.73°, Z = 2. Mo radiation, R = 0.024 for 1628 reflexions.

Atomic positions (x 10^5)

	x	y	z
Ca	72442	0	66593
Mn(1)	0	0	50000
Mn(2)	0	25914	0
O(1)	10059	22158	39171
O(2)	59726	50000	90202
O(3)	60385	0	96258

The structure contains (100) layers of edge-sharing MnO_6 octahedra, linked by Ca ions which have trigonal prismatic coordination. Mn-O = 1.863-2.008, Ca-O = 2.259-2.418(1) Å.

HOLLANDITE
$(Ba,Pb,Na,K)(Mn,Fe,Al)_8(O,OH)_{16}$

CRYPTOMELANE
$(K,Na,Sr,Ba)_{1.4}(Mn,Fe,Al)_8(O,OH)_{16}$

PRIDERITE
$(K,Ba)_{1.25}(Ti,Fe,Mg)_8O_{16}$

J.E. POST, R.B. VON DREELE and P.R. BUSECK, 1982. Acta Cryst., B38, 1056-1065.

Hollandite, cryptomelane
Monoclinic, I2/m, a = 10.026, 9.956, b = 2.878, 2.871, c = 9.729, 9.706 Å, β = 91.03, 90.95°, Z = 1. Mo radiation, R = 0.017, 0.030 for 599, 623 reflexions.

Priderite
Tetragonal, I4/m, a = 10.139, c = 2.966 Å, Z = 1. Mo radiation, R = 0.010 for 316 reflexions.

Hollandite-type structures (1), with distortions from tetragonal symmetry (Fig. 1) depending on cation sizes.

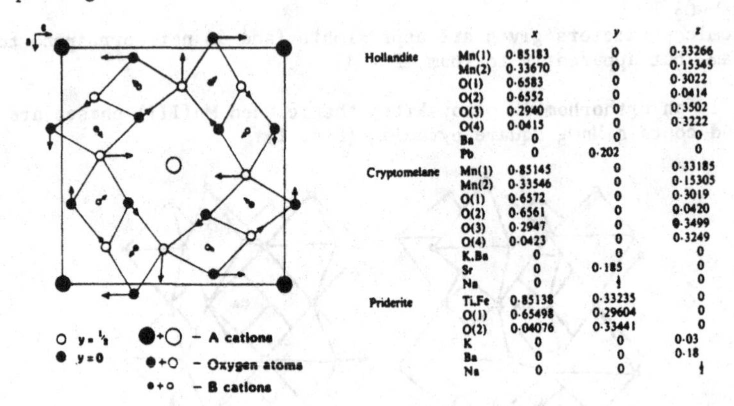

		x	y	z
Hollandite	Mn(1)	0·85183	0	0·33266
	Mn(2)	0·33670	0	0·15345
	O(1)	0·6583	0	0·3022
	O(2)	0·6552	0	0·0414
	O(3)	0·2940	0	0·3502
	O(4)	0·0415	0	0·3222
	Ba	0	0	0
	Pb	0	0·202	0
Cryptomelane	Mn(1)	0·85145	0	0·33185
	Mn(2)	0·33546	0	0·15305
	O(1)	0·6572	0	0·3019
	O(2)	0·6561	0	0·0420
	O(3)	0·2947	0	0·3499
	O(4)	0·0423	0	0·3249
	K.Ba	0	0	0
	Sr	0	0·185	0
	Na	0	½	0
Priderite	Ti,Fe	0·85138	0·33235	0
	O(1)	0·65498	0·29604	0
	O(2)	0·04076	0·33441	0
	K	0	0	0·03
	Ba	0	0	0·18
	Na	0	0	½

Fig. 1. Structure of hollandite (arrows indicate direction of displacement from an ideal tetragonal cell), and atomic positional parameters in hollandite, cryptomelane, and priderite.

1. Structure Reports, 13, 188; 22, 339; 42A, 269; 44A, 208; 46A, 243.

LITHIUM RHENATES
$LiReO_3$, Li_2ReO_3

LITHIUM IRON TRIVANADATE
$Li_2FeV_3O_8$

I. R.J. CAVA, A. SANTORO, D.W. MURPHY, S. ZAHURAK and R.S. ROTH, 1982. J. Solid State Chem., 42, 251-262.
II. Idem, 1981. Solid State Ionics, 5, 323-326.

$LiReO_3$, Li_2ReO_3 (I and II)
Rhombohedral, R3c, a = 5.092, 4.971, c = 13.403, 14.788 Å, Z = 6. Neutron powder data. Re in 6(a): 0,0,z, z = 0; Li in 6(a): z = 0.273, 0.312; O in 18(b): x,y,z, x = -0.3801, -0.3580, y = 0.012, -0.008, z = 0.2460, 0.2551; for Li_2ReO_3, additional Li in 6(a): z = 0.139.

The structures are based on a ReO_3 lattice of corner-sharing octahedra, distorted by insertion of Li into two octahedral sites, as in $LiNbO_3$ (1).

$Li_2FeV_3O_8$
The FeV_3O_8 host lattice is extensively edge-shared and changes little on Li insertion. Li has 5-coordination.

1. Structure Reports, 31A, 147.

MERCURY RHENIUM OXIDE
$Hg_5Re_2O_{10}$

J.-P. PICARD, G. BAUD, J.-P. BESSE, R. CHEVALIER and M. GASPÉRIN, 1982. Acta Cryst., B38, 2242-2245.

Monoclinic, $P2_1/b$, a = 6.401, b = 7.981, c = 11.538 Å, γ = 98.87°, Z = 2. Mo radiation, R = 0.062 for 1791 reflexions.

The structure (Fig. 1) contains $(Hg_2)_4O_4$ rings formed by $(Hg_2)^{2+}$ pairs linked by O atoms. These rings are linked by Hg^{2+} ions to form a two-dimensional network in which ReO_4^- tetrahedra are located. Hg-Hg = 2.546(2), Hg-O = 2.05-2.19(2) (next nearest at 2.66 Å), Re-O = 1.67-1.77(3) Å.

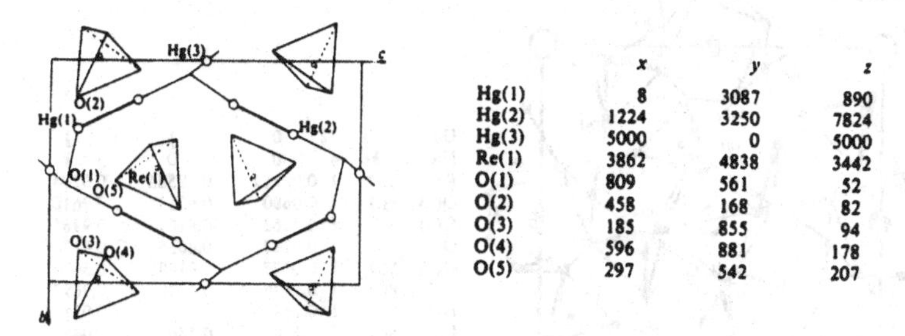

	x	y	z
Hg(1)	8	3087	890
Hg(2)	1224	3250	7824
Hg(3)	5000	0	5000
Re(1)	3862	4838	3442
O(1)	809	561	52
O(2)	458	168	82
O(3)	185	855	94
O(4)	596	881	178
O(5)	297	542	207

Fig. 1. Structure of $Hg_5Re_2O_{10}$, and atomic positional parameters (x 10^4 for Hg, Re; x 10^3 for O).

NEODYMIUM MOLYBDATE PERRHENATE
$Nd(MoO_4)(ReO_4)$

D. ARGELÈS, J.-P. SILVESTRE and M. QUARTON, 1982. Acta Cryst., B38, 1690-1693.

Monoclinic, $P2_1/c$, a = 6.169, b = 9.822, c = 13.139 Å, β = 111.65°, Z = 4. Mo radiation, R = 0.036 for 1766 reflexions.

The structure (Fig. 1) contains MoO_4^{2-} and ReO_4^- tetrahedra, linked by 8-coordinate Nd^{3+} ions. Mo-O = 1.75-1.79, Re-O = 1.72-1.75, Nd-O = 2.31-2.68(1) Å.

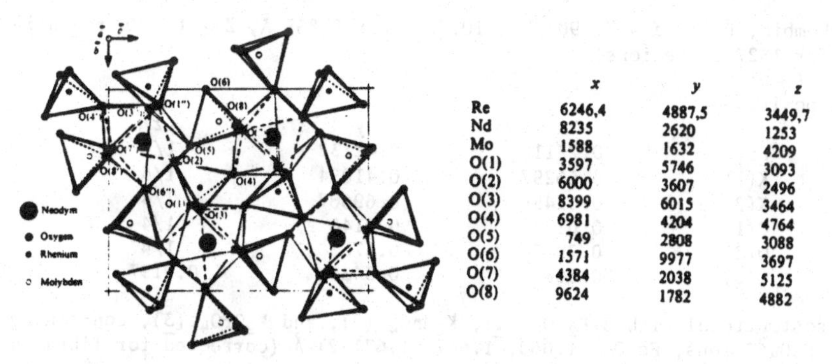

	x	y	z
Re	6246,4	4887,5	3449,7
Nd	8235	2620	1253
Mo	1588	1632	4209
O(1)	3597	5746	3093
O(2)	6000	3607	2496
O(3)	8399	6015	3464
O(4)	6981	4204	4764
O(5)	749	2808	3088
O(6)	1571	9977	3697
O(7)	4384	2038	5125
O(8)	9624	1782	4882

Fig. 1. Structure of neodymium molybdate perrhenate and atomic positional parameters (x 10^4).

SODIUM GADOLINIUM PERRHENATE TETRAHYDRATE
$NaGd(ReO_4)_4.4H_2O$

Z. AÏT ALI SLIMANE, J.-P. SILVESTRE, W. FREUNDLICH and A. RIMSKY, 1982. Acta Cryst., B38, 1070-1074.

Tetragonal, $P\bar{4}n2$, a = 12.483, c = 5.728 Å, D_m = 4.68, Z = 2. Mo radiation, R = 0.040 for 1605 reflexions.

The structure (Fig. 1) contains ReO_4^- tetrahedra, linked by GdO_8 antiprisms, NaO_8 polyhedra, and hydrogen bonds via the water molecules. Re-O = 1.71-1.74, Gd-O = 2.36, 2.40, Na-O = 2.51, 2.64(1), O-H...O = 2.71-2.81(2) Å.

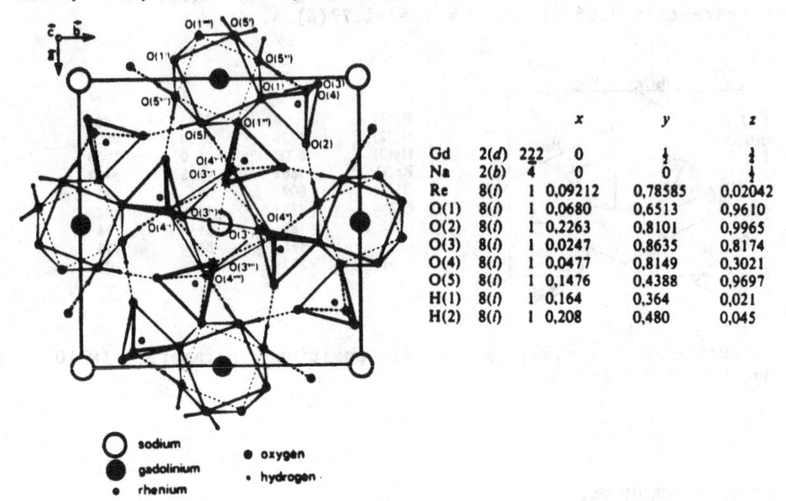

			x	y	z
Gd	2(d)	222	0	½	¾
Na	2(b)	$\bar{4}$	0	0	½
Re	8(i)	1	0,09212	0,78585	0,02042
O(1)	8(i)	1	0,0680	0,6513	0,9610
O(2)	8(i)	1	0,2263	0,8101	0,9965
O(3)	8(i)	1	0,0247	0,8635	0,8174
O(4)	8(i)	1	0,0477	0,8149	0,3021
O(5)	8(i)	1	0,1476	0,4388	0,9697
H(1)	8(i)	1	0,164	0,364	0,021
H(2)	8(i)	1	0,208	0,480	0,045

Fig. 1. Structure of sodium gadolinium perrhenate tetrahydrate.

POTASSIUM FERRATE(VI)
K_2FeO_4

M.L. HOPPE, E.O. SCHLEMPER and R.K. MURMANN, 1982. Acta Cryst., B38, 2237-2239.

Orthorhombic, Pnam, a = 7.690, b = 10.328, c = 5.855 Å, Z = 4. Mo radiation, R = 0.029 for 1527 reflexions.

Atomic positions

	x	y	z
Fe	0.23110	0.42132	1/4
K(1)	0.66297	0.41054	1/4
K(2)	-0.01456	0.69667	1/4
O(1)	0.0173	0.4148	1/4
O(2)	0.2997	0.5733	1/4
O(3)	0.3084	0.3488	0.0197

Isostructural with β-K_2SO_4 (1), K_2MnO_4 (2), and K_2CrO_4 (3), containing tetrahedral FeO_4^{2-} ions, Fe-O = 1.660, 1.667, 1.671(2) Å (corrected for libration).

1. Strukturbericht, 2, 86, 423; Structure Reports, 22, 447; 33A, 367; 38A, 330; 48A, 304.
2. Structure Reports, 32A, 284.
3. Strukturbericht, 2, 88, 446; Structure Reports, 38A, 330; 44A, 207.

CALCIUM FERRITE
$Ca_4Fe_9O_{17}$

B. MALAMAN, H. ALEBOUYEH, A. COURTOIS, R. GÉRARDIN and O. EVRARD, 1982. Mater. Res. Bull., 17, 795-800.

Monoclinic, C2, a = 10.441, b = 6.025, c = 11.384 Å, β = 98.80°, D_m = 4.42, Z = 2. Mo radiation, R = 0.050 for 731 reflexions.

The structure (Fig. 1) contains layers of FeO_6 octahedra and FeO_5 trigonal bipyramids, linked by FeO_4 tetrahedra and 7-coordinate Ca ions. Fe-O = 2.00-2.12 (octahedra), 1.96-1.98, 2.10, 2.11 (bipyramids), 1.84-1.90 (tetrahedra), Ca-O = 2.28-2.83(2) Å.

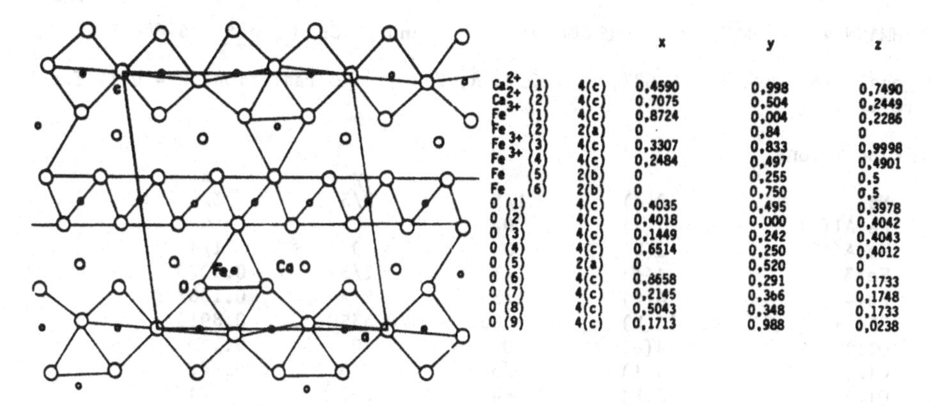

		x	y	z
Ca^{2+} (1)	4(c)	0,4590	0,998	0,7490
Ca^{2+} (2)	4(c)	0,7075	0,504	0,2449
Fe^{3+} (1)	4(c)	0,8724	0,004	0,2286
Fe (2)	2(a)	0	0,84	0
Fe^{3+} (3)	4(c)	0,3307	0,833	0,9998
Fe^{3+} (4)	4(c)	0,2484	0,497	0,4901
Fe (5)	2(b)	0	0,255	0,5
Fe (6)	2(b)	0	0,750	0,5
O (1)	4(c)	0,4035	0,495	0,3978
O (2)	4(c)	0,4018	0,000	0,4042
O (3)	4(c)	0,1449	0,242	0,4043
O (4)	4(c)	0,6514	0,250	0,4012
O (5)	2(a)	0	0,520	0
O (6)	4(c)	0,8658	0,291	0,1733
O (7)	4(c)	0,2145	0,366	0,1748
O (8)	4(c)	0,5043	0,348	0,1733
O (9)	4(c)	0,1713	0,988	0,0238

Fig. 1. Structure of $Ca_4Fe_9O_{17}$.

LANTHANON IRON ALUMINUM OXIDES
$LaFe_{11}AlO_{19}$, $SmFe_4Al_8O_{19}$, $Eu_{0.83}Fe_2Al_{10}O_{19}$

U. SCHWANITZ-SCHÜLLER and H. MÜLLER-BUSCHBAUM, 1982. Mh. Chem., 113, 1079-1085.

Hexagonal, $P6_3/mmc$, a = 5.826, 5.634, 5.658, c = 22.83, 22.08, 22.18 Å, Z = 2. Mo radiation, R = 0.086, 0.10, 0.061 for 471, 318, 282 reflexions.

Atomic positions (parameters for La compound)

		x	y	z
Ln	in 2(d)	1/3	2/3	3/4
M(1)	2(a)	0	0	0
M(2)	2(b)	0	0	1/4
M(3)	4(f)	1/3	2/3	0.0273
M(4)	4(f)	1/3	2/3	0.1895
M(5)	12(k)	0.1680	0.3360	0.8911
O(1)	4(e)	0	0	0.151
O(2)	4(f)	1/3	2/3	0.946
O(3)	6(h)	0.186	0.372	1/4
O(4)	12(k)	0.156	0.311	0.055
O(5)	12(k)	0.506	0.011	0.154

Occupancies

Ln	2 La	2 Sm	1.66 Eu
M(1)	3/2 Fe + 1/2 Al	2 Al	2 Al
M(2)	3/2 Fe + 1/2 Al	1 Fe + 1 Al	1 Fe + 1 Al
M(3)	4 Fe	3 Fe + 1 Al	2 Fe + 2 Al
M(4)	4 Fe	2 Fe + 2 Al	1 Fe + 3 Al
M(5)	11 Fe + 1 Al	2 Fe + 10 Al	3 Fe + 9 Al

[occupancies for the Eu compound do not correspond to the stoichiometry given]

Magnetoplumbite-type structures (1). Ln-12 O = 2.68-2.92, M-4, 5, or 6 O = 1.82-2.21 Å.

1. Strukturbericht, 6, 74; Structure Reports, 43A, 179.

NEODYMIUM IRON ALUMINUM OXIDE

$Nd_2Fe_{15}Al_9O_{38}$ $Nd_2Fe(II)_2Fe(III)_{13}Al_9O_{38}$

U. LEHMANN and H. MÜLLER-BUSCHBAUM, 1982. Z. anorg. Chem., 486, 45-48.

Hexagonal, $P6_3/mmc$, a = 5.687, c = 22.23 Å, Z = 1. Mo radiation, R = 0.11 for 413 reflexions.

Atomic positions

			x	y	z
Nd	in	2(d)	1/3	2/3	3/4
Fe,Al(1)		2(a)	0	0	0
Fe,Al(2)		2(b)	0	0	1/4
Fe(3)		4(f)	1/3	2/3	0.0279
Fe(4)		4(f)	1/3	2/3	0.1887
5Fe + 7Al(5)		12(k)	0.1680	0.3359	0.8916
O(1)		4(e)	0	0	0.149
O(2)		4(f)	1/3	2/3	0.942
O(3)		6(h)	0.184	0.367	1/4
O(4)		12(k)	0.151	0.302	0.053
O(5)		12(k)	0.509	0.018	0.151

Magnetoplumbite structure (1). Nd-O = 2.70, 2.85 (12-coordination), Fe,Al(1)-O = 1.90 (6), Fe,Al(2)-O = 1.81, 2.23 (5), Fe(3)-O = 1.89, 1.91 (4), Fe(4)-O = 1.92, 2.01 (6), Fe,Al(5) = 1.85-2.00 (6-coordination) Å.

1. Strukturbericht, 6, 74.

IRON MANGANESE OXIDE

$(Mn_xFe_{1-x})_yO$ (x = 0-0.975, y = 0.910-0.998)

D.A.O. HOPE, A.K. CHEETHAM and G.J. LONG, 1982. Inorg. Chem., 21, 2804-2809.

Rhombohedral, $R\bar{3}$, unit cell parameters not given. Neutron powder data. Mn,Fe in 1(a) (octahedral site); 0.023-0.036 Fe in 2(c) (tetrahedral site): x taken as 0.25; O in 1(b). Defect sodium chloride structure, with interstitial tetrahedrally-coordinated Fe.

LITHIUM PLATINUM OXIDE

$Li_{0.64}Pt_3O_4$

COBALT SODIUM PLATINUM OXIDE

$Co_{0.37}Na_{0.14}Pt_3O_4$

K.B. SCHWARTZ, J.B. PARISE, C.T. PREWITT and R.D. SHANNON, 1982. Acta Cryst., B38, 2109-2116.

Li compound, cubic, $P\bar{4}3n$, a = 5.6242 Å, Z = 2. Neutron powder data. 1.28 Li in 2(a): 0,0,0; Pt in 6(d): 1/4,0,1/2; O in 8(e): x,x,x, x = 0.2717.

Co/Na compound, cubic, Pm3, a = 5.6321 Å, Z = 2. Neutron powder data. 0.3 Co + 0.3 Na in 1(a): 0,0,0; site 1(b): 1/2,1/2,1/2, nearly empty; 0.4 Co + 5.6 Pt in 6(f): x,0,1/2, x = 0.251; O in 8(i): x,x,x, x = 0.2425.

The structures are distortions of that of $NaPt_3O_4$ (1, 2). Li has a coordination of two interpenetrating tetrahedra, Li-O = 2.22 (x 4), 2.65 (x 4) Å. Co/Na has cubic coordination, Co/Na-O = 2.36 (x 8) Å, with the other 8-coordination site vacant.

1. Structure Reports, 13, 273.
2. Following report.

SODIUM PLATINUM OXIDES
$NaPt_3O_4$, $Na_{0.73}Pt_3O_4$

K.B. SCHWARTZ, C.T. PREWITT, R.D. SHANNON, L.M. CORLISS, J.M. HASTINGS and B.L. CHAMBERLAND, 1982. Acta Cryst., B38, 363-368.

Cubic, Pm3n, a = 5.6868, 5.675 Å, Z = 2. X-ray and neutron powder data. Na in 2(a); Pt in 6(c); O in 8(e); Na occupancies = 1.0, 0.73.

Structure as previously described (1).

1. Structure Reports, 13, 273.

RUBIDIUM OXOCUPRATE(I)
$Rb_3Cu_5O_4$

H. KLASSEN and R. HOPPE, 1982. Z. anorg. Chem., 494, 20-30.

Monoclinic, $P2_1/c$, a = 9.886, b = 7.508, c = 14.401 Å, β = 106.85°, Z = 4. Mo radiation, R = 0.106 for 1453 reflexions.

The structure contains spiral -O-Cu-O-Cu- chains along b linked by additional Cu^+ ions to give sheets parallel to (001); the sheets are linked by Rb ions. Cu ions have approximately linear coordinations, Cu-O = 1.78-1.90(2) Å.

EUROPIUM STRONTIUM CUPRATE
$Eu_{1.3}Sr_{1.7}Cu_2O_{5.65}$

N. NGUYEN, J. CHOISNET and B. RAVEAU, 1982. Mater. Res. Bull., 17, 567-573.

Orthorhombic, Immm, a = 3.744, b = 11.337, c = 20.047 Å, Z = 6. Powder data.

A structure is derived similar to that of $La_2SrCu_2O_6$ (1), but with a tripled b axis. The structure contains an ordered Eu/Sr distribution and oxygen vacancies, with double perovskite and SrO layers.

1. Structure Reports, 46A, 291.

BARIUM LANTHANON CUPRATES
$BaLn_2CuO_5$ (Ln = Y, Gd)

C. MICHEL and B. RAVEAU, 1982. J. Solid State Chem., 43, 73-80.

Orthorhombic, Pbnm, a = 7.132, 7.226, b = 12.181, 12.321, c = 5.658, 5.724 Å, D_m = 6.22, 7.75, Z = 4. Powder data.

The structures (Fig. 1) contain a framework of LnO_7 distorted monocapped trigonal prisms sharing a triangular face to form Ln_2O_{11} blocks; the blocks share edges to form a three-dimensional framework with cavities which contain 11-coordinate Ba^{2+} and 5-coordinate Cu (CuO_5 distorted square pyramid). Y-O = 2.22-2.44, Gd-O = 2.23-2.58, Ba-O = 2.46-3.33, Cu-O = 1.98-2.34(3) Å.

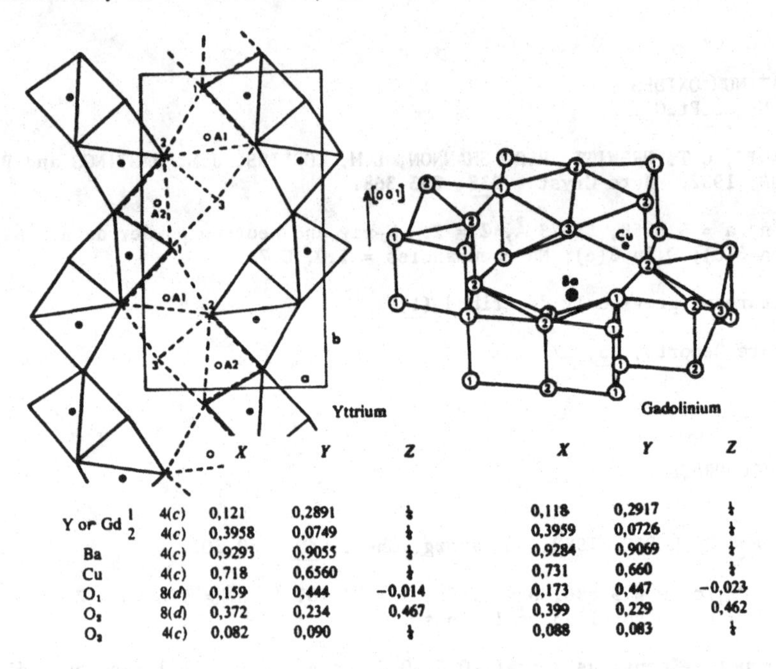

		Yttrium X	Y	Z	Gadolinium X	Y	Z
Y or Gd 1	4(c)	0,121	0,2891	¼	0,118	0,2917	¼
Y or Gd 2	4(c)	0,3958	0,0749	¼	0,3959	0,0726	¼
Ba	4(c)	0,9293	0,9055	¼	0,9284	0,9069	¼
Cu	4(c)	0,718	0,6560	¼	0,731	0,660	¼
O_1	8(d)	0,159	0,444	−0,014	0,173	0,447	−0,023
O_2	8(d)	0,372	0,234	0,467	0,399	0,229	0,462
O_3	4(c)	0,082	0,090	¼	0,088	0,083	¼

Fig. 1. Structure of $BaLn_2CuO_5$ (A = Ln).

SODIUM ARGENTATE(I)
Na_3AgO_2

H. KLASSEN and R.HOPPE, 1982. Z. anorg. Chem., 485, 92-100.

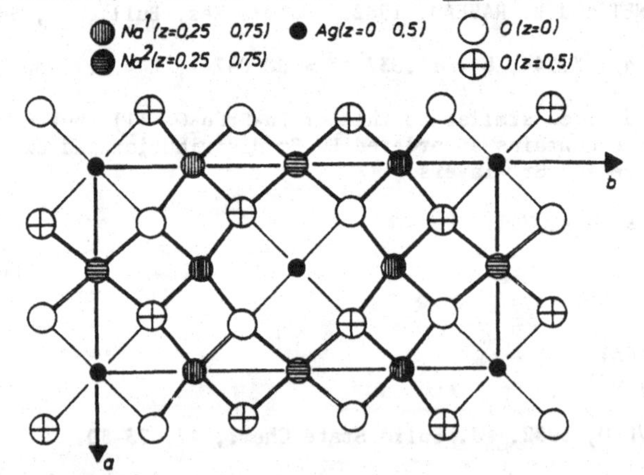

Fig. 1. Structure of Na_3AgO_2.

Orthorhombic, Ibam, a = 5.463, b = 10.926, c = 5.926 Å, D_m = 3.89, Z = 4. Mo radiation, R = 0.029 for 311 reflexions. Na(1) in 8(g): 0,0.2405,1/4; Na(2) in 4(b): 1/2,0,1/4; Ag in 4(c): 0,0,0; O in 8(j): 0.2735,0.1344,0.

The structure (Fig. 1) contains linear AgO_2 groups linked by 4-coordinate Na ions. Ag-O = 2.10, Na-O = 2.37-2.43 Å.

POTASSIUM ARGENTATE(I)

KAgO $\qquad$ 0.25 x $K_4[Ag_4O_4]$

H. KLASSEN and R. HOPPE, 1982. Z. anorg. Chem., 485, 101-114.

Tetragonal, I$\bar{4}$m2, a = 9.893, c = 5.445 Å, D_m = 4.00, Z = 8. Mo radiation, R = 0.067 for 406 reflexions. K in 8(i): 0.1780,0,0.7661; Ag in 8(h): 0.3520,0.1480, 1/4; O in 8(i): 0.2058,0,0.2538. Previous study in 1.

The structure (Fig. 1) contains slightly non-planar Ag_4O_4 rings; Ag has linear coordination, Ag-O = 2.06(1) Å, O-Ag-O = 179, Ag-O-Ag = 91°. K has 4 O neighbours at 2.67-2.80 Å.

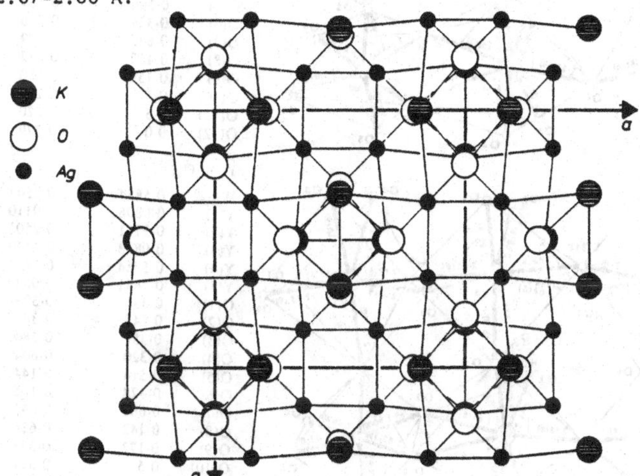

Fig. 1. Structure of $K_4[Ag_4O_4]$.

1. Structure Reports, 33A, 280.

LANTHANUM BARIUM ZINCATES

Ln_2BaZnO_5 (Ln = La, Nd)

C. MICHEL, L. ER-RAHKO and B. RAVEAU, 1982. J. Solid State Chem., 42, 176-182.

Tetragonal, I4/mcm, a = 6.914, 6.756, c = 11.594, 11.540 Å, D_m = 6.66, 7.16, Z = 4. Powder data. (Ln,Ba) in 4(a): 0,0,1/4 and 8(h): x,1/2+x,0, x = 0.1744, 0.1711; Zn in 4(b): 0,1/2,1/4; O in 4(c): 0,0,0 and 16(ℓ): x,1/2+x,z, x = 0.351, 0.352, z = 0.135, 0.135 (bond distances suggest that the 4(a) position contains Ba, and 8(h) contains Ln).

The structures contain layers of face- and edge-sharing LnO_8 bicapped trigonal prisms, with 10-coordinate Ba (bicapped antiprism) and tetrahedrally-coordinated Zn between the layers. Ln-O = 2.32-2.74, Ba-O = 2.89-2.95, Zn-O = 1.97(3) Å.

ERBIUM TUNGSTEN OXIDE
$Er_{10}W_2O_{21}$

YTTRIUM TUNGSTEN OXIDE
$Y_{10}W_2O_{21}$

D.J.M. BEVAN, J. DRENNAN and H.J. ROSSELL, 1982. Acta Cryst., B38, 2991-2997.

Orthorhombic, Pbcn, a = 15.8221, 15.8761, b = 10.4814, 10.5232, c = 10.5407, 10.5778 Å, Z = 4. Powder data and electron diffraction patterns.

Fluorite superstructure (3 x 2 x 2) with cation and anion-vacancies ordered. One Er (Y) has 7-coordination, and the other cations have six-coordination (Fig. 1), with pairs of corner-sharing WO_6 octahedra.

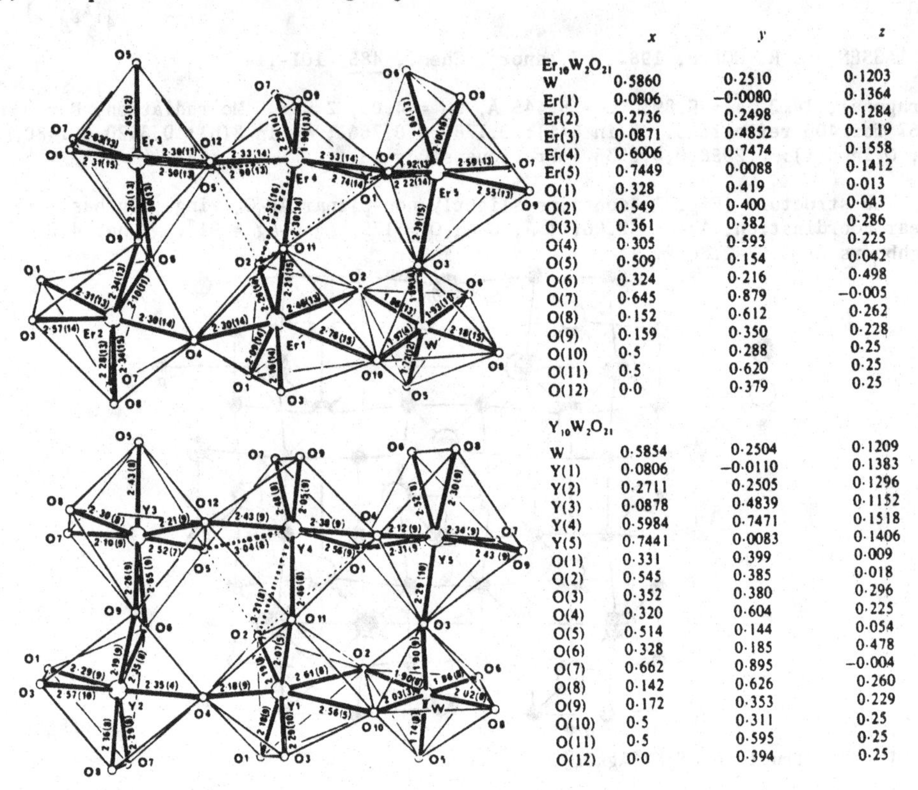

$Er_{10}W_2O_{21}$	x	y	z
W	0·5860	0·2510	0·1203
Er(1)	0·0806	−0·0080	0·1364
Er(2)	0·2736	0·2498	0·1284
Er(3)	0·0871	0·4852	0·1119
Er(4)	0·6006	0·7474	0·1558
Er(5)	0·7449	0·0088	0·1412
O(1)	0·328	0·419	0·013
O(2)	0·549	0·400	0·043
O(3)	0·361	0·382	0·286
O(4)	0·305	0·593	0·225
O(5)	0·509	0·154	0·042
O(6)	0·324	0·216	0·498
O(7)	0·645	0·879	−0·005
O(8)	0·152	0·612	0·262
O(9)	0·159	0·350	0·228
O(10)	0·5	0·288	0·25
O(11)	0·5	0·620	0·25
O(12)	0·0	0·379	0·25

$Y_{10}W_2O_{21}$	x	y	z
W	0·5854	0·2504	0·1209
Y(1)	0·0806	−0·0110	0·1383
Y(2)	0·2711	0·2505	0·1296
Y(3)	0·0878	0·4839	0·1152
Y(4)	0·5984	0·7471	0·1518
Y(5)	0·7441	0·0083	0·1406
O(1)	0·331	0·399	0·009
O(2)	0·545	0·385	0·018
O(3)	0·352	0·380	0·296
O(4)	0·320	0·604	0·225
O(5)	0·514	0·144	0·054
O(6)	0·328	0·185	0·478
O(7)	0·662	0·895	−0·004
O(8)	0·142	0·626	0·260
O(9)	0·172	0·353	0·229
O(10)	0·5	0·311	0·25
O(11)	0·5	0·595	0·25
O(12)	0·0	0·394	0·25

Fig. 1. Structures of $Er_{10}W_2O_{21}$ and $Y_{10}W_2O_{21}$.

LITHIUM TRIURANATE
$Li_2U_3O_{10}$

V.I. SPICYN, L.M. KOVBA, V.V. TABAČENKO, N.V. TABAČENKO and Ju.N. MIKHAILOV, 1982. Izv. Akad. Nauk SSSR, Ser. Khim., No. 4, 807-812.

Monoclinic, P2$_1$/b, a = 6.818, b = 7.298, c = 18.914 Å, γ = 121.58°, Z = 4. R = 0.058 Previous study in 1.

The structure contains $(UO_2)O_4$ octahedra and $(UO_2)O_5$ pentagonal bipyramids, linked by bridging O into a $(UO_2)_3O_4^{2-}$ group.

1. Structure Reports, 38A, 286.

BARIUM URANIUM OXIDE
$BaUO_{3+x}$

S.A. BARRETT, A.J. JACOBSON, B.C. TOFIELD and B.E.F. FENDER, 1982. Acta Cryst., B38, 2775-2781.

$BaUO_{3.30}$ ($Ba_{0.9}U_{0.9}O_3$), $Ba_{0.98}UO_3$
Orthorhombic, Pnma, a = 6.209, 6.200, b = 8.799, 8.764, c = 6.237, 6.208 Å, Z = 4. Neutron powder data. Some UO_2 also present, Fm3m, a = 5.468 Å, fluorite structure.

Atomic positions

$BaUO_{3.30}$

		x	y	z	Occupation
Ba	4(c)	0·0042	0·25	−0·0053	0·911
U	4(b)	0·5	0	0	0·909
O(1)	4(c)	0·5007	0·25	0·0512	1·0
O(2)	8(d)	0·2694	0·0400	−0·2618	2·0

$Ba_{0.98}UO_3$

		x	y	z	Occupation
Ba	4(c)	0·0240	0·25	0·0060	0·976
U	4(b)	0·5	0	0	1·001
O(1)	4(c)	−0·0235	0·25	0·5761	1·0
O(2)	8(d)	0·7700	0·0288	0·2351	2·0

Perovskite-type structures, with the non-stoichiometry arising from Ba and U vacancies.

COBALT URANIUM OXIDE
$Co_3U_2O_8$

M. BACMANN and B. LAMBERT-ANDRON, 1982. Phys. Status Solidi, A, 72, 833-837.

Orthorhombic, Pnnm above 143K, $Pn2_1m$ below 143K, a = 5.111, b = 10.300, c = 6.151 Å, at 300K, Z = 2. Ag radiation, R = 0.058, 0.054 for 258, 291 reflexions at 300, 110K; also neutron powder data.

Atomic positions

300K, Pnnm		x	y	z
U	in 4(g)	0.9694	0.2703	0
Co(1)	4(e)	0	0	0.2638
Co(2)	2(d)	0	1/2	1/2
O(1)	8(h)	0.2161	0.3437	0.2612
O(2)	4(g)	0.2841	0.5818	0
O(3)	4(g)	0.1884	0.0981	0
110K, $Pn2_1m$				
U(1)	in 2(a)	0.9931	0.2703	0
U(2)	2(a)	0.0415	0.7285	0
Co(1)	4(b)	-0.0179	-0.0048	0.2628
Co(2)	2(a)	0.5385	0.0294	0
O(11)	4(b)	0.2391	0.3369	0.2638
O(12)	4(b)	0.7849	0.6543	0.7372
O(21)	2(a)	0.2708	0.5811	0
O(22)	2(a)	0.7079	0.4160	0
O(31)	2(a)	0.1971	0.0878	0
O(32)	2(a)	0.8132	0.8989	0

The high-temperature structure is as previously described (1), with octahedrally-coordinated cations (very flattened for Co(2)); U-O = 1.91-2.26, Co(1)-O = 2.07-2.16, Co(2)-O = 1.88 (x 2), 2.47(2) (x 4) Å. In the low-temperature form, the principal change is in the Co(2) coordination; U-O = 1.92-2.37, Co(1)-O = 1.94-2.18, Co(2)-O = 1.85, 1.94, 2.15(x 2), 2.84(3) (x 2) Å.

1. Structure Reports, 39A, 254.

LITHIUM HYDROXIDE MONOHYDRATE
$LiOH.H_2O$

I. K. HERMANSSON and J.O. THOMAS, 1982. Acta Cryst., B38, 2555-2563.
II. K. HERMANSSON and S. LUNELL, 1982. Ibid., B38, 2563-2569.

Monoclinic, C2/m, a = 7.4153, b = 8.3054, c = 3.1950 Å, β = 110.107°, Z = 4. X-ray and neutron radiations, R = 0.037-0.076 for 277-947 reflexions.

Atomic positions (neutron)

	x	y	z
Li	0	0.3482	1/2
O(1)	0.2861	0	0.3952
O(W)	0	0.2067	0
H(1)	0.2651	0	0.6738
H(W)	0.1113	0.1329	0.1391

Structure as previously described (1). Deformation electron density is found on O-H bonds and in regions corresponding to oxygen lone pairs.

1. Strukturbericht, 7, 4; Structure Reports, 21, 240; 37A, 268.

SODIUM HYDROXIDE
NaOH

H.-J. BLEIF and H. DACHS, 1982. Acta Cryst., A38, 470-476.

298K, orthorhombic, Bmmb, a = 3.401, b = 3.401, c = 11.382 Å, Z = 4. Structure as previously described (1): atoms in 4(c): (1/2,0,1/2)±(1/4,1/4,z), z = -0.087, 0.117 0.197 for Na, O, H.

535K, monoclinic, $P2_1/m$, a = 3.435, b = 3.445, c = 6.080 Å, β = 109.9°, Z = 2. Neutron radiation, R = 0.06 for 140 reflexions. Atoms in 2(e): x,1/4,z, x = 0.141, 0.400, 0.462, z = -0.179, 0.236, 0.389 for Na, O, H.

578K, cubic, Fm3m, a = 5.10 Å, Z = 4. X-ray film data. Na in 4(a): 0,0,0; OH in 4(b); 1/2,1/2,1/2.

The orthorhombic and monoclinic structures (Fig. 1) are as previously described (1); in the cubic form the OH group behaves as a spherically symmetric ion.

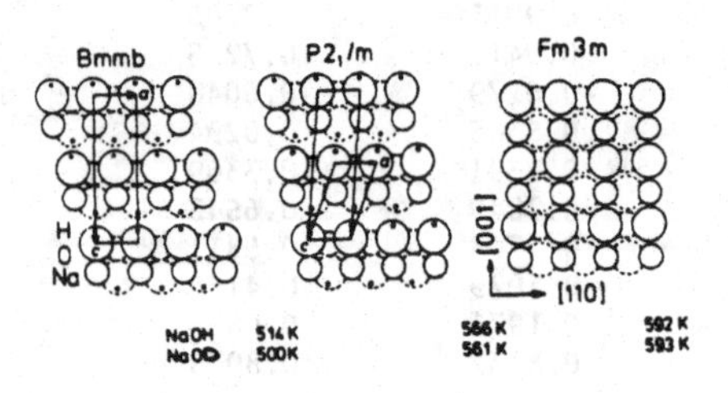

Fig. 1. Structures of the three forms of sodium hydroxide, and transition temperatures.

1. Structure Reports, 11, 252; 32A, 241.

CAESIUM HYDROXIDE MONOHYDRATE (HIGH-TEMPERATURE)
$CsOH.H_2O$

H. JACOBS, B. HARBRECHT, P. MÜLLER and W. BRONGER, 1982. Z. anorg. Chem., 491, 154-162.

Hexagonal, P6/mmm, a = 4.574, 4.576, c = 4.440, 4.436 Å, at 355, 400K, Z = 1. Mo radiation, R = 0.039, 0.040 for 31, 23 reflexions. Cs in 1(a): 0,0,0; O in 2(d): 1/3,2/3,1/2.

The structure contains an infinite hydrogen-bonded sheet of composition $H_3O_2^-$. At room temperature the space group is probably P3ml, with distinguishable OH^- and H_2O; below 225K the structure is monoclinic.

ALUMINUM OXIDE HYDROXIDE [BOEHMITE]
IRON(III) OXIDE HYDROXIDE [LEPIDOCROCITE]
γ-MOOH (M = Al, Fe)

A.N. CHRISTENSEN, M.S. LEHMANN and P. CONVERT, 1982. Acta Chem. Scand., A36, 303-308.

Orthorhombic, Cmcm, a = 2.876, 3.070, b = 12.24, 12.53, c = 3.709, 3.876 Å, for $AlOOH_{0.33}D_{0.67}$, $FeOOH_{0.04}D_{0.96}$, Z = 4. Neutron powder data. M, O(1), O(2) in 4(c): 0,y,1/4, y = -0.3172, 0.2902, 0.0820 for Al compound, -0.3137, 0.2842, 0.0724 for Fe compound; 0.5 H,D in 8(f): 1/2,y,z, y = 0.519, z = 0.392 for Al compound, y = 0.514, z = 0.366 for Fe compound.

Structures as previously described (1, 2), the higher-symmetry space group now preferred, with disordered H positions.

1. Structure Reports, 10, 99, 20, 284; 43A, 219; 44A, 321; 45A, 399.
2. Strukturbericht, 3, 66, 375; Structure Reports, 40A, 304; 44A, 224.

BARIUM TETRAHYDROXODIOXOSTANNATE(IV) DECAHYDRATE
$Ba_2SnO_2(OH)_4.10H_2O$

S. GRIMVALL, 1982. Acta Chem. Scand., A36, 309-311.

Triclinic, P$\bar{1}$, a = 8.771, b = 8.816, c = 6.175 Å, α = 91.33, β = 67.70, γ = 116.16°, Z = 1. Mo radiation, R = 0.049 for 3069 reflexions.

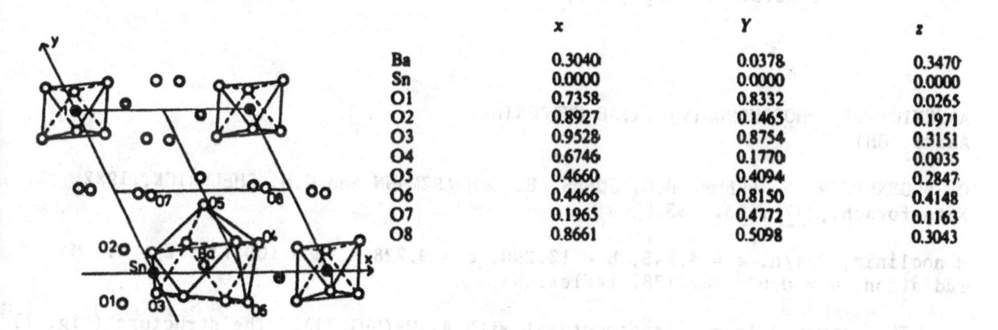

	x	y	z
Ba	0.3040	0.0378	0.3470
Sn	0.0000	0.0000	0.0000
O1	0.7358	0.8332	0.0265
O2	0.8927	0.1465	0.1971
O3	0.9528	0.8754	0.3151
O4	0.6746	0.1770	0.0035
O5	0.4660	0.4094	0.2847
O6	0.4466	0.8150	0.4148
O7	0.1965	0.4772	0.1163
O8	0.8661	0.5098	0.3043

Fig. 1. Structure of $Ba_2SnO_2(OH)_4.10H_2O$.

The structure (Fig. 1) contains $SnO_2(OH)_4{}^{4-}$ trans-octahedra and Ba^{2+} ions arranged in layers, with alternate layers of water molecules. Bonding within and between layers is via Ba-O bonds (9-coordination) and O-H...O hydrogen bonds. Sn-O = 2.04-2.07, Ba-O = 2.76-2.91 Å.

ARSENIC(III) ARSENIC(V) OXIDE HYDROXIDE
$As_3O_5(OH)$ $As(III)_2As(V)O_5(OH)$

D. BODENSTEIN, A. BREHM, P.G. JONES, E. SCHWARZMANN and G.M. SHELDRICK, 1982. Z. Naturforsch., 37B, 138-140.

Monoclinic, $P2_1/c$, a = 12.504, b = 4.593, c = 10.976 Å, β = 118.08°, Z = 4. Mo radiation, R = 0.058 for 1303 reflexions.

The structure (Fig. 1) contains infinite layers, with trigonal pyramidal $As(III)O_3$ and tetrahedral $As(V)O_3(OH)$ groups, and stabilized by hydrogen bonding. As(III)-O = 1.78-1.84, As(V)-O = 1.59-1.67(1) Å, O-As(III)-O = 94-98, O-As(V)-O = 102-113, As-O-As = 126-132°, O-H...O = 2.44 Å (very short). Shortest distances between sheets are As(III)...O = 3.21, 3.26 Å.

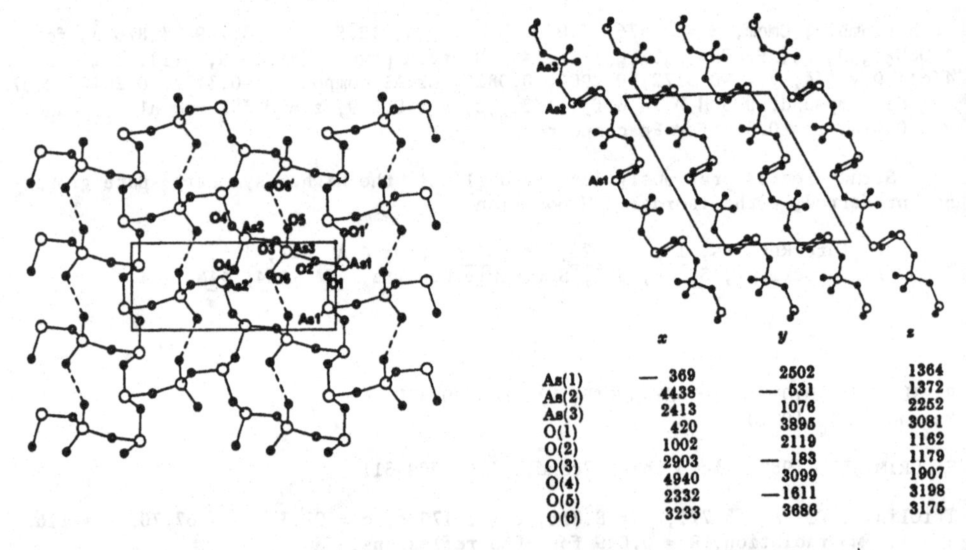

	x	y	z
As(1)	— 369	2502	1364
As(2)	4438	— 531	1372
As(3)	2413	1076	2252
O(1)	420	3895	3081
O(2)	1002	2119	1162
O(3)	2903	— 183	1179
O(4)	4940	3099	1907
O(5)	2332	—1611	3198
O(6)	3233	3686	3175

Fig. 1. Structure of $As_3O_5(OH)$, and atomic positional parameters (x 10^4).

ARSENIC(III) PHOSPHORUS(V) OXIDE HYDROXIDE
$As_2PO_5(OH)$

D. BODENSTEIN, A. BREHM, P.G. JONES, E. SCHWARZMANN and G.M. SHELDRICK, 1982. Z. Naturforsch., 37B, 531-533.

Monoclinic, $P2_1/n$, a = 4.545, b = 12.290, c = 9.728 Å, β = 100.13°, Z = 4. Mo radiation, R = 0.028 for 1387 reflexions.

The compound is not isostructural with $As_3O_5(OH)$ (1). The structure (Fig. 1) contains (001) layers of AsO_3 trigonal pyramids and PO_4 tetrahedra sharing corners, and stabilized by P-O-H...O-P hydrogen bonds. As-O = 1.77-1.87, P-O = 1.50-1.56 Å, O-As-O = 93-98, O-P-O = 103-112, As-O-As = 125, 126, As-O-P = 132, 134°.

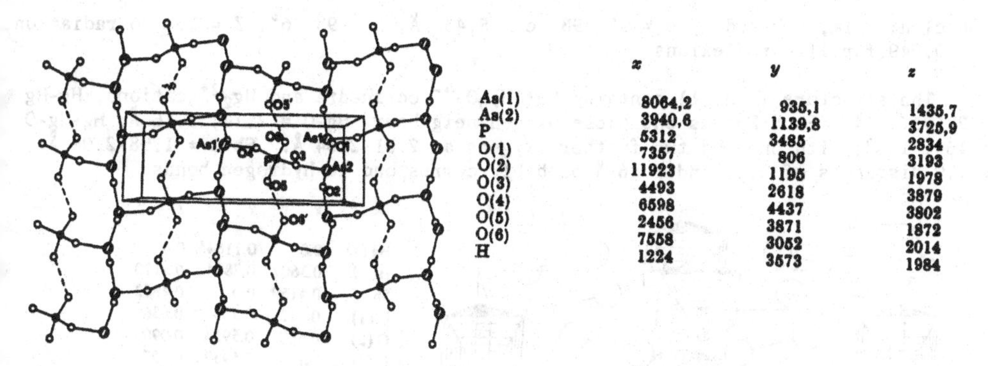

	x	y	z
As(1)	8064,2	935,1	1435,7
As(2)	3940,6	1139,8	3725,9
P	5312	3485	2834
O(1)	7357	806	3193
O(2)	11923	1195	1978
O(3)	4493	2618	3879
O(4)	6598	4437	3802
O(5)	2456	3871	1872
O(6)	7558	3052	2014
H	1224	3573	1984

Fig. 1. Structure of $As_2PO_5(OH)$ viewed along c, and atomic positional parameters (x 10^4).

1. Preceding report.

AMMONIUM HEXAHYDROXOPLATINATE(IV)
POTASSIUM HEXAHYDROXOPLATINATE(IV)
$M_2Pt(OH)_6$ (M = NH_4, K)

G. BANDEL, C. PLATTE and M. TRÖMEL, 1982. Acta Cryst., B38, 1544-1546.

Rhombohedral, $R\bar{3}$, a = 5.668, 5.658, Å, α = 77.01, 69.04°, Z = 1 (hexagonal cells, a = 7.057, 6.413, c = 11.820, 12.836 Å, Z = 3). Mo radiation, R = 0.033, 0.038 for 316, 275 reflexions. Pt in 1(a): 0,0,0; N and K in 2(c): x,x,x, x = 0.328, 0.2859; O in 6(f): (0.835,0.258,0.209) and (0.799,0.294,0.176).

The structures (Fig. 1) contain $Pt(OH)_6^{2-}$ octahedra, Pt-O = 2.02(1) Å, linked by K^+ or NH_4^+ ions. K has 9 oxygen neighbours, K-O = 2.76, 2.89, 3.02 (each x 3), but NH_4 has three close hydrogen-bonded oxygens at 2.76 Å, with next nearest at 3.14 Å.

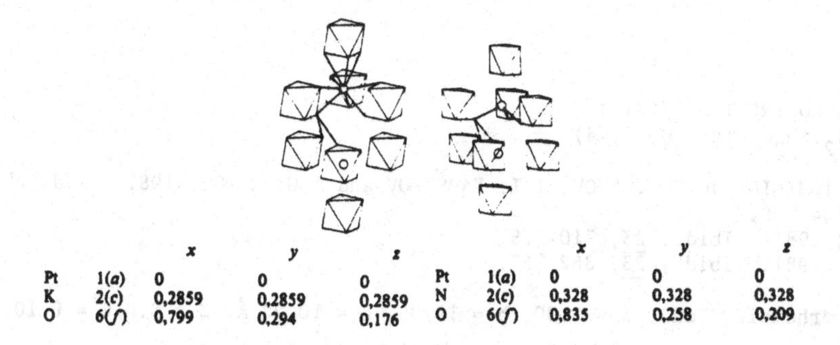

		x	y	z
Pt	1(a)	0	0	0
K	2(c)	0,2859	0,2859	0,2859
O	6(f)	0,799	0,294	0,176

		x	y	z
Pt	1(a)	0	0	0
N	2(c)	0,328	0,328	0,328
O	6(f)	0,835	0,258	0,209

Fig. 1. Structures of $M_2Pt(OH)_6$, M = K (left), and NH_4 (right).

SHAKHOVITE (SHAHOVITE)
$Hg_4Sb(OH)_3O_3$

E. TILLMANNS, R. KRUPP and K. ABRAHAM, 1982. Tschermaks Min. Petr. Mitt., 30, 227-235.

Monoclinic, Im, a = 4.871, b = 15.098, c = 5.433 Å, β = 98.86°, Z = 2. Mo radiation, R = 0.049 for 1126 reflexions.

The structure (Fig. 1) contains $Sb(OH)_3O_3^{4-}$ octahedra and Hg_2^{2+} cations, Hg-Hg = 2.543(1) Å. Each Hg has one close oxygen neighbour, Hg-O = 2.14, 2.16 Å, Hg-Hg-O = 166, 155°, with one and two further oxygens at 2.51-2.54 Å. Sb-O = 1.98-2.04 Å. O...O distances of 2.64 and 2.86 Å probably correspond to hydrogen bonds.

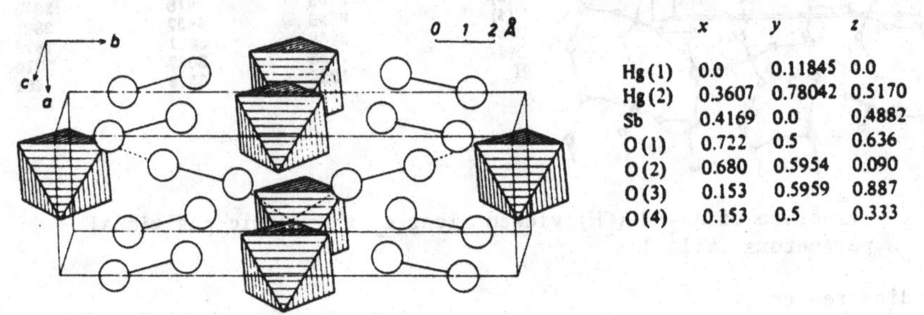

	x	y	z
Hg(1)	0.0	0.11845	0.0
Hg(2)	0.3607	0.78042	0.5170
Sb	0.4169	0.0	0.4882
O(1)	0.722	0.5	0.636
O(2)	0.680	0.5954	0.090
O(3)	0.153	0.5959	0.887
O(4)	0.153	0.5	0.333

Fig. 1. Structure of shakhovite.

BECQUERELITE

$Ca[(UO_2)_6O_4(OH)_6].8H_2O$ $CaO.6UO_3.11H_2O$

J. PIRET-MEUNIER and P. PIRET, 1982. Bull. Minéral., 105, 606-610.

Orthorhombic, $Pn2_1a$, a = 13.86, b = 12.30, c = 14.92 Å, D_m = 5.12, Z = 4. Mo radiation, R = 0.070 for 1983 reflexions. Previous study in 1.

The structure contains $[(UO_2)_6O_4(OH)_6{}^{2-}]_n$ layers connected by Ca^{2+} ions and water molecules. U ions have pentagonal bipyramidal coordinations, U-O = 1.76-1.93 (2 uranyl O), 2.18-2.30 (2 O), 2.28-2.59 (2 OH), 2.58-2.77 (1 OH); Ca has 8-coordination (square antiprism) to 4 O and 4 H_2O, Ca-O = 2.25-2.83 Å.

1. A. COURTOIS, 1968. Thesis, Nancy.

SODIUM DIHYDROXYNEPTUNATE(VII)

$Na_3NpO_4(OH)_2.xH_2O$ (x = 0, 2, 4)

I. S.V. TOMILIN, Ju.F. VOLKOV, I.I. KAPŠUKOV and A.G. RYKOV, 1981. Radiokhimija, 23, 704-709.
II. Idem, 1981. Ibid., 23, 710-715.
III. Idem, 1981. Ibid., 23, 862-867.

x = 0, orthorhombic, Fdd2, a = 5.90, b = 19.95, c = 10.22 Å, Z = 8. R = 0.10.

x = 2, monoclinic, $P2_1/b$, a = 7.82, b = 6.85, c = 7.79 Å, γ = 113.1°, Z = 2. R = 0.084.

x = 4, triclinic, $P\bar{1}$, a = 7.68, b = 5.94, c = 5.10 Å, α = 90.2, β = 78.5, γ = 103.2°, Z = 1. R = 0.11.

All three structures contain isolated trans-octahedral $NpO_4(OH)_2{}^{3-}$ anions, linked by Na^+ cations and water molecules. Np-O = 1.90, Np-OH = 2.30 Å.

CARBON DISULPHIDE
CS_2

B.M. POWELL, G. DOLLING and B.H. TORRIE, 1982. Acta Cryst., B38, 28-32.

Orthorhombic, Cmca, a = 6.086, 6.414, b = 5.287, 5.579, c = 9.462, 8.893 Å, at 5.3, 150K (data at four intermediate temperatures also given), Z = 4. Neutron powder data.

Structure as previously described (1).

1. Structure Reports, 33A, 161.

CARBONYL SULPHIDE
COS

J.S.W. OVERELL, G.S. PAWLEY and B.M. POWELL, 1982. Acta Cryst., B38, 1121-1123.

Rhombohedral, R3m, a = 4.063, α = 98.81° at 90K, Z = 1. Neutron powder data. Atoms in 1(a): x,x,x, [x = 0, -0.2056, 0.2576 for C, O, S].

Structure as previously described (1). C-O = 1.205(6), C-S = 1.510(7) Å.

1. Strukturbericht, 2, 62, 373.

SODIUM SULPHIDE ENNEAHYDRATE
$Na_2S.9D_2O$

K. MEREITER, O. BAUMGARTNER, G. HEGER, W. MIKENDA and H. STEIDL, 1982. Inorg. Chim. Acta, 57, 237-246.

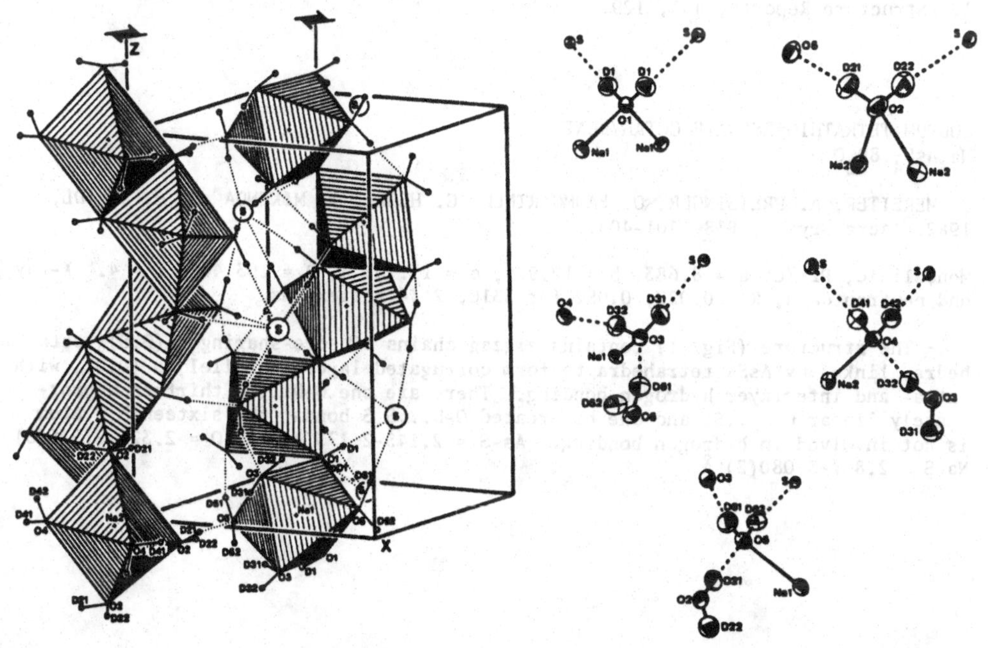

Fig. 1. Structure of $Na_2S.9D_2O$.

Tetragonal, $P4_322$, a = 9.337, c = 12.783 Å, Z = 4. Mo and neutron radiations, R = 0.025, 0.053 for 866, 930 reflexions.

The structure is as previously described (1), and the hydrogen positions have now been determined (Fig. 1).

1. Structure Reports, 30A, 358.

DISILYL SULPHIDE
$H_3Si\text{-}S\text{-}SiH_3$

DISILYL SELENIDE
$H_3Si\text{-}Se\text{-}SiH_3$

M.J. BARROW and E.A.V. EBSWORTH, 1982. J. Chem. Soc., Dalton, 211-216.

Sulphide, orthorhombic, Pbcn, a = 8.14, b = 14.76, c = 8.68 Å, at 120K, Z = 8. Cu radiation, R = 0.038 for 410 reflexions (films, densitometer intemsities).

Selenide, tetragonal, $P4_32_12$, a = 8.36, c = 15.4 Å, at 125K, Z = 8. Cu radiation, R = 0.055 for 263 reflexions.

Atomic positions

	X = S			X = Se		
	x	y	z	x	y	z
Si(1)	0.2576	0.0313	0.1058	0.2275	0.3399	0.3386
X(2)	0.2751	0.1243	-0.0831	0.2098	0.1357	0.2435
Si(3)	0.0896	0.2181	-0.0116	0.3967	0.2337	0.1487

Both structures contain isolated molecules but with intermolecular Si...X (X = S, Se) contacts about 0.4 Å less than van der Waals distances; these contacts result in 4 + 1 coordination for each Si and 2 + 2 coordination for S and Se. Si-S = 2.14(1), Si-Se = 2.27(2) Å, Si-X-Si = 98.4, 95.7° for X = S, Se; Si...S = 3.55, 3.56, Si...Se = 3.58, 3.62 Å. The oxide exhibits only one Si...O intermolecular contact (1).

1. Structure Reports, 45A, 129.

SODIUM TETRATHIOARSENATE OCTAHYDRATE
$Na_3AsS_4.8D_2O$

K. MEREITER, A. PREISINGER, O. BAUMGARTNER, G. HEGER, W. MIKENDA and H. STEIDL, 1982. Acta Cryst., B38, 401-408.

Monoclinic, $P2_1/c$, a = 8.683, b = 12.979, c = 13.702 Å, β = 103.40°, Z = 4. X-ray and neutron data, R = 0.022, 0.052 for 3310, 2144 reflexions.

The structure (Fig. 1) contains zigzag chains of edge-sharing $Na(O,S)_6$ octahedra, linked by AsS_4 tetrahedra to form corrugated layers parallel to (010), with intra- and interlayer hydrogen bonding. There are one O-D...O, thirteen approximately linear O-D...S, and one bifurcated O-D...S,S bonds; the sixteenth D atom is not involved in hydrogen bonding. As-S = 2.145-2.173(1), Na-O = 2.326-2.536(3), Na-S = 2.837-3.080(2) Å.

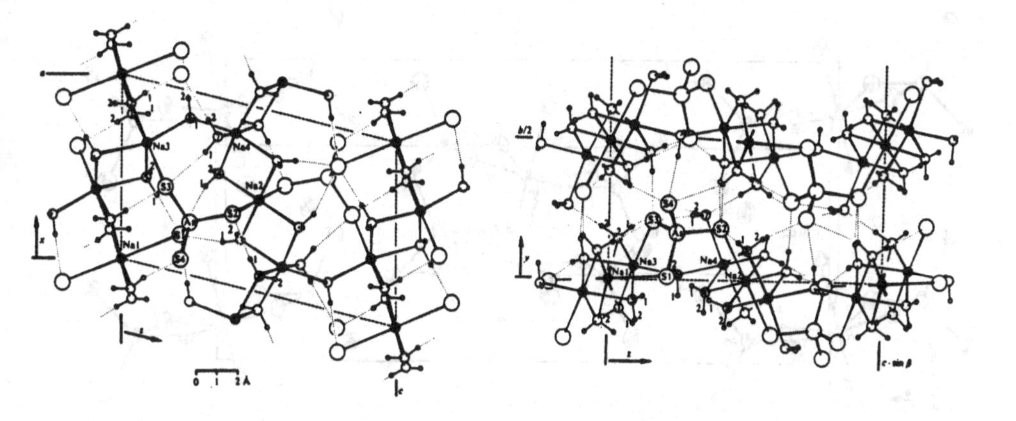

Fig. 1. Structure of $Na_3AsS_4.8D_2O$.

SELENIUM SULPHUR
Se_3S_5

R. LAITINEN, N. RAUTENBERG, J. STEIDEL and R. STEUDEL, 1982. Z. anorg. Chem., 486, 116-128.

Monoclinic, P2/c, a = 8.550, b = 13.340, c = 9.336 Å, β = 124.17°, Z = 4. Mo radiation, R = 0.056 for 1817 reflexions.

The material is the 1,2,3-Se_3S_5 compound, and is isostructural with the other Se_nS_{8-n} phases (1), with a crown-shaped eight-membered ring of statistically distributed Se and S atoms; the structure contains two independent molecules, each having C_2 symmetry.

1. Structure Reports, 43A, 354; 44A, 97, 226; 45A, 275; 48A, 251.

CAESIUM THIOTUNGSTATE HYDRATE
$Cs_2[(WS_4)WO(WS_4)H_2O].2H_2O$

A. MÜLLER, H. BÖGGE, E. KRICKEMEYER, G. HENKEL and B. KREBS, 1982. Z. Naturforsch., 37B, 1014-1019.

Monoclinic, $P2_1/n$, a = 6.716, b = 20.185, c = 7.144 Å, β = 101.42°, Z = 2. Mo radiation, R = 0.069 for 1220 reflexions.

The structure (Fig. 1) contains an isopoly-thioanion disordered about a centre of symmetry, linked by Cs ions and hydrogen bonds via the water molecules.

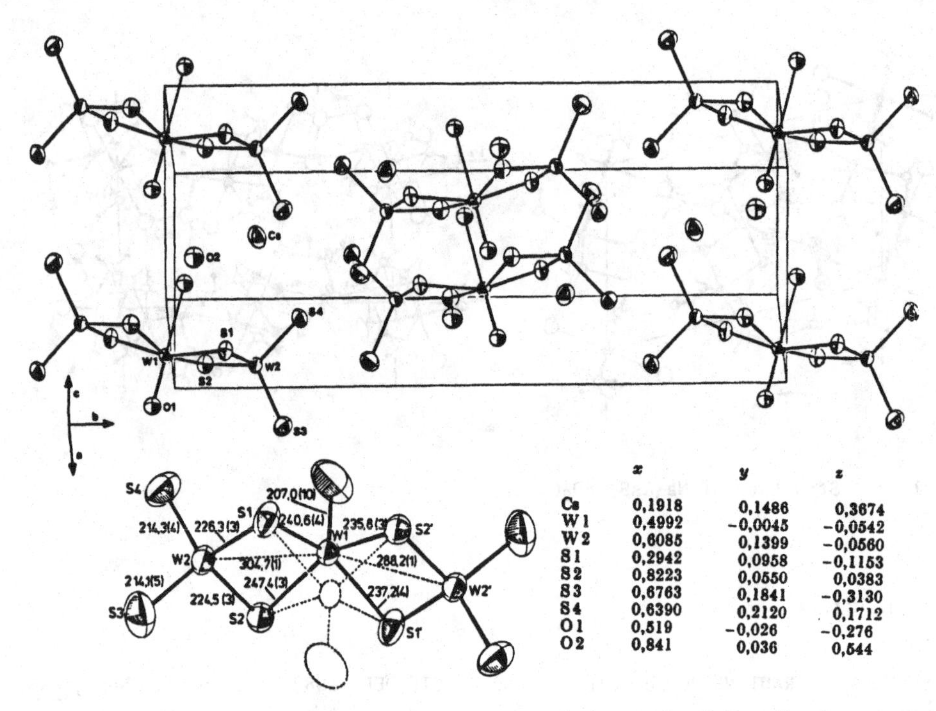

	x	y	z
Cs	0,1918	0,1486	0,3674
W1	0,4992	-0,0045	-0,0542
W2	0,6085	0,1399	-0,0560
S1	0,2942	0,0958	-0,1153
S2	0,8223	0,0550	0,0383
S3	0,6763	0,1841	-0,3130
S4	0,6390	0,2120	0,1712
O1	0,519	-0,026	-0,276
O2	0,841	0,036	0,544

Fig. 1. Structure of caesium thiotungstate hydrate (bond lengths in Å x 10^2).

LANTHANUM GALLIUM OXYSULPHIDE

$La_{3.33}Ga_6O_2S_{12}$

A. MAZURIER, M. GUITTARD and S. JAULMES, 1982. Acta Cryst., B38, 379-382.

Tetragonal, $P\bar{4}2_1m$, a = 9.351, c = 6.049 Å, D_m = 4.12, Z = 1. Mo radiation, R = 0.053 for 619 reflexions.

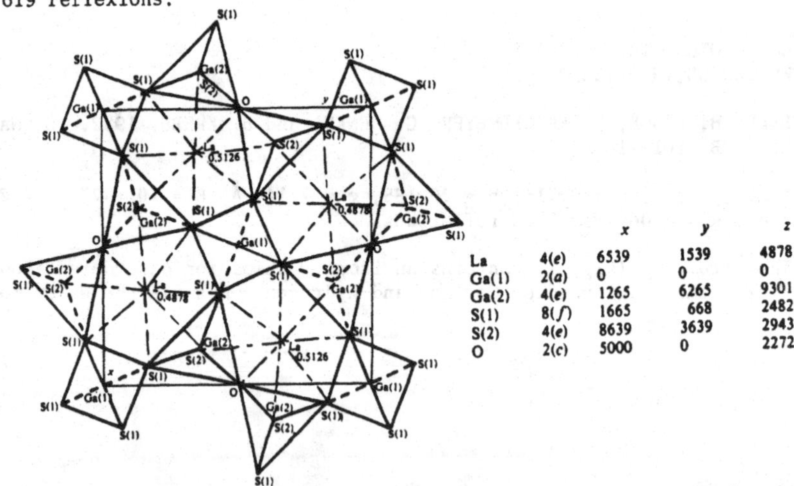

		x	y	z
La	4(e)	6539	1539	4878
Ga(1)	2(a)	0	0	0
Ga(2)	4(e)	1265	6265	9301
S(1)	8(f)	1665	668	2482
S(2)	4(e)	8639	3639	2943
O	2(c)	5000	0	2272

Fig. 1. Structure of lanthanum gallium oxysulphide, and atomic positional parameters (x 10^4).

The structure (Fig. 1) contains sheets of corner-sharing GaS_4 and GaS_3O tetrahedra, linked by eight-coordinate La (occupancy = 0.832). Ga-S = 2.21-2.29, Ga-O = 1.93, La-S = 3.02-3.06, La-O = 2.57(1) Å.

LANTHANUM INDIUM OXYSULPHIDE
$La_{10}In_6O_6S_{17}$

L. GASTALDI, D. CARRÉ and M.P. PARDO, 1982. Acta Cryst., B38, 2365-2367.

Orthorhombic, Immm, a = 26.45, b = 15.81, c = 4.060 Å, Z = 2. Mo radiation, R = 0.037 for 1193 reflexions.

Indium atoms have 4- and 6-, and La atoms 7- and 8-coordinations (Fig. 1); one In atom is distributed over two positions separated by 0.55 Å. In-S = 2.42-3.11, La-S = 2.91-3.21, La-O = 2.34-2.50, In(2)-In(2') = 2.86 Å.

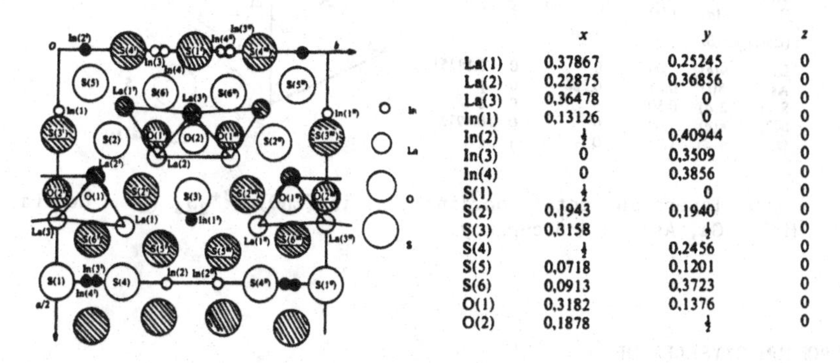

	x	y	z
La(1)	0,37867	0,25245	0
La(2)	0,22875	0,36856	0
La(3)	0,36478	0	0
In(1)	0,13126	0	0
In(2)	½	0,40944	0
In(3)	0	0,3509	0
In(4)	0	0,3856	0
S(1)	½	0	0
S(2)	0,1943	0,1940	0
S(3)	0,3158	½	0
S(4)	½	0,2456	0
S(5)	0,0718	0,1201	0
S(6)	0,0913	0,3723	0
O(1)	0,3182	0,1376	0
O(2)	0,1878	½	0

Fig. 1. Structure of lanthanum indium oxysulphide.

LANTHANUM COPPER OXYSULPHIDE
LANTHANUM SILVER OXYSULPHIDE
(LaO)CuS, (LaO)AgS

M. PALAZZI, C. CARCALY, P. LARUELLE and J. FLAHAUT, 1982. Rare Earths Mod. Sci. Technol., 3, 347-350.

Structures as previously described (1).

1. Structure Reports, 48A, 253.

CERIUM GALLIUM OXYSULPHIDE
$(CeO)_4Ga_2S_5$

LANTHANUM ARSENIC OXYSULPHIDE
$(LaO)_4As_2S_5$

S. JAULMES, E. GODLEWSKI, M. PALAZZI and J. ETIENNE, 1982. Acta Cryst., B38, 1707-1710.

Tetragonal, I4/mmm, a = 3.986, 4.092, c = 18.331, 18.046 Å, Z = 1. Mo radiation, R = 0.054, 0.048 for 172, 175 reflexions.

The structures (Fig. 1) contain (LnO) layers parallel to (001), Ln ions having 9-coordination to 4 O and 5 S. Partially-occupied Ga (As) sites have tetrahedral coordination to 4 S. Ce-O = 2.36, La-O = 2.38, Ce-S = 3.08, 3.31, La-S = 3.23, 3.29, Ga-S = 2.19, 2.36, As-S = 2.19, 2.23 Å.

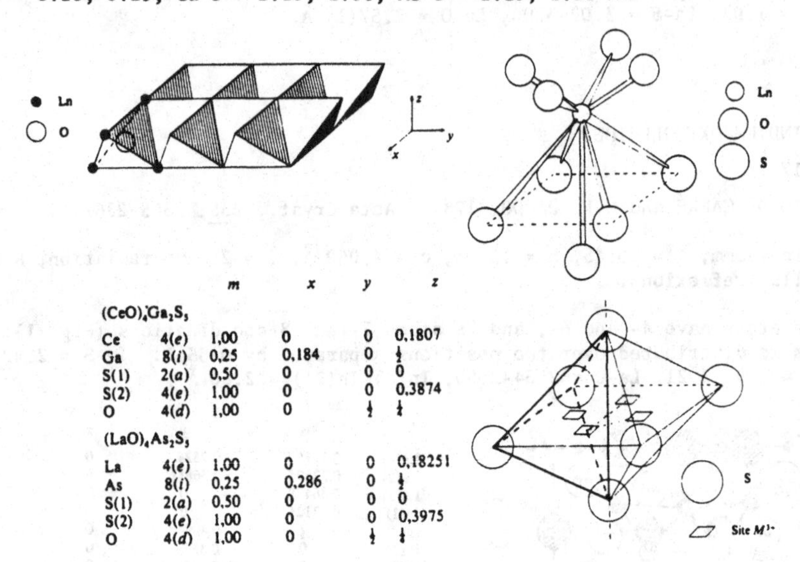

		m	x	y	z
$(CeO)_4Ga_2S_5$					
Ce	4(e)	1,00	0	0	0,1807
Ga	8(i)	0,25	0,184	0	½
S(1)	2(a)	0,50	0	0	0
S(2)	4(e)	1,00	0	0	0,3874
O	4(d)	1,00	0	½	¼
$(LaO)_4As_2S_5$					
La	4(e)	1,00	0	0	0,18251
As	8(i)	0,25	0,286	0	½
S(1)	2(a)	0,50	0	0	0
S(2)	4(e)	1,00	0	0	0,3975
O	4(d)	1,00	0	½	¼

Fig. 1. (LnO) layers and metal coordinations in $(LnO)_4M^{3+}{}_2S_5$ (Ln = Ce, La; M^{3+} = Ga, As); m = occupancy.

CERIUM CHROMIUM OXYSELENIDE
$CeCrSe_2O$

TIEN VO VAN and DUNG NGUYEN HUY, 1981. C.R. Acad. Sci. Paris, Ser. II, 293, 933-936.

Monoclinic, B2/m, a = 11.698, b = 8.248, c = 3.845 Å, γ = 90.25°, Dm = 6.60, Z = 4. Mo radiation, R = 0.068 for 578 reflexions.

Atomic positions

			x	y	z
Ce	in	4(i)	0.2713	0.1963	1/2
Se(1)		4(i)	0.0468	0.1991	0
Se(2)		4(i)	0.3603	0.4711	0
Cr(1)		2(b)	1/2	0	0
Cr(2)		2(c)	0	1/2	0
O		4(i)	0.334	0.058	0

Isostructural with the oxysulphide (1). Ce has tricapped-trigonal-prismatic 9-coordination to 6 S at 3.15-3.26 Å and $\overline{3}$ O at 2.36-2.43 Å. Cr atoms have octahedral coordinations, Cr(1) to 4 S at 2.59 Å and 2 O at 2.00 Å, and Cr(2) to 6 S at 2.53 and 2.54 Å.

1. Structure Reports, 46A, 299.

NEODYMIUM GALLIUM OXYSULPHIDE
$(NdO)_4Ga_2S_5$

J. DUGUÉ and M. GUITTARD, 1982. Acta Cryst., B38, 2368-2371.

Orthorhombic, Pbca, a = 18.293, b = 22.586, c = 5.737 Å, Z = 8. Mo radiation, R = 0.052 for 2035 reflexions.

The structure (Fig. 1) contains (NdO) layers of Nd_4O tetrahedra parallel to (100), with intervening Ga_2S_5 layers of GaS_4 tetrahedra. Nd atoms have 4 O and 2, 3, or 4 S neighbours. Ga-S = 2.21-2.34(1), Nd-O = 2.28-2.39(2), Nd-S = 2.87-3.27(1) Å.

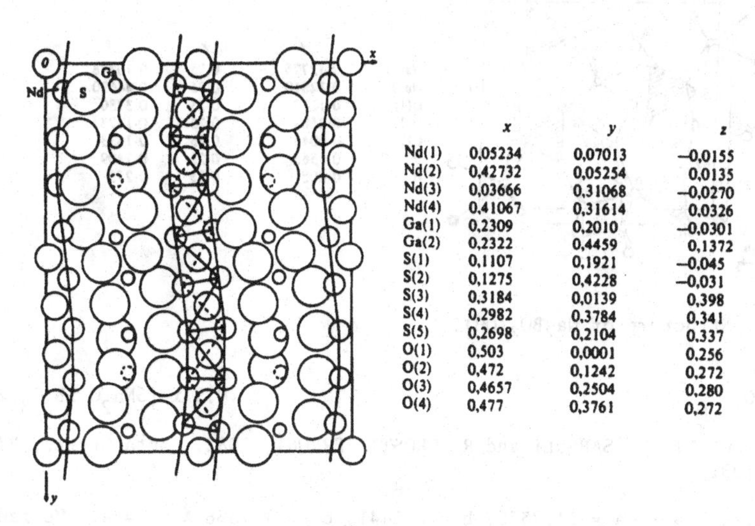

	x	y	z
Nd(1)	0,05234	0,07013	−0,0155
Nd(2)	0,42732	0,05254	0,0135
Nd(3)	0,03666	0,31068	−0,0270
Nd(4)	0,41067	0,31614	0,0326
Ga(1)	0,2309	0,2010	−0,0301
Ga(2)	0,2322	0,4459	0,1372
S(1)	0,1107	0,1921	−0,045
S(2)	0,1275	0,4228	−0,031
S(3)	0,3184	0,0139	0,398
S(4)	0,2982	0,3784	0,341
S(5)	0,2698	0,2104	0,337
O(1)	0,503	0,0001	0,256
O(2)	0,472	0,1242	0,272
O(3)	0,4657	0,2504	0,280
O(4)	0,477	0,3761	0,272

Fig. 1. Structure of neodymium gallium oxysulphide.

LITHIUM METABORATE (MONOCLINIC)

$LiBO_2$

G. WILL, A. KIRFEL and B. JOSTEN, 1981. J. Less-Common Metals, 82, 255-267.

Monoclinic, $P2_1/c$, a = 5.845, b = 4.353, c = 6.454 Å, β = 115.09°, D_m = 2.18, Z = 4. Mo radiation, R = 0.036 for 1226 reflexions.

Atomic positions (high-angle data)

	x	y	z
Li	0.4332	0.2142	0.3441
B	0.1242	0.6753	0.2722
O(1)	0.0845	0.3556	0.2588
O(2)	0.3546	0.7743	0.3168

The structure is as previously described (1). B-O = 1.324, 1.392, 1.407(1), Li-O = 1.944-2.007 (4 distances), 2.473 Å. The charge distribution is described.

1. Structure Reports, 29, 386.

SODIUM BORATE HYDRATES

$Na_2BO_2(OH)$ 0.5 x $[2Na_2O.B_2O_3.H_2O]$

I. S. MENCHETTI and C. SABELLI, 1982. Acta Cryst., B38, 1282-1285.

Orthorhombic, Pnma, a = 8.627, b = 3.512, c = 9.863 Å, Z = 4. Mo radiation, R = 0.028 for 531 reflexions.

The structure (Fig. 1) contains dense sheets parallel to (001) of edge-sharing NaO_5 square pyramids and NaO_6 trigonal prisms; sheets are linked by isolated $BO_2(OH)$ triangles and by hydrogen bonds. B-O = 1.353, B-OH = 1.439, Na-O = 2.275-2.623, O-H...O = 2.566(2) Å.

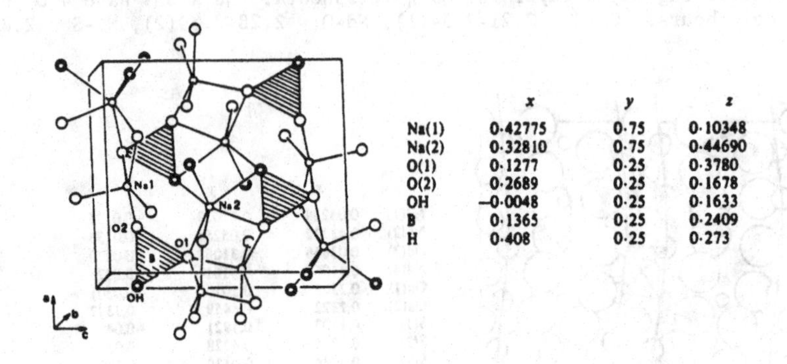

	x	y	z
Na(1)	0·42775	0·75	0·10348
Na(2)	0·32810	0·75	0·44690
O(1)	0·1277	0·25	0·3780
O(2)	0·2689	0·25	0·1678
OH	−0·0048	0·25	0·1633
B	0·1365	0·25	0·2409
H	0·408	0·25	0·273

Fig. 1. Structure of $Na_2BO_2(OH)$.

$Na_3B_5O_9.H_2O$ 0.5 x $[3Na_2O.5B_2O_3.2H_2O]$

II. S. MENCHETTI, C. SABELLI and R. TROSTI-FERRONI, 1982. Acta Cryst., B38, 2987-2991.

Orthorhombic, $Pca2_1$, a = 11.2373, b = 6.0441, c = 11.1336 Å, Z = 4. Mo radiation, R = 0.030 for 1060 reflexions.

The structure (Fig. 2) contains an infinite $B_5O_9^{3-}$ polyanion, with three tetrahedra and two triangles; channels parallel to the three crystal axes contain corner-sharing NaO_5 and NaO_6 polyhedra. B-O = 1.435-1.510 (tetrahedra), 1.367-1.377 (triangles), Na-O = 2.269-2.655(3) Å.

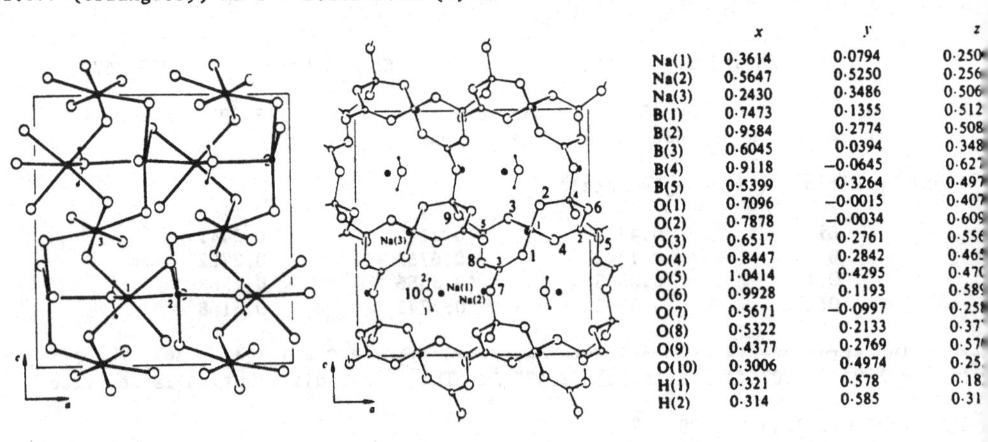

	x	y	z
Na(1)	0·3614	0·0794	0·250
Na(2)	0·5647	0·5250	0·256
Na(3)	0·2430	0·3486	0·506
B(1)	0·7473	0·1355	0·512
B(2)	0·9584	0·2774	0·508
B(3)	0·6045	0·0394	0·348
B(4)	0·9118	−0·0645	0·627
B(5)	0·5399	0·3264	0·497
O(1)	0·7096	−0·0015	0·407
O(2)	0·7878	−0·0034	0·609
O(3)	0·6517	0·2761	0·556
O(4)	0·8447	0·2842	0·465
O(5)	1·0414	0·4295	0·470
O(6)	0·9928	0·1193	0·589
O(7)	0·5671	−0·0997	0·258
O(8)	0·5322	0·2133	0·37
O(9)	0·4377	0·2769	0·57
O(10)	0·3006	0·4974	0·25
H(1)	0·321	0·578	0·18
H(2)	0·314	0·585	0·31

Fig. 2. Na-O and B-O frameworks and atomic positional parameters in $Na_3B_5O_9.H_2O$.

TEEPLEITE
$Na_2Cl[B(OH)_4]$

H. EFFENBERGER, 1982. Acta Cryst., B38, 82-85.

Tetragonal, P4/nmm, a = 7.260, c = 4.847 Å, D_m = 2.076, Z = 2. Mo radiation, R = 0.037 for 218 reflexions.

The structure (Fig. 1) contains tetrahedral $B(OH)_4^-$ groups, connected by edge-sharing $Na(OH)_4Cl_2$ octahedra, with four very weak O-H...Cl hydrogen bonds. B-O = 1.481(2), Na-Cl = 2.805(1), Na-O = 2.454(1), O-H...Cl = 3.292(2) Å.

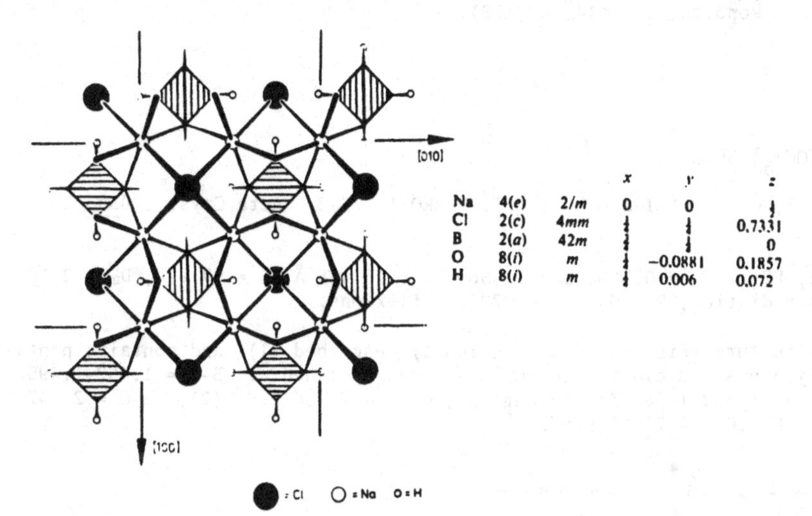

			x	y	z
Na	4(e)	2/m	0	0	½
Cl	2(c)	4mm	¼	¼	0.7331
B	2(a)	42m	¾	¼	0
O	8(i)	m	¼	−0.0881	0.1857
H	8(i)	m	¼	0.006	0.072

Fig. 1. Structure of teepleite.

KOTOITE
$Mg_3(BO_3)_2$

J. ZEMANN, H. EFFENBERGER and F. PERTLIK, 1982. Oesterr. Akad. Wiss., Math.-Naturwiss. Kl., Anz., 61-62.

Orthorhombic, Pnmn, a = 5.398, b = 8.416, c = 4.497 Å, Z = 2. Mo radiation, R = 0.047 for 1121 reflexions.

Atomic positions

		x	y	z
Mg(1)	2a	0	0	0
Mg(2)	4f	0	0.3128	1/2
B	4g	0.2546	0	0.5453
O(1)	4g	0.3222	0	0.2502
O(2)	8h	0.2045	0.1387	0.7029

The structure is as previously described (1), with slightly non-planar borate ions, B-O = 1.376(6), 1.392(4) (x 2) Å, linked by MgO_6 octahedra, Mg-O = 2.083, 2.110 Å.

1. Structure Reports, 11, 425; 30A, 415.

INYOITE
$Ca[B_3O_3(OH)_5].4H_2O$

I.M. RUMANOVA and E.A. GENKINA, 1981. Latv. PSR Zinat. Akad. Vestis, Kim. Ser., No. 6, 643-653.

Monoclinic, $P2_1/n$, a = 10.530, b = 12.073, c = 8.409 Å, β = 112°52', Z = 4. R = 0.031.

The structure is as previously described (1), and hydrogen atoms are now located.

1. Structure Reports, 23, 414; 29, 391.

PROBERTITE

$CaNa[B_5O_7(OH)_4].3H_2O$

S. MENCHETTI, C. SABELLI and R. TROSTI-FERRONI, 1982. Acta Cryst., B38, 3072-3075.

Monoclinic, $P2_1/c$, a = 6.588, b = 12.560, c = 13.428 Å, β = 99.97°, D_m = 2.14, Z = 4. Mo radiation, R = 0.036 for 2235 reflexions.

The structure (Fig. 1) is as previously described (1), and contains pentaborate polyanions and clusters of CaO_9 and NaO_6 polyhedra. B-O = 1.452-1.495 (tetrahedra), 1.352-1.381(3) (triangles), Ca-O = 2.400-2.838(2), Na-O = 2.347-2.653(2), O-H...O = 2.732-2.986(3) Å.

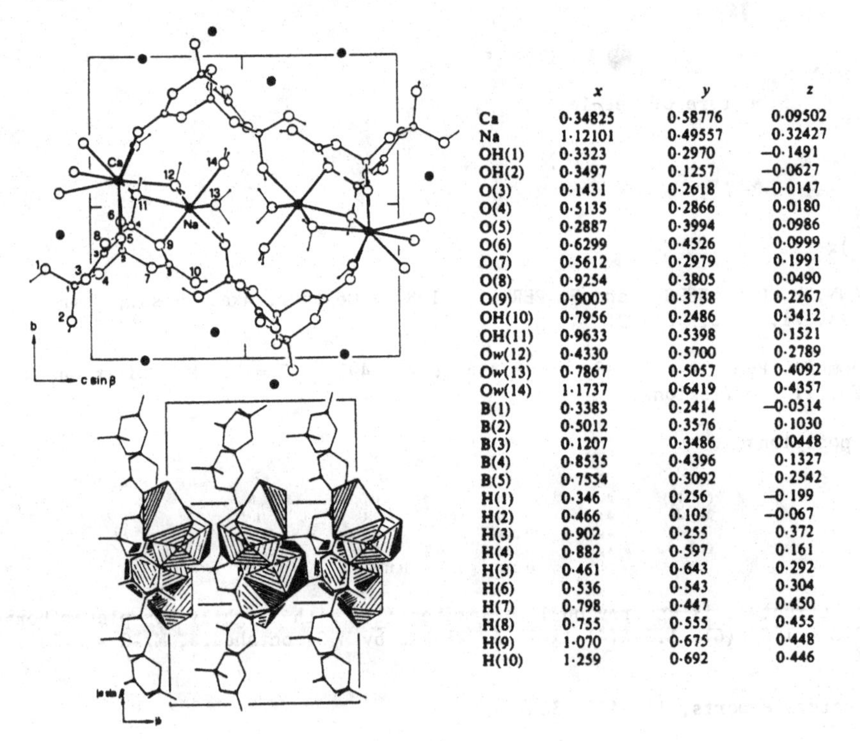

	x	y	z
Ca	0·34825	0·58776	0·09502
Na	1·12101	0·49557	0·32427
OH(1)	0·3323	0·2970	−0·1491
OH(2)	0·3497	0·1257	−0·0627
O(3)	0·1431	0·2618	−0·0147
O(4)	0·5135	0·2866	0·0180
O(5)	0·2887	0·3994	0·0986
O(6)	0·6299	0·4526	0·0999
O(7)	0·5612	0·2979	0·1991
O(8)	0·9254	0·3805	0·0490
O(9)	0·9003	0·3738	0·2267
OH(10)	0·7956	0·2486	0·3412
OH(11)	0·9633	0·5398	0·1521
Ow(12)	0·4330	0·5700	0·2789
Ow(13)	0·7867	0·5057	0·4092
Ow(14)	1·1737	0·6419	0·4357
B(1)	0·3383	0·2414	−0·0514
B(2)	0·5012	0·3576	0·1030
B(3)	0·1207	0·3486	0·0448
B(4)	0·8535	0·4396	0·1327
B(5)	0·7554	0·3092	0·2542
H(1)	0·346	0·256	−0·199
H(2)	0·466	0·105	−0·067
H(3)	0·902	0·255	0·372
H(4)	0·882	0·597	0·161
H(5)	0·461	0·643	0·292
H(6)	0·536	0·543	0·304
H(7)	0·798	0·447	0·450
H(8)	0·755	0·555	0·455
H(9)	1·070	0·675	0·448
H(10)	1·259	0·692	0·446

Fig. 1. Structure of probertite.

1. Structure Reports, 28, 172; 30A, 419.

CALCIUM POTASSIUM TETRABORATE HYDRATE
$CaK_2[B_4O_5(OH)_4]_2.8H_2O$

X. SOLANS, M. FONT-ALTABA, J. SOLANS and M.V. DOMENECH, 1982. Acta Cryst., B38, 2438-2441.

Orthorhombic, $P2_12_12_1$, a = 16.597, b = 12.469, c = 11.569 Å, Z = 4. Mo radiation, R = 0.052 for 3096 reflexions.

The structure (Fig. 1, two anions per asymmetric unit) contains layers of tetraborate anions and water molecules linked by hydrogen bonding, with intervening layers of cations and water molecules; Ca has 7-, and K ions have 6- and 8-coordinations. Ca-O = 2.338-2.483, K-O = 2.743-2.990 and 3.021-3.197, O-H...O = 2.605-2.998(6) Å.

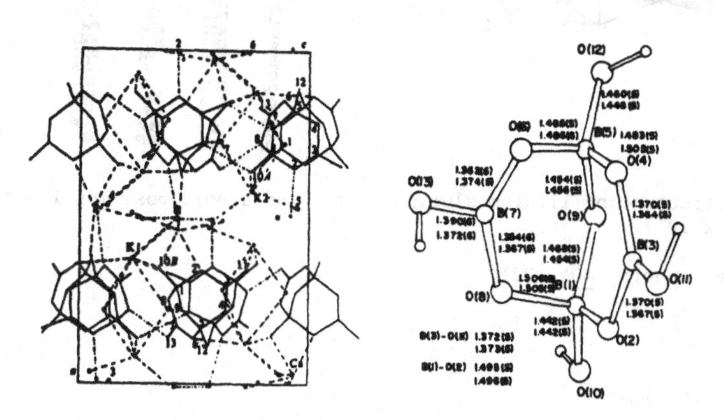

Fig. 1. Structure of calcium potassium tetraborate hydrate.

BARIUM BORATE
$Ba_3(B_3O_6)_2$

S. LU, M. HO and J. HUANG, 1982. Wuli Xuebao, 31, 948-955.

Rhombohedral, R3, a = 12.532, c = 12.717 Å, Z = 6. R = 0.046 for 693 reflexions.

The structure contains alternate layers of Ba^{2+} ions and nearly-planar cyclic $B_3O_6{}^{3-}$ ions.

LITHIUM ALUMINOBORATE
$Li_6[Al_2(BO_3)_4]$

G.K. ABDULLAEV and Kh.S. MAMEDOV, 1982. Kristallografija, 27, 381-383 [Soviet Physics - Crystallography, 27, 229-230].

Triclinic, $P\bar{1}$, a = 6.131, b = 4.819, c = 8.227 Å, α = 90.26, β = 117.03, γ = 89.87°, Z = 1. Mo radiation, R = 0.053 for 1567 reflexions.

The structure (Fig. 1) is as previously described (1), with infinite aluminoborate anions linked by tetrahedrally-coordinated Li ions. Al-4 O = 1.85-1.98, B-3 O = 1.34-1.41, Li-4 O = 1.82-2.13(1) Å.

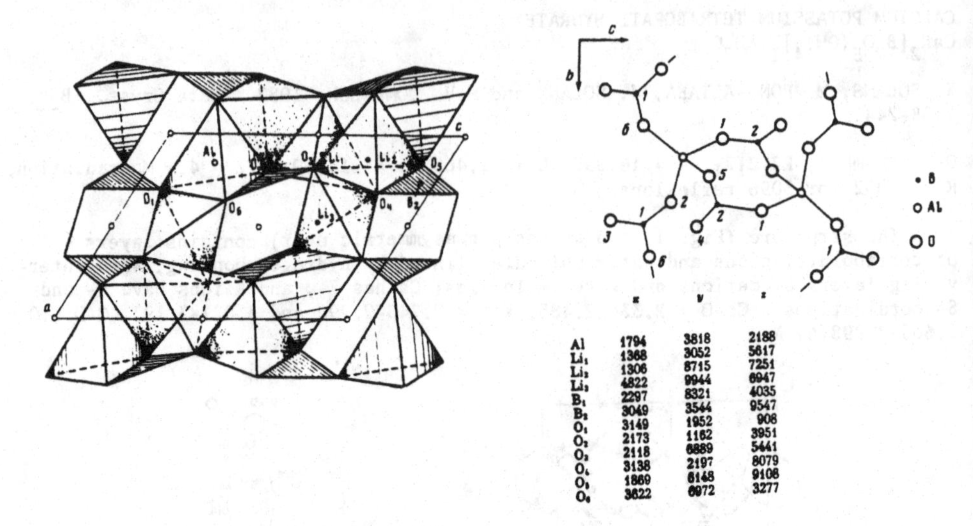

	x	y	z
Al	1794	3818	2188
Li_1	1368	3052	5617
Li_2	1306	8715	7251
Li_3	4822	9944	6947
B_1	2297	8321	4035
B_2	3049	3544	9547
O_1	3149	1952	908
O_2	2173	1162	3951
O_3	2118	6889	5441
O_4	3138	2197	8079
O_5	1869	6148	9108
O_6	3822	6972	3277

Fig. 1. Structure of lithium aluminoborate, and atomic positional parameters (x 10^4).

1. Structure Reports, 40A, 221.

NICKEL NIOBIUM(V) BORATE
Ni_2NbBO_6

G.B. ANSELL, M.E. LEONOWICZ, M.A. MODRICK, B.M. WANKLYN and F.R. WONDRE, 1982. Acta Cryst., B38, 892-893.

Orthorhombic, Pnma, a = 10.057, b = 8.618, c = 4.490 Å, Z = 4. Mo radiation, R = 0.050 for 1285 reflexions.

Isostructural with Fe_3BO_6 (1) and with norbergite (2). The structure (Fig. 1) contains layers with zigzag chains of edge-sharing NiO_6 and NbO_6 octahedra cross-linked by corner sharing with BO_4 tetrahedra; layers are linked by corner-sharing.

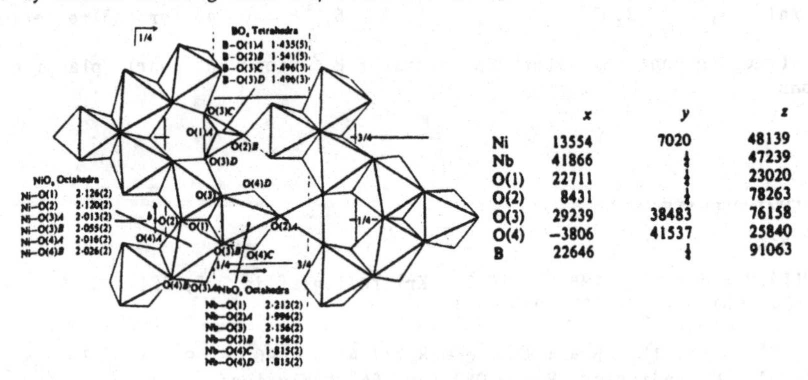

	x	y	z
Ni	13554	7020	48139
Nb	41866	1/4	47239
O(1)	22711	1/4	23020
O(2)	8431	1/4	78263
O(3)	29239	38483	76158
O(4)	−3806	41537	25840
B	22646	1/4	91063

Fig. 1. Structure of Ni_2NbBO_6, and atomic positional parameters (x 10^5).

1. Structure Reports, 30A, 417; 41A, 289.
2. Strukturbericht, 2, 121, 516; Structure Reports, 34A, 366.

BORACITES (COBALT IODINE, COPPER BROMINE, CADMIUM SULPHUR)

$Co_3B_7O_{13}I$, $Cu_3B_7O_{13}Br$

I. R.J. NELMES and W.J. HAY, 1981. J. Phys. C, 14, 5247-5257.

Cubic, F$\bar{4}$3c, a = 12.119, 11.955 Å, Z = 8. Mo radiation, R = 0.032, 0.044 for 338, 313 reflexions. Co or Cu in 24(c): 1/4,0,1/4; B(1) in 24(d): 1/4,0,0; B(2) in 32(e): x,x,x, x = 0.0797, 0.0804; O(1) in 8(a): 0,0,0; O(2) in 96(h): x,y,z, x = 0.1803, 0.1804, y = 0.0196, 0.0178, z = 0.0966, 0.0997; I or Br in 8(b): 1/4,1/4,1/4.

Boracite structures (1).

$Cd_3B_7O_{12.5}S$

II. R.O. GOULD, R.J. NELMES and S.E.B. GOULD, 1981. J. Phys. C, 14, 5259-5267.

Cubic, F$\bar{4}$3c, a = 12.491 Å, Z = 8. R = 0.044.

Atomic positions

				x	y	z
0.47	Cd(1)	in	48(g)	1/4	0.0456	1/4
0.06	Cd(2)		24(c)	1/4	0	1/4
	B(1)		24(d)	1/4	0	0
	B(2)		32(e)	0.0799	0.0799	0.0799
0.77	O(1)		8(a)	0	0	0
	O(2)		96(h)	0.1814	0.0257	0.0924
0.94	S(1)		8(b)	1/4	1/4	1/4
0.24	S(2)		32(e)	0.3466	0.3466	0.3466

Boracite structure (1), with disordered Cd positions, and with the S site occupied by a disordered S_2 ion.

1. Strukturbericht, 2, 407; Structure Reports, 15, 282; 39A, 264; 43A, 223.

PALLADIUM METABORATE

PdB_2O_4

W. DEPMEIER and H. SCHMID, 1982. Acta Cryst., B38, 605-606.

Tetragonal, I$\bar{4}$2d, a = 11.672, c = 5.698 Å, Z = 12. Mo radiation, R = 0.020 for 476 reflexions.

Atomic positions (x 10^4)

			x	y	z
Pd(1)	4(h)	$\bar{4}$	0	0	5000
Pd(2)	8(d)	2	802	2500	1250
O(1)	16(e)	1	1619	702	4978
O(2)	8(d)	2	2498	2500	6250
O(3)	8(d)	2	2500	893	8750
O(4)	16(e)	1	737	1878	7922
B(1)	16(e)	1	1843	1490	6979
B(2)	8(d)	2	-8	2500	6250

Isostructural with the Cu analogue (1), with a tetrahedral B-O framework and square-planar coordinated Pd ions. B-O = 1.45-1.49, Pd-O = 1.98-2.06(1) Å.

1. Structure Reports, 37A, 277.

COPPER BORATE DIBORATE OXIDE
$Cu_{15}(BO_3)_6(B_2O_5)_2O_2$ $5 \times 3CuO.B_2O_3$

H. BEHM, 1982. Acta Cryst., B38, 2781-2784.

Triclinic, $P\bar{1}$, a = 3.353, b = 19.665, c = 19.627 Å, α = 88.77, β = 69.71, γ = 69.24°, Dm = 4.4, Z = 2. Mo radiation, R = 0.03 for 3407 reflexions.

The structure (Fig. 1) contains nearly planar B_2O_5 groups, planar BO_3 groups, isolated oxide anions, and Cu ions with 4, 4+1, and 4+2 coordinations. B-O = 1.32-1.42(2) Å, O-B-O = 108-131, B-O-B = 130°, Cu-O = 1.86-1.95, 2.42-2.84(1) Å.

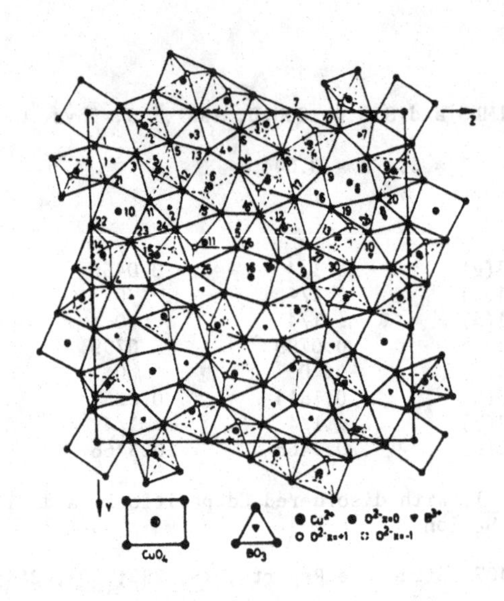

Fig. 1. Structure of copper borate diborate oxide.

SILVER ORTHOBORATE (FORM II)
Ag_3BO_3

M. JANSEN and G. BRACHTEL, 1982. Z. anorg. Chem., 489, 42-46.

Rhombohedral, $R\bar{3}c$, a = 9.878, c = 13.512 Å, Z = 12. Mo radiation, R = 0.035 for 376 reflexions. Ag in 36(f): 0.4971,0.0019,0.3761; B in 12(c): 0,0,0.3784; O in 36(f): -0.0063,0.1362,0.3768.

Stacking variant of form I (1), with a quadrupled c axis. Ag-O = 2.111(2) (x 2) Å, O-Ag-O = 178°, B-O = 1.378(5) (x 3) Å, O-B-O = 120°.

1. Structure Reports, 48A, 260.

STRONTIUM NEODYMIUM BORATE
$Sr_3Nd_2(BO_3)_4$

G.K. ABDULLAEV and K.S. MAMEDOV, 1982. Kristallografija, 27, 795-797 [Soviet Physics - Crystallography, 27, 478-480].

Orthorhombic, $Pc2_1n$, a = 8.791, b = 16.176, c = 7.386 Å, Z = 4. Mo radiation, R = 0.065 for 1734 reflexions.

The structure is as previously described (1), containing trigonal planar BO_3^{3-} ions linked by edge- and face-sharing cation coordination polyhedra, NdO_8, SrO_8, SrO_9, SrO_{10}. B-O = 1.33-1.40(1), Nd-O = 2.37-3.01(1), Sr-O = 2.31-3.08(1) Å.

1. Structure Reports, 40A, 225.

SODIUM CARBONATE HEPTAHYDRATE
β-$Na_2CO_3.7H_2O$

C. BETZEL, W. SAENGER and D. LOEWUS, 1982. Acta Cryst., B38, 2802-2804.

Orthorhombic, Pbca, a = 14.492, b = 19.490, c = 7.017 Å, Z = 8. Cu radiation, R = 0.067 for 1230 reflexions.

The structure (Fig. 1) contains chains of edge-sharing $Na(H_2O)_6^+$ octahedra, cross-linked by $Na_2(H_2O)_{10}^{2+}$ units into layers parallel to (100); interstices contain carbonate anions, the oxygens of which are hydrogen bonded to three or four water molecules. Na-O = 2.370-2.500(4), C-O = 1.262-1.296(6), O-H...O = 2.672-3.060(5) Å, 148-178°.

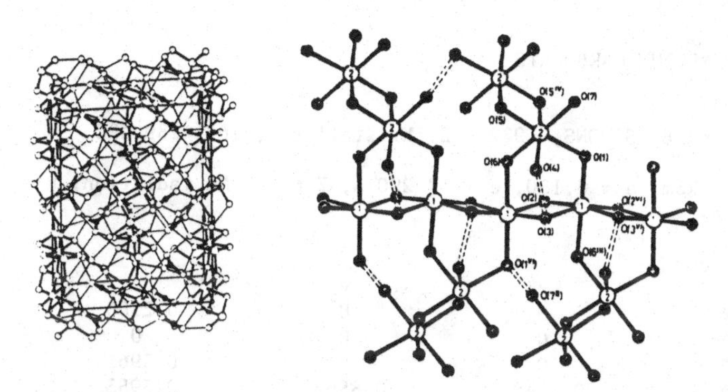

Fig. 1. Structure of sodium carbonate heptahydrate.

SODIUM SESQUICARBONATE DIHYDRATE
$Na_2CO_3.NaHCO_3.2H_2O$

C.S. CHOI and A.D. MIGHELL, 1982. Acta Cryst., B38, 2874-2876.

Monoclinic, C2/c, a = 20.36, b = 3.48, c = 10.29 Å, β = 106.48°, Z = 4. Neutron radiation, R = 0.040 for 754 reflexions.

The structure (Fig. 1) is as previously described (1); the thermal parameters of atom H(3) in the symmetrical hydrogen bond are interpreted in terms of two disordered atoms with separation 0.21(1) Å.

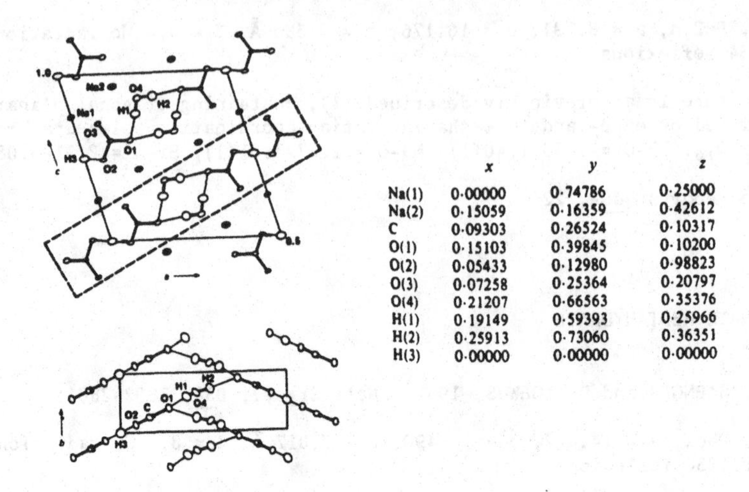

	x	y	z
Na(1)	0·00000	0·74786	0·25000
Na(2)	0·15059	0·16359	0·42612
C	0·09303	0·26524	0·10317
O(1)	0·15103	0·39845	0·10200
O(2)	0·05433	0·12980	0·98823
O(3)	0·07258	0·25364	0·20797
O(4)	0·21207	0·66563	0·35376
H(1)	0·19149	0·59393	0·25966
H(2)	0·25913	0·73060	0·36351
H(3)	0·00000	0·00000	0·00000

Fig. 1. Structure of sodium sesquicarbonate dihydrate.

1. Structure Reports, 12, 238; 20, 389.

POTASSIUM MAGNESIUM CARBONATE
$K_2Mg(CO_3)_2$

K.-F. HESSE and B. SIMONS, 1982. Z. Kristallogr., 161, 289-292.

Rhombohedral, $R\bar{3}m$, a = 5.150, c = 17.290 Å, Z = 3. Mo radiation, R = 0.057 for 280 reflexions.

Atomic positions

	x	y	z
K	0	0	0.2103
Mg	0	0	0
C	0	0	0.5961
O	0.1438	0.8562	0.5953

Isostructural with buetschliite, $K_2Ca(CO_3)_2$ (1). The structure contains slightly non-planar carbonate ions, linked by MgO_6 octahedra and KO_9 polyhedra. C-O = 1.283, Mg-O = 2.093, K-O = 2.733, 2.938(1) Å.

1. Structure Reports, 40A, 228; 46A, 305.

HYDRODRESSERITE
$BaAl_2(CO_3)_2(OH)_4.3H_2O$

J.T. SZYMAŃSKI, 1982. Canad. Miner., 20, 253-262.

Triclinic, $P\bar{1}$, a = 9.7545, b = 10.4069, c = 5.6322 Å, α = 95.695, β = 92.273, γ = 115.643°, Dm = 2.80, Z = 2. Mo radiation, R = 0.024 for 4544 reflexions.

The structure (Fig. 1) contains $AlO_2(OH)_4$ distorted trans-octahedra, linked by carbonate ions into zigzag ribbons. Ba ions have 9-coordination, and the structure contains an extensive network of hydrogen bonds.

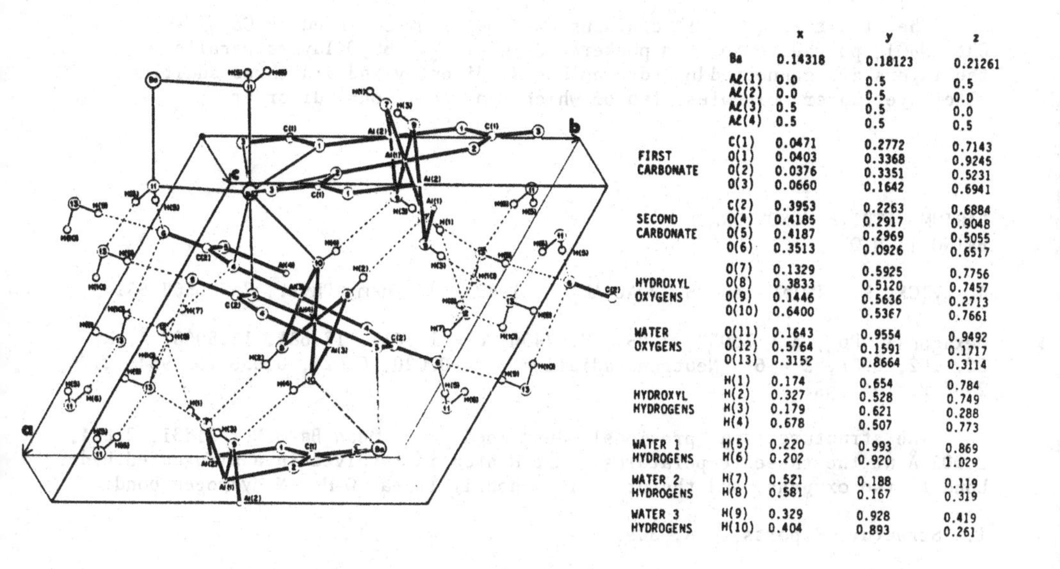

		x	y	z
	Ba	0.14318	0.18123	0.21261
	Aℓ(1)	0.0	0.5	0.5
	Aℓ(2)	0.0	0.5	0.0
	Aℓ(3)	0.5	0.5	0.0
	Aℓ(4)	0.5	0.5	0.5
FIRST CARBONATE	C(1)	0.0471	0.2772	0.7143
	O(1)	0.0403	0.3368	0.9245
	O(2)	0.0376	0.3351	0.5231
	O(3)	0.0660	0.1642	0.6941
SECOND CARBONATE	C(2)	0.3953	0.2263	0.6884
	O(4)	0.4185	0.2917	0.9048
	O(5)	0.4187	0.2969	0.5055
	O(6)	0.3513	0.0926	0.6517
HYDROXYL OXYGENS	O(7)	0.1329	0.5925	0.7756
	O(8)	0.3833	0.5120	0.7457
	O(9)	0.1446	0.5636	0.2713
	O(10)	0.6400	0.5367	0.7661
WATER OXYGENS	O(11)	0.1643	0.9554	0.9492
	O(12)	0.5764	0.1591	0.1717
	O(13)	0.3152	0.8664	0.3114
HYDROXYL HYDROGENS	H(1)	0.174	0.654	0.784
	H(2)	0.327	0.528	0.749
	H(3)	0.179	0.621	0.288
	H(4)	0.678	0.507	0.773
WATER 1 HYDROGENS	H(5)	0.220	0.993	0.869
	H(6)	0.202	0.920	1.029
WATER 2 HYDROGENS	H(7)	0.521	0.188	0.119
	H(8)	0.581	0.167	0.319
WATER 3 HYDROGENS	H(9)	0.329	0.928	0.419
	H(10)	0.404	0.893	0.261

Fig. 1. Structure of hydrodresserite.

LIEBIGITE

$Ca_2UO_2(CO_3)_3 \cdot {\sim}11H_2O$

K. MEREITER, 1982. Tschermaks Min. Petr. Mitt., 30, 277-288.

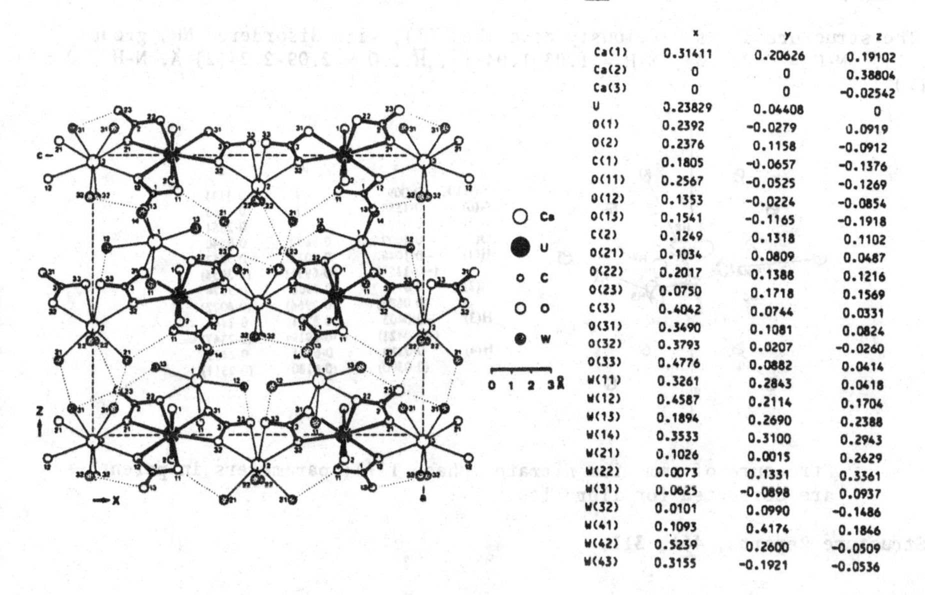

	x	y	z
Ca(1)	0.31411	0.20626	0.19102
Ca(2)	0	0	0.38804
Ca(3)	0	0	-0.02542
U	0.23829	0.04408	0
O(1)	0.2392	-0.0279	0.0919
O(2)	0.2376	0.1158	-0.0912
C(1)	0.1805	-0.0657	-0.1376
O(11)	0.2567	-0.0525	-0.1269
O(12)	0.1353	-0.0224	-0.0854
O(13)	0.1541	-0.1165	-0.1918
C(2)	0.1249	0.1318	0.1102
O(21)	0.1034	0.0809	0.0487
O(22)	0.2017	0.1388	0.1216
O(23)	0.0750	0.1718	0.1569
C(3)	0.4042	0.0744	0.0331
O(31)	0.3490	0.1081	0.0824
O(32)	0.3793	0.0207	-0.0260
O(33)	0.4776	0.0882	0.0414
W(11)	0.3261	0.2843	0.0418
W(12)	0.4587	0.2114	0.1704
W(13)	0.1894	0.2690	0.2388
W(14)	0.3533	0.3100	0.2943
W(21)	0.1026	0.0015	0.2629
W(22)	0.0073	0.1331	0.3361
W(31)	0.0625	-0.0898	0.0937
W(32)	0.0101	0.0990	-0.1486
W(41)	0.1093	0.4174	0.1846
W(42)	0.5239	0.2600	-0.0509
W(43)	0.3155	-0.1921	-0.0536

Fig. 1. Structure of liebigite.

Orthorhombic, Bba2, a = 16.699, b = 17.557, c = 13.697 Å, D_m = 2.41, Z = 8. Mo radiation, R = 0.030 for 3005 reflexions.

The structure (Fig. 1) contains $UO_2(CO_3)_3$ units linked by $CaO_4(H_2O)_4$ and $CaO_3(H_2O)_4$ polyhedra to form puckered $Ca_2UO_2(CO_3)_3.8H_2O$ layers parallel to (010); the layers are connected by hydrogen bonds, directly and via three additional interlayer water molecules, two of which show positional disorder.

BARIUM NITRITE MONOHYDRATE
$Ba(NO_2)_2.H_2O$

Å. KVICK, R. LIMINGA and S.C. ABRAHAMS, 1982. J. Chem. Phys., 76, 5508-5514.

Hexagonal, $P6_5$, a = 7.052, 7.056, 7.07490, c = 17.637, 17.681, 17.89087 Å, at 20, 102, 298K, Z = 6. Neutron radiation, R = 0.020, 0.024, 0.050 for 954, 2179, 2393 reflexions.

The structure is as previously described (1). Mean Ba-O,N = 2.881, 2.884, 2.902 Å at the three temperatures. One H atom is involved in a bifurcated hydrogen bond to two oxygens, and the other in a nearly linear O-H...N hydrogen bond.

1. Structure Reports, 46A, 309.

AMMONIUM NITRATE (PHASE III)
NH_4NO_3

C.S. CHOI and H.J. PRASK, 1982. Acta Cryst., B38, 2324-2328.

Orthorhombic, Pnma, a = 7.6772, b = 5.8208, c = 7.1396 Å, Z = 4. Neutron radiation, R = 0.042 for 348 reflexions, for crystal containing 5% KNO_3; rigid-body refinement of NH_4 group.

The structure is as previously described (1), with disordered NH_4 groups (Fig. 1). N-O = 1.239(3), N-H = 1.03-1.04(1), H...O = 2.09-2.24(2) Å, N-H...O = 127-170°.

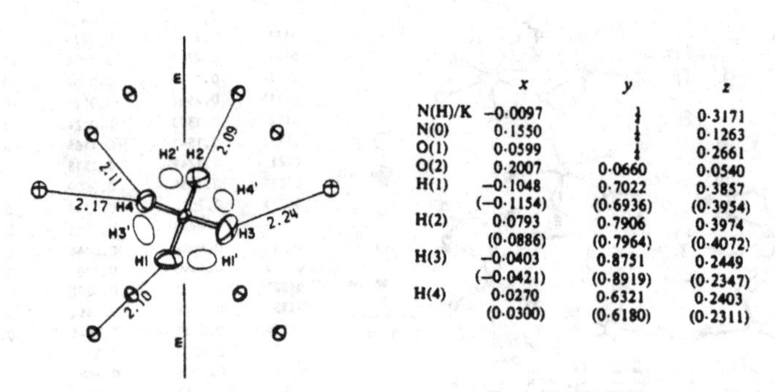

	x	y	z
N(H)/K	−0·0097	1/4	0·3171
N(0)	0·1550	1/4	0·1263
O(1)	0·0599	1/4	0·2661
O(2)	0·2007	0·0660	0·0540
H(1)	−0·1048	0·7022	0·3857
	(−0·1154)	(0·6936)	(0·3954)
H(2)	0·0793	0·7906	0·3974
	(0·0886)	(0·7964)	(0·4072)
H(3)	−0·0403	0·8751	0·2449
	(−0·0421)	(0·8919)	(0·2347)
H(4)	0·0270	0·6321	0·2403
	(0·0300)	(0·6180)	(0·2311)

Fig. 1. Structure of ammonium nitrate (phase III); parameters in parentheses are corrected for libration.

1. Structure Reports, 46A, 311.

TRISODIUM ORTHONITRATE
Na_3NO_4

M. JANSEN, 1982. Z. anorg. Chem., 491, 175-183.

Orthorhombic, Pbca, a = 8.632, b = 9.731, c = 9.042 Å, Z = 8. Mo radiation, R = 0.050 for 1092 reflexions.

The structure (Fig. 1) contains tetrahedral NO_4^{3-} anions, linked by Na ions which have 4- and 5-coordinations. N-O = 1.39 Å, O-N-O = 108.9-110.0°, Na-O = 2.27-2.57 Å.

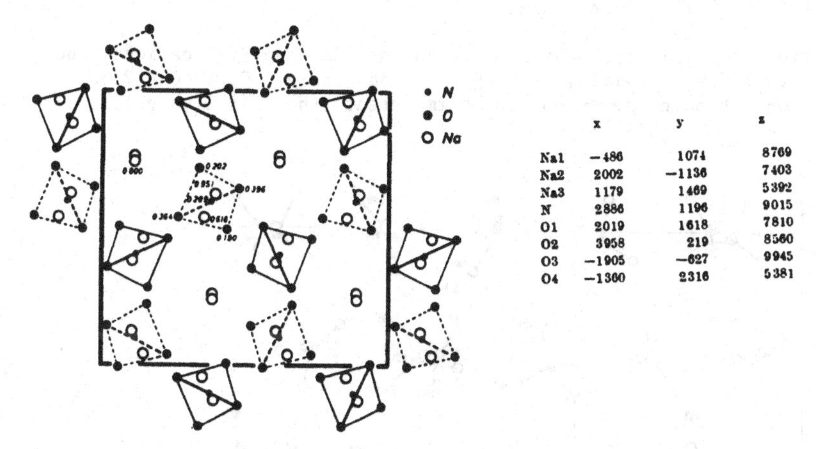

	x	y	z
Na1	−486	1074	8769
Na2	2002	−1136	7403
Na3	1179	1469	5392
N	2886	1196	9015
O1	2019	1618	7810
O2	3958	219	8560
O3	−1905	−627	9945
O4	−1360	2316	5381

Fig. 1. Structure of Na_3NO_4, and atomic positional parameters (x 10^4).

RUBIDIUM NITRATE (PHASE IV)
$RbNO_3$

M. SHAMSUZZOHA and B.W. LUCAS, 1982. Acta Cryst., B38, 2353-2357.

Trigonal, $P3_1$, a = 10.55, 10.61, c = 7.47, 7.55 Å, at 298, 403K, Z = 9. Neutron radiation, R = 0.068, 0.078 for 1080 (including symmetry equivalents) and 451 (independent) reflexions. Previous study in 1.

The structure (Fig. 1) contains a pseudocubic Rb sublattice, with nine pseudocubes per cell, with nitrate groups parallel to pseudocube faces.

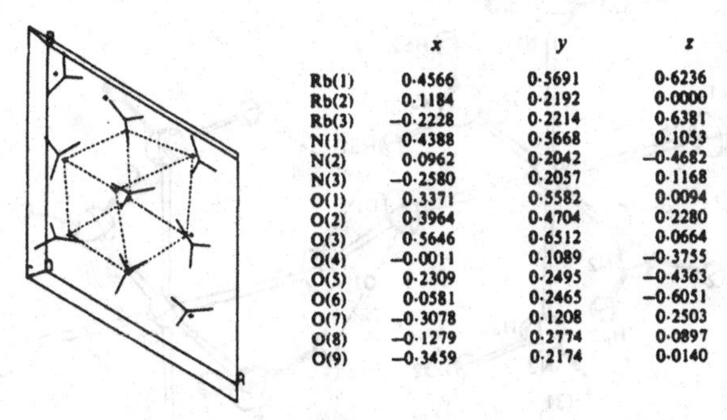

	x	y	z
Rb(1)	0·4566	0·5691	0·6236
Rb(2)	0·1184	0·2192	0·0000
Rb(3)	−0·2228	0·2214	0·6381
N(1)	0·4388	0·5668	0·1053
N(2)	0·0962	0·2042	−0·4682
N(3)	−0·2580	0·2057	0·1168
O(1)	0·3371	0·5582	0·0094
O(2)	0·3964	0·4704	0·2280
O(3)	0·5646	0·6512	0·0664
O(4)	−0·0011	0·1089	−0·3755
O(5)	0·2309	0·2495	−0·4363
O(6)	0·0581	0·2465	−0·6051
O(7)	−0·3078	0·1208	0·2503
O(8)	−0·1279	0·2774	0·0897
O(9)	−0·3459	0·2174	0·0140

Fig. 1. Structure of $RbNO_3$-IV, and atomic positional parameters at 298K (slightly different values at 403K).

1. Strukturbericht, 2, 381.

CAESIUM BISMUTH(III) NITRATE MONOHYDRATE
$Cs_2Bi(NO_3)_5.H_2O$

F. LAZARINI and I. LEBAN, 1982. Cryst. Struct. Comm., 11, 653-657.

Orthorhombic, $Pna2_1$, a = 26.433, b = 7.139, c = 8.044 Å, D_m = 3.45, Z = 4. Mo radiation, R = 0.064 for 1224 reflexions.

The structure (Fig. 1) contains nitrate anions, Cs^+ and Bi^{3+} cations, and water molecules. N-O = 1.14-1.35(5), Bi-9 O = 2.38-2.69(3), Cs-11 O = 2.99-3.26(4) Å; hydrogen bonding seems not to be important (shortest O(water)...O = 2.93 Å).

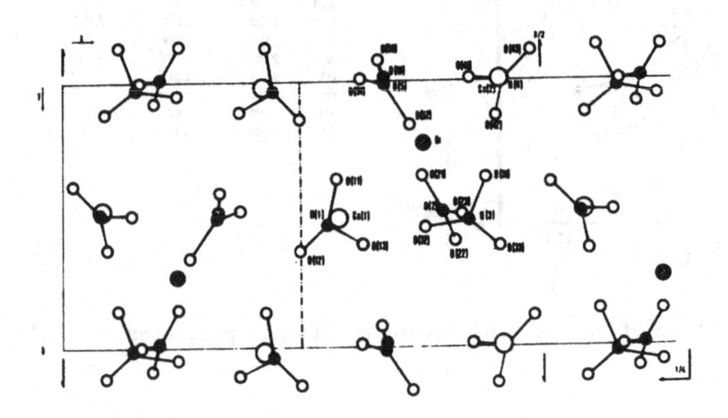

Fig. 1. Structure of $Cs_2Bi(NO_3)_5.H_2O$.

μ-PEROXO-BIS[PENTAAMMINECOBALT(III)] TETRANITRATE DIHYDRATE
$[(NH_3)_5CoO_2Co(NH_3)_5](NO_3)_4.2H_2O$

U. THEWALT, 1982. Z. anorg. Chem., 485, 122-128.

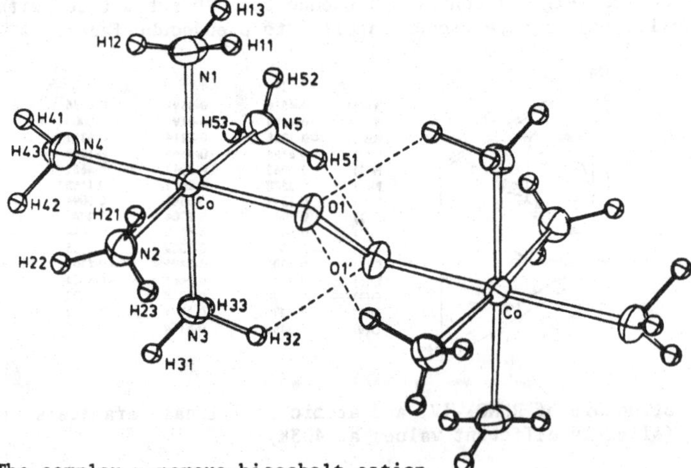

Fig. 1. The complex μ-peroxo-biscobalt cation.

Monoclinic, $P2_1/n$, a = 11.657, b = 11.977, c = 8.082 Å, β = 91.58°, D_m = 1.75, Z = 2. Mo radiation, R = 0.051 for 1978 reflexions.

The centrosymmetric cation (Fig. 1) contains octahedrally-coordinated Co ions linked by a planar Co-O-O-Co bridge; Co-O = 1.89, O-O = 1.47, Co-N = 1.96-2.00 Å, Co-O-O = 111°. There are also intramolecular N-H...O hydrogen bonds, 2.77 and 2.80 Å. Two of the three independent nitrate groups are disordered.

GERHARDTITE

$Cu_2(OH)_3NO_3$

B. BOVIO and S. LOCCHI, 1982. J. Cryst. Spect. Res., 12, 507-517.

Orthorhombic, $P2_12_12_1$, a = 6.087, b = 13.813, c = 5.597 Å, Z = 4. Mo radiation, R = 0.097 for 532 reflexions.

The structure (Fig. 1) contains $Cu(OH)_4O_2$ and $Cu(OH)_5O$ distorted octahedra sharing edges, to form layers which are linked by hydrogen bonds. Cu-O = 1.93-2.02 (4 distances), 2.31-2.48(2) (2 distances), N-O = 1.24-1.28(3), O-H...O = 2.88-2.95(3) Å. A monoclinic form is also known (1).

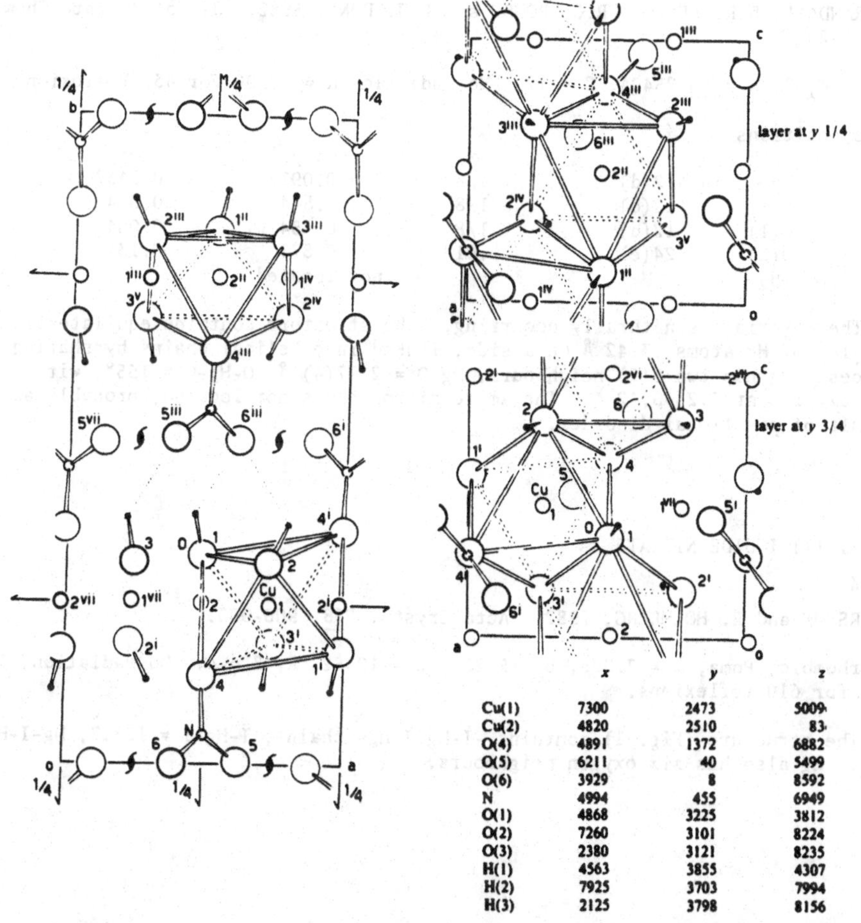

	x	y	z
Cu(1)	7300	2473	5009
Cu(2)	4820	2510	83
O(4)	4891	1372	6882
O(5)	6211	40	5499
O(6)	3929	8	8592
N	4994	455	6949
O(1)	4868	3225	3812
O(2)	7260	3101	8224
O(3)	2380	3121	8235
H(1)	4563	3855	4307
H(2)	7925	3703	7994
H(3)	2125	3798	8156

Fig. 1. Structure of gerhardtite, and atomic positional parameters (x 10^4).

1. Structure Reports, 13, 302; 16, 321.

SILVER NITRATE (HIGH-TEMPERATURE)
$AgNO_3$

P. MEYER and J.-J. CAPPONI, 1982. Acta Cryst., B38, 2543-2546.

Rhombohedral, R3m, a = 4.125 Å, α = 78.1°, at 444K, Z = 1 (hexagonal cell, a = 5.196, c = 8.49 Å, Z = 3). Ag radiation, R = 0.047 for 149 reflexions. Ag in 3(a): 0,0,0; N in 3(a): 0,0,0.4269; O in 9(b): 0.1334,-0.1334,0.4239 (hexagonal axes).

The Ag ions are at the cell corners, with the nitrate ions close to the cell centres. A metastable rhombohedral phase has a doubled c axis.

MERCURY(II) AMIDE NITRATE
$HgNH_2NO_3$

C.J. RANDALL, D.R. PEACOR, R.C. ROUSE and P.J. DUNN, 1982. J. Solid State Chem., 42, 221-226.

Cubic, $P4_132$, a = 10.254 Å, Z = 12. Mo radiation R = 0.090 for 437 reflexions.

Atomic positions

		x	y	z
Hg	in 12(d)	1/8	0.0937	0.3437
N	12(d)	1/8	0.844	0.094
O(1)	12(d)	1/8	0.754	0.004
O(2)	24(e)	0.018	0.873	0.131
NH_2			not located	

The material is naturally occurring. The structure contains equilateral triangles of Hg atoms, 3.42 Å on a side, linked into helical chains by sharing vertices. Hg has two O(2) neighbours, Hg-O = 2.77(4) Å, O-Hg-O = 155°, with other oxygens at 3.23-3.42 Å. The amide nitrogen was not located, probably as a result of positional disorder.

MERCURY(II) IODIDE NITRATE
$HgINO_3$

K. PERSSON and B. HOLMBERG, 1982. Acta Cryst., B38, 900-903.

Orthorhombic, Pnma, a = 7.385, b = 5.257, c = 12.528 Å, Z = 4. Mo radiation, R = 0.044 for 610 reflexions.

The structure (Fig. 1) contains -I-Hg-I-Hg- chains, I-Hg-I = 158.7, Hg-I-Hg = 90.3°. Hg also has six oxygen neighbours.

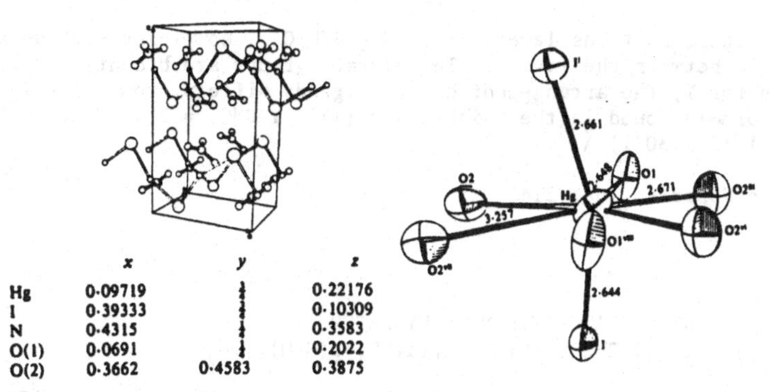

	x	y	z
Hg	0·09719	¼	0·22176
I	0·39333	¼	0·10309
N	0·4315	¼	0·3583
O(1)	0·0691	¼	0·2022
O(2)	0·3662	0·4583	0·3875

Fig. 1. Structure of $HgINO_3$.

MERCURY SILVER IODIDE NITRATE MONOHYDRATE
$Ag_2HgI_2(NO_3)_2.H_2O$

K. PERSSON and B. HOLMBERG, 1982. Acta Cryst., B38, 904-907.

Orthorhombic, Pbam, a = 10.947, b = 18.968, c = 5.313 Å, Z = 4. Mo radiation, R = 0.06 for 817 reflexions.

The structure (Fig. 1) contains sheets of Ag_2I-Hg-IAg_2 units sharing Ag corners; cavities between the sheets contain the nitrate ions and water molecules. There is disorder of Ag, nitrate, and water positions. Hg-I = 2.632(3), Ag-I = 2.71, 2.74 (very short), 2.81, 3.11, Hg-O = 2.70-2.86 (6 distances), Ag-O = 1.62-3.65 Å (coordination depends on the occupancy, and the 1.62 Å distance cannot be real), I-Hg-I = 177°.

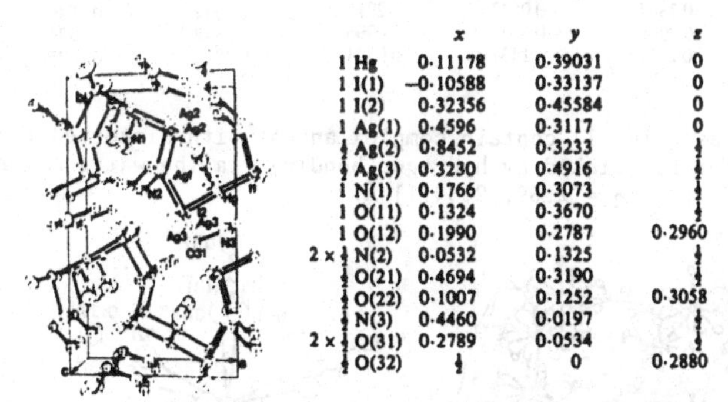

		x	y	z
1	Hg	0·11178	0·39031	0
1	I(1)	−0·10588	0·33137	0
1	I(2)	0·32356	0·45584	0
1	Ag(1)	0·4596	0·3117	0
½	Ag(2)	0·8452	0·3233	½
½	Ag(3)	0·3230	0·4916	½
1	N(1)	0·1766	0·3073	½
1	O(11)	0·1324	0·3670	½
1	O(12)	0·1990	0·2787	0·2960
2 × ½	N(2)	0·0532	0·1325	½
½	O(21)	0·4694	0·3190	½
½	O(22)	0·1007	0·1252	0·3058
½	N(3)	0·4460	0·0197	½
2 × ½	O(31)	0·2789	0·0534	½
½	O(32)	½	0	0·2880

Fig. 1. Structure of $Ag_2HgI_2(NO_3)_2.H_2O$.

YTTRIUM(III) NITRATE PENTAHYDRATE (TETRAAQUOTRINITRATOYTTRIUM(III) MONOHYDRATE)
$Y(NO_3)_3.5H_2O$ $[Y(H_2O)_4(NO_3)_3].H_2O$

B. ERIKSSON, 1982. Acta Chem. Scand., A36, 186-188.

Triclinic, P$\bar{1}$, a = 6.652, b = 9.558, c = 10.563 Å, α = 63.59, β = 84.62, γ = 76.06°, Z = 2. Mo radiation, R = 0.074 for 2880 reflexions.

The structure contains layers of $[Y(NO_3)_3(H_2O)_4]$ complexes with an additional water molecule between the layers. The nitrate groups are bidentate giving 10-coordination for Y; the arrangement of the ligands differs from that of the complex of similar formula found in the hexahydrate (1). Y-ONO_2 = 2.43-2.60, Y-OH_2 = 2.36-2.44, N-O = 1.19-1.30(1) Å.

1. Structure Reports, 46A, 314.

AMMONIUM DIAQUOPENTANITRATOLANTHANATE HYDRATES
$(NH_4)_2[La(NO_3)_5(H_2O)_2].2H_2O$, $(NH_4)_2[La(NO_3)_5(H_2O)_2].H_2O$

B. ERIKSSON, L.O. LARSSON and L. NIINISTO, 1982. Acta Chem. Scand., A36, 465-470.

Monoclinic, C2/c, a = 11.152, 10.969, b = 8.966, 9.012, c = 17.881, 17.439 Å, β = 101.6, 100.1°, Z = 4. Mo radiation, R = 0.065, 0.041 for 2389, 1866 reflexions.

Atomic positions (O(2) has occupancy 0.5 in the second compound)

	x	y	z			
	$(NH_4)_2[La(NO_3)_5(H_2O_2)] \cdot 2H_2O$			$(NH_4)_2[La(NO_3)_5(H_2O)_2] \cdot H_2O$		
La	0	0.0567	0.25	0	0.0572	0.25
O(11)	0.0925	−0.2222	0.2401	0.0947	−0.2195	0.2413
O(12)	0	−0.4296	0.25	0	−0.4261	0.25
O(21)	−0.0791	0.2519	0.1395	−0.0789	0.2518	0.1362
O(22)	0.1075	0.2925	0.1957	0.1067	0.2935	0.1945
O(23)	0.0272	0.4213	0.0961	0.0310	0.4220	0.0921
O(31)	0.0991	−0.0017	0.1277	0.1106	0.0053	0.1264
O(32)	−0.0852	−0.0864	0.1174	−0.0728	−0.0861	0.1131
O(33)	0.0235	−0.1495	0.0350	0.0466	−0.1514	0.0330
N(1)	0	−0.2959	0.25	0	−0.2899	0.25
N(2)	0.0186	0.3244	0.1426	0.0190	0.3246	0.1398
N(3)	0.0144	−0.0781	0.0912	0.0287	−0.0789	0.0891
O(1)	0.2364	0.0542	0.2914	0.2342	0.0546	0.2902
O(2)	0.2962	−0.1157	−0.0306	0.2842	−0.0984	−0.0213
NH_4	0.2757	0.1431	0.0573	0.2924	0.1423	0.0566

The structures (Fig. 1) contain complex anions with distorted icosahedral 12-coordination for La, linked by hydrogen bonding via the water molecules. La-ON = 2.66-2.73, La-OH_2 = 2.59, 2.54(1) Å.

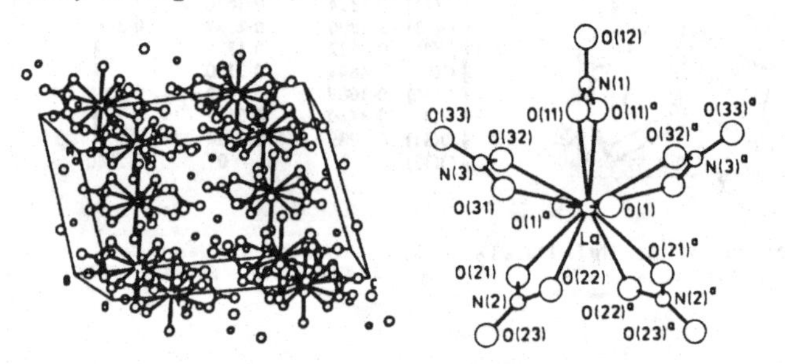

Fig. 1. Structure of the pentanitratolanthanates.

CERIUM(III) NITRATE TETRAHYDRATE
$Ce(NO_3)_3.4H_2O$

N. MILINSKI, P. RADIVOJEVIČ and B. RIBÁR, 1982. Cryst. Struct. Comm., 11, 1241-1244

Orthorhombic, Pbca, a = 11.756, b = 12.901, c = 13.522 Å, Z = 8. Mo radiation, R = 0.050 for 2225 reflexions.

The structure contains chains along c of $CeO_6(H_2O)_4$ bicapped square anti-prisms bridged by one nitrate group, the other two nitrate groups chelating one Ce ion (Fig. 1); the chains are linked by hydrogen bonds. Ce-O = 2.59-2.69, $Ce-OH_2$ = 2.49-2.56(1) Å. The hexahydrate has been described previously (1).

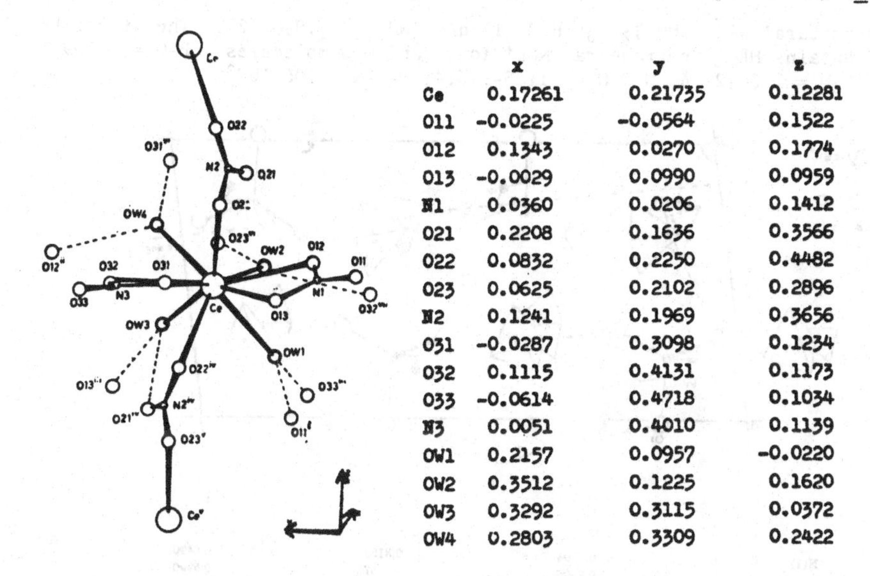

	x	y	z
Ce	0.17261	0.21735	0.12281
O11	-0.0225	-0.0564	0.1522
O12	0.1343	0.0270	0.1774
O13	-0.0029	0.0990	0.0959
N1	0.0360	0.0206	0.1412
O21	0.2208	0.1636	0.3566
O22	0.0832	0.2250	0.4482
O23	0.0625	0.2102	0.2896
N2	0.1241	0.1969	0.3656
O31	-0.0287	0.3098	0.1234
O32	0.1115	0.4131	0.1173
O33	-0.0614	0.4718	0.1034
N3	0.0051	0.4010	0.1139
OW1	0.2157	0.0957	-0.0220
OW2	0.3512	0.1225	0.1620
OW3	0.3292	0.3115	0.0372
OW4	0.2803	0.3309	0.2422

Fig. 1. Structure of cerium(III) nitrate tetrahydrate.

1. Structure Reports, 46A, 316.

RUBIDIUM TRIS(NITRATO)DIOXONEPTUNATE(VI)
$Rb[NpO_2(NO_3)_3]$

N.W. ALCOCK, M.M. ROBERTS and D. BROWN, 1982. J. Chem. Soc., Dalton, 33-36.

Rhombohedral, R$\bar{3}$c, a = 9.281, c = 19.033 Å, Z = 6. Mo radiation, R = 0.067 for 305 reflexions.

Atomic positions

		x	y	z
Np	in 6(a)	0	0	1/4
O(1)	12(c)	0	0	0.3410
O(2)	18(e)	-0.4426	0	1/4
O(3)	36(f)	-0.1729	0.1332	0.2552
N	18(e)	-0.3414	0	1/4
Rb	6(b)	0	0	0

Isostructural with the U analogue (1). Np has hexagonal bipyramidal coordination to six equatorial O(3) atoms from three bidentate nitrate groups, and two apical neptunyl O(1) atoms; Np-O(1) = 1.73(2), Np-O(3) = 2.47(3), N-O(3) = 1.28(2), N-O(2) = 1.20(3) Å. Rb-O = 3.03-3.26(2) Å.

1. Structure Reports, 14, 36; 30A, 406.

AMMONIUM PHOSPHITE MONOHYDRATE
$(NH_4)_2HPO_3.H_2O$

M. RAFIQ, J. DURAND and L. COT, 1982. Z. anorg. Chem., 484, 187-194.

Monoclinic, $P2_1/c$, a = 6.322, b = 8.323, c = 12.676 Å, β = 98.84°, D_m = 1.37, Z = 4. Mo radiation, R = 0.022 for 853 reflexions.

Isostructural with $(NH_4)_2PO_3F.H_2O$ (1) and $(NH_4)_2SO_3.H_2O$ (2). The structure (Fig. 1) contains HPO_3^- tetrahedra, NH_4^+ ions, and H_2O molecules. P-O = 1.509-1.519(2), P-H = 1.34(2) Å, O-P-O = 111.5-113.4, O-P-H = 106-107°.

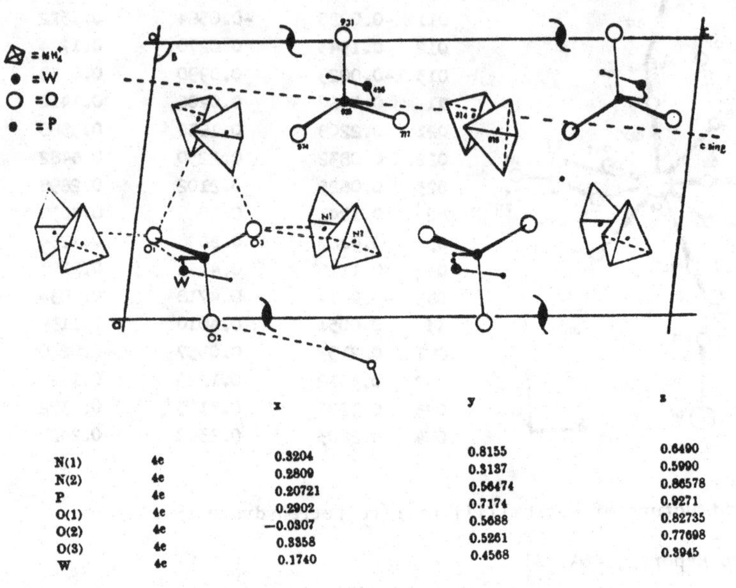

		x	y	z
N(1)	4e	0.3204	0.8155	0.6490
N(2)	4e	0.2809	0.3137	0.5990
P	4e	0.20721	0.56474	0.86578
O(1)	4e	0.2902	0.7174	0.9271
O(2)	4e	−0.0307	0.5688	0.82735
O(3)	4e	0.3358	0.5261	0.77698
W	4e	0.1740	0.4568	0.3945

Fig. 1. Structure of ammonium phosphite monohydrate.

1. Structure Reports, 38A, 317; 44A, 239.
2. Ibid., 30A, 378; 43A, 276.

SODIUM METAPHOSPHATE (KURROL SALT) (TYPE C)
$NaPO_3$

A. IMMIRZI and W. PORZIO, 1982. Acta Cryst., B38, 2788-2792.

Tetragonal, $I4_1/a$, a = 13.18, c = 5.93 Å, Z = 16. Powder data.

Atomic positions

	x	y	z
Na	0.402	-0.162	0.351
P	0.327	-0.100	-0.062
O(1)	0.366	0.015	-0.083
O(2)	0.414	-0.171	-0.069
O(3)	0.262	-0.126	-0.257

The proposed structure is similar to that of the B-form (1), the only difference being the relative position of layers of helices parallel to [101].

1. Structure Reports, 28, 186.

SODIUM HYDROGENPHOSPHITE
$Na_{0.5}H_{2.5}PO_3$

R.G. HAZELL, A.C. HAZELL and B. KRATOCHVÍL, 1982. Acta Cryst., B38, 1267-1269.

Trigonal, P$\bar{3}$c1, a = 5.908, c = 9.554 Å, D_m = 2.135, Z = 4. Mo, Ag radiations, R = 0.022, 0.022 for 884, 843 reflexions.

The structure (Fig. 1) contains tetrahedral $HP(OH_{0.5})_3{}^{1/2-}$ ions connected into sheets by disordered O...H(2)...O hydrogen bonds; the sheets are held together by octahedrally-coordinated Na^+ ions. P-O = 1.529(4), P-H = 1.26(1), Na-O = 2.423(5) Å.

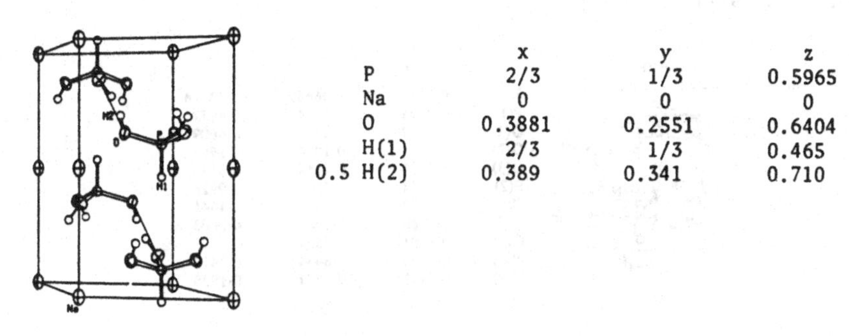

	x	y	z
P	2/3	1/3	0.5965
Na	0	0	0
O	0.3881	0.2551	0.6404
H(1)	2/3	1/3	0.465
0.5 H(2)	0.389	0.341	0.710

Fig. 1. Structure of sodium hydrogenphosphite.

SODIUM THIOPHOSPHATE DODECAHYDRATE
$Na_3PO_3S.12H_2O$

B.M. GOLDSTEIN, 1982. Acta Cryst., B38, 1116-1120.

Rhombohedral, R$\bar{3}$c, a = 9.061, c = 34.34 Å, Z = 6. Mo radiation, R = 0.051 for 326 reflexions.

The structure (Fig. 1) contains disordered PO_3S^{3-} tetrahedra and $Na_3(H_2O)_{12}{}^{3+}$ complexes of three face-sharing octahedra, linked by hydrogen bonding. P-S = 1.99, P-O = 1.52(1) Å, S-P-O = 110.6, O-P-O = 108.3°, Na-O = 2.38-2.40 Å.

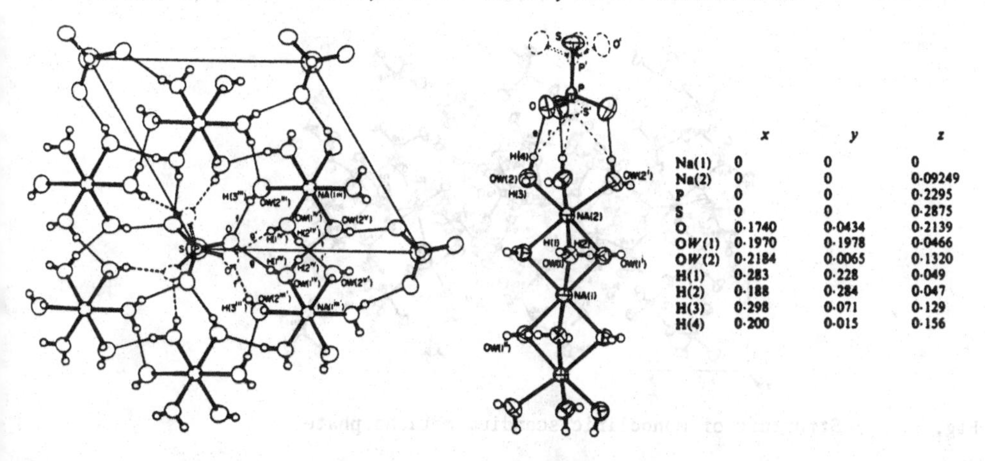

	x	y	z
Na(1)	0	0	0
Na(2)	0	0	0·09249
P	0	0	0·2295
S	0	0	0·2875
O	0·1740	0·0434	0·2139
OW(1)	0·1970	0·1978	0·0466
OW(2)	0·2184	0·0065	0·1320
H(1)	0·283	0·228	0·049
H(2)	0·188	0·284	0·047
H(3)	0·298	0·071	0·129
H(4)	0·200	0·015	0·156

Fig. 1. Structure of sodium thiophosphate dodecahydrate (occupancy = 0.5 for P, S, O).

AMMONIUM TIN(II) PHOSPHITE
$(NH_4)_2Sn(HPO_3)_2$

T. YAMAGUCHI and O. LINDQVIST, 1982. Acta Cryst., B38, 1441-1445.

Triclinic, P$\bar{1}$, a = 4.799, b = 10.060, c = 10.338 Å, α = 107.07, β = 82.18, γ = 90.32°, Z = 2. Mo radiation, R = 0.046 for 2364 reflexions.

The structure (Fig. 1) contains tetrahedral HPO_3^{2-} anions (H atoms not located), linked by Sn^{2+} ions with 3+1+1 coordination, and by ammonium ions which each form three strong hydrogen bonds. P-O = 1.50-1.54, Sn-O = 2.11, 2.12, 2.16, 2.70, 3.26, N-H...O = 2.77-2.89(1) Å.

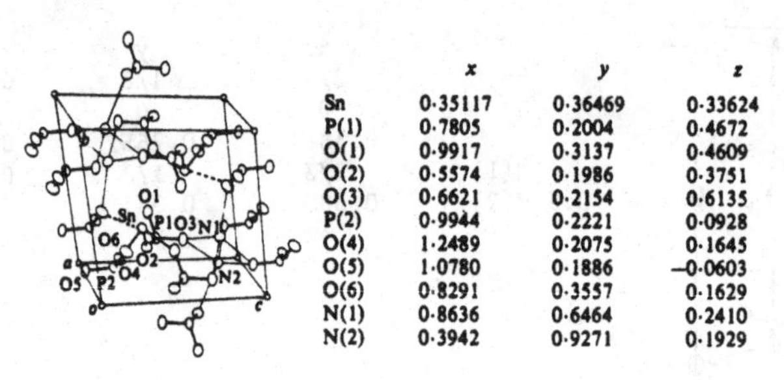

	x	y	z
Sn	0·35117	0·36469	0·33624
P(1)	0·7805	0·2004	0·4672
O(1)	0·9917	0·3137	0·4609
O(2)	0·5574	0·1986	0·3751
O(3)	0·6621	0·2154	0·6135
P(2)	0·9944	0·2221	0·0928
O(4)	1·2489	0·2075	0·1645
O(5)	1·0780	0·1886	−0·0603
O(6)	0·8291	0·3557	0·1629
N(1)	0·8636	0·6464	0·2410
N(2)	0·3942	0·9271	0·1929

Fig. 1. Structure of ammonium tin(II) phosphite.

SCANDIUM METAPHOSPHATE (MONOCLINIC)
$Sc(PO_3)_3$

A.I. DOMANSKIJ, Ju.F. ŠEPELEV, Ju.I. SMOLIN and B.N. LITVIN, 1982. Kristallografija, 27, 229-232 [Soviet Physics - Crystallography, 27, 140-142].

Monoclinic, Cc, a = 13.558, b = 19.588, c = 9.690 Å, β = 127.11°, Z = 12. Mo radiation, R = 0.02 for 3760 reflexions.

Isostructural with the monoclinic form of $Al(PO_3)_3$ (1), with infinite polyphosphate chains linked by octahedrally-coordinated Sc ions (Fig. 1). A high-temperature cubic form, with cyclic $P_4O_{12}^{4-}$ ions, is also known (2).

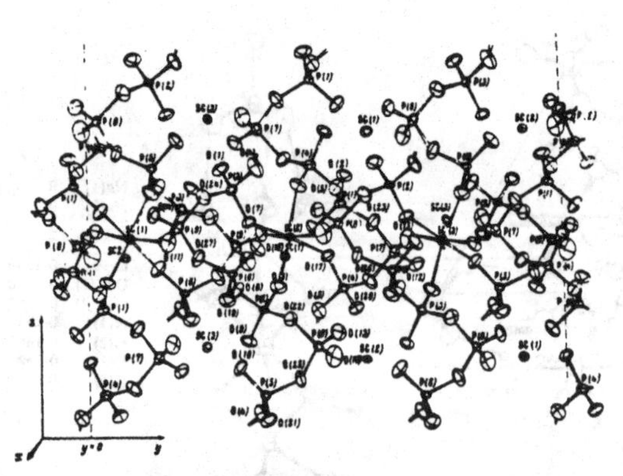

Fig. 1. Structure of monoclinic scandium metaphosphate.

1. Structure Reports, 42A, 332.
2. Strukturbericht, 5, 15, 80; Structure Reports, 44A, 241.

MANGANESE(III) HYDROGENBIS(ORTHOPHOSPHITE) DIHYDRATE

$MnH_3P_2O_6.2H_2O$ $MnH(HPO_3)_2.2H_2O$

I. CÍSAŘOVÁ, C. NOVÁK, V. PETŘÍČEK, B. KRATOCHVÍL and J. LOUB, 1982. Acta Cryst., B38, 1687-1689.

Monoclinic, $P2_1/b$, a = 7.618, b = 8.839, c = 6.703 Å, γ = 125.46°, D_m = 2.24, Z = 2. Mo radiation, R = 0.049 for 903 reflexions (twinned crystal).

The structure (Fig. 1) contains tetrahedral HPO_3^{2-} ions probably connected by H^+ in a linear symmetric hydrogen bond to form a $H_3P_2O_6^{3-}$ anion (H atoms not located). The anions are linked by $MnO_4(OH_2)_2$ octahedra. P-O = 1.52-1.55, Mn-O = 1.90, 1.91, Mn-OH_2 = 2.23(1) Å, O-P-O = 109-112°, O-H...O = 2.44 (anion), 2.83, 2.84 (water) Å.

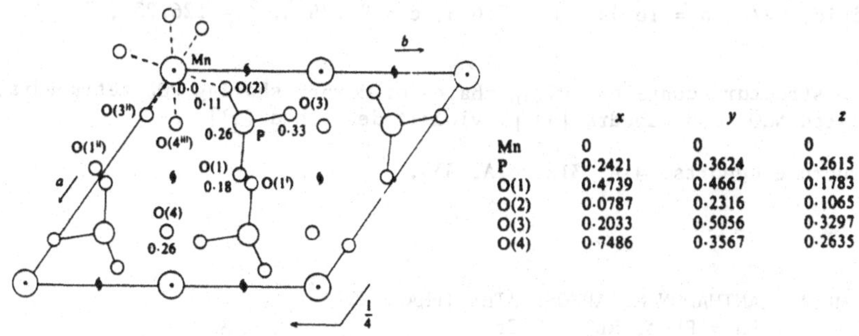

	x	y	z
Mn	0	0	0
P	0·2421	0·3624	0·2615
O(1)	0·4739	0·4667	0·1783
O(2)	0·0787	0·2316	0·1065
O(3)	0·2033	0·5056	0·3297
O(4)	0·7486	0·3567	0·2635

Fig. 1. Structure of manganese(III) hydrogenbis(orthophosphite) dihydrate.

SODIUM COBALT(II) DIHYDROGENPHOSPHITE MONOHYDRATE

$NaCo(H_2PO_3)_3.H_2O$

B. KRATOCHVÍL, J. PODLAHOVÁ, S. HABIBPUR, V. PETŘÍČEK and K. MALÝ, 1982. Acta Cryst., B38, 2436-2438.

Orthorhombic, Pbca, a = 9.054, b = 14.706, c = 14.759 Å, D_m = 2.30, Z = 8. Mo radiation, R = 0.046 for 1980 reflexions.

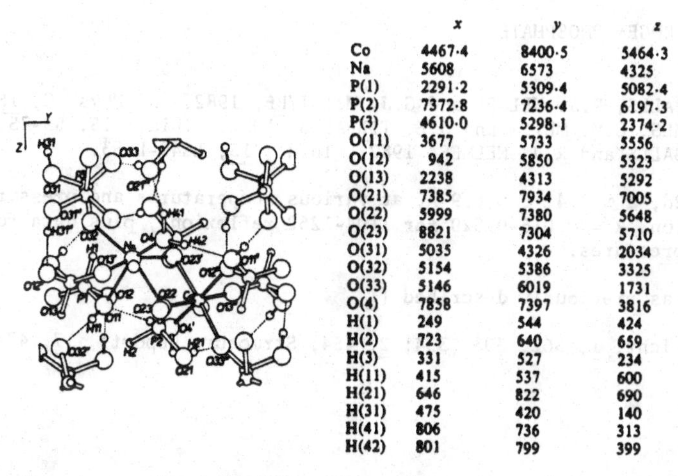

	x	y	z
Co	4467·4	8400·5	5464·3
Na	5608	6573	4325
P(1)	2291·2	5309·4	5082·4
P(2)	7372·8	7236·4	6197·3
P(3)	4610·0	5298·1	2374·2
O(11)	3677	5733	5556
O(12)	942	5850	5323
O(13)	2238	4313	5292
O(21)	7385	7934	7005
O(22)	5999	7380	5648
O(23)	8821	7304	5710
O(31)	5035	4326	2034
O(32)	5154	5386	3325
O(33)	5146	6019	1731
O(4)	7858	7397	3816
H(1)	249	544	424
H(2)	723	640	659
H(3)	331	527	234
H(11)	415	537	600
H(21)	646	822	690
H(31)	475	420	140
H(41)	806	736	313
H(42)	801	799	399

Fig. 1. Structure of $NaCo(H_2PO_3)_3.H_2O$, and atomic positional parameters (x 10^3 for H, x 10^4 for others).

The structure (Fig. 1) contains three independent $HPO_2(OH)^-$ tetrahedra, linked by octahedrally-coordinated Co^{2+} and Na^+ ions, and by hydrogen bonding which also involves the water molecule. P-OH = 1.563-1.573(3), P-O = 1.493-1.504(3), P-H = 1.18-1.37(5), Co-O = 2.061-2.151(3), Na-O = 2.312-2.809(3), O-H...O = 2.560-2.995(4) Å.

LITHIUM NEODYMIUM METAPHOSPHATE
$LiNd(PO_3)_4$

J. LIU, 1982. Wuli Xuebao, 31, 537-542.

Monoclinic, C2/c, a = 16.486, b = 7.073, c = 9.775 Å, β = 126.23°, Z = 4. R = 0.054.

The structure contains $(PO_3)_n$ chains of corner-sharing PO_4 tetrahedra, linked by isolated NdO_8 dodecahedra [as previously described (1)].

1. Structure Reports, 41A, 312; 42A, 335.

ALKALI-METAL LANTHANON METAPHOSPHATES (PHASE IV)
$MLn(PO_3)_4$ (MLn = RbHo, RbTm, CsEr)

S.I. MAKSIMOVA, K.K. PALKINA and N.T. ČIBISKOVA, 1982. Izv. Akad. Nauk SSSR, Neorg. Mater., 18, 653-659.

Monoclinic, $P2_1/n$, a = 10.266, 10.217, 10.215, b = 8.853, 8.803, 8.833, c = 10.953, 10.928, 11.136 Å, β = 106.28, 106.28, 106.32°, Z = 4. R = 0.049, 0.040, 0.032.

The structures contain chains of corner-sharing PO_4 tetrahedra, as previously described for related materials (1).

1. Structure Reports, 45A, 402.

POTASSIUM DIHYDROGEN PHOSPHATE
KH_2PO_4

I. J.E. TIBBALLS, R.J. NELMES and G.J. McINTYRE, 1982. J. Phys. C, 15, 37-58.
II. R.J. NELMES, G.M. MEYER and J.E. TIBBALLS, 1982. Ibid., 15, 59-75.
III. J.E. TIBBALLS and R.J. NELMES, 1982. Ibid., 15, L849-L853.

Tetragonal, $I\bar{4}2d$, a ∿ 7.4, c ∿ 6.9 Å, at various temperatures and pressures. Neutron radiation, R = 0.016-0.029 for 1104-1262 reflexions, plus data for various zones at high pressures.

Structure as previously described (1).

1. Strukturbericht, 1, 362, 393, 394; 2, 454; Structure Reports, 17, 478; 46A, 416.

RUBIDIUM DIHYDROGEN PHOSPHATE
$Rb(H,D)PO_4$

J.E. TIBBALLS, W.-L. ZHONG and R.J. NELMES, 1982. J. Phys. C, 15, 4431-4436.

Tetragonal, I$\bar{4}$2d, a = 7.618, c = 7.310 Å, Z = 4. Neutron radiation, R_W = 0.032 for 136 reflexions.

Atomic positions

	x	y	z
Rb	0	0	1/2
P	0	0	0
O	0.1427	0.0853	0.1206
0.55 D	0.1391	0.2199	0.1210
0.45 H	0.1392	0.2244	0.1233

Structure as previously described (1). Extrapolation predicts O...O = 2.522 Å for a fully-deuterated sample.

1. Structure Reports, 44A, 245; 45A, 300; 48A, 280.

CAESIUM PENTAHYDROGEN PHOSPHATE
$CsH_5(PO_4)_2$

V.A. EFREMOV, V.K. TRUNOV, I. MACIČEK, E.N. GUDINICA and A.A. FAKEEV, 1981. Ž. Neorg. Khim., 26, 3213-3216 [Russian J. Inorg. Chem., 26, 1721-1723].

Monoclinic, $P2_1/c$, a = 10.879, b = 7.768, c = 9.526 Å, β = 96.60°, D_m = 2.71, Z = 4. Mo radiation, R = 0.037 for 1803 reflexions.

The structure contains layers of $H_2PO_4^{2-}$ tetrahedra linked by four O-H...O hydrogen bonds (O...O = 2.53-2.59 Å); the layers are linked by a fifth shorter hydrogen bond (O...O = 2.43 Å) and by 10-coordinated Cs ions, Cs-O = 3.17-3.79 Å. The formula can best be written $CsH(H_2PO_4)_2$ (as in the K salt (1)).

1. Structure Reports, 37A, 292; 38A, 307.

LITHIUM MAGNESIUM PHOSPHATE
$LiMgPO_4$

F. HANIC, M. HANDLOVIĆ, K. BURDOVÁ and J. MAJLING, 1982. J. Cryst. Spect. Res., 12, 99-127.

Orthorhombic, Pnma, a = 10.147, b = 5.909, c = 4.692 Å, D_m = 3.00, Z = 4. Mo radiation, R = 0.048 for 1262 reflexions.

Atomic positions

	x	y	z
Li	0	0	0
Mg	0.27832	1/4	0.98200
P	0.09515	1/4	0.41693
O(1)	0.10172	1/4	0.74100
O(2)	0.45367	1/4	0.19862
O(3)	0.16636	0.04424	0.28039

Olivine-type structure (1), with an ordered arrangement of Li and Mg. P-4 O = 1.522-1.553, Li-6 O = 2.098-2.174, Mg-6 O = 2.049-2.175(1) Å.

1. Strukturbericht, 1, 352.

MAGNESIUM SODIUM PHOSPHATE HEPTAHYDRATE
$MgNaPO_4.7H_2O$

M. MATHEW, P. KINGSBURY, S. TAKAGI and W.E. BROWN, 1982. Acta Cryst., B38, 40-44.

Tetragonal, $P4_2/mmc$, a = 6.731, c = 10.982 Å, D_m = 1.77, Z = 2. Mo radiation, R = 0.038 for 418 reflexions. Structure related to that of struvite (1).

The structure (Fig. 1) contains PO_4 tetrahedra and $Mg(OH_2)_6$ octahedra linked by hydrogen bonding, with an open channel along c which contains a column of alternating Na ions (disordered, six coordination) and water molecules. P-O = 1.531(1), Mg-O = 2.051, 2.105, Na-O = 2.384, 2.746, O-H...O = 2.624, 2.669 Å.

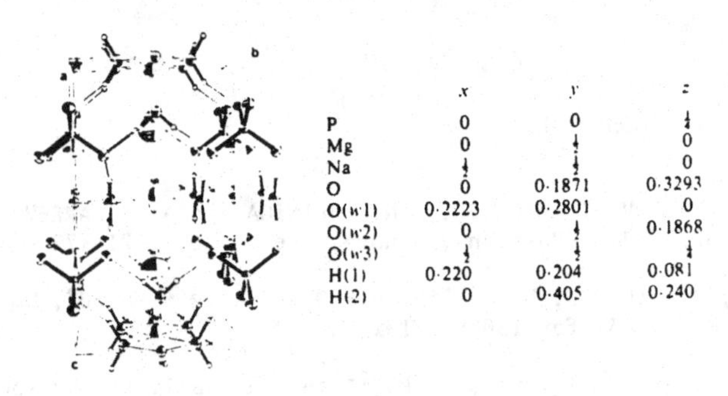

	x	y	z
P	0	0	1/4
Mg	0	1/2	0
Na	1/2	1/2	0
O	0	0·1871	0·3293
O(w1)	0·2223	0·2801	0
O(w2)	0	1/2	0·1868
O(w3)	1/2	1/2	1/4
H(1)	0·220	0·204	0·081
H(2)	0	0·405	0·240

Fig. 1. Structure of $MgNaPO_4.7H_2O$ and atomic positional parameters (large thermal parameters indicate disorder for Na, O(w2), and especially O(w3)).

1. Structure Reports, 19, 440; 35A, 329.

MAGNESIUM CALCIUM SODIUM PHOSPHATE
$Mg_{21}Ca_4Na_4(PO_4)_{18}$

A.I. DOMANSKIJ, Ju.I. SMOLIN, Ju.F. ŠEPELEV and J. MAJLING, 1982. Kristallografija, 27, 891-895 [Soviet Physics - Crystallography, 27, 535-537].

Rhombohedral, $R\bar{3}$, a = 14.974, c = 42.74 Å, Z = 6. Mo radiation, R = 0.038 for 3800 reflexions.

The structure (Fig. 1) contains phosphate tetrahedra, linked by Mg, Ca, and Na ions with 5- to 9-coordinations.

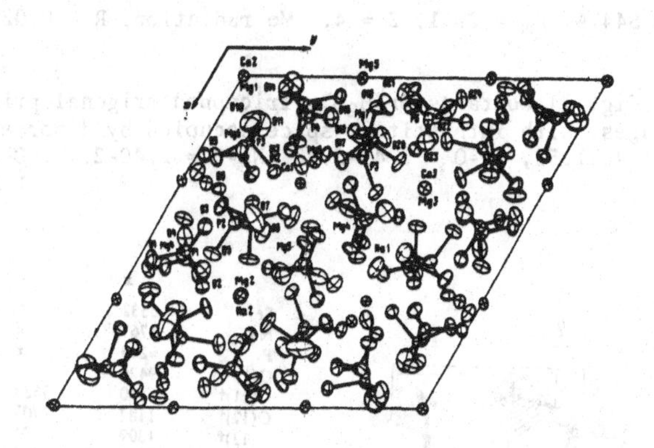

Fig. 1. Structure of magnesium calcium sodium phosphate.

THADEUITE
$Mg(Ca,Mn)(Mg,Fe,Mn)_2(PO_4)_2(OH,F)_2$

A.M. ISAACS and D.R. PEACOR, 1982. Amer. Min., 67, 120-125.

Orthorhombic, $C222_1$, a = 6.412, b = 13.563, c = 8.545 Å, Z = 4. Mo radiation, R = 0.032 for 591 reflexions.

The structure (Fig. 1) contains three cation coordination octahedra, $MO_4(OH)_2$. Helical chains along c of edge-sharing M(3) octahedra are cross-linked by PO_4 tetrahedra to form channels parallel to a into which are fitted chains of alternating edge-sharing M(1) and M(2) octahedra. Cation site occupancies are: M(1) = Mg, M(2) = $Ca_{0.95}Mn_{0.05}$, M(3) = $Mg_{0.59}Fe_{0.29}Mn_{0.12}$. P-O = 1.528-1.536(3), M(1)-O = 2.01-2.06, M(2)-O = 2.32-2.47, M(3)-O = 2.06-2.21 Å.

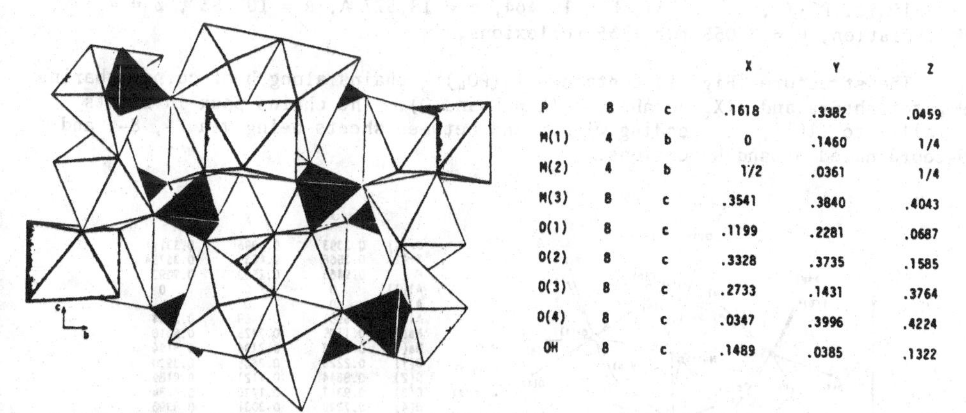

			X	Y	Z
P	8	c	.1618	.3382	.0459
M(1)	4	b	0	.1460	1/4
M(2)	4	b	1/2	.0361	1/4
M(3)	8	c	.3541	.3840	.4043
O(1)	8	c	.1199	.2281	.0687
O(2)	8	c	.3328	.3735	.1585
O(3)	8	c	.2733	.1431	.3764
O(4)	8	c	.0347	.3996	.4224
OH	8	c	.1489	.0385	.1322

Fig. 1. Structure of thadeuite.

SODIUM STRONTIUM PHOSPHATE NONAHYDRATE
$NaSrPO_4.9H_2O$

S. TAKAGI, M. MATHEW and W.E. BROWN, 1982. Acta Cryst., B38, 1408-1413.

Cubic, $P2_13$, a = 10.544 Å, D_m = 2.11, Z = 4. Mo radiation, R = 0.025 for 323 reflexions.

The structure (Fig. 1) contains $Sr(H_2O)_9$ tricapped trigonal prisms and $Na(H_2O)_6$ octahedra sharing edges, with interstitial space occupied by disordered phosphate tetrahedra. P-O = 1.48-1.52, Sr-O = 2.40-2.73, Na-O = 2.40-2.41, O-H...O = 2.67-2.90(2) Å.

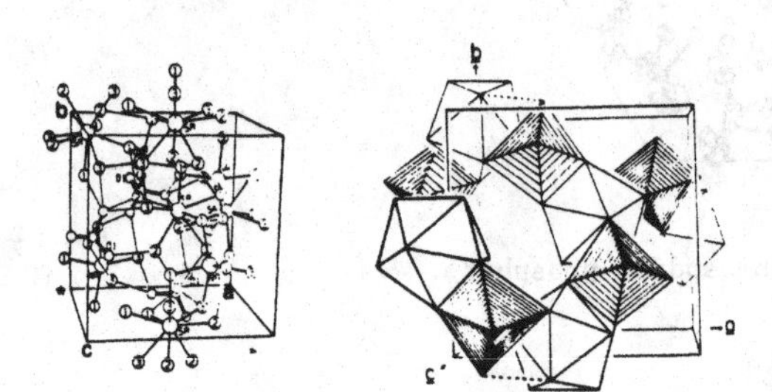

	x	y	z
Sr	5532	x	x
Na	1767	x	x
P	9243	x	x
O(1)	9426	x	x
O(11)†	590	3628	4465
O(12)†	1187	4703	4499
O(13)†	1809	3971	4814
O(w1)	3674	456	7088
O(w2)	1012	808	6517
O(w3)	2081	1485	4004

† The O atom in the general position is statistically disordered to occupy three different positions, O(11), O(12) and O(13) (not related by the threefold operation) with occupancy factors of 50, 29 and 21%, respectively.

Fig. 1. Structure of $NaSrPO_4.9H_2O$ and atomic positional parameters (x 10^4).

BØGGILDITE
$Sr_2Na_2[Al_2(PO_4)F_9]$

F.C. HAWTHORNE, 1982. Canad. Miner., 20, 263-270.

Monoclinic, $P2_1/c$, a = 5.251, b = 10.464, c = 18.577 Å, β = 107.53°, Z = 4. Mo radiation, R = 0.055 for 1736 reflexions.

The structure (Fig. 1) contains $Al_2(PO_4)F_9$ chains along b of corner-sharing PO_4 tetrahedra and AlX_6 octahedra (X = F and O). The chains pack in sheets parallel to (100), bonding within and between sheets being via 7-, 8-, and 9-coordinated Sr and Na cations.

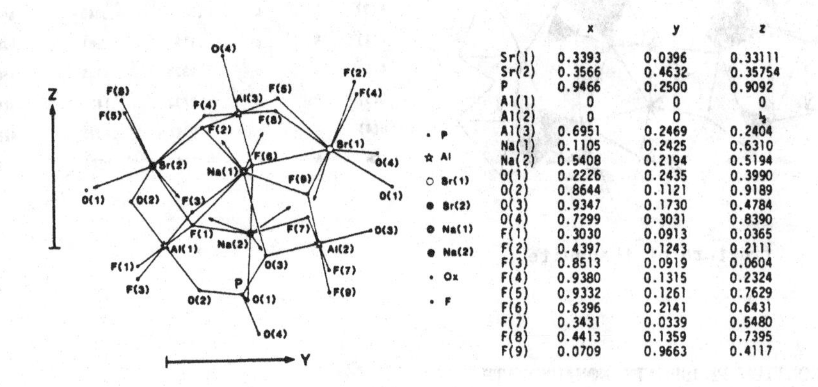

	x	y	z
Sr(1)	0.3393	0.0396	0.33111
Sr(2)	0.3566	0.4632	0.35754
P	0.9466	0.2500	0.9092
Al(1)	0	0	0
Al(2)	0	0	½
Al(3)	0.6951	0.2469	0.2404
Na(1)	0.1105	0.2425	0.6310
Na(2)	0.5408	0.2194	0.5194
O(1)	0.2226	0.2435	0.3990
O(2)	0.8644	0.1121	0.9189
O(3)	0.9347	0.1730	0.4784
O(4)	0.7299	0.3031	0.8390
F(1)	0.3030	0.0913	0.0365
F(2)	0.4397	0.1243	0.2111
F(3)	0.8513	0.0919	0.0604
F(4)	0.9380	0.1315	0.2324
F(5)	0.9332	0.1261	0.7629
F(6)	0.6396	0.2141	0.6431
F(7)	0.3431	0.0339	0.5480
F(8)	0.4413	0.1359	0.7395
F(9)	0.0709	0.9663	0.4117

Fig. 1. Structure of bøggildite.

GORCEIXITE
$BaAl_3(PO_4)(HPO_4)(OH)_6$

E.W. RADOSLOVICH, 1982. Neues Jb. Miner., Mh., 446-464.

Monoclinic, Cm, a = 12.915, b = 7.040, c = 7.055 Å, β = 125.10°, Z = 2. Mo radiation, R = 0.038 for 2646 reflexions (0.031 for 720 reflexions used in the refinement).

Alunite-type structure (1), but with a reduction from rhombohedral to monoclinic symmetry, as previously described (2). There are two independent phosphate groups, one of which is protonated (compare crandallite (3)). P-OH = 1.63, P-O = 1.45-1.59, Al-6 O = 1.80-2.02, Ba- 12 O = 2.79-2.92 Å.

1. Strukturbericht, 5, 92; Structure Reports, 30A, 376; 42A, 369.
2. Structure Reports, 46A, 322.
3. Ibid., 40A, 243.

LEAD(II) DIHYDROGENPHOSPHATE
$Pb(H_2PO_4)_2$

LEAD(II) HYDROGENPHOSPHATE
$PbHPO_4$

H. WORZALA and K.H. JOST, 1982. Z. anorg. Chem., 486, 165-176.

$Pb(H_2PO_4)_2$
Triclinic, P$\bar{1}$, a = 7.823, b = 8.315, c = 5.856 Å, α = 108.24, β = 96.90, γ = 108.61°, Dm = 3.94, Z = 2. Mo radiation, R = 0.09 for 1927 reflexions. Independent study in 1.

$PbHPO_4$
Monoclinic, Pc, a = 4.684, b = 6.642, c = 5.781 Å, β = 97.18°, D_m = 5.57, Z = 2. Mo radiation, R = 0.04 for 751 reflexions (refinement in P2/c; H atom positions deviate from centrosymmetry).

Atomic positions

	x	y	z
		$Pb(H_2PO_4)_2$	
Pb	0.1880	0.0768	0.3223
P(1)	0.2480	0.3493	0.9898
P(2)	0.7757	0.1947	0.3313
O(1)	0.084	0.190	-0.013
O(2)	0.319	0.321	0.759
O(3)	0.795	0.470	0.943
O(4)	0.402	0.392	0.224
O(5)	0.920	0.135	0.439
O(6)	0.623	0.186	0.488
O(7)	0.669	0.066	0.065
O(8)	0.864	0.392	0.340
		$PbHPO_4$	
Pb	0	0.3002	1/4
P	1/2	0.7062	1/4
O(1)	0.225	0.576	0.127
O(2)	0.619	0.843	0.065

The structures (Fig. 1) contain phosphate tetrahedra, linked by hydrogen bonds (some of which are symmetric in the higher-symmetry space groups), and by 7- and 6-coordinate Pb^{2+} ions. $Pb(H_2PO_4)_2$ decomposes topotactically to $PbHPO_4$ and amorphous H_3PO_4.

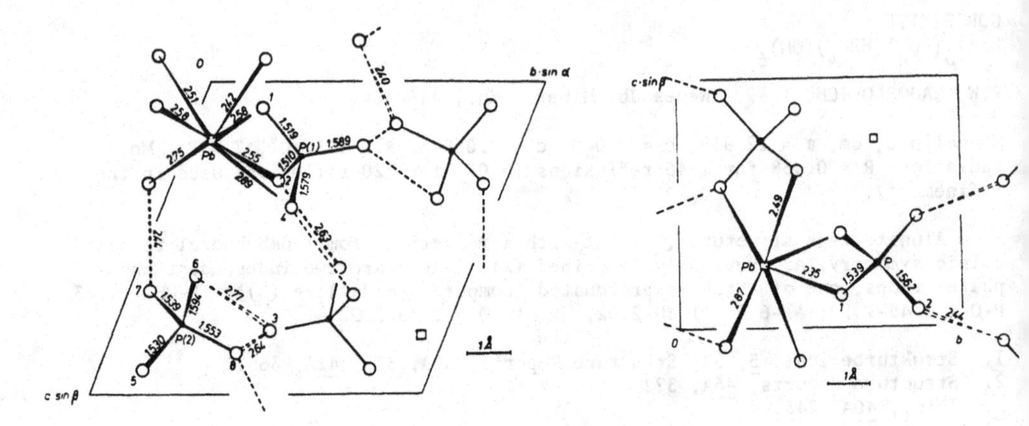

Fig. 1. Structures of $Pb(H_2PO_4)_2$ (left) and $PbHPO_4$ (right).

1. Structure Reports, 48A, 283.

FLUOROPYROMORPHITE
$Pb_5(PO_4)_3F$

E.L. BELOKONEVA, E.A. TRONEVA, L.N. DEM'JANEC, N.G. DUDEROV and N.V. BELOV, 1982. Kristallografija, 27, 793-794 [Soviet Physics - Crystallography, 27, 476-477].

Hexagonal, $P6_3/m$, a = 9.760, c = 7.300 Å, Z = 2. Mo radiation, R = 0.049 for 500 reflexions, for synthetic material.

Atomic positions

	x	y	z
Pb(1)	2/3	1/3	-0.0029
Pb(2)	-0.0026	0.2376	1/4
P	0.4004	0.3813	1/4
O(1)	0.323	0.487	1/4
O(2)	0.582	0.486	1/4
O(3)	0.347	0.267	0.422
F	0	0	1/2 [in text, 1/4 in Table]

Apatite structure as for pyromorphite (1),

1. Strukturbericht, 2, 101.

LEAD BISMUTH PHOSPHATE OXIDE
$Pb_8Bi_2(PO_4)_6O_2$

E.P. MOORE, H.-Y. CHEN, L.H. BRIXNER and C.M. FORIS, 1982. Mater. Res. Bull., 17, 653-660.

Orthorhombic, Pnma, a = 13.313, b = 10.284, c = 9.219 Å, Z = 2. Mo radiation, R = 0.082 for 1180 reflexions.

The material does not have an apatite structure. The structure contains PO_4 tetrahedra, and four independent heavy metal sites, one (Pb) with distorted octahedral coordination and three with 7-fold plus lone-pair coordinations; Bi is probably ordered in one of the latter sites. Oxide ions are located at the centres of Pb_4 tetrahedra, so that the formula is most appropriately written $Bi(Pb_4O)(PO_4)_3$.

TELLURIUM PHOSPHATE
$Te_2O_3.HPO_4$

N.W. ALCOCK and W.D. HARRISON, 1982. Acta Cryst., B38, 1809-1811.

Orthorhombic, $Pca2_1$, a = 10.239, b = 7.018, c = 7.933 Å, Z = 4. Mo radiation, R = 0.022 for 696 reflexions.

The structure (Fig. 1) contains PO_4 tetrahedra, linked by two Te atoms which have trigonal bipyramidal coordinations (including equatorial lone-pairs) and several longer Te...O interactions. P-O = 1.50-1.58, Te-O = 1.90-2.18, Te...O = 2.73-3.26(1) Å. O(7)...O(1) and O(7)...O(2) contacts of 2.90 and 2.99 Å may be hydrogen bonds.

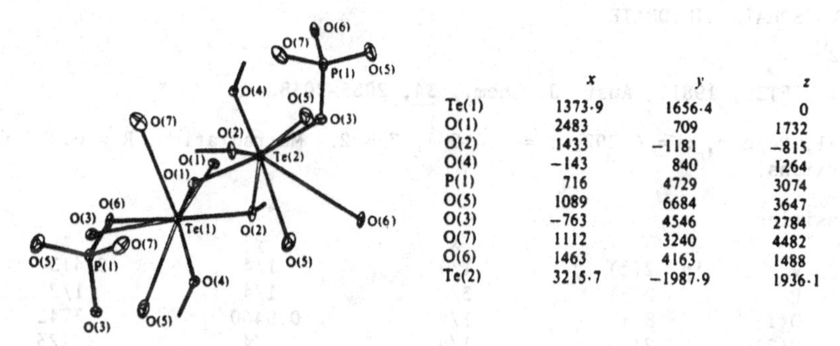

	x	y	z
Te(1)	1373·9	1656·4	0
O(1)	2483	709	1732
O(2)	1433	−1181	−815
O(4)	−143	840	1264
P(1)	716	4729	3074
O(5)	1089	6684	3647
O(3)	−763	4546	2784
O(7)	1112	3240	4482
O(6)	1463	4163	1488
Te(2)	3215·7	−1987·9	1936·1

Fig. 1. Structure of $Te_2O_3.HPO_4$, and atomic positional parameters (x 10^4).

SCANDIUM DIHYDROGENPHOSPHATE
$Sc(H_2PO_4)_3$

Ju.I. SMOLIN, Ju.F. ŠEPELEV and A.I. DOMANSKIJ, 1982. Kristallografija, 27, 239-241 [Soviet Physics - Crystallography, 27, 146-147].

Rhombohedral, R3c, a = 8.274, c = 25.98 Å, Z = 6. Mo radiation, R = 0.038 for 735 reflexions.

The structure (Fig. 1) contains $PO_2(OH)_2^-$ tetrahedra, linked by hydrogen bonds and by octahedrally-coordinated Sc ions. P-OH = 1.57, 1.58, P-O = 1.46, 1.49, Sc-O = 2.06, 2.08(1) Å.

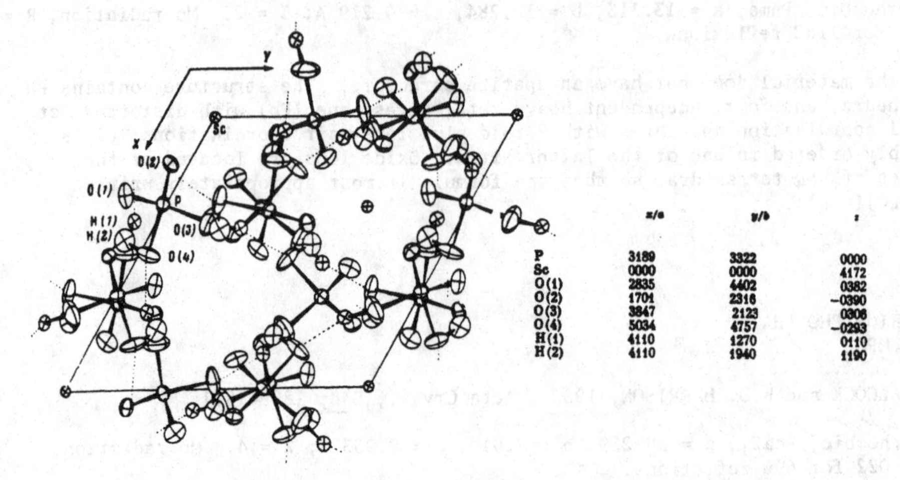

	x/a	y/b	z
P	3189	3322	0000
Sc	0000	0000	4172
O(1)	2835	4402	0382
O(2)	1701	2316	−0390
O(3)	3847	2123	0306
O(4)	5034	4757	−0293
H(1)	4110	1270	0110
H(2)	4110	1940	1190

Fig. 1. Structure of scandium dihydrogenphosphate and atomic positional parameters (x 10^4).

VANADYL PHOSPHATE DIHYDRATE
$VOPO_4.2H_2O$

I. H.R. TIETZE, 1981. Aust. J. Chem., 34, 2035-2038.

Tetragonal, P4/nmm, a = 6.202, c = 7.410 Å, Z = 2. Mo radiation, R = 0.053 for 269 reflexions.

Atomic positions

		x	y	z
V	in 2(c)	1/4	1/4	0.4137
P	2(b)	3/4	1/4	1/2
O(1)	8(i)	1/4	0.9460	0.3742
O(2)	2(c)	1/4	1/4	0.1123
$H_2O(3)$	2(a)	3/4	1/4	0
$H_2O(4)$	2(c)	1/4	1/4	0.6251

The structure contains layers of corner-sharing PO_4 tetrahedra and VO_6 octahedra. P-O = 1.532(2) Å, O-P-O = 105.0, 111.7°, V-O = 1.567(5) (vanadyl), 1.908(2) (x 4, phosphate), 2.233(5) (water) Å.

$VOPO_4.2D_2O$

II. M. TACHEZ, F. THEOBALD, J. BERNARD and A.W. HEWAT, 1982. Rev. Chim. Minér., 19 291-300.

Tetragonal, P4/n, a = 6.2154, c = 7.4029 Å, Z = 2. Neutron powder data.

Atomic positions (occupancy = 0.5 for D(1), D(2), O(4))

		x	y	z
2(c)	V	0	0.5	0.568
2(c)	O(1)	0	0.5	0.375
2(b)	P	0	0	0.5
8(g)	O(2)	0.498	0.6988	0.3757
2(c)	O(3)	0	0.5	0.905
8(g)	D(1)	0.148	0.48	0.944
8(g)	D(2)	0.551	0.429	0.062
4(f)	O(4)	0	0	0.092

The structure is assumed to be similar to that of α_I-$VOPO_4$ (1), with two water molecules inserted between the layers, one bonded to V and the other (disordered and poorly defined) hydrogen bonded to two layers. V has distorted octahedral coordination, V=O = 1.42, V-OP = 1.92 (x 4), V-OH_2 = 2.50; P-O = 1.54 Å.

1. Structure Reports, 48A, 284.

IRON(II) PHOSPHATE OXIDE
$Fe_4(PO_4)_2O$

M. BOUCHDOUG, A. COURTOIS, R. GERARDIN, J. STEINMETZ and C. GLEITZER, 1982. J. Solid State Chem., 42, 149-157.

Monoclinic, $P2_1/c$, a = 6.564, b = 11.271, c = 9.383 Å, β = 103.95°, D_m = 4.22, Z = 4. Mo radiation, R = 0.033 for 1672 reflexions.

The structure (Fig. 1) contains PO_4 tetrahedra linked by Fe atoms, three of which have distorted octahedral and one of which has trigonal bipyramidal coordination; the oxide ion is bonded to four Fe atoms. P-O = 1.514-1.562, Fe-O = 1.939-2.354 Å.

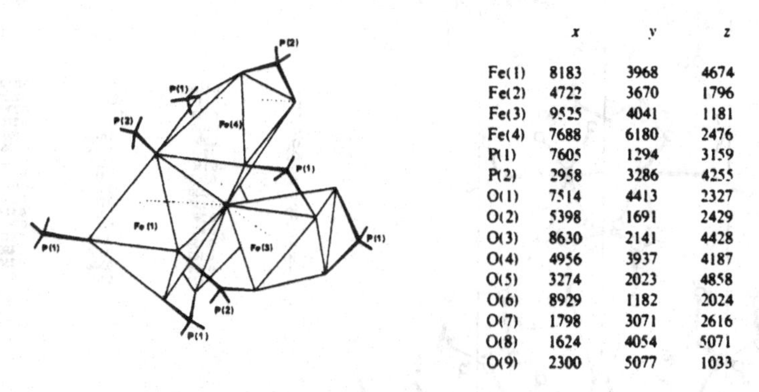

	x	y	z
Fe(1)	8183	3968	4674
Fe(2)	4722	3670	1796
Fe(3)	9525	4041	1181
Fe(4)	7688	6180	2476
P(1)	7605	1294	3159
P(2)	2958	3286	4255
O(1)	7514	4413	2327
O(2)	5398	1691	2429
O(3)	8630	2141	4428
O(4)	4956	3937	4187
O(5)	3274	2023	4858
O(6)	8929	1182	2024
O(7)	1798	3071	2616
O(8)	1624	4054	5071
O(9)	2300	5077	1033

Fig. 1. Structure of $Fe_4(PO_4)_2O$, and atomic positional parameters (x 10^4).

METAVIVIANITE
$Fe_3(PO_4)_2.8H_2O$

J.-L. DORMANN, M. GASPÉRIN and J.-F. POULLEN, 1982. Bull. Minéral., 105, 147-160.

Triclinic, $P\bar{1}$, a = 7.84, b = 9.11, c = 4.67 Å, α = 95.04, β = 96.94, γ = 107.72°, Z = 1. Mo radiation, R = 0.13 for 838 reflexions (twinned crystal).

The structure is similar to that of monoclinic vivianite (1), with PO_4 tetrahedra and FeO_6 octahedra. There is some oxidation to Fe^{3+} and formation of OH^-. P-O = 1.52-1.53, Fe-O = 1.98-2.01 and 2.02-2.15(2) Å.

1. Structure Reports, 46A, 326.

RHODIUM(II) PHOSPHATE HYDRATE
$Rh_2(H_2PO_4)_4(H_2O)_2$

L.M. DIKAREVA, G.G. SADIKOV, M.A. PORAJ-KOŠIC, I.B. BARANOVSKIJ, S.S. ABDULLAEV and R.N. ŠČELOKOV, 1982. Ž. Neorg. Khim., 27, 417-423 [Russian J. Inorg. Chem., 27, 236-240].

Monoclinic, C2, a = 7.285, b = 13.564, c = 7.225 Å, β = 93.59°, Z = 2 dimers. Mo radiation, R = 0.039 for 996 reflexions. Previous study in 1.

The structure (Fig. 1) contains molecular dimers, linked by hydrogen bonds. Rh-Rh = 2.487(1) (single bond), Rh-O = 2.02-2.08(1) (phosphate), 2.29(1) (water), P-O = 1.48-1.59(2) Å.

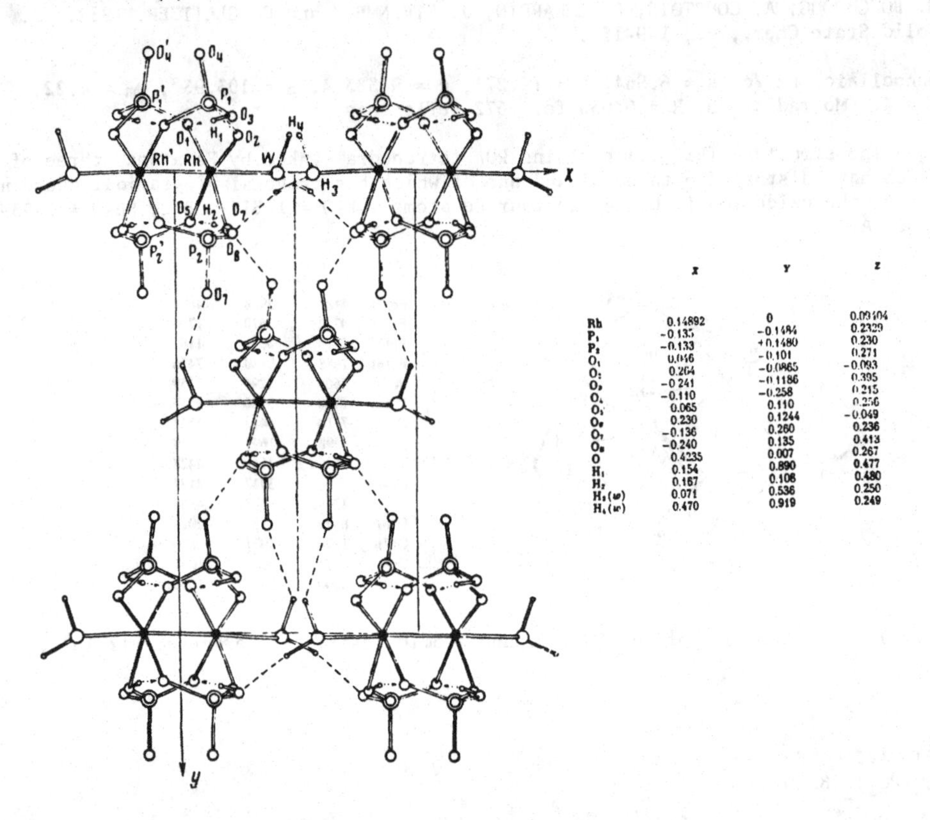

	x	y	z
Rh	0.14892	0	0.09404
P_1	-0.135	-0.1484	0.2329
P_2	-0.133	+0.1480	0.230
O_1	0.046	-0.101	0.271
O_2	0.264	-0.0865	-0.093
O_3	-0.241	-0.1186	0.395
O_4	-0.110	-0.258	0.215
O_5	0.065	0.110	0.256
O_6	0.230	0.1244	-0.049
O_7	-0.136	0.260	0.236
O_8	-0.240	0.135	0.413
O	0.4235	0.007	0.267
H_1	0.154	0.890	0.477
H_2	0.167	0.108	0.480
H_3(w)	0.071	0.536	0.250
H_4(w)	0.470	0.919	0.249

Fig. 1. Structure of rhodium(II) phosphate hydrate.

1. Structure Reports, 46A, 329.

NICKEL ZINC PHOSPHATE
$(Ni_{0.75}Zn_{0.25})_3(PO_4)_2$

A.G. NORD, 1982. Neues Jb. Miner., Mh., 422-432.

Monoclinic, $P2_1/a$, a = 10.150, b = 4.707, c = 5.870 Å, β = 91.11°, Z = 4. Neutron powder data.

Atomic positions

		x	y	z
M(1)	$Ni_{0.90}Zn_{0.10}$	0	0	1/2
M(2)	$Ni_{0.68}Zn_{0.32}$	0.2771	-0.0223	0.2401
M(3)	vacant	0	0	0
P		0.0968	0.4194	0.2558
O(1)		0.1033	0.7444	0.2704
O(2)		0.4606	0.1910	0.2541
O(3)		0.1774	0.3045	0.0555
O(4)		0.1635	0.2628	0.4735

Isostructural with $Ni_3(PO_4)_2$ (1) and with sarcopside, $(Fe,Mn,Mg)_3(PO_4)_2$ (2). Ni and Zn are octahedrally coordinated, with partial ordering of Ni in M(1) and Zn in M(2), and the M(3) site vacant. M(1)-O = 2.069-2.098, M(2)-O = 1.999-2.190, P-O = 1.521-1.595(2) Å.

1. Structure Reports, 41A, 323.
2. Ibid., 38A, 313.

SODIUM PLATINUM PHOSPHATE HYDRATE (SODIUM TETRA-μ-HYDROGENPHOSPHATO-DIAQUO-DIPLATINUM(Pt-Pt))

$Na_2[Pt_2(HPO_4)_4(H_2O)_2]$

F.A. COTTON, L.R. FALVELLO and S. HAN, 1982. Inorg. Chem., 21, 1709-1710.

Triclinic, $P\bar{1}$, a = 7.912, b = 7.983, c = 13.732 Å, α = 82.66, β = 98.29, γ = 114.27°, Z = 2. Mo radiation, R = 0.045 for 1673 reflexions.

The structure contains two independent anions on symmetry centres, each containing a Pt-Pt unit bridged by four $PO_3(OH)^{2-}$ ions with two axial water molecules (Fig. 1). Mean Pt-Pt = 2.486(2), Pt-O = 2.006(7), $Pt-OH_2$ = 2.151(11), P-O = 1.543(5) (coordinated), 1.505(6), P-OH = 1.549(6) Å. Na ions have distorted octahedral coordination to phosphate oxygens, Na-O = 2.79-3.16(2) Å.

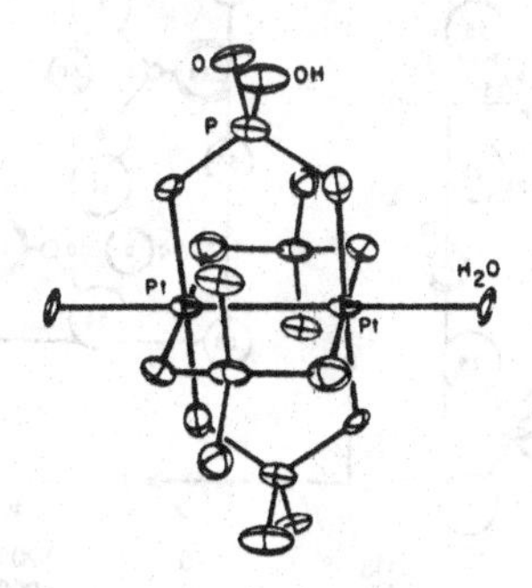

Fig. 1. The $[Pt_2(HPO_4)_4(H_2O)_2]^{2-}$ ion.

ZINC MANGANESE PHOSPHATE

γ-$(Zn_{0.75}Mn_{0.25})_3(PO_4)_2$

A.G. NORD and T. STEFANIDIS, 1982. Polyhedron, 1, 349-353.

Monoclinic, $P2_1/n$, a = 7.563, b = 8.553, c = 5.056 Å, β = 94.78°, Z = 2. Neutron powder data.

Atomic positions

	x	y	z
M(1)	0.622	0.134	0.081
M(2)	0	0	1/2
P	0.206	0.192	0.036
O(1)	0.051	0.136	0.836
O(2)	0.135	0.205	0.304
O(3)	0.254	0.354	0.936
O(4)	0.362	0.079	0.046

M(1) = Zn; M(2) = (Zn + 3 Mn)/4

Structure as previously described (1). The cations are strongly ordered, with ZnO_5 distorted trigonal bipyramids and $(Zn_{0.25}Mn_{0.75})O_6$ octahedra.

1. Structure Reports, 28, 189.

SODIUM MERCURY(II) PHOSPHATE
$NaHgPO_4$

M. HATA and F. MARUMO, 1982. Acta Cryst., B38, 239-241.

Orthorhombic, Cmcm, a = 5.883, b = 9.401, c = 6.448 Å, Z = 4. Mo radiation, R = 0.054 for 722 reflexions.

The structure (Fig. 1) is related to that of olivine (1), but with a different orientation of the tetrahedral phosphate anions, so that Hg has strongly-distorted tetrahedral coordination; Na has octahedral coordination, with NaO_6 octahedra sharing edges to form ribbons along c. P-O = 1.55, Hg-O = 2.08 (x 2), 2.56 (x 2), Na-O = 2.37, 2.43(1) Å, O(1)-Hg-O(1') = 152°.

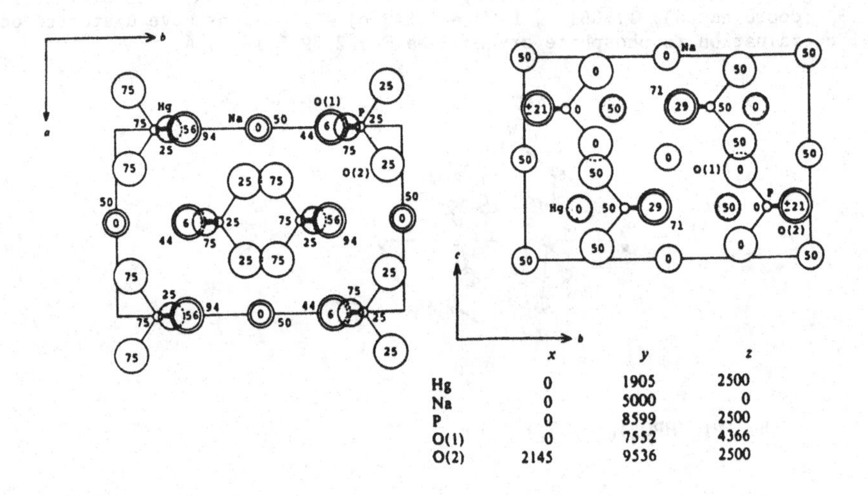

	x	y	z
Hg	0	1905	2500
Na	0	5000	0
P	0	8599	2500
O(1)	0	7552	4366
O(2)	2145	9536	2500

Fig. 1. Structure of $NaHgPO_4$, and atomic positional parameters (x 10^4).

1. Strukturbericht, 1, 352.

YTTRIUM, SCANDIUM, and LUTETIUM PHOSPHATES
MPO_4 (M = Y, Sc, Lu)

W.O. MILLIGAN, D.F. MULLICA, G.W. BEALL and L.A. BOATNER, 1982. Inorg. Chim. Acta, 60, 39-43.

Tetragonal, $I4_1/amd$, a = 6.882, 6.574, 6.792, c = 6.018, 5.791, 5.954 Å, Z = 4. Mo radiation, R = 0.024, 0.028, 0.038 for 241, 287, 153 reflexions. M in 4(a): 0,3/4,1/8; P in 4(b): 0,1/4,3/8; O in 16(h): 0,y,z, y = 0.4251, 0.4305, 0.4262, z = 0.2147, 0.2071, 0.2126 (origin at 2/m).

Zircon structure (1). P-O = 1.543(3), 1.534(1), 1.534(10), M-O = 2.300, 2.373(3); 2.153, 2.260($\overline{1}$); 2.262, 2.344(10) Å.

1. Strukturbericht, 1, 345.

CALCIUM NEODYMIUM POTASSIUM PHOSPHATE
$CaKNd(PO_4)_2$

M. VLASSE, P. BOCHU, C. PARENT, J.P. CHAMINADE, A. DAOUDI, G. LE FLEM and P. HAGENMULLER, 1982. Acta Cryst., B38, 2328-2331.

Hexagonal, $P6_222$, a = 7.033, c = 6.397 Å, D_m = 3.74, Z = 1.5. Mo radiation, R = 0.042 for 407 reflexions. (Nd,Ca) in 3(c): 1/2,0,0; 0.5 K in 3(b): 0,0,1/2; P in 3(d): 1/2,0,1/2; O in 12(k): 0.4397,0.1384,0.3535.

The structure (Fig. 1) contains PO_4 tetrahedra, linked by eight-coordinate disordered Nd,Ca and half-occupied K sites. P-O = 1.56, Nd,Ca-O = 2.36, 2.58, K-O = 2.89, 2.99(1) Å.

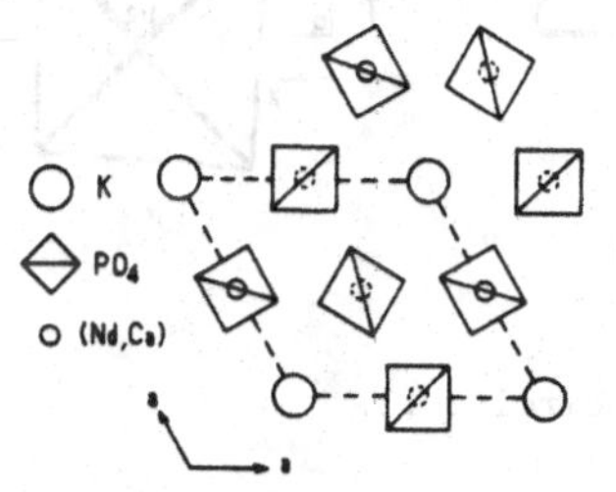

Fig. 1. Structure of $CaKNd(PO_4)_2$.

VANMEERSSCHEITE
$U(UO_2)_3(PO_4)_2(OH)_6.4H_2O$

P. PIRET and M. DELIENS, 1982. Bull. Minéral., 105, 125-128.

Orthorhombic, $P2_1mn$, a = 17.06, b = 16.76, c = 7.023 Å, Z = 4. Mo radiation, R = 0.126 for 429 reflexions (oxygen positions not refined, water molecules not located). The dihydrate, meta-vanmeersscheite, has space group Fddd, 34 x 34 x 14 Å, Z = 32.

Phosphuranylite-type structure (1), with $(UO_2)_3(PO_4)_2(OH)_2$ layers.

1. Structure Reports, 41A, 410; 44A, 255; 45A, 313.

META-URANOCIRCITE II
$Ba(UO_2)_2(PO_4)_2.6H_2O$

F. KHOSRAWAN-SAZEDJ, 1982. Tschermaks Min. Petr. Mitt., 29, 193-204.

Monoclinic, $P2_1/a$, a = 9.789, b = 9.822, c (unique axis) = 16.868 Å, γ = 89.95°, Z = 4. Mo radiation, R = 0.071 for 1743 reflexions, for a synthetic specimen.

The structure (Fig. 1) contains slightly corrugated UO_2PO_4 layers parallel to (001), connected by 9-coordinated Ba ions and water molecules. U has distorted octahedral coordination, mean U-O = 1.78 (uranyl), 2.28(2) Å; P-O = 1.47-1.55, Ba-O = 2.74-3.11(3) Å.

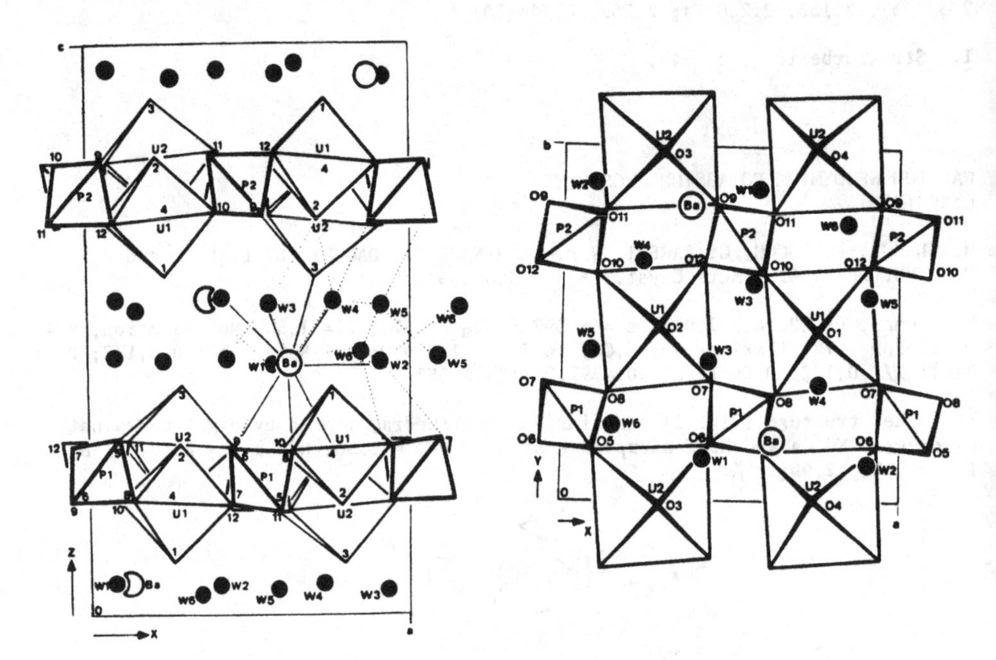

Fig. 1. Structure of meta-uranocircite II.

THREADGOLDITE
$Al(UO_2)_2(PO_4)_2(OH).8H_2O$

F. KHOSRAWAN-SAZEDJ, 1982. Tschermaks Min. Petr. Mitt., 30, 111-115.

The structure is as previously determined (1), but space group C2/c (R = 0.077) is preferred rather than Cc (R = 0.076 for 853 reflexions). The bond distances show considerably smaller scatter than in 1; U-O = 1.70-1.87, 2.25-2.31, P-O = 1.40-1.68, Al-O = 1.79-1.99 Å.

1. Structure Reports, 45A, 313.

POTASSIUM HYDROGEN PHOSPHATE PYROPHOSPHATE
$K_2H_8(PO_4)_2P_2O_7$

A. LARBOT, A. NORBERT and J. DURAND, 1982. Z. anorg. Chem., 486, 200-206.

Orthorhombic, $Pca2_1$, a = 9.364, b = 7.458, c = 19.560 Å, D_m = 2.17, Z = 4. Mo radiation, R = 0.025 for 416 reflexions.

The structure (Fig. 1) contains phosphate and pyrophosphate anions, linked by hydrogen bonds and by 8- and 9-coordinate K ions. P-O = 1.51-1.59 (phosphate), 1.46-1.54 (pyrophosphate, terminal), 1.58(2) Å (pyrophosphate, bridging), P-O-P = 135°, K-O = 2.70-3.25(2) Å.

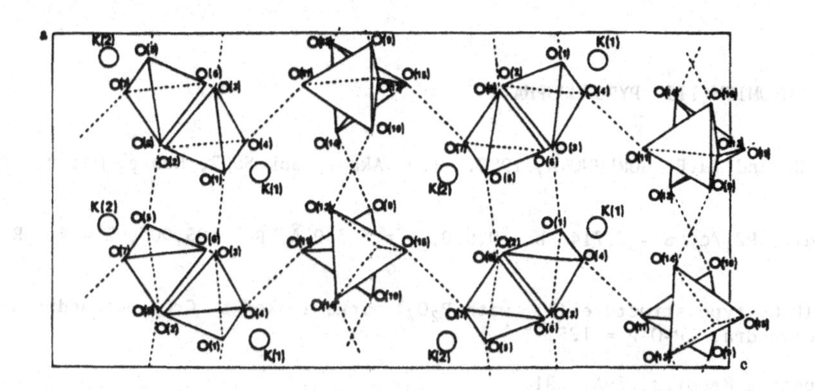

Fig. 1. Structure of $K_2H_8(PO_4)_2P_2O_7$.

MAGNESIUM PYROPHOSPHATE DIHYDRATE
$Mg_2P_2O_7.2H_2O$

J. OKA and A. KAWAHARA, 1982. Acta Cryst., B38, 3-5.

Monoclinic, $P2_1/n$, a = 7.367, b = 13.906, c = 6.277 Å, β = 94.37°, D_m = 2.60, Z = 4. Mo radiation, R = 0.043 for 1051 reflexions.

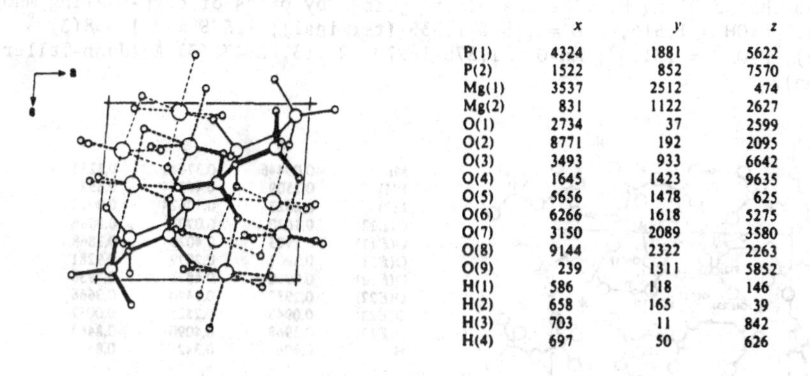

	x	y	z
P(1)	4324	1881	5622
P(2)	1522	852	7570
Mg(1)	3537	2512	474
Mg(2)	831	1122	2627
O(1)	2734	37	2599
O(2)	8771	192	2095
O(3)	3493	933	6642
O(4)	1645	1423	9635
O(5)	5656	1478	625
O(6)	6266	1618	5275
O(7)	3150	2089	3580
O(8)	9144	2322	2263
O(9)	239	1311	5852
H(1)	586	118	146
H(2)	658	165	39
H(3)	703	11	842
H(4)	697	50	626

Fig. 1. Structure of $Mg_2P_2O_7.2H_2O$, and atomic positional parameters (x 10^4 for P, Mg, O; x 10^3 for H).

The structure (Fig. 1) contains $O_3P\text{-}O\text{-}PO_3^{4-}$ anions, linked by chains of edge-sharing MgO_6 octahedra. Mg-O = 2.002-2.220(3), P-O = 1.485-1.526 (terminal), 1.608 Å (bridging), P-O-P = 125.7°. There is one hydrogen bond, and possibly three other bifurcated bonds.

SODIUM CHROMIUM(III) PYROPHOSPHATE
$NaCrP_2O_7$

L. BOHATÝ, J. LIEBERTZ and R. FRÖHLICH, 1982. Z. Kristallogr., 161, 53-59.

Monoclinic, $P2_1/c$, a = 7.294, b = 7.838, c = 9.484 Å, β = 111.72°, D_m = 3.26, Z = 4. Mo radiation, R = 0.029 for 2381 reflexions.

The structure contains nearly eclipsed $O_3P\text{-}O\text{-}PO_3^{4-}$ ions, linked by octahedrally coordinated Cr^{3+} ions and by Na^+ ions with irregular eight-coordination. P-O = 1.605, 1.617(2) (bridging), 1.497-1.524(2) (terminal) Å, P-O-P = 133.0°, Cr-O = 1.930-2.019(2), Na-O = 2.361-3.029(2) Å.

CAESIUM CHROMIUM(III) PYROPHOSPHATE
$CsCrP_2O_7$

S.A. LINDE and Ju.E. GORBUNOVA, 1982. Izv. Akad. Nauk SSSR, Neorg. Mater., 18, 464-467.

Monoclinic, $P2_1/c$, a = 7.714, b = 9.920, c = 8.359 Å, β = 105.00°, Z = 4. R = 0.025.

$KAlP_2O_7$-type structure (1), with $P_2O_7^{4-}$ ions linked by CrO_6 octahedra and CsO_{10} polyhedra. P-O-P = 125.9°.

1. Structure Reports, 39A, 291.

MANGANESE(III) HYDROGEN PYROPHOSPHATE
$MnHP_2O_7$

A. DURIF and M.T. AVERBUCH-POUCHOT, 1982. Acta Cryst., B38, 2883-2885.

Monoclinic, $P2_1/n$, a = 7.951, b = 12.645, c = 4.922 Å, β = 100.92°, Z = 4. Mo radiation, R = 0.028 for 765 reflexions.

The structure (Fig. 1) contains $O_3P\text{-}O\text{-}PO_3(OH)^{3-}$ anions linked by hydrogen bonds into chains along b. The chains are joined by pairs of edge-sharing MnO_6 octahedra. P-OH = 1.516, P-O = 1.506-1.535 (terminal), 1.575 and 1.608(3) Å (bridging), P-O-P = 131.5°; Mn-O = 1.876-1.970, 2.113, 2.433(3) Å (Jahn-Teller distortion).

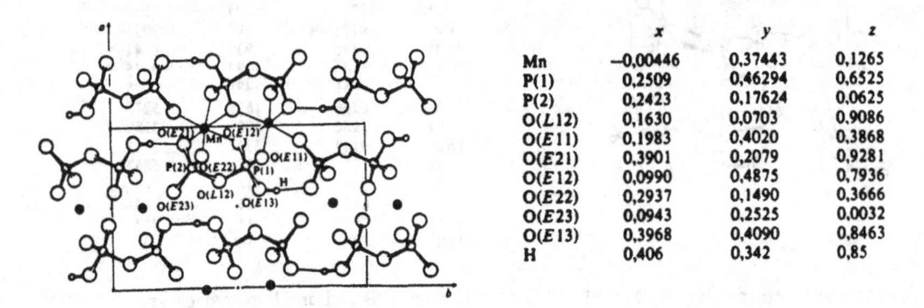

	x	y	z
Mn	−0,00446	0,37443	0,1265
P(1)	0,2509	0,46294	0,6525
P(2)	0,2423	0,17624	0,0625
O(L12)	0,1630	0,0703	0,9086
O(E11)	0,1983	0,4020	0,3868
O(E21)	0,3901	0,2079	0,9281
O(E12)	0,0990	0,4875	0,7936
O(E22)	0,2937	0,1490	0,3666
O(E23)	0,0943	0,2525	0,0032
O(E13)	0,3968	0,4090	0,8463
H	0,406	0,342	0,85

Fig. 1. Structure of $MnHP_2O_7$.

IRON(II) DIPHOSPHATE
$Fe_2P_2O_7$

T. STEFANIDIS and A.G. NORD, 1982. Z. Kristallogr., 159, 255-264.

Triclinic, P1, a = 5.517, b = 5.255, c = 4.488 Å, α = 98.73, β = 98.33, γ = 103.81°, Z = 1. Mo radiation, R = 0.049 for 722 reflexions.

The structure (Fig. 1) contains edge-sharing FeO_6 octahedra, which share corners with staggered $O_3P\text{-}O\text{-}PO_3^{4-}$ anions. Fe-O = 1.98-2.63(1), P-O = 1.58, 1.62 (bridging), 1.48-1.53(1) Å (terminal), P-O-P = 153°.

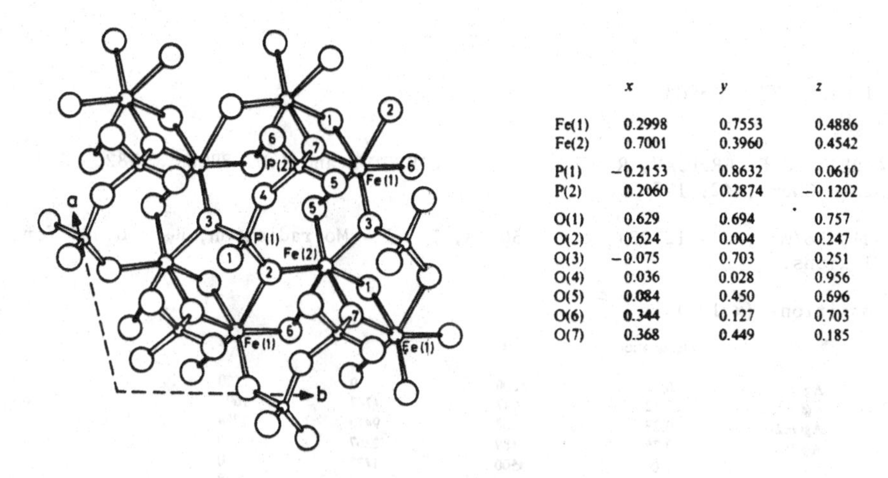

	x	y	z
Fe(1)	0.2998	0.7553	0.4886
Fe(2)	0.7001	0.3960	0.4542
P(1)	−0.2153	0.8632	0.0610
P(2)	0.2060	0.2874	−0.1202
O(1)	0.629	0.694	0.757
O(2)	0.624	0.004	0.247
O(3)	−0.075	0.703	0.251
O(4)	0.036	0.028	0.956
O(5)	0.084	0.450	0.696
O(6)	0.344	0.127	0.703
O(7)	0.368	0.449	0.185

Fig. 1. Structure of iron(II) diphosphate.

SODIUM IRON(III) PYROPHOSPHATE (HIGH-TEMPERATURE FORM II)
$NaFeP_2O_7$

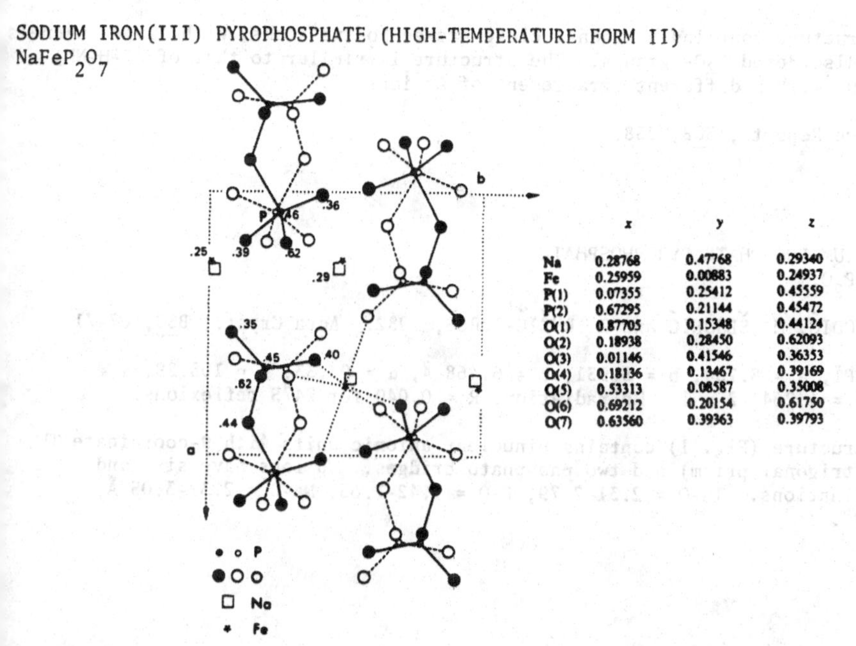

	x	y	z
Na	0.28768	0.47768	0.29340
Fe	0.25959	0.00883	0.24937
P(1)	0.07355	0.25412	0.45559
P(2)	0.67295	0.21144	0.45472
O(1)	0.87705	0.15348	0.44124
O(2)	0.18938	0.28450	0.62093
O(3)	0.01146	0.41546	0.36353
O(4)	0.18136	0.13467	0.39169
O(5)	0.53313	0.08587	0.35008
O(6)	0.69212	0.20154	0.61750
O(7)	0.63560	0.39363	0.39793

Fig. 1. Structure of $NaFeP_2O_7$-II.

M. GABELICA-ROBERT, M. GOREAUD, P. LABBE and B. RAVEAU, 1982. J. Solid State Chem., 45, 389-395.

Monoclinic, $P2_1/c$, a = 7.324, b = 7.905, c = 9.575 Å, β = 111.86°, Z = 4. Mo radiation, R = 0.040 for 3842 reflexions.

The structure (Fig. 1) contains (001) layers of P_2O_7 groups alternating with layers of corner-sharing FeO_6 octahedra. Na^+ ions are located in elongated cages. The pyrophosphate groups have a nearly-eclipsed conformation, P-O = 1.61 (bridging), 1.50-1.53 (terminal) Å, P-O-P = 133°; Fe-O = 1.94-2.05 Å; Na-O = 2.38-2.66 (6 distances), 2.98, 3.06 Å (σ = 0.001-0.002 Å).

SILVER IODIDE PYROPHOSPHATE

$Ag_{16}I_{12}P_2O_7$

J.D. GARRETT, J.E. GREEDAN, R. FAGGIANI, S. CARBOTTE and I.D. BROWN, 1982. J. Solid State Chem., 42, 183-190.

Hexagonal, P6/mcc, a = 12.054, c = 7.504 Å, Z = 1. Mo radiation, R_w = 0.084 for 270 reflexions.

Atomic positions (x 10^4)

	Occupation	*x*	*y*	*z*
Ag(*f*)	0.52	5000	0	2500
Ag(*m*1)	0.12	947	4747	1605
Ag(*m*2)	0.29	2488	9470	1754
Ag(*l*)	0.26	489	2467	0
I	1.0	4500	1479	0
P	0.5	0	0	2018
O(*b*)	0.083	500	200	0
O(1)	0.083	1000	500	3500
O(2)	0.083	−800	−1500	2000
O(3)	0.083	−1000	400	2000

The structure contains hexagonal close-packed iodine atoms, with large channels containing disordered P_2O_7 groups. The structure is similar to that of (C_5H_5NH)-Ag_5I_6 (1), but with a different arrangement of Ag ions.

1. Structure Reports, 38B, 258.

SODIUM THORIUM PHOSPHATE PYROPHOSPHATE

$Na_3Th(PO_4)(P_2O_7)$

B. KOJIĆ-PRODIĆ, M. ŠLJUKIĆ and Ž. RUŽIĆ-TOROŠ, 1982. Acta Cryst., B38, 67-71.

Triclinic, $P\bar{1}$, a = 8.734, b = 8.931, c = 6.468 Å, α = 93.33, β = 108.29, γ = 110.10°, Dm = 4.264, Z = 2. Mo radiation, R = 0.040 for 2475 reflexions.

The structure (Fig. 1) contains binuclear anionic units with 9-coordinate Th (tricapped trigonal prism) and two phosphato bridges. Na ions have six- and seven-coordinations. Th-O = 2.31-2.79, P-O = 1.42-1.63, Na-O = 2.31-3.05 Å, P-O-P = 129°.

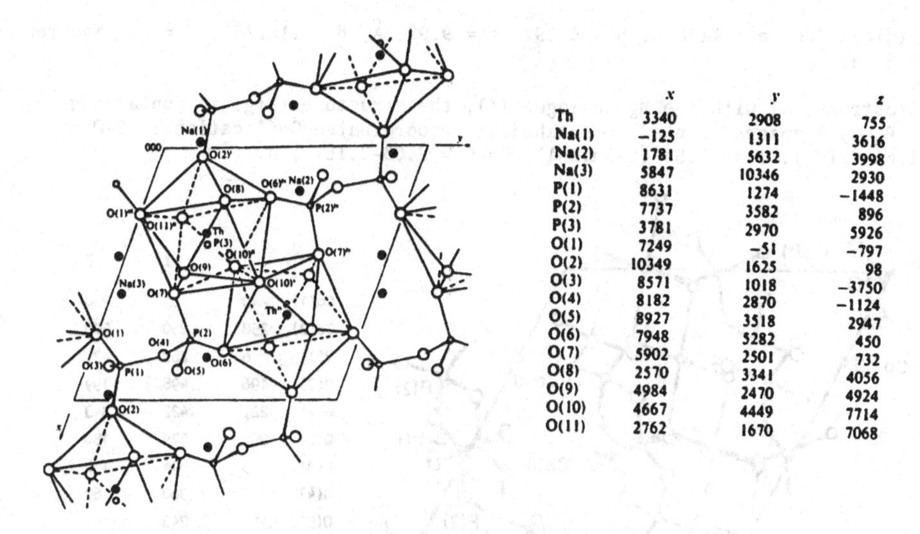

	x	y	z
Th	3340	2908	755
Na(1)	−125	1311	3616
Na(2)	1781	5632	3998
Na(3)	5847	10346	2930
P(1)	8631	1274	−1448
P(2)	7737	3582	896
P(3)	3781	2970	5926
O(1)	7249	−51	−797
O(2)	10349	1625	98
O(3)	8571	1018	−3750
O(4)	8182	2870	−1124
O(5)	8927	3518	2947
O(6)	7948	5282	450
O(7)	5902	2501	732
O(8)	2570	3341	4056
O(9)	4984	2470	4924
O(10)	4667	4449	7714
O(11)	2762	1670	7068

Fig. 1. Structure of sodium thorium phosphate pyrophosphate, and atomic positional parameters (x 10^4).

CAESIUM URANYL PYROPHOSPHATE
$Cs_2UO_2P_2O_7$

S.A. LINDE, Ju.E. GORBUNOVA, A.V. LAVROV and A.B. POBEDINA, 1981. Izv. Akad. Nauk SSSR, Neorg. Mater., 17, 1062-1066.

Orthorhombic, Pmmn, a = 12.670, b = 12.807, c = 6.152 Å, Z = 4. R = 0.061.

The structure contains $(UO_2P_2O_7^{2-})_n$ layers linked by Cs^+ ions; U has flattened octahedral coordination.

SODIUM TRIPOLYPHOSPHATE HEXAHYDRATE
$Na_5P_3O_{10}.6H_2O$

D.M. WIENCH, M. JANSEN and R. HOPPE, 1982. Z. anorg. Chem., 488, 80-86.

Triclinic, $P\bar{1}$, a = 10.370, b = 9.848, c = 7.615 Å, α = 92.24, β = 94.55, γ = 90.87°, Z = 2. R = 0.053 for 2089 reflexions.

The structure contains discrete $P_3O_{10}^{5-}$ anions, O_3P-O-PO_2-O-PO_3, 5- and 6-coordinate Na^+ ions (to phosphate and water oxygens), and water molecules which are involved in hydrogen bonding. P-O = 1.60-1.67 (bridging), 1.50-1.53(1) Å (terminal), P-O-P = 123.4, 124.4°, Na-O = 2.32-2.55, O-H...O = 2.73-2.99 Å.

COBALT(II) TETRAMETAPHOSPHATE
$Co_2P_4O_{12}$

A.G. NORD, 1982. Cryst. Struct. Comm., 11, 1467-1474.

Monoclinic, C2/c, a = 11.809, b = 8.297, c = 9.923 Å, β = 118.72°, Z = 4. Neutron powder data.

Isostructural with the Mg analogue (1), the structure (Fig. 1) containing cyclic $P_4O_{12}^{4-}$ anions, linked by octahedrally coordinated Co^{2+} cations. P-O = 1.58-1.66 (ring), 1.45-1.54 (terminal), Co-O = 2.03-2.16(2) Å.

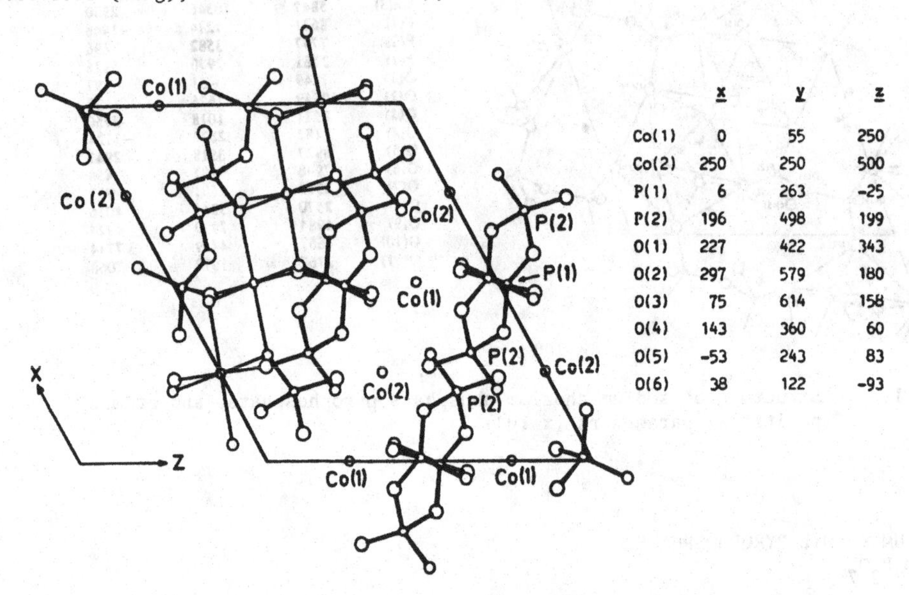

	x	y	z
Co(1)	0	55	250
Co(2)	250	250	500
P(1)	6	263	-25
P(2)	196	498	199
O(1)	227	422	343
O(2)	297	579	180
O(3)	75	614	158
O(4)	143	360	60
O(5)	-53	243	83
O(6)	38	122	-93

Fig. 1. Structure of cobalt(II) tetrametaphosphate, and atomic positional parameters (x 10^3).

1. Structure Reports, 41A, 334.

POTASSIUM THULIUM TETRAMETAPHOSPHIMATE HYDRATE

$K_4(H_3O)[Tm\{(PO_2NH)_4\}_2].17H_2O$

V.I. SOKOL, M.A. PORAJ-KOŠIC, D.A. MURAŠOV and G.G. SADIKOV, 1982. Koord. Khim., 8, 1408-1414.

Monoclinic, C2/c, a = 18.866, b = 11.063, c = 19.165 Å, β = 94.84°, Dm = 2.05, Z = 4. R = 0.053.

Tm is coordinated to eight O atoms from two tetrametaphosphimate groups.

TERBIUM PENTAPHOSPHATE

TbP_5O_{14}

Y. LIN, N. HU, M. WANG and E. SHI, 1982. Huaxue Xuebao, 40, 211-216.

Monoclinic, $P2_1/c$, a = 8.721, b = 8.877, c = 12.911 Å, β = 90.52°, Z = 4. R = 0.075

The structure contains polyphosphate chains, linked by eight-coordinate Tb cations (square antiprism). Isostructural compounds have been described previously (1).

1. Structure Reports, 40A, 257, 258.

ERBIUM PENTAPHOSPHATE
ErP_5O_{14}

I. Y. LIN, N. HU, Q. ZHOU, E. SHI, M. WANG, S. LIU and S. WU, 1982. Kexue Tongbao, 27, 281-284.
II. Idem, 1982. Ibid., 27, 1126-1129.

Monoclinic, C2/c, a = 12.835, b = 12.705, c = 12.363 Å, β = 88.75°, Z = 8. R = 0.08.

[Structure as previously described (1).]

1. Structure Reports, 46A, 335.

THULIUM ULTRAPHOSPHATE
TmP_5O_{14}

K. ZHOU, Y. QIAN and G. HONG, 1982. Cryst. Struct. Comm., 11, 1695-1699.

Monoclinic, C2/c, a = 12.822, b = 12.709, c = 12.358 Å, β = 91.25°, Dm = 3.55, Z = 8. Mo radiation, R = 0.061 for 1812 reflexions.

Isostructural with the Ho, Yb, and Er compounds (1, 2). Tm-O = 2.24-2.36 Å.

1. Structure Reports, 39A, 294; 40A, 258; 46A, 335.
2. Preceding report.

BARIUM ZINC DECAMETAPHOSPHATE
$Ba_2Zn_3P_{10}O_{30}$

M. BAGIEU-BEUCHER, A. DURIF and J.C. GUITEL, 1982. J. Solid State Chem., 45, 159-163.

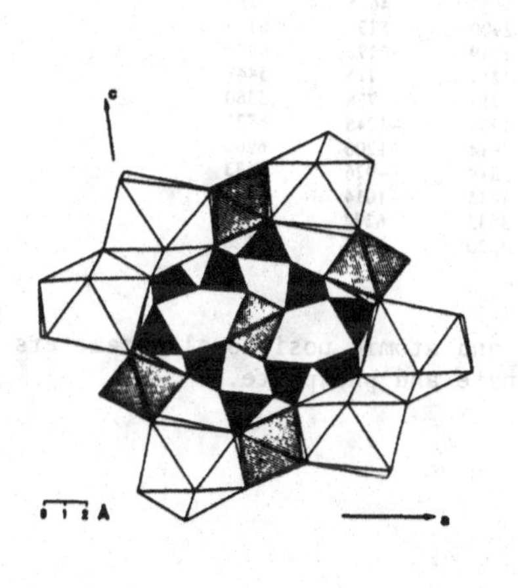

	x	y	z
Ba	0.57580	0.4983	0.18736
Zn(1)	0.5000	0.5000	0.5000
Zn(2)	0.2500	0.4514	0.7500
Zn(3)	0.2500	0.5307	0.2500
P(1)	0.42580	0.0939	0.0552
P(2)	0.47814	−0.0158	0.3173
P(3)	0.37042	−0.2145	0.4134
P(4)	0.30431	−0.0444	0.6222
P(5)	0.20406	0.2234	0.4789
O(L12)	0.4595	−0.078	0 1694
O(L23)	0.4140	−0.007	0.3666
O(L34)	0.3504	−0.056	0.5249
O(L45)	0.2356	−0.032	0.5395
O(L51)	0.3586	0.116	0.0979
O(E11)	0.4544	0.348	0.0586
O(E12)	0.4211	−0.050	−0.0653
O(E21)	0.5051	0.240	0.3365
O(E22)	0.5168	−0.234	0.3692
O(E31)	0.3162	−0.245	0.3108
O(E32)	0.4066	−0.439	0.4620
O(E41)	0.3180	0.195	0.6918
O(E42)	0.3093	−0.283	0.6950
O(E51)	0.2433	0.317	0.3896
O(E52)	0.1908	0.394	0.5809

Fig. 1. Structure of $Ba_2Zn_3P_{10}O_{30}$.

Monoclinic, P2/n, a = 21.738, b = 5.356, c = 10.748 Å, β = 99.65°, Z = 2. R = 0.041 for 2759 reflexions.

The structure (Fig. 1) contains rings of ten corner-sharing phosphate tetrahedra, linked along b by corner sharing with ZnO_4 tetrahedra; linkage in the other directions is via double chains of BaO_9 polyhedra connected by ZnO_6 octahedra. P-O = 1.475-1.495 (terminal), 1.566-1.619(3) (ring) Å, P-O-P = 123-144°, Zn-O = 1.904-1.911 (tetrahedral), 2.029-2.260 (octahedral), Ba-O = 2.742-3.237 Å.

MAGNESIUM POTASSIUM HYDROGENBIS(ARSENATE) 15-HYDRATE
MAGNESIUM POTASSIUM HYDROGENBIS(PHOSPHATE) 15-HYDRATE
$Mg_2KH(XO_4)_2.15H_2O$ (X = As, P)

S. TAKAGI, M. MATHEW and W.E. BROWN, 1982. Acta Cryst., B38, 44-50.

Triclinic, P$\bar{1}$, a = 6.390, 6.288, b = 12.477, 12.228, c = 6.659, 6.554 Å, α = 93.54, 93.64, β = 88.71, 89.18, γ = 94.51, 94.69°, Z = 1. Mo radiation, R = 0.079, 0.035 for 964, 2633 reflexions.

The structures (Fig. 1) contain $H(XO_4)_2$ units with symmetrical hydrogen bonds, hydrogen bonded to $Mg(OH_2)_6$ octahedra, with channels occupied by disordered K/H_2O. As-O = 1.68-1.70(1), P-O = 1.527-1.571(1), Mg-O = 2.04-2.09, O...H...O = 2.52, 2.50 Å.

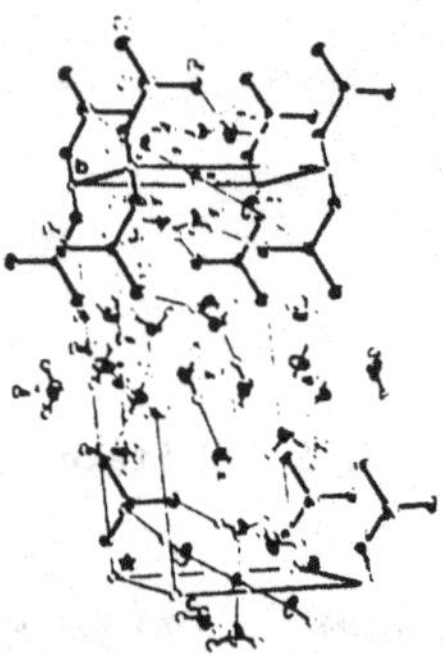

	x	y	z
Mg(1)	0	5000	5000
	0	5000	5000
Mg(2)	5000	0	5000
	5000	0	5000
As	1195	1933	472
P	1144	1913	434
M	4741	3116	5724
	4668	3101	5598
O(1)	301	1828	2865
	322	1819	2635
O(2)	394	3053	−486
	383	2931	−470
O(3)	132	852	−926
	160	881	−883
O(4)	3820	1939	328
	3575	1933	331
O(w1)	232	6579	4349
	336	6628	4391
O(w2)	1596	4636	2304
	1593	4614	2289
O(w3)	2900	5220	6375
	2949	5192	6448
O(w4)	2211	713	5449
	2212	728	5360
O(w5)	3901	−1246	6735
	3914	−1200	6902
O(w6)	3819	−926	2523
	3835	−1014	2538
O(w7)	3533	6297	53
	3520	6331	120

Fig. 1. Structure of $Mg_2KH(PO_4)_2.15H_2O$, and atomic positional parameters ($x\ 10^4$, M = K/H_2O) for the arsenate and phosphate.

MACHATSCHKIITE
$(Ca,Na)_6(AsO_4)(AsO_3OH)_3[(P,S)O_4].15H_2O$

H. EFFENBERGER, K. MEREITER, M. PIMMINGER and J. ZEMANN, 1982. Tschermaks Min. Petr. Mitt., 30, 145-155.

Rhombohedral, R3c, a = 11.506 Å, α = 82.20°, Z = 2 (hexagonal cell, a = 15.127, c = 22.471 Å, Z = 6). Mo radiation, R = 0.04 for 645 reflexions.

The structure (Fig. 1) contains an AsO_4 anion hydrogen bonded to three $AsO_3(OH)$ anions, and an XO_4 anion (X = P, S). The anions are linked by two edge-sharing polyhedra, CaO_7 and CaO_8, and by hydrogen bonding. As-OH = 1.73, As-O = 1.65-1.70, X-O = 1.52, Ca-O = 2.36-2.80, O-H...O = 2.66-2.87 Å.

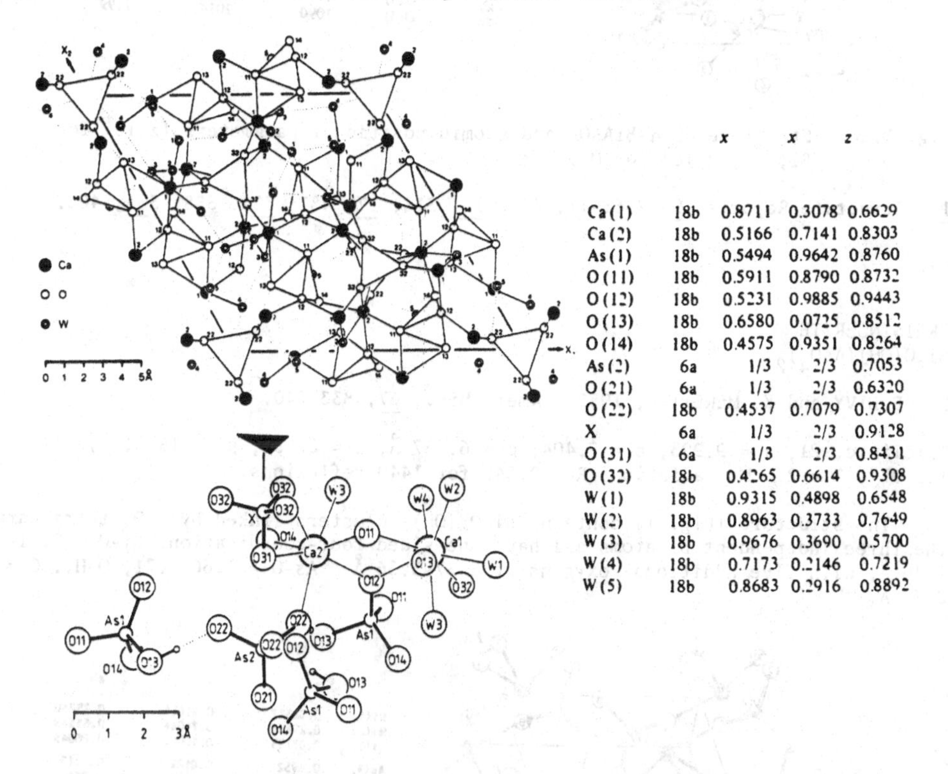

		x	x	z
Ca(1)	18b	0.8711	0.3078	0.6629
Ca(2)	18b	0.5166	0.7141	0.8303
As(1)	18b	0.5494	0.9642	0.8760
O(11)	18b	0.5911	0.8790	0.8732
O(12)	18b	0.5231	0.9885	0.9443
O(13)	18b	0.6580	0.0725	0.8512
O(14)	18b	0.4575	0.9351	0.8264
As(2)	6a	1/3	2/3	0.7053
O(21)	6a	1/3	2/3	0.6320
O(22)	18b	0.4537	0.7079	0.7307
X	6a	1/3	2/3	0.9128
O(31)	6a	1/3	2/3	0.8431
O(32)	18b	0.4265	0.6614	0.9308
W(1)	18b	0.9315	0.4898	0.6548
W(2)	18b	0.8963	0.3733	0.7649
W(3)	18b	0.9676	0.3690	0.5700
W(4)	18b	0.7173	0.2146	0.7219
W(5)	18b	0.8683	0.2916	0.8892

Fig. 1. Structure of machatschkiite.

ROOSEVELTITE
α-$BiAsO_4$

D. BEDLIVY and K. MEREITER, 1982. Acta Cryst., B38, 1559-1561.

Monoclinic, $P2_1/n$, a = 6.879, b = 7.159, c = 6.732 Å, β = 104.84°, Z = 4. Mo radiation, R = 0.027 for 772 reflexions.

Monazite-type ($CePO_4$) structure (1), but with distortions resulting from the Bi^{3+} lone pair, Bi having irregular one-sided eight coordination (Fig. 1). As-O = 1.67-1.70, Bi-O = 2.33-2.65 (1) Å (next nearest Bi-O = 3.24 Å).

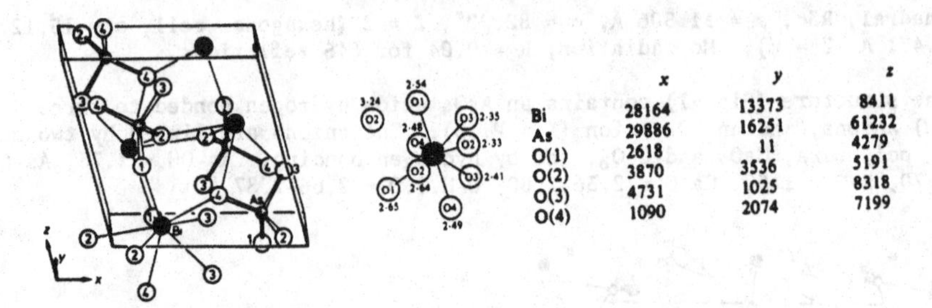

	x	y	z
Bi	28164	13373	8411
As	29886	16251	61232
O(1)	2618	11	4279
O(2)	3870	3536	5191
O(3)	4731	1025	8318
O(4)	1090	2074	7199

Fig. 1. Structure of α-$BiAsO_4$ and atomic positional parameters (x 10^5 for Bi, As; x 10^4 for O).

1. Structure Reports, 9, 236; 11, 374; 13, 316; 17, 487; 32A, 358; 33A, 402.

PREISINGERITE

$Bi_3O(OH)(AsO_4)_2$

D. BEDLIVY and K. MEREITER, 1982. Amer. Min., 67, 833-840.

Triclinic, P$\bar{1}$, a = 9.993, b = 7.404, c = 6.937 Å, α = 87.82, β = 115.01, γ = 111.07°, Z = 2. Mo radiation, R = 0.046 for 1440 reflexions.

The structure (Fig. 1) contains $Bi_6O_2(OH)_2$ clusters linked by AsO_4 tetrahedra. The three independent Bi atoms all have one-sided four-coordination, Bi-O = 2.11-2.46 Å, with five additional oxygens at 2.57-3.36 Å. As-O = 1.66-1.71, O-H...O = 2.72 Å.

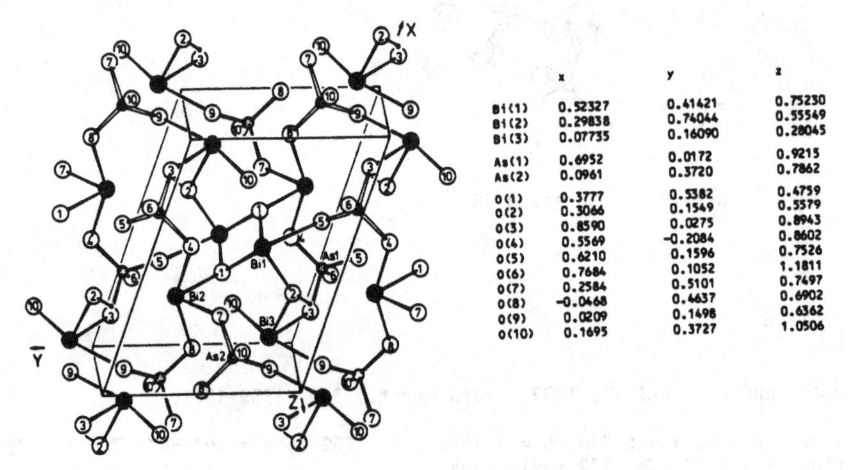

	x	y	z
Bi(1)	0.52327	0.41421	0.75230
Bi(2)	0.29838	0.74044	0.55549
Bi(3)	0.07735	0.16090	0.28045
As(1)	0.6952	0.0172	0.9215
As(2)	0.0961	0.3720	0.7862
O(1)	0.3777	0.5382	0.4759
O(2)	0.3066	0.1549	0.5579
O(3)	0.8590	0.0275	0.8943
O(4)	0.5569	-0.2084	0.8602
O(5)	0.6210	0.1596	0.7526
O(6)	0.7684	0.1052	1.1811
O(7)	0.2584	0.5101	0.7497
O(8)	-0.0468	0.4637	0.6902
O(9)	0.0209	0.1498	0.6362
O(10)	0.1695	0.3727	1.0506

Fig. 1. Structure of preisingerite.

CABRERITE

$(Ni,Mg)_3(AsO_4)_2.8H_2O$

G. GIUSEPPETTI and C. TADINI, 1982. Bull. Minéral., 105, 333-337.

Monoclinic, C2/m, a = 10.211, b = 13.335, c = 4.728 Å, β = 104.97°, D_m = 3.11, Z = 2. Mo radiation R = 0.046 for 609 reflexions.

Atomic positions

	x	y	z
M(1)	0	0	0
M(2)	0	0.3861	0
As	0.3161	0	0.3754
O(1)	0.1500	0	0.3775
O(2)	0.4066	0	0.7272
O(3)	0.3434	0.1078	0.2095
W(1)	0.0955	0.1135	-0.1875
W(2)	0.4009	0.2251	0.7189

Isostructural with vivianite (1), with AsO_4 tetrahedra, linked by $M(1)O_2W_4$ octahedra and pairs of edge-sharing $M(2)O_4W_2$ octahedra. As-O = 1.68-1.70, M-O = 2.03-2.11, O-H...O = 2.72-2.89(1) Å.

1. Structure Reports, 46A, 326.

COPPER(II) SILVER TRIHYDROGENBIS(ARSENATE)
$CuAgH_3(AsO_4)_2$

A. BOUDJADA, R. MASSE and J.C. GUITEL, 1982. Acta Cryst., B38, 710-713.

Monoclinic, $P2_1/a$, a = 9.716, b = 7.704, c = 9.209 Å, β = 103.73°, Z = 4. Ag radiation, R = 0.031 for 1118 reflexions.

The structure (Fig. 1) contains AsO_4 tetrahedra, linked by 5-coordinate Cu (square-pyramidal Cu_2O_8 dimers), 6-coordinate Ag, and hydrogen bonds. As-O = 1.71, 1.71, 1.74 (probably OH), 1.65-1.69, Cu-O = 1.93-2.02, 2.36, Ag-O = 2.42-2.68, O-H...O = 2.61, 2.63, 2.79 Å.

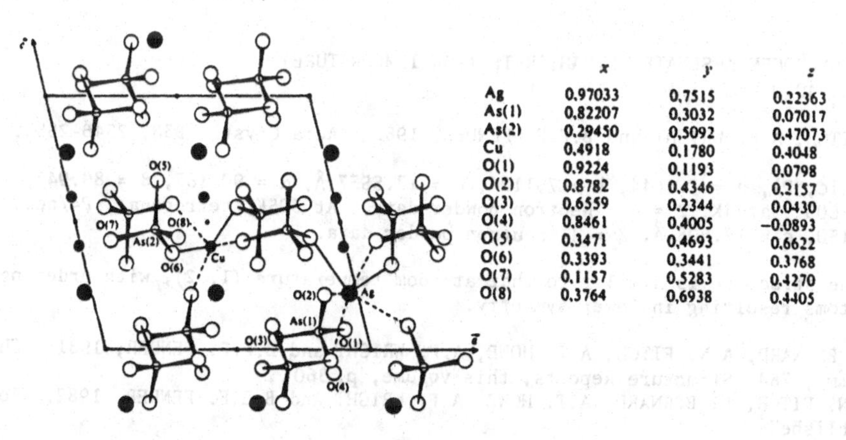

	x	y	z
Ag	0,97033	0,7515	0,22363
As(1)	0,82207	0,3032	0,07017
As(2)	0,29450	0,5092	0,47073
Cu	0,4918	0,1780	0,4048
O(1)	0,9224	0,1193	0,0798
O(2)	0,8782	0,4346	0,2157
O(3)	0,6559	0,2344	0,0430
O(4)	0,8461	0,4005	−0,0893
O(5)	0,3471	0,4693	0,6622
O(6)	0,3393	0,3441	0,3768
O(7)	0,1157	0,5283	0,4270
O(8)	0,3764	0,6938	0,4405

Fig. 1. Structure of $CuAgH_3(AsO_4)_2$.

PROSPERITE
$Ca_2Zn_4(AsO_4)_4.H_2O$

P. KELLER, H. RIFFEL and H. HESS, 1982. Z. Kristallogr., 158, 33-42.

Monoclinic, C2/c, a = 19.238, b = 7.731, c = 9.765 Å, β = 104.47°, Z = 4. Mo radiation, R = 0.034 for 1528 reflexions.

The structure (Fig. 1) contains AsO_4 tetrahedra and a framework of face-, edge-, and corner-sharing cation polyhedra, 9-coordination for Ca and 5-coordination for Zn ions; the H_2O molecule forms hydrogen bonds. As-O = 1.677-1.696, Ca-O = 2.333-3.160, Zn-O = 1.935-2.159(4) Å.

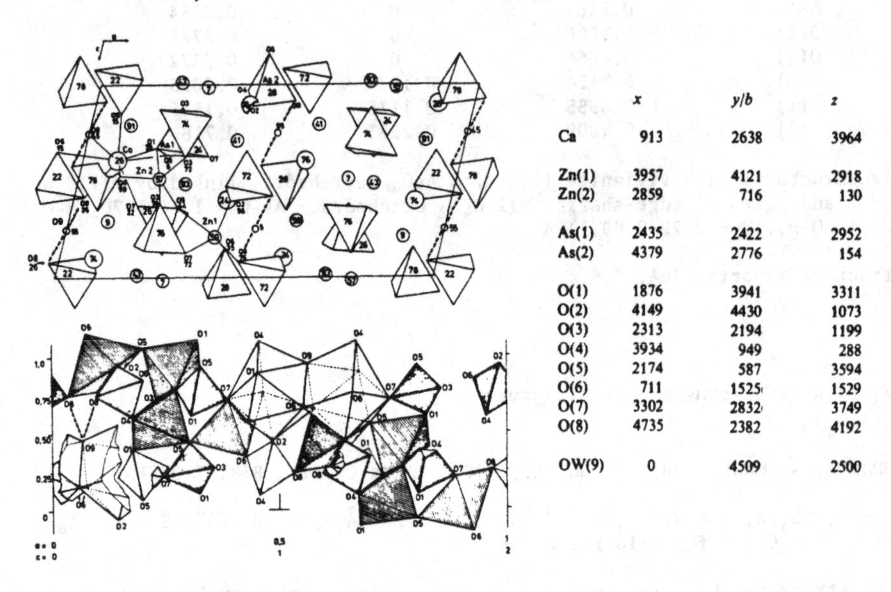

	x	y/b	z
Ca	913	2638	3964
Zn(1)	3957	4121	2918
Zn(2)	2859	716	130
As(1)	2435	2422	2952
As(2)	4379	2776	154
O(1)	1876	3941	3311
O(2)	4149	4430	1073
O(3)	2313	2194	1199
O(4)	3934	949	288
O(5)	2174	587	3594
O(6)	711	1525	1529
O(7)	3302	2832	3749
O(8)	4735	2382	4192
OW(9)	0	4509	2500

Fig. 1. Structure of prosperite, and atomic positional parameters (x 10^4).

URANYL HYDROGEN ARSENATE TETRAHYDRATE (LOW-TEMPERATURE)

$UO_2DAsO_4 \cdot 4D_2O$

A.N. FITCH, A.F. WRIGHT and B.E.F. FENDER, 1982. Acta Cryst., B38, 2546-2554.

Triclinic, P$\bar{1}$, a = 7.1644, b = 7.1124, c = 17.5537 Å, α = 90.187, β = 89.947, γ = 90.003°, at 4K, Z = 4. Neutron powder data. At 305K, tetragonal, P4/ncc, a = 7.150, c = 17.608 Å, Z = 4, neutron powder data.

The structure is similar to that at room temperature (1, 2), with ordering of H atoms resulting in lower symmetry.

1. L. BERNARD, A.N. FITCH, A.T. HOWE, A.F. WRIGHT and B.E.F. FENDER, 1981. Chem. Comm., 784; Structure Reports, this volume, p. 360.
2. A.N. FITCH, L. BERNARD, A.T. HOWE, A.F. WRIGHT and B.E.F. FENDER, 1982. To be published.

LITHIUM URANYL ARSENATE TETRAHYDRATE

$LiUO_2AsO_4 \cdot 4D_2O$

A.N. FITCH, B.E.F. FENDER and A.F. WRIGHT, 1982. Acta Cryst., B38, 1108-1112.

Tetragonal, P4/n, a = 7.0969, c = 9.1903 Å, at 294K, Z = 2. Neutron powder data.

The structure (Fig. 1) contains layers of tetrahedral AsO_4^- and linear UO_2^{2+} ions and water molecules; four arsenate oxygen atoms complete octahedral coordination at U. The layers are linked by Li coordinated to four water molecules. As-O = 1.67, U-O = 1.79 (uranyl), 2.31, Li-O = 1.96, O-H...O = 2.78, 2.95 Å.

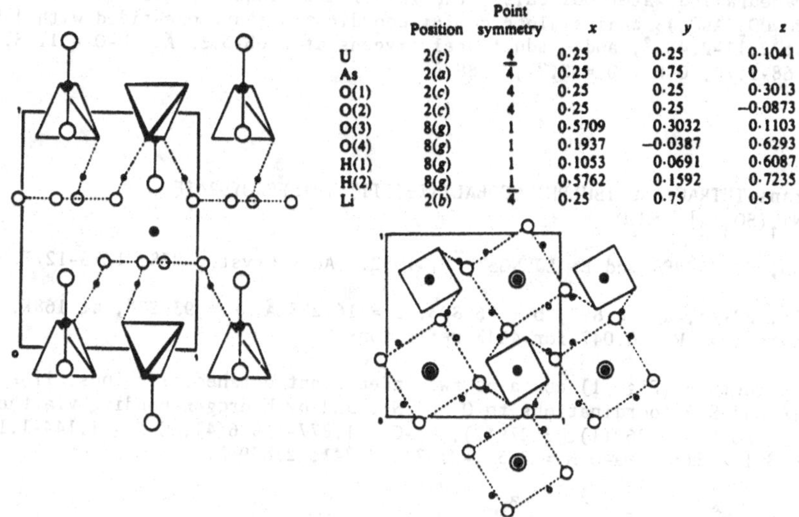

	Position	Point symmetry	x	y	z
U	2(c)	$\bar{4}$	0·25	0·25	0·1041
As	2(a)	$\bar{4}$	0·25	0·75	0
O(1)	2(c)	4	0·25	0·25	0·3013
O(2)	2(c)	4	0·25	0·25	−0·0873
O(3)	8(g)	1	0·5709	0·3032	0·1103
O(4)	8(g)	1	0·1937	−0·0387	0·6293
H(1)	8(g)	1	0·1053	0·0691	0·6087
H(2)	8(g)	$\bar{1}$	0·5762	0·1592	0·7235
Li	2(b)	$\bar{4}$	0·25	0·75	0·5

Fig. 1. Structure of lithium uranyl arsenate tetrahydrate.

WALPURGITE

$(UO_2)Bi_4O_4(AsO_4)_2.2H_2O$

K. MEREITER, 1982. Tschermaks Min. Petr. Mitt., 30, 129-139.

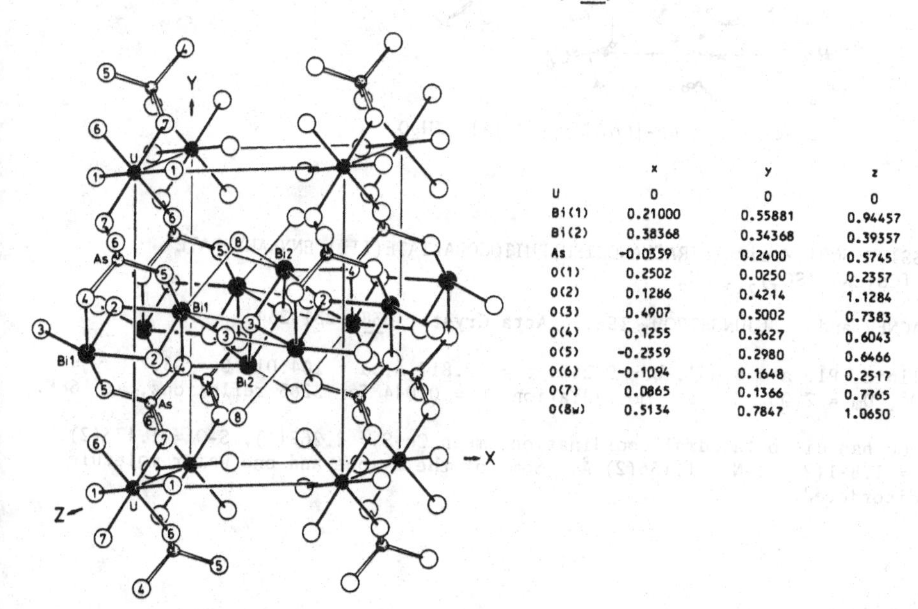

	x	y	z
U	0	0	0
Bi(1)	0.21000	0.55881	0.94457
Bi(2)	0.38368	0.34368	0.39357
As	-0.0359	0.2400	0.5745
O(1)	0.2502	0.0250	0.2357
O(2)	0.1286	0.4214	1.1284
O(3)	0.4907	0.5002	0.7383
O(4)	0.1255	0.3627	0.6043
O(5)	-0.2359	0.2980	0.6466
O(6)	-0.1094	0.1648	0.2517
O(7)	0.0865	0.1366	0.7765
O(8w)	0.5134	0.7847	1.0650

Fig. 1. Structure of walpurgite.

Triclinic, P$\bar{1}$, a = 7.135, b = 10.426, c = 5.494 Å, α = 101.47, β = 110.82, γ = 88.20°, Z = 1. Mo radiation, R = 0.041 for 1381 reflexions.

The structure (Fig. 1) contains (010) layers of Bi and O atoms with attached AsO_4 tetrahedra and water molecules; the layers are linked by $(UO_2)O_4$ octahedra, which form $UO_2(AsO_4)_2$ chains along c. Bi coordinations are one-sided with 4-5 oxygens at 2.11-2.48 Å, and 4 additional oxygens at 2.63-3.35 Å. U-O = 1.78, 2.28, As-O = 1.68-1.70, O-H...O = 2.77, 2.88 Å.

SODIUM trans-TETRACYANODISULPHITOCOBALTATE(III) TRIDECAHYDRATE
$Na_5[Co(CN)_4(SO_3)_2].13H_2O$

M. ASPLUND, S. JAGNER and E. LJUNGSTRÖM, 1982. Acta Cryst., B38, 1275-1277.

Monoclinic, $P2_1/c$, a = 8.893, b = 16.893, c = 16.258 Å, β = 93.27°, at 168K, Z = 4. Mo radiation, R = 0.047 for 4143 reflexions.

The structure (Fig. 1) contains two independent octahedral anions, linked by Na ions (6 and 5+2 coordinations to O and N), and by hydrogen bonding via the water molecules. Co-S = 2.262(1), 2.279(2), Co-C = 1.877-1.896(4), C-N = 1.144-1.159(5), S-O = 1.478-1.498(3), Na-O,N = 2.301-2.674, 2.741, 2.839 Å.

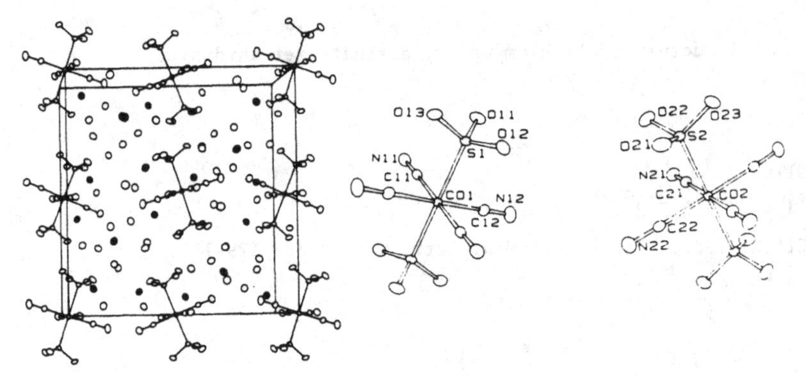

Fig. 1. Structure of $Na_5[Co(CN)_4(SO_3)_2].13H_2O$.

POTASSIUM SODIUM cis-TETRACYANODISULPHITOCOBALTATE(III) ENNEAHYDRATE
$K_5Na_5[Co(CN)_4(SO_3)_2]_2.9H_2O$

S. JAGNER and E. LJUNGSTRÖM, 1982. Acta Cryst., B38, 231-234.

Triclinic, P$\bar{1}$, a = 8.057, b = 9.253, c = 12.817 Å, α = 104.01, β = 98.77, γ = 96.39°, D_m = 2.00, Z = 1. Mo radiation, R = 0.044 for 3285 reflexions, at 168K.

Co has cis-octahedral coordination, mean Co-S = 2.262(1), S-O = 1.473(2), Co-C = 1.891(2), C-N = 1.156(2) Å. Some of the K ions and one water molecule are disordered.

POTASSIUM cis-DIBROMO(HYDROGENBISSULPHITO)PLATINATE(II) MONOHYDRATE
$K_3[Pt\{(SO_3)_2H\}Br_2].H_2O$

D.K. BREITINGER, G. PETRIKOWSKI and G. BAUER, 1982. Acta Cryst., B38, 2997-3000.

Monoclinic, $P2_1/c$, a = 8.764, b = 6.934, c = 21.544 Å, β = 99.32°, D_m = 3.47, Z = 4. Ag radiation, R = 0.025 for 1801 reflexions.

The structure (Fig. 1) contains a square-planar complex anion, with a very short intra-ion O-H...O hydrogen bond. These anions are arranged in layers held together by three K^+ ions and by hydrogen-bonded water molecules. A related chloro compound has been described previously (1).

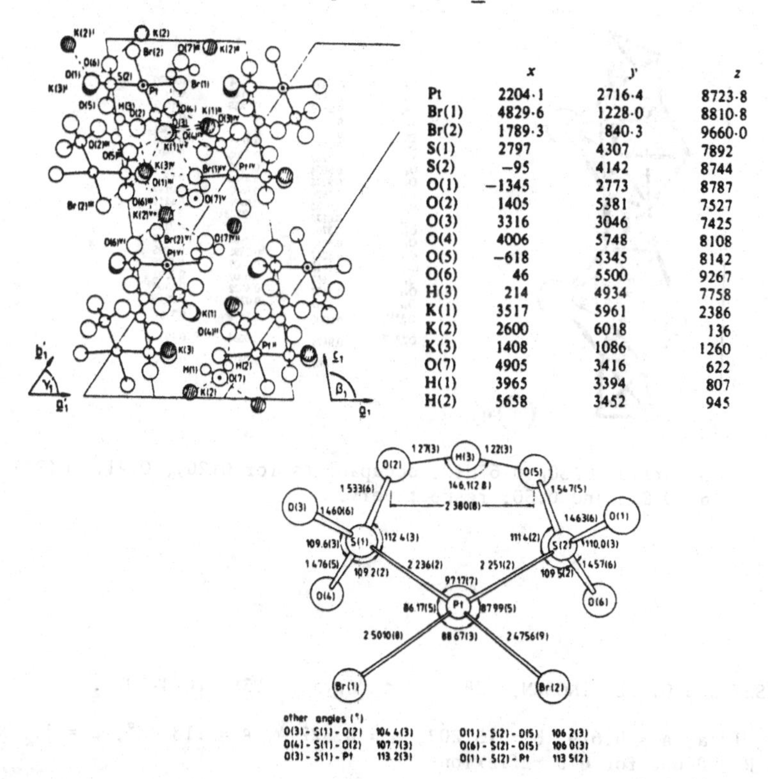

	x	y	z
Pt	2204·1	2716·4	8723·8
Br(1)	4829·6	1228·0	8810·8
Br(2)	1789·3	840·3	9660·0
S(1)	2797	4307	7892
S(2)	−95	4142	8744
O(1)	−1345	2773	8787
O(2)	1405	5381	7527
O(3)	3316	3046	7425
O(4)	4006	5748	8108
O(5)	−618	5345	8142
O(6)	46	5500	9267
H(3)	214	4934	7758
K(1)	3517	5961	2386
K(2)	2600	6018	136
K(3)	1408	1086	1260
O(7)	4905	3416	622
H(1)	3965	3394	807
H(2)	5658	3452	945

Fig. 1. Structure of $K_3[Pt\{(SO_3)_2H\}Br_2].H_2O$, and atomic positional parameters ($\times 10^4$).

1. Structure Reports, 46A, 346.

TRISODIUM HYDROGEN BISSULPHATE
$Na_3H(SO_4)_2$

W. JOSWIG, H. FUESS and G. FERRARIS, 1982. Acta Cryst., B38, 2798-2801.

Monoclinic, $P2_1/c$, a = 8.648, b = 9.648, c = 9.143 Å, β = 108.77°, Z = 4. Neutron radiation, R = 0.049 for 2055 reflexions.

The structure is as previously described (1), with a short, asymmetric hydrogen bond, O...O = 2.432(2), O-H = 1.156(3), H...O = 1.276(3) Å, O-H...O = 179.1°.

1. Structure Reports, 45A, 331.

CALCIUM SULPHATE HYDRATE
$CaSO_4.0{\cdot}67H_2O$

N.N. BUŠUEV, 1982. Ž. Neorg. Khim., 27, 610-615 [Russ. J. Inorg. Chem., 27, 344-347].

Monoclinic, I2, a = 12.028, b = 38.022, c (unique axis) = 6.927 Å, γ = 90.21°, Dm = 2.77, Z = 36 (subcell with b' = b/3, Z = 12). Mo radiation, R = 0.079 for 1014 subcell reflexions.

The structure (Fig. 1) contains chains of SO_4 tetrahedra and CaO_9 polyhedra; the water sites are only partially occupied. S-O = 1.35-1.59, Ca-O = 2.26-2.87(2) Å.

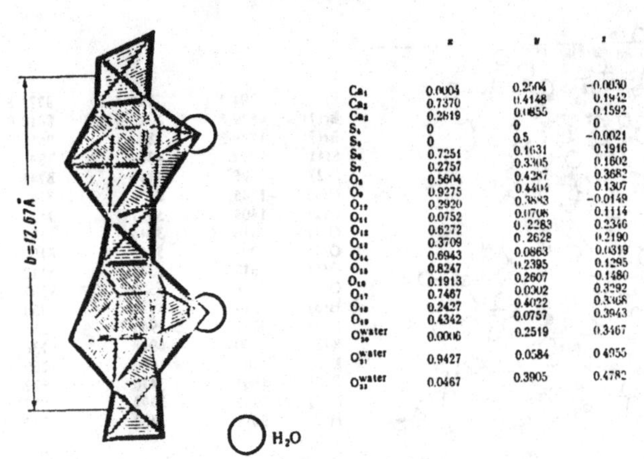

	x	y	z
Ca_1	0.0004	0.2504	-0.0030
Ca_2	0.7370	0.4148	0.1942
Ca_3	0.2819	0.0855	0.1592
S_4	0	0	0
S_5	0	0.5	-0.0021
S_6	0.7251	0.1631	0.1916
S_7	0.2757	0.3305	0.1602
O_8	0.5604	0.4287	0.3682
O_9	0.9275	0.4404	0.1307
O_{10}	0.2920	0.3883	-0.0149
O_{11}	0.0752	0.0708	0.1114
O_{12}	0.6272	0.2283	0.2346
O_{13}	0.3709	0.2628	0.2190
O_{14}	0.6943	0.0863	0.0319
O_{15}	0.8247	0.2395	0.1295
O_{16}	0.1913	0.2607	0.1480
O_{17}	0.7467	0.0302	0.3292
O_{18}	0.2427	0.4022	0.3368
O_{19}	0.4342	0.0757	0.3943
O_{20}^{water}	0.0006	0.2519	0.3467
O_{21}^{water}	0.9427	0.0584	0.4955
O_{22}^{water}	0.0467	0.3905	0.4782

Fig. 1. Structure of $CaSO_4.0{\cdot}67H_2O$; occupancies for O(20), O(21), O(22) are 0.66, 0.82, and 0.50, respectively.

GYPSUM
$CaSO_4.2H_2O$

B.F. PEDERSEN and D. SEMMINGSEN, 1982. Acta Cryst., B38, 1074-1077.

Monoclinic, I2/a, a = 5.679, b = 15.202, c = 6.522 Å, β = 118.43°, Z = 4. Neutron radiation, R = 0.036 for 610 reflexions.

Atomic positions

	x	y	z
Ca	0·5	0·07967	0·25
S	0·0	0·07705	0·75
O(1)	0·96320	0·13190	0·55047
O(2)	0·75822	0·02226	0·66709
O(W)	0·37960	0·18212	0·45881
H(1)	0·25112	0·16158	0·50372
H(2)	0·40458	0·24275	0·49217

The structure is as previously described (1), containing slightly-distorted sulphate tetrahedra linked by CaO_8 polyhedra and O-H...O hydrogen bonds via the water molecules. O-H = 0.959, 0.942(3), H...O = 1.856, 1.941, O...O = 2.807, 2.882 Å, O-H...O = 170.9, 177.2°.

1. Strukturbericht, 2, 97, 430; 4, 47, 179; Structure Reports, 22, 449; 40A, 263.

NATROALUNITE
$(K,Na)Al_3(SO_4)_2(OH)_6$

K. OKADA, J.-I. HIRABAYASHI and J. OSSAKA, 1982. Neues Jb. Miner., Mh., 534-540.

Rhombohedral, R$\bar{3}$m, a = 6.990, c = 16.905 Å, Z = 3. Mo radiation, R = 0.031 for 446 reflexions.

Atomic positions

	x	y	z
M	0	0	0
Al	0	1/2	1/2
S	0	0	0.3059
O(1)	0	0	0.3921
O(2)	0.2181	-0.2181	-0.0573
OH	0.1253	-0.1253	0.1390
H	0.178	-0.178	0.116

M = 0.42 K + 0.58 Na

Isostructural with alunite (1). S-O = 1.457, 1.484 (x 3), Al-2 O = 1.951, Al-4 OH = 1.879, M-6 O = 2.812, M-6 OH = 2.797(2) Å.

1. Strukturbericht, 5, 92; Structure Reports, 30A, 376; 42A, 369.

MINAMIITE
$(Na,Ca,K)Al_3(SO_4)_2(OH)_6$

J. OSSAKA, J.-I. HIRABAYASHI, K. OKADA, R. KOBAYASHI and T. HAYASHI, 1982. Amer. Min., 67, 114-119.

Rhombohedral, R$\bar{3}$m, a = 6.981, c = 33.490 Å, Z = 6. Mo radiation, R = 0.045 for 389 reflexions.

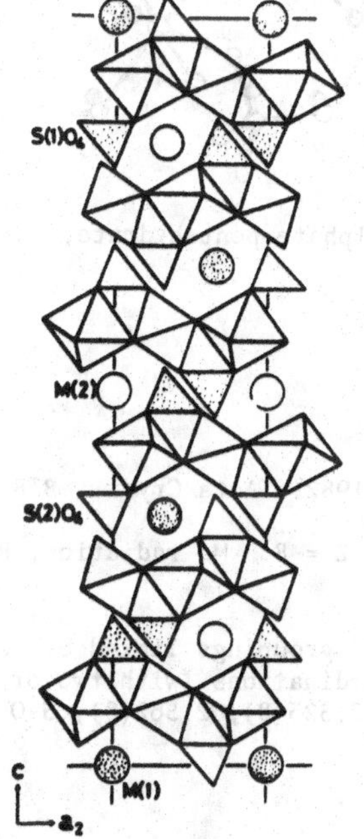

	multipl.	x	y	z
M(1) {Na / K	0.036 / 0.017	0	0	0
M(2) {Na / Ca	0.024 / 0.045	0	0	0.5
S(1)	0.167	0	0	0.8458
S(2)	0.167	0	0	0.3473
Al	0.5	0.5001	-0.5001	0.2496
O(1)	0.167	0	0	0.8022
O(2)	0.167	0	0	0.3043
O(3)	0.5	0.2181	-0.2181	0.0290
O(4)	0.5	0.2183	-0.2183	0.5274
OH(1)	0.5	0.1268	-0.1268	0.9294
OH(2)	0.5	0.1239	-0.1239	0.4321
H(1)	0.5	0.180	-0.180	0.9483
H(2)	0.5	0.180	-0.180	0.4477

Fig. 1. Structure of minamiite.

Alunite-type structure (1, 2), but with a doubled c-axis resulting from partial cation ordering in the M sites (Fig. 1). The structure contains sheets of corner-sharing $AlO_2(OH)_4$ octahedra and SO_4 tetrahedra, with M(1) and M(2) in large cavities, each coordinated to 6 O and 6 OH. S-O = 1.44-1.48, Al-O = 1.88-1.96, M(1)-O = 2.81, 2.82, M(2)-O = 2.72, 2.79(1) Å.

1. Strukturbericht, 5, 92; Structure Reports, 30A, 376; 42A, 369.
2. Preceding report.

OXONIUM DIAQUODISULPHATOINDATE(III) DIHYDRATE (INDIUM(III) HYDROGEN SULPHATE PENTAHYDRATE)

$[H_3O]^+[In(H_2O)_2(SO_4)_2]^-\cdot 2H_2O$ $InH(SO_4)_2\cdot 5H_2O$

R. CAMINITI, G. MARONGIU and G. PASCHINA, 1982. Cryst. Struct. Comm., 11, 955-958.

Monoclinic, $P2_1/c$, a = 10.57, b = 10.77, c = 9.25 Å, β = 102.6°, D_m = 2.48, Z = 4. Mo radiation, R = 0.034 for 1524 reflexions.

The structure (Fig. 1) consists of $InO_4(OH_2)_2$ octahedra and SO_4 tetrahedra sharing corners, with additional water molecules in cavities (W(4) is probably H_3O^+). In-O = 2.10-2.16, S-O = 1.43-1.50(1) Å.

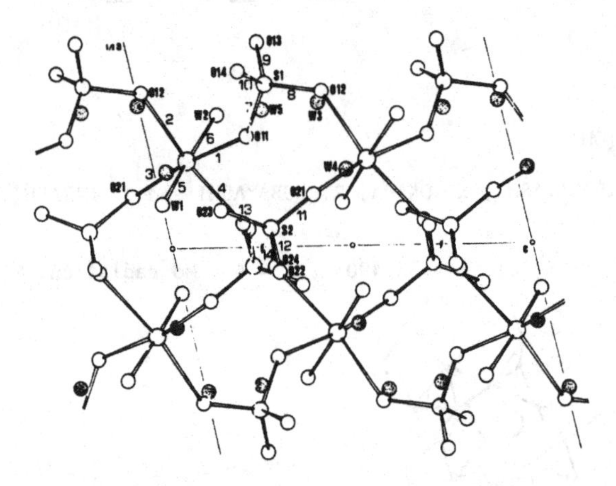

Fig. 1. Structure of indium hydrogen sulphate pentahydrate.

TIN(II) OXIDE SULPHATE

Sn_2OSO_4

G. LUNDGREN, G. WERNFORS and T. YAMAGUCHI, 1982. Acta Cryst., B38, 2357-2361.

Tetragonal, $P\bar{4}2_1c$, a = 10.930, c = 8.931 Å, Z = 8. Mo radiation, R = 0.049 for 1796 reflexions.

The structure (Fig. 1) contains $Sn_8O_4^{8+}$ groupings linked by sulphate ions. The two independent Sn atoms have 3 + 1 coordinations (with two or three longer Sn...O distances). Sn-O = 2.137-2.323(6), 2.523(8), 2.562(8), S-O = 1.448-1.488(8) Å.

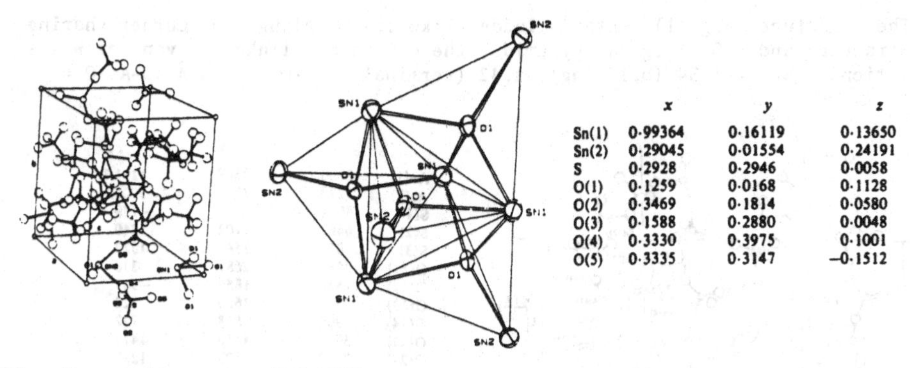

	x	y	z
Sn(1)	0·99364	0·16119	0·13650
Sn(2)	0·29045	0·01554	0·24191
S	0·2928	0·2946	0·0058
O(1)	0·1259	0·0168	0·1128
O(2)	0·3469	0·1814	0·0580
O(3)	0·1588	0·2880	0·0048
O(4)	0·3330	0·3975	0·1001
O(5)	0·3335	0·3147	−0·1512

Fig. 1. Structure of Sn_2OSO_4.

TIN HYDROXIDE SULPHATE
$Sn_7(OH)_{12}(SO_4)_2$

S. GRIMVALL, 1982. Acta Chem. Scand., A36, 361-364.

Orthorhombic, Pbca, a = 12.472, b = 12.649, c = 12.676 Å, Z = 4. Cu radiation, R = 0.089 for 277 reflexions (films, densitometer intensities).

The structure (Fig. 1) contains a network of Sn(II), Sn(IV), and O, strengthened by sulphate groups. Sn(IV) (Sn(4)) has octahedral coordination, Sn-O = 1.89-2.06(9) Å, and Sn(II) ions have 3 + 1 coordinations, Sn-O = 2.05-2.28 and 2.56-2.77(9) Å, O-Sn(II)-O = 81-94°; S-O = 1.43-1.49(8) Å. Short O...O distances are probably hydrogen bonds.

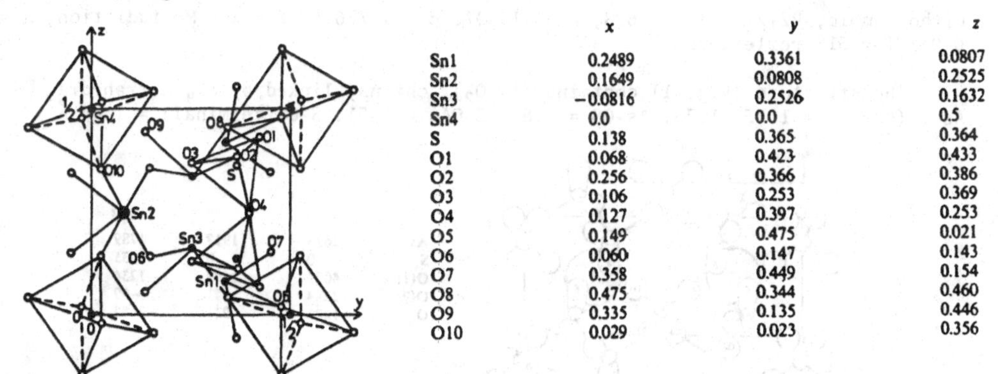

	x	y	z
Sn1	0.2489	0.3361	0.0807
Sn2	0.1649	0.0808	0.2525
Sn3	−0.0816	0.2526	0.1632
Sn4	0.0	0.0	0.5
S	0.138	0.365	0.364
O1	0.068	0.423	0.433
O2	0.256	0.366	0.386
O3	0.106	0.253	0.369
O4	0.127	0.397	0.253
O5	0.149	0.475	0.021
O6	0.060	0.147	0.143
O7	0.358	0.449	0.154
O8	0.475	0.344	0.460
O9	0.335	0.135	0.446
O10	0.029	0.023	0.356

Fig. 1. Structure of $Sn_7(OH)_{12}(SO_4)_2$.

ARSENIC(III) SULPHATE
$As_2(SO_4)_3$

J. DOUGLADE and R. MERCIER, 1982. Acta Cryst., B38, 720-723.

Monoclinic, $P2_1/c$, a = 9.389, b = 5.255, c = 19.355 Å, β = 91.88°, D_m = 3.00, Z = 4. Mo radiation, R_w = 0.040 for 1922 reflexions.

The structure (Fig. 1) contains ladder-like chains along $\underline{a}$ of corner-sharing SO_4 tetrahedra and AsO_3 trigonal pyramids; the chains are linked by van der Waals interactions. S-O = 1.54 (bridging), 1.42 (terminal), As-O = 1.83 Å, O-As-O = 84-94°.

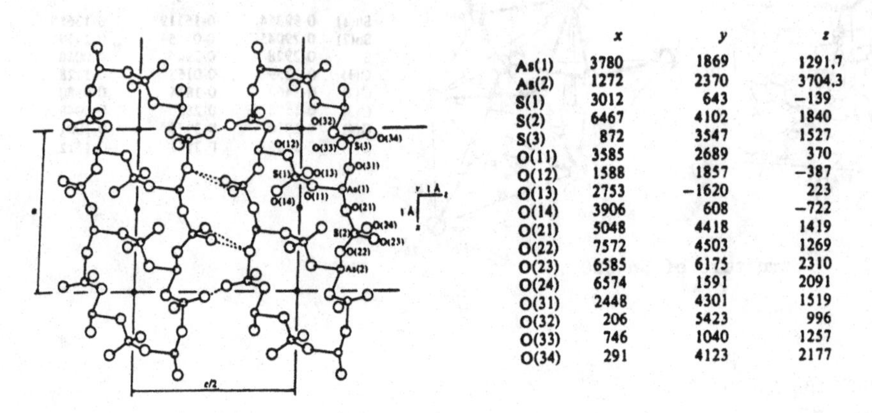

	x	y	z
As(1)	3780	1869	1291.7
As(2)	1272	2370	3704.3
S(1)	3012	643	−139
S(2)	6467	4102	1840
S(3)	872	3547	1527
O(11)	3585	2689	370
O(12)	1588	1857	−387
O(13)	2753	−1620	223
O(14)	3906	608	−722
O(21)	5048	4418	1419
O(22)	7572	4503	1269
O(23)	6585	6175	2310
O(24)	6574	1591	2091
O(31)	2448	4301	1519
O(32)	206	5423	996
O(33)	746	1040	1257
O(34)	291	4123	2177

Fig. 1. Structure of arsenic(III) sulphate, and atomic positional parameters (x 10^4).

ARSENIC OXIDE SULPHATE

$As_2O_2SO_4$

I. R. MERCIER and J. DOUGLADE, 1982. Acta Cryst., B38, 896-898.

Orthorhombic, $P2_12_12$, a = 4.628, b = 11.497, c = 4.776 Å, Z = 2. Mo radiation, R = 0.031 for 518 reflexions.

The structure (Fig. 1) contains $(As_2O_2)_\infty$ chains, linked by SO_4 tetrahedra. As-O (chain) = 1.75, 1.78, As-OS = 1.89, S-OAs = 1.51, S-O (terminal) = 1.44 Å.

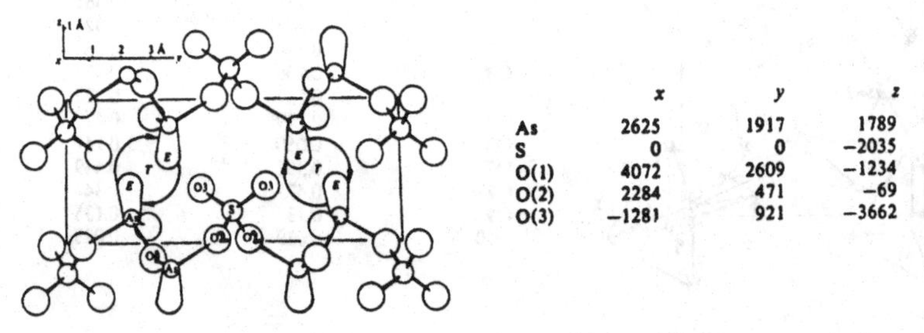

	x	y	z
As	2625	1917	1789
S	0	0	−2035
O(1)	4072	2609	−1234
O(2)	2284	471	−69
O(3)	−1281	921	−3662

Fig. 1. Structure of $As_2O_2SO_4$, and atomic positional parameters (x 10^4) [O1 and O2 labels are incorrect in the original paper].

$As_2O(SO_4)_2$ $As_2O_3.2SO_3$

II. R. MERCIER and J. DOUGLADE, 1982. Acta Cryst., B38, 1731-1735.

Monoclinic, Pc, a = 6.650, b = 6.671, c = 16.612 Å, β = 94.34°, Z = 4. Mo radiatio
R_w = 0.044 for 1066 reflexions.

The structure (Fig. 2) contains two independent isolated molecules in which an As-O-As group is linked to two SO_4 tetrahedra to give two six-membered rings. Mean As-OAs = 1.75, As-OS = 1.86, S-OAs = 1.53, S-O (terminal) = 1.42(3) Å, O-As-O = 95 (trigonal pyramid, with lone pair completing a tetrahedron), As-O-As = 129, As-O-S = 133°.

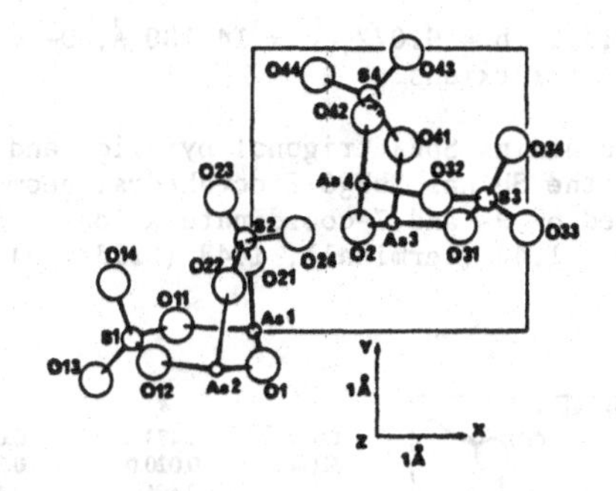

Fig. 2. Structure of $As_2O(SO_4)_2$.

μ-HYDROGENSULPHATO-μ-HYDROXO-μ-OXO-BIS[TRICHLOROANTIMONY(V)] HYDRATES
$Cl_3Sb(HOSO_3)(OH)(O)SbCl_3.xH_2O$ (x = 1.5 and 2)

S. BLÖSL, W. SCHWARZ and A. SCHMIDT, 1982. Z. anorg. Chem., 495, 165-176.

x = 1.5
Triclinic, P$\bar{1}$, a = 9.072, b = 11.382, c = 15.074 Å, α = 89.17, β = 85.06, γ = 82.94°, at 173K, Z = 4. Mo radiation, R = 0.054 for 4048 reflexions.

x = 2
Orthorhombic, Pban, a = 19.040, b = 18.024, c = 9.752 Å, Z = 8. Mo radiation, R = 0.061 for 1660 reflexions.

Both hydrates contain molecular complexes linked by hydrogen bonding which involves the water molecules (Fig. 1).

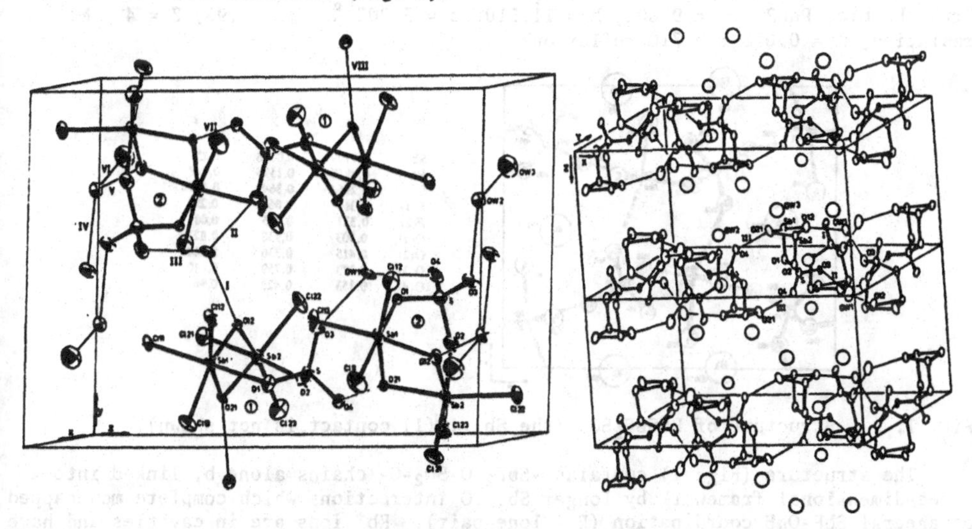

Fig. 1. Structures of hydrogensulphato-bridged antimony chloride hydrates, 1.5 H_2O (left) and $2H_2O$ (right).

POTASSIUM SULPHATOFLUOROANTIMONATE(III)
$K_2SO_4.SbF_3$

T. BIRCHALL, B. DUCOURANT, R. FOURCADE and G. MASCHERPA, 1982. J. Chem. Soc., Dalton, 2313-2316.

Orthorhombic, $P2_12_12_1$, a = 5.601, b = 9.072, c = 14.180 Å, D_m = 3.22, Z = 4. Mo radiation, R = 0.035 for 1016 reflexions.

The structure (Fig. 1) contains SbF_3 trigonal pyramids and SO_4 tetrahedra linked into a helix, so that the Sb has SbF_3O_2E octahedral geometry (E = lone-pair). The helices are linked by 9- and 7-coordinate K ions. Sb-F = 1.94, 1.98, 2.00, Sb-O = 2.35, 2.45, S-O = 1.44 (terminal), 1.48 (bridging) Å, Sb-O-S = 123°, K-F,O = 2.64-3.08 Å.

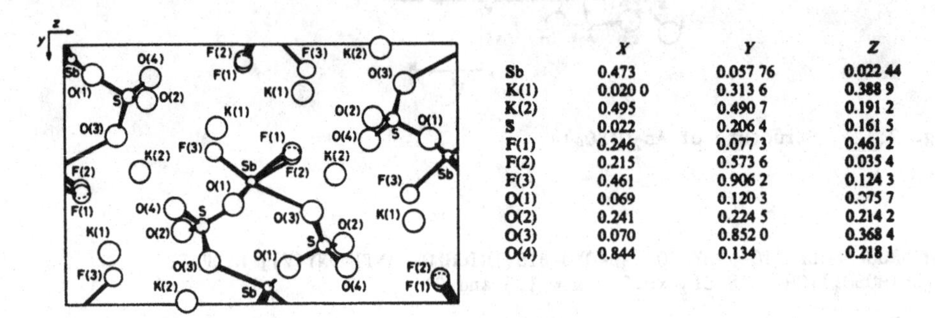

	X	Y	Z
Sb	0.473	0.057 76	0.022 44
K(1)	0.020 0	0.313 6	0.388 9
K(2)	0.495	0.490 7	0.191 2
S	0.022	0.206 4	0.161 5
F(1)	0.246	0.077 3	0.461 2
F(2)	0.215	0.573 6	0.035 4
F(3)	0.461	0.906 2	0.124 3
O(1)	0.069	0.120 3	0.375 7
O(2)	0.241	0.224 5	0.214 2
O(3)	0.070	0.852 0	0.368 4
O(4)	0.844	0.134	0.218 1

Fig. 1. Structure of $K_2SO_4.SbF_3$.

RUBIDIUM ANTIMONY(III) FLUORIDE SULPHATE
$RbSbF_2SO_4$

R. FOURCADE, M. BOURGAULT, B. BONNET and B. DUCOURANT, 1982. J. Solid State Chem., 43, 81-86.

Orthorhombic, $Pna2_1$, a = 9.601, b = 11.510, c = 5.202 Å, D_m = 3.93, Z = 4. Mo radiation, R = 0.078 for 710 reflexions.

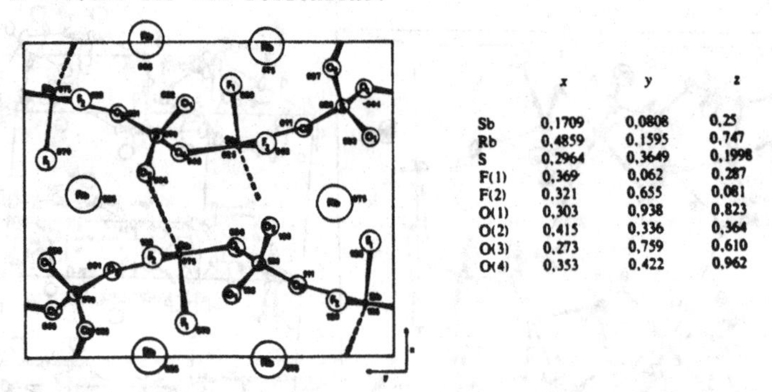

	x	y	z
Sb	0,1709	0,0808	0,25
Rb	0,4859	0,1595	0,747
S	0,2964	0,3649	0,1998
F(1)	0,369	0,062	0,287
F(2)	0,321	0,655	0,081
O(1)	0,303	0,938	0,823
O(2)	0,415	0,336	0,364
O(3)	0,273	0,759	0,610
O(4)	0,353	0,422	0,962

Fig. 1. Structure of $RbSbF_2SO_4$ (the Sb...O(1) contact is not shown).

The structure (Fig. 1) contains $-SbF_2-O-SO_2-O-$ chains along b, linked into a three-dimensional framework by longer Sb...O interactions which complete monocapped octahedral SbF_2O_4E coordination (E = lone-pair). Rb^+ ions are in cavities and have irregular 10-coordination to F and O. Sb-F = 1.93, Sb-O = 2.14, 2.25, Sb...O =

2.71, 3.04, S-O = 1.43-1.50(2) Å, F-Sb-F = 86, O-Sb-O = 166, Sb-O-S = 140°, Rb-F,O = 2.87-3.32 Å.

DI-μ-HYDROXO-BIS[AQUOSULPHATOBISMUTH(III)]
$Bi_2(H_2O)_2(SO_4)_2(OH)_2$ (I)

catena-DI-μ-HYDROXO-μ3-OXO-DIBISMUTH(III) SULPHATE
$[Bi_2O(OH)_2]SO_4$ (II)

I. M. GRAUNAR and F. LAZARINI, 1982. Acta Cryst., B38, 2879-2881.
II. L. GOLIČ, M. GRAUNAR and F. LAZARINI, 1982. Ibid., B38, 2881-2883.

I. Monoclinic, $P2_1/n$, a = 6.021, b = 13.363, c = 6.495 Å, β = 112.94°, D_m = 4.62, Z = 2. Mo radiation, R = 0.063 for 853 reflexions.

II. Monoclinic, $P2_1/c$, a = 7.692, b = 13.87, c = 5.688 Å, β = 109.01°, D_m = 6.5, Z = 4. Mo radiation, R = 0.039 for 1356 reflexions.

The structure of I (Fig. 1) contains a planar $Bi_2(OH)_2^{4+}$ group; each Bi is bonded to one sulphate oxygen atom and one water molecule, with the coordination sphere (6 + lone-pair) completed by two more-distant sulphate oxygens of neighbouring units. Bi-OH = 2.24, Bi-OS = 2.29, Bi-OH_2 = 2.44, Bi...OS = 2.58, 2.69, S-O = 1.43-1.54(3) Å. There is also a hydrogen bond, O-H...O = 2.65 Å.

The structure of II (Fig. 1) contains sulphate anions and $[Bi_2O(OH)_2^{2+}]_n$ double chains; Bi atoms have trigonal pyramidal and square pyramidal coordinations, Bi-O = 2.16-2.36 Å, with more-distant oxygens at 2.54-2.74 Å.

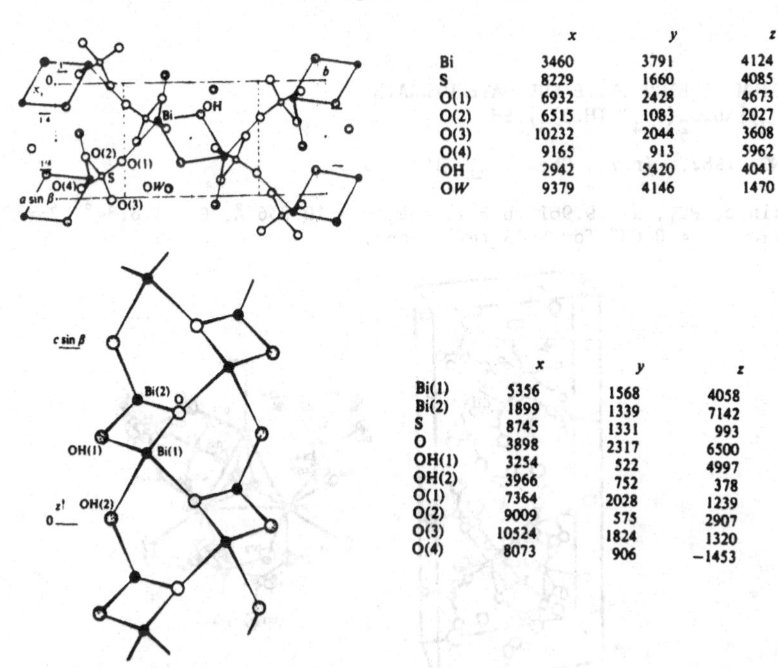

	x	y	z
Bi	3460	3791	4124
S	8229	1660	4085
O(1)	6932	2428	4673
O(2)	6515	1083	2027
O(3)	10232	2044	3608
O(4)	9165	913	5962
OH	2942	5420	4041
OW	9379	4146	1470

	x	y	z
Bi(1)	5356	1568	4058
Bi(2)	1899	1339	7142
S	8745	1331	993
O	3898	2317	6500
OH(1)	3254	522	4997
OH(2)	3966	752	378
O(1)	7364	2028	1239
O(2)	9009	575	2907
O(3)	10524	1824	1320
O(4)	8073	906	−1453

Fig. 1. Structures of $Bi_2(H_2O)_2(SO_4)_2(OH)_2$ (top) and $[Bi_2O(OH)_2]SO_4$ (bottom), with atomic positional parameters (x 10^4).

AMMONIUM HAFNIUM SULPHATE TETRAHYDRATE
$(NH_4)_4Hf(SO_4)_4.4H_2O$

D.L. ROGAČEV, L.M. DIKAREVA, V. Ja. KUZNECOV, V.V. FOMENKO and M.A. PORAJ-KOŠIC, 1982. Ž. Strukt. Khim., 23, No. 5, 130-133.

Monoclinic, C2/c, a = 22.264, b = 7.356, c = 18.065 Å, β = 138.22° [in text, 138.33° in abstract], Dm = 2.40, Z = 4. Mo radiation, R = 0.052 for 2330 reflexions.

The structure contains an $[Hf(SO_4)_4(H_2O)_2]^{4-}$ anion, NH_4^+ cations, and additional water molecules. Hf has 8-coordination to six sulphate oxygens (two bidentate and two unidentate groups) and two water molecules, Hf-O = 2.07, 2.22, 2.28, Hf-OH_2 = 2.20 Å. S-O = 1.43-1.52, N...O = 2.82-3.50 Å.

VANADYL SULPHATE TRIHYDRATE
$VOSO_4.3D_2O$

M. TACHEZ, F. THÉOBALD and A.W. HEWAT, 1982. Acta Cryst., B38, 1807-1809.

Monoclinic, $P2_1/n$, a = 7.397, b = 7.419, c = 12.076 Å, β = 106.51°, Z = 4. Neutron powder data.

The structure is as previously determined by X-ray methods (1). There are four normal and two weak hydrogen bonds, O...O = 2.69-2.78; 2.90, 3.00 Å; one water has tetrahedral, and two have trigonal environments.

1. Structure Reports, 39A, 311; 46A, 353.

POTASSIUM NIOBIUM OXIDE SULPHATE HYDRATE
$K_4(H_5O_2)[Nb_3O_2(SO_4)_6(H_2O)_3].5H_2O$

A. BINO, 1982. Inorg. Chem., 21, 1917-1920.

Monoclinic, $P2_1$, a = 9.961, b = 18.088, c = 10.036 Å, β = 118.84°, Z = 2. Mo radiation, R = 0.037 for 2645 reflexions.

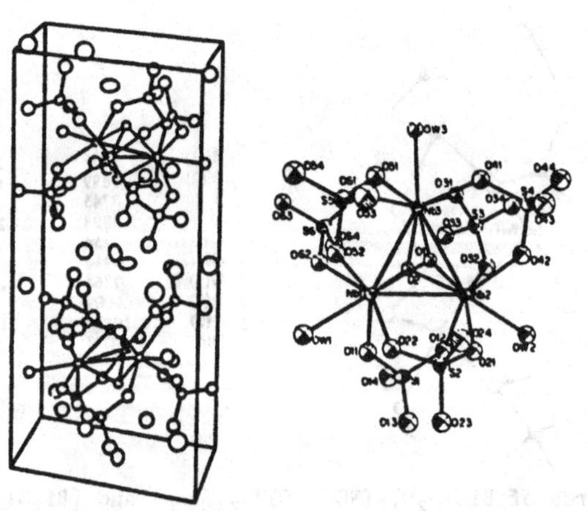

Fig. 1. Structure of the complex niobium sulphate.

The structure (Fig. 1) contains a discrete $Nb_3X_{17}^{5-}$ cluster, with a Nb_3 triangle triply bridged by two μ_3-O atoms, and with each pair of Nb atoms bridged by two μ-SO_4 groups. Nb-Nb = 2.875, 2.889, 2.892(1), Nb-O(μ_3) = 2.20-2.08, Nb-OS = 2.11-2.15, Nb-OH_2 = 2.22-2.25(1) Å. K ions have 6-8 oxygen neighbours, and the $H_5O_2^+$ ion and the water molecules participate in hydrogen bonding.

MANGANESE(II) SULPHATE PENTAHYDRATE
$MnSO_4.5H_2O$

R. CAMINITI, G. MARONGIU and G. PASCHINA, 1982. Z. Naturforsch., 37A, 581-586.

Triclinic, $P\bar{1}$, a = 6.36, b = 10.77, c = 6.16 Å, α = 80.3, β = 110.1, γ = 106.0°, Dm = 2.07, Z = 2. Mo radiation, R = 0.034 for 1334 reflexions.

Isostructural with the Mg (1) and Cu (2) salts. S-O = 1.456-1.480(5), Mn-O = 2.136-2.191(8) Å.

1. Structure Reports, 38A, 331.
2. Strukturbericht, 3, 102, 449; Structure Reports, 27, 614; 41A, 353.

μ-HYPEROXO-BIS[PENTAAMMINECOBALT(III)] HYDROGENSULPHATE BISSULPHATE TRIHYDRATE
$[Co_2(NH_3)_{10}(O_2)](HSO_4)(SO_4)_2.3H_2O$

W.P. SCHAEFER, S.E. EALICK, D. FINLEY and R.E. MARSH, 1982. Acta Cryst., B38, 2232-2235.

Monoclinic, $P2_1/c$, a = 13.392, b = 9.742, c = 17.709 Å, β = 100.98°, Z = 4. Mo radiation, R = 0.109 for 3640 reflexions.

The structure (Fig. 1) contains two independent centrosymmetric cations, and disordered sulphate anions and water molecules. Co-O = 1.90(1), Co-N = 1.93-1.98(1), O-O = 1.27(1) Å, Co-O-O = 117°. Many H_2O/NH_3...O contacts probably correspond to hydrogen bonds, including O(4)...O(8) = 2.43, 2.53 Å involving the hydrogensulphate proton.

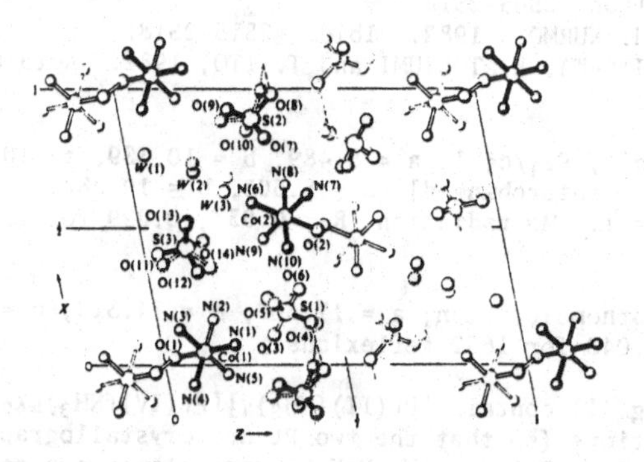

Fig. 1. Structure of the μ-hyperoxo-cobalt sulphate.

TETRAAMMINEPLATINUM(II) TETRAAMMINEDICHLOROPLATINUM(IV) HYDROGENSULPHATE
$[Pt(NH_3)_4][Pt(NH_3)_4Cl_2](HSO_4)_4$

P.E. FANWICK and J.L. HUCKABY, 1982. Inorg. Chem., 21, 3067-3071.

Monoclinic, $P2_1/c$, a = 5.466, b = 10.280, c = 10.633 Å, β = 93.16°, Z = 1. Mo radiation, R = 0.034 for 636 reflexions. [The bromo compound is isostructural (1).]

Atomic position (Cl occupancy = 0.5)

	x	y	z
Pt	0	0	0
Cl	0.423	0.0084	0.0027
S	0.5346	-0.1385	0.3561
O1	0.720	-0.1986	0.2849
O2	0.556	0.0098	0.330
O3	0.289	-0.1773	0.3214
O4	0.589	-0.1570	0.4925
N1	0.008	0.057	0.1854
N2	0.043	-0.1917	0.0540

The structure contains chains along a of disordered square-planar $Pt(II)(NH_3)_4^{2+}$ and octahedral $Pt(IV)(NH_3)_4Cl_2^{2+}$ ions with eclipsed ammines in neighbouring ions; these chains are linked by N-H...O hydrogen bonding via pairs of O-H...O bonded bisulphate anions. Pt-N = 2.06(1), Pt(IV)-Cl = 2.31(1), Pt(II)...Cl = 3.15(1), S-OH = 1.56(1), S-O = 1.43-1.48(1), N-H...O = 2.92-3.47, O-H...O = 2.58 Å.

1. Following report.

TETRAAMMINEPLATINUM(II) TETRAAMMINEDIBROMOPLATINUM(IV) HYDROGENSULPHATE
$[Pt(NH_3)_4][Pt(NH_3)_4Br_2](HSO_4)_4$

TETRAAMMINEPLATINUM(II) TETRAAMMINEDIIODOPLATINUM(IV) HYDROGENSULPHATE HYDROXIDE MONOHYDRATE
$[Pt(NH_3)_4][Pt(NH_3)_4I_2](HSO_4)_3(OH).H_2O$

I. R.J.H. CLARK, M. KURMOO, A.M.R. GALAS and M.B. HURSTHOUSE, 1982. J. Chem. Soc., Dalton, 2505-2513.
II. R.J.H. CLARK and M. KURMOO, 1982. Ibid., 2515-2518.
III. M. TANAKA, I. TSUJIKAWA, K. TORIUMI and T. ITO, 1982. Acta Cryst., B38, 2793-2797.

Bromo compound, monoclinic, $P2_1/c$; I, a = 5.489, b = 10.339, c = 10.648 Å, β = 92.57°; III [with a and c interchanged], a = 5.505, b = 10.282, c = 10.654 Å, β = 93.17°, at 269K; Z = 1. Mo radiation, R = 0.037, 0.029 for 1288, 1448 reflexions.

Iodo compound (I), orthorhombic, Pmcn, a = 11.667, b = 14.319, c = 15.374 Å, Z = 4. Mo radiation, R = 0.048 for 1652 reflexions.

The structures (Fig. 1) contain $[Pt(II)(NH_3)_4][Pt(IV)(NH_3)_4X_2]$ chains with disordered Br positions (so that the two Pt are crystallographically equivalent) but ordered I positions; the PtN_4 square planes are eclipsed in the bromo compound, but staggered in the iodo compound. The HSO_4^{2-} anions are hydrogen bonded in pairs in the bromo compound; in the iodo compound there is some anion disorder, and probably OH^- ions and water molecules. Pt(IV)-Br = 2.467, Pt(II)...Br = 3.022, Pt(IV)-I = 2.686, Pt(II)...I = 3.148(4), Pt-N = 2.01-2.11(2) Å.

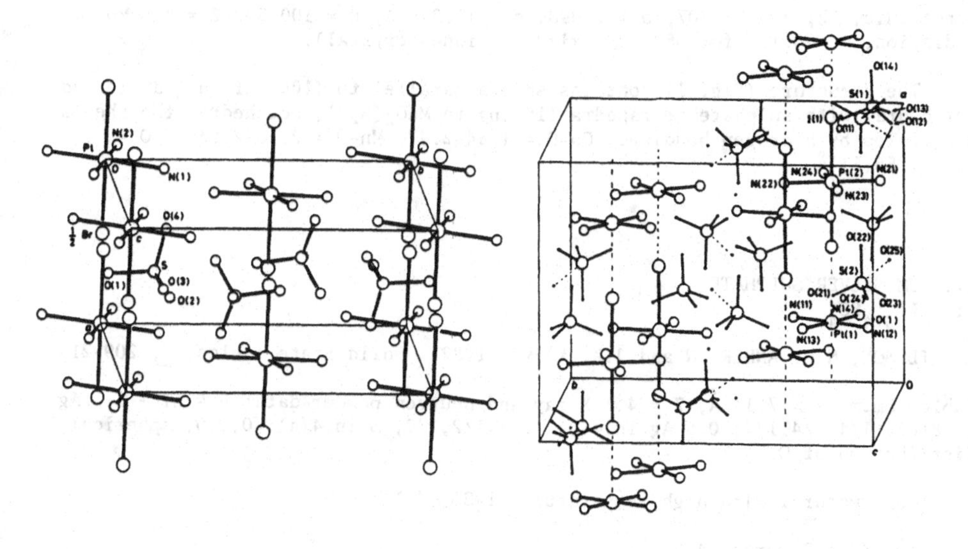

Fig. 1. Structures of the bromo (left) and iodo (right) platinum complexes.

CAMPIGLIAITE

$Cu_4Mn(SO_4)_2(OH)_6.4H_2O$

S. MENCHETTI and C. SABELLI, 1982. Amer. Min., 67, 385-393.

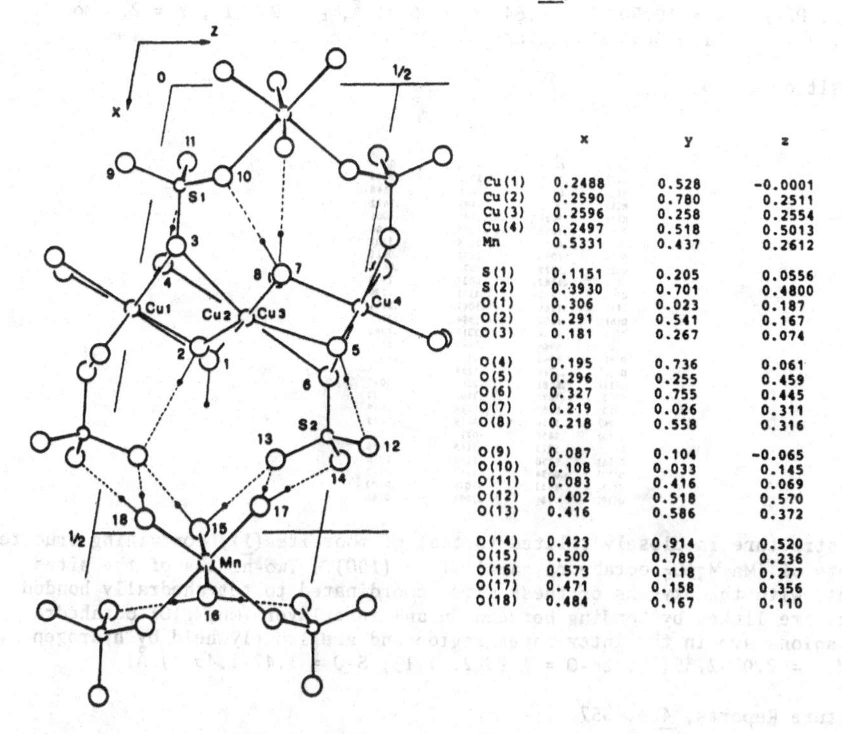

	x	y	z
Cu(1)	0.2488	0.528	-0.0001
Cu(2)	0.2590	0.780	0.2511
Cu(3)	0.2596	0.258	0.2554
Cu(4)	0.2497	0.518	0.5013
Mn	0.5331	0.437	0.2612
S(1)	0.1151	0.205	0.0556
S(2)	0.3930	0.701	0.4800
O(1)	0.306	0.023	0.187
O(2)	0.291	0.541	0.167
O(3)	0.181	0.267	0.074
O(4)	0.195	0.736	0.061
O(5)	0.296	0.255	0.459
O(6)	0.327	0.755	0.445
O(7)	0.219	0.026	0.311
O(8)	0.218	0.558	0.316
O(9)	0.087	0.104	-0.065
O(10)	0.108	0.033	0.145
O(11)	0.083	0.416	0.069
O(12)	0.402	0.518	0.570
O(13)	0.416	0.586	0.372
O(14)	0.423	0.914	0.520
O(15)	0.500	0.789	0.236
O(16)	0.573	0.118	0.277
O(17)	0.471	0.158	0.356
O(18)	0.484	0.167	0.110

Fig. 1. Structure of campigliaite.

Monoclinic, C2, a = 21.707, b = 6.098, c = 11.245 Å, β = 100.3°, Z = 4. Mo radiation, R = 0.124 for 689 reflexions (twinned crystal).

The structure (Fig. 1) contains sheets parallel to (100) of CuO_6 distorted octahedra, with sulphate tetrahedra linking to $MnO_2(H_2O)_4$ octahedra; the sheets are linked by hydrogen bonding. Cu-O = 1.84-2.47, Mn-O = 2.13-2.53, S-O = 1.45-1.56(5) Å.

LITHIUM SILVER SULPHATE
$Li_{1.6}Ag_{0.4}SO_4$

L. NILSSON, N.H. ANDERSEN and J.K. KJEMS, 1982. Solid State Ionics, 6, 209-214.

Cubic, Fm3m, a = 7.14 Å, Z = 4. X-ray and neutron powder data; 6.4 Li + 1.1 Ag in 8(c): 1/4,1/4,1/4; 0.5 Ag in 4(b): 1/2,1/2,1/2; S in 4(a): 0,0,0; spherical distribution of O.

Isostructural with high-temperature Li_2SO_4 (1).

1. Structure Reports, 48A, 372.

LAWSONBAUERITE
$(Mn,Mg)_9Zn_4(SO_4)_2(OH)_{22}.8H_2O$

A.H. TREIMAN and D.R. PEACOR, 1982. Amer. Min., 67, 1029-1034.

Monoclinic, $P2_1/c$, a = 10.50, b = 9.64, c = 16.41 Å, β = 95.21°, Z = 2. Mo radiation, R = 0.12 for 3045 reflexions.

Atomic positions

	x	y	z
M(1)	0	0	0
M(2)	.1551	.3340	.9999
M(3)	.0123	.1676	.1688
M(4)	.0025	.4998	.1672
M(5)	.5021	.4582	.3006
Zn(1)	.1828	.3355	.3466
Zn(2)	.8372	.3329	.3237
O(1)	.1060	.1602	.0611
O(2)	.1029	.5077	.0613
O(3)	.1182	.0037	.2303
O(4)	.1270	.3351	.2290
O(5)	.1164	.5067	.3951
O(6)	.1290	.1594	.3965
O(7)	.9136	.3314	.1080
O(8)	.8964	.1646	.2690
O(9)	.8943	.3321	.4404
O(10)	.3714	.3269	.3536
O(11)	.6481	.3083	.3070
O(12)	.3895	.3966	.1733
O(13)	.6250	.1284	.1939
O(14)	.3950	.0419	.0825
O(15)	.3850	.0723	.2870
S	.5973	.3148	.0138
O(16)	.6250	.1260	.4350
O(17)	.6656	.1804	.0240
O(18)	.6453	.4074	.0183
O(19)	.4586	.2888	.0164

The structure is closely related to that of mooreite (1), containing brucite-like sheets of $(Mn,Mg)O_6$ octahedra parallel to (100). Two-ninths of the sites are vacant, with the oxygens of these sites coordinated to tetrahedrally bonded Zn; sheets are linked by bonding between Zn and interlayer $(Mn,Mg)O_6$ octahedra. Sulphate anions are in the inter-sheet region and are loosely held by hydrogen bonds. M-O = 2.07-2.39(1), Zn-O = 1.95-2.01(1), S-O = 1.47-1.49(1) Å.

1. Structure Reports, 46A, 357.

CADMIUM HYDROXIDE SULPHATE
α-$Cd_2(OH)_2SO_4$

M. LOUËR, D. LOUËR and D. GRANDJEAN, 1982. Acta Cryst., B38, 909-912.

Monoclinic, P2/n (pseudo P4/n), a = 10.020, b = 4.858, c = 10.019 Å, β = 90.07°, D_m = 4.81, Z = 4. Mo radiation, R = 0.069 for 549 reflexions.

The structure (Fig. 1) contains sulphate tetrahedra linked by sharing corners with distorted edge- and corner-sharing CdO_6 octahedra; the sulphate groups are disordered. S-O = 1.45-1.53, Cd-O = 2.20-2.63, O-H...O = 2.90 Å, The β- and γ-forms have been described previously (1).

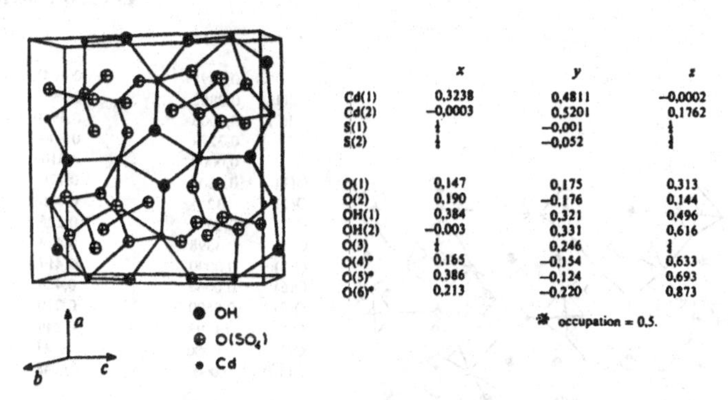

	x	y	z
Cd(1)	0,3238	0,4811	−0,0002
Cd(2)	−0,0003	0,5201	0,1762
S(1)	$\frac{1}{4}$	−0,001	$\frac{1}{4}$
S(2)	$\frac{1}{4}$	−0,052	$\frac{1}{4}$
O(1)	0,147	0,175	0,313
O(2)	0,190	−0,176	0,144
OH(1)	0,384	0,321	0,496
OH(2)	−0,003	0,331	0,616
O(3)	$\frac{1}{4}$	0,246	$\frac{1}{4}$
O(4)*	0,165	−0,154	0,633
O(5)*	0,386	−0,124	0,693
O(6)*	0,213	−0,220	0,873

* occupation = 0.5.

Fig. 1. Structure of cadmium hydroxide sulphate.

1. Structure Reports, 41A, 355; 42A. 375.

CADMIUM HYDROXIDE SULPHATE HYDRATE
$Cd_8(OH)_{12}(SO_4)_2.H_2O$

D. LOUËR, J. LABARRE, J.-P. AUFFREDIC and M. LOUËR, 1982. Acta Cryst., B38, 1079-1084.

Monoclinic, C2/c, a = 27.180, b = 5.855, c = 14.825 Å, β = 124.96°, D_m = 4.51, Z = 4. Mo radiation, R = 0.051 for 2174 reflexions.

The structure (Fig. 1) is built up from SO_4 tetrahedra, CdO_6 octahedra, and one CdO_5 pyramid, sharing edges and corners. S-O = 1.45-1.50, Cd-O = 2.15-2.75(1) Å.

	x	y	z
Cd(1)	0,34725	0,49209	−0,02782
Cd(2)	0,26555	0,01422	−0,14702
Cd(3)	0,07086	0,49490	−0,09034
Cd(4)	0	0	0
Cd(5)	0	0,05326	$\frac{1}{4}$
S	0,1310	0,3279	−0,2975
O(1)	0,1845	0,2495	−0,1907
O(2)	0,3558	0,1804	0.3829
O(3)	−0,0791	0,1875	−0,1677
O(4)	−0,1204	0,4330	0,2834
O(5)	0,2604	0,3190	0,4332
O(6)	0,2069	0,1753	0,1985
O(7)	0,3347	0,1442	0,0220
O(8)	−0,0006	0,2502	0,1172
O(9)	0,0969	0,1735	0,3689
O(10)	−0,0719	0,2009	0,4937
O(11)	0	−0,3129	$\frac{1}{4}$

Fig. 1. Structure of $Cd_8(OH)_{12}(SO_4)_2.H_2O$.

POTASSIUM CADMIUM HYDROXIDE SULPHATE DIHYDRATE
CAESIUM CADMIUM HYDROXIDE SULPHATE DIHYDRATE
$M_2Cd_3(OH)_2(SO_4)_3.2H_2O$ (M = K, Cs)

M. LOUËR and D. LOUËR, 1982. Rev. Chim. Minér., 19, 162-171.

Orthorhombic, $Cmc2_1$, a = 18.657, 19.376, b = 7.918, 8.114, c = 9.94, 10.132 Å, Z = 4. Mo radiation, R = 0.093, 0.056 for 334, 756 reflexions.

The structures (Fig. 1) contain SO_4 tetrahedra linked by $CdO_3(OH)_2(H_2O)$ octahedra into layers normal to a; the layers are linked by 10-coordinate M^+ ions and hydrogen bonds. S-O = 1.42-1.56, Cd-O = 2.21-2.40, Cs-O = 3.19-3.43(2) Å.

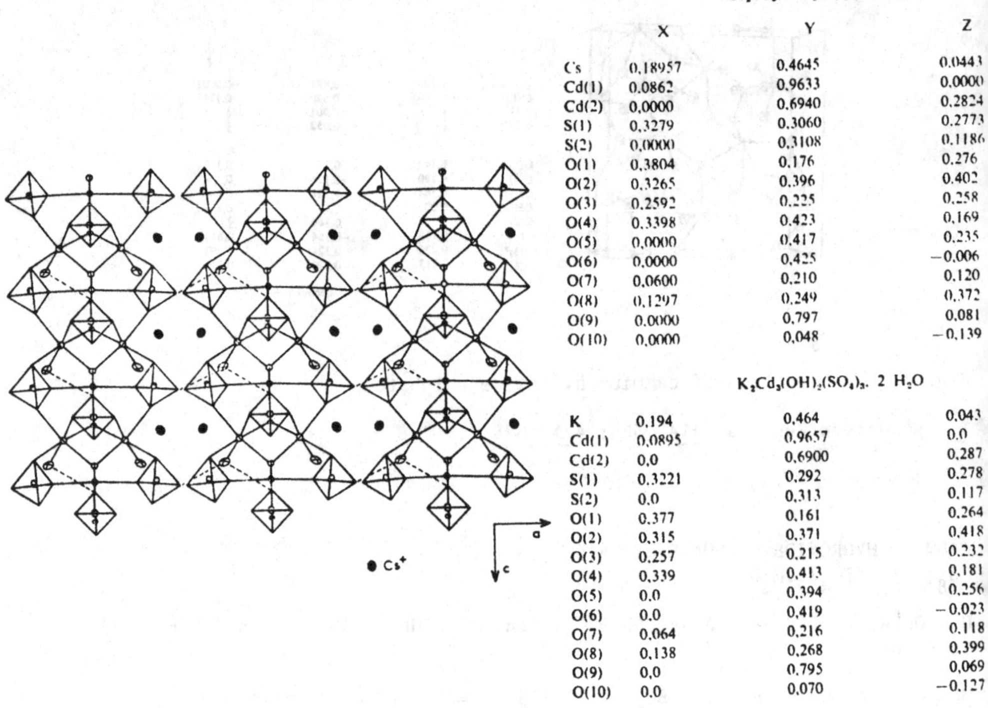

$Cs_2Cd_3(OH)_2(SO_4)_3.\ 2\ H_2O$

	X	Y	Z
Cs	0,18957	0,4645	0,0443
Cd(1)	0.0862	0,9633	0,0000
Cd(2)	0,0000	0,6940	0,2824
S(1)	0,3279	0,3060	0,2773
S(2)	0,0000	0,3108	0,1186
O(1)	0,3804	0,176	0,276
O(2)	0,3265	0,396	0,402
O(3)	0,2592	0,225	0,258
O(4)	0,3398	0,423	0,169
O(5)	0,0000	0,417	0,235
O(6)	0,0000	0,425	−0,006
O(7)	0,0600	0,210	0,120
O(8)	0,1297	0,249	0,372
O(9)	0,0000	0,797	0,081
O(10)	0,0000	0,048	−0,139

$K_2Cd_3(OH)_2(SO_4)_3.\ 2\ H_2O$

	X	Y	Z
K	0,194	0,464	0,043
Cd(1)	0,0895	0,9657	0,0
Cd(2)	0,0	0,6900	0,287
S(1)	0,3221	0,292	0,278
S(2)	0,0	0,313	0,117
O(1)	0,377	0,161	0,264
O(2)	0,315	0,371	0,418
O(3)	0,257	0,215	0,232
O(4)	0,339	0,413	0,181
O(5)	0,0	0,394	0,256
O(6)	0,0	0,419	−0,023
O(7)	0,064	0,216	0,118
O(8)	0,138	0,268	0,399
O(9)	0,0	0,795	0,069
O(10)	0,0	0,070	−0,127

Fig. 1. Structure of $Cs_2Cd_3(OH)_2(SO_4)_3.2H_2O$, and atomic positional parameters for Cs and K salts.

RUBIDIUM PRASEODYMIUM SULPHATE TETRAHYDRATE
$RbPr(SO_4)_2.4H_2O$

L.D. ISKHAKOVA, Z.A. STARIKOVA and V.K. TRUNOV, 1981. Koord. Khim,, 7, 1713-1718.

Monoclinic, $P2_1/c$, a = 6.622, b = 18.997, c = 8.740 Å, β = 96.16°, Z = 4. R = 0.032.

Isostructural with related materials (1). Pr has 9-coordination (tri-capped trigonal prism) to six sulphate and three water oxygens; these polyhedra are linked by the sulphate groups and by 13-coordinate Rb ions.

1. Structure Reports, 40A, 267; 42A, 377; 46A, 417.

SODIUM THULIUM SULPHATE
α-$NaTm(SO_4)_2$

S.M. ČIŽOV, A.N. POKROVSKIJ and L.M. KOVBA, 1982. Kristallografija, 27, 997-998 [Soviet Physics - Crystallography, 27, 598-599].

Monoclinic, $P2_1/m$, a = 4.669, b = 10.143, c = 6.837 Å, γ = 110.40°, Z = 2. Mo radiation, R = 0.060 for 803 reflexions. High-temperature form.

Atomic positions (x 10^4)

		x	y	z
Tm	2e	3297	1789	2500
S_1	2e	0770	4077	2500
S_2	2e	4300	0723	7500
Na	2e	2124	3486	7500
O_1	2e	2631	9383	2500
O_2	2e	7864	0758	2500
O_3	4f	1601	3380	4206
O_4	4f	3880	1435	5761
O_5	2e	7439	3766	2500
O_6	2e	2289	5572	2500

Isostructural with the Er salt (1). S-O = 1.43-1.49, Tm-O = 2.25-2.68 (8-coordination), Na-O = 2.26-2.98(1) Å (8-coordination).

1. Structure Reports, 44A, 277.

URANYL SULPHATE HYDRATE
$UO_2SO_4.H_2SO_4.5H_2O$

NEPTUNYL SULPHATE HYDRATE
$2NpO_2SO_4.H_2SO_4.4H_2O$

N.W. ALCOCK, M.M. ROBERTS and D. BROWN, 1982. J. Chem. Soc., Dalton, 869-873.

Uranyl compound, monoclinic, C2/c, a = 15.619, b = 8.242, c = 11.008 Å, β = 113.71°, Z = 4. Mo radiation, R = 0.037 for 985 reflexions.

Neptunyl compound, orthorhombic, $P2_12_12$, a = 9.474, b = 10.065, c = 8.409 Å, Z = 2. Mo radiation, R = 0.038 for 1007 reflexions.

Both structures (Fig. 1) exhibit pentagonal-bipyramidal coordination at the metal atom, with bidentate bridging sulphates in the uranyl compound, and tri- and quadridentate sulphates in the neptunyl compound. U-O = 1.78(uranyl), 2.36-2.46, Np-O = 1.74, 2.36-2.44(1) Å.

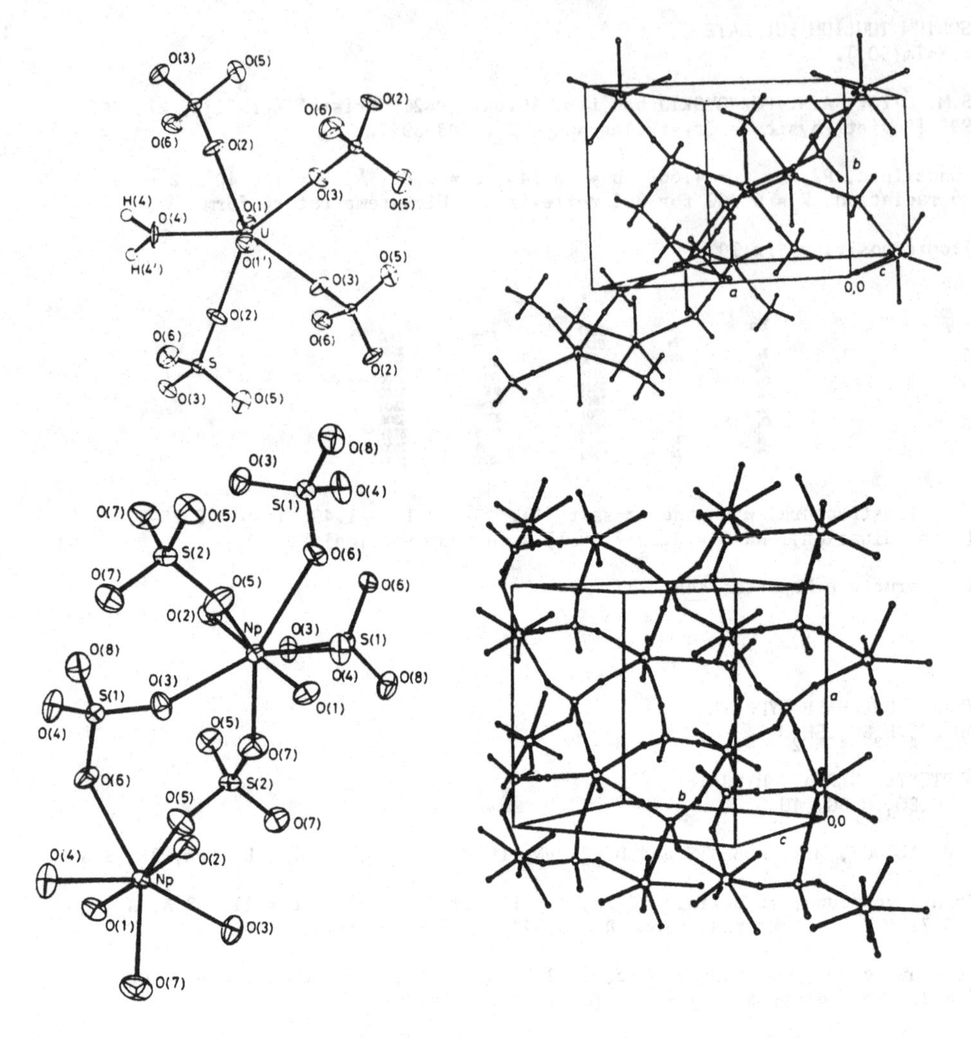

Fig. 1. Structures of uranyl (top) and neptunyl (bottom) sulphate hydrates.

JOHANNITE
$Cu(UO_2)_2(OH)_2(SO_4)_2.8H_2O$

K. MEREITER, 1982. Tschermaks Min. Petr. Mitt., 30, 47-57.

Triclinic, $P\bar{1}$, a = 8.903, b = 9.499, c = 6.812 Å, α = 109.87, β = 112.01, γ = 100.40°, Z = 1. Mo radiation, R = 0.039 for 2005 reflexions.

The structure (Fig. 1) contains pairs of OH-edge-sharing $UO_2(OH)_2O_3$ pentagonal bipyramids, linked by SO_4 tetrahedra to form $(UO_2)_2(OH)_2(SO_4)_2$ layers parallel to (100). The layers are connected by elongated $Cu(H_2O)_4O_2$ octahedra and by hydrogen bonding via water molecules. U-O = 1.78 (uranyl), 2.34-2.39, Cu-O = 1.97 (x 4), 2.40 (x 2), mean S-O = 1.47, O-H...O = 2.71-2.91, 3.30 Å.

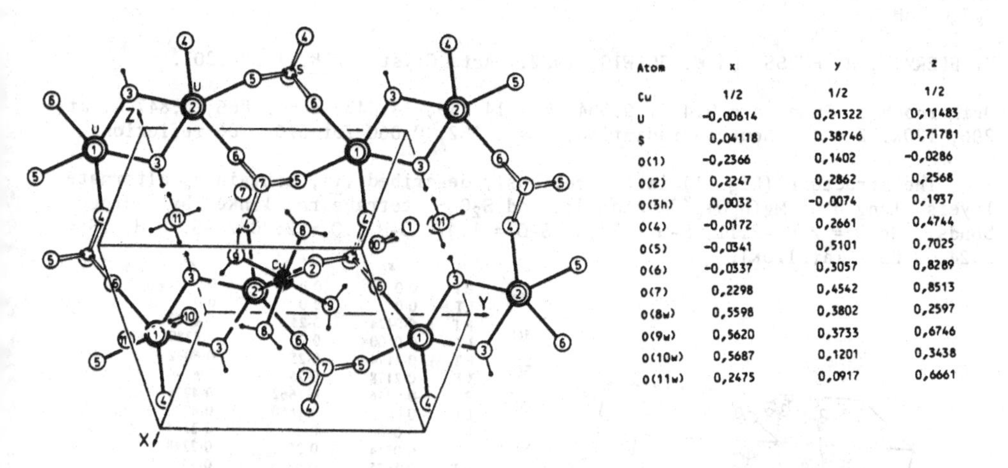

Atom	x	y	z
Cu	1/2	1/2	1/2
U	-0,00614	0,21322	0,11483
S	0,04118	0,38746	0,71781
O(1)	-0,2366	0,1402	-0,0286
O(2)	0,2247	0,2862	0,2568
O(3h)	0,0032	-0,0074	0,1937
O(4)	-0,0172	0,2661	0,4744
O(5)	-0,0341	0,5101	0,7025
O(6)	-0,0337	0,3057	0,8289
O(7)	0,2298	0,4542	0,8513
O(8w)	0,5598	0,3802	0,2597
O(9w)	0,5620	0,3733	0,6746
O(10w)	0,5687	0,1201	0,3438
O(11w)	0,2475	0,0917	0,6661

Fig. 1. Structure of johannite.

ZINC URANYL SULPHATE HYDRATE
$Zn(UO_2)_2SO_4(OH)_4.1\cdot 5H_2O$

V.I. SPICYN, L.M. KOVBA, V.V. TABAČENKO, N.V. TABAČENKO and Ju.N. MIKHAILOV, 1982. Izv. Akad. Nauk SSSR, Ser. Khim., No. 4, 807-812.

Monoclinic, B2/m, a = 8.654, b = 17.714, c = 14.182 Å, γ = 103.92°, Z = 8. R = 0.078.

The structure contains $UO_2(OH)_3O_2$ pentagonal bipyramids, SO_4 tetrahedra, and tridentate OH^- groups. Zn has distorted octahedral coordination to 2 OH, 2 H_2O, and 2 uranyl O atoms.

PLUTONIUM(IV) SULPHATE TETRAHYDRATE
$Pu(SO_4)_2.4H_2O$

N.C. JAYADEVAN, K.D. SINGH MUDHER and D.M. CHACKRABURTTY, 1982. Z. Kristallogr., 161, 7-13.

α-Form, orthorhombic, Fddd, a = 26.527, b = 11.995, c = 5.687 Å, Z = 8. No structure analysis.

β-Form, orthorhombic, Pnma, a = 14.544, b = 10.980, c = 5.667 Å, Z = 4. Cu radiation, R = 0.089 for 476 reflexions (films, visual intensities).

The structure of the β-form contains sulphate tetrahedra, linked by $PuO_4(OH_2)_4$ square antiprisms, and by hydrogen bonds via the water molecules. S-O = 1.48-1.52(3), Pu-OS = 2.31-2.34, Pu-OH_2 = 2.31-2.41(4), O-H...O = 2.61-2.83(5) Å. The α-form is isostructural with $Zr(SO_4)_2.4H_2O$ (1).

1. Structure Reports, 23, 448; 24, 382.

MAGNESIUM THIOSULPHATE HEXAHYDRATE
$MgS_2O_3.6H_2O$

Y. ELERMAN, H. FUESS and W. JOSWIG, 1982. Acta Cryst., B38, 1799-1801.

Orthorhombic, Pnma, a = 9.405, 9.304, b = 14.449, 14.447, c = 6.866, 6.847 Å, at 296, 120K, Z = 4. Neutron radiation, R = 0.062, 0.060 for 516, 706 reflexions.

The structure (Fig. 1) is as previously described (1), containing alternate layers along b of $Mg(H_2O)_6^{2+}$ octahedra and $S_2O_3^{2-}$ tetrahedra, linked by hydrogen bonds. Mg-O = 2.04-2.11, S-S = 2.05, S-O = 1.47, O-H...O = 2.70-2.95, O-H...S = 3.24, 3.25 Å (at 120K).

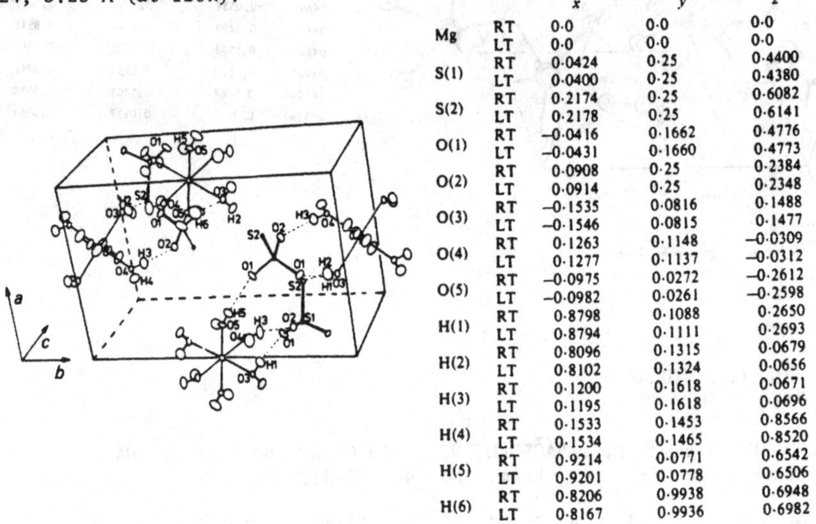

		x	y	z
Mg	RT	0·0	0·0	0·0
	LT	0·0	0·0	0·0
S(1)	RT	0·0424	0·25	0·4400
	LT	0·0400	0·25	0·4380
S(2)	RT	0·2174	0·25	0·6082
	LT	0·2178	0·25	0·6141
O(1)	RT	−0·0416	0·1662	0·4776
	LT	−0·0431	0·1660	0·4773
O(2)	RT	0·0908	0·25	0·2384
	LT	0·0914	0·25	0·2348
O(3)	RT	−0·1535	0·0816	0·1488
	LT	−0·1546	0·0815	0·1477
O(4)	RT	0·1263	0·1148	−0·0309
	LT	0·1277	0·1137	−0·0312
O(5)	RT	−0·0975	0·0272	−0·2612
	LT	−0·0982	0·0261	−0·2598
H(1)	RT	0·8798	0·1088	0·2650
	LT	0·8794	0·1111	0·2693
H(2)	RT	0·8096	0·1315	0·0679
	LT	0·8102	0·1324	0·0656
H(3)	RT	0·1200	0·1618	0·0671
	LT	0·1195	0·1618	0·0696
H(4)	RT	0·1533	0·1453	0·8566
	LT	0·1534	0·1465	0·8520
H(5)	RT	0·9214	0·0771	0·6542
	LT	0·9201	0·0778	0·6506
H(6)	RT	0·8206	0·9938	0·6948
	LT	0·8167	0·9936	0·6982

Fig. 1. Structure of magnesium thiosulphate hexahydrate, and atomic positional parameters at 296K (RT) and 120K (LT).

1. Structure Reports, 10, 148; 27, 631; 34A, 314.

TIN(II) DITHIONITE
$Sn_2(S_2O_4)_2$

A. MAGNUSSON and L.-G. JOHANSSON, 1982. Acta Chem. Scand., A36, 429-433.

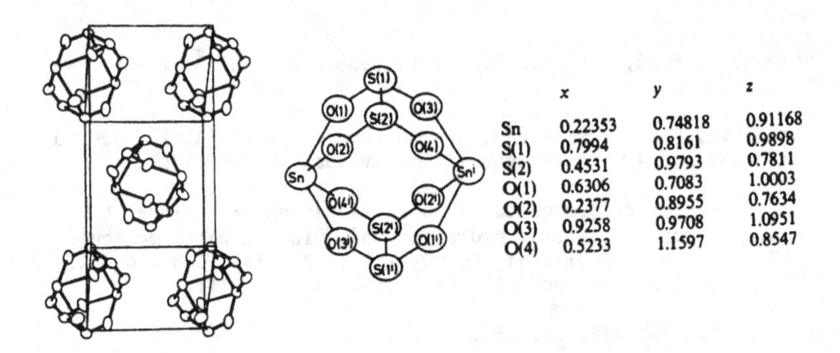

	x	y	z
Sn	0.22353	0.74818	0.91168
S(1)	0.7994	0.8161	0.9898
S(2)	0.4531	0.9793	0.7811
O(1)	0.6306	0.7083	1.0003
O(2)	0.2377	0.8955	0.7634
O(3)	0.9258	0.9708	1.0951
O(4)	0.5233	1.1597	0.8547

Fig. 1. Structure of tin(II) dithionite.

Monoclinic, $P2_1/c$, a = 7.015, b = 7.480, c = 12.652 Å, β = 133.59°, Z = 2. Mo radiation, R = 0.037 for 1264 reflexions.

The structure (Fig. 1) contains a dimeric complex with approximate D_{2h} symmetry, Sn atoms having square pyramidal coordination with a lone-pair at the apex. Sn-O = 2.237-2.323(3) Å.

POTASSIUM μ-(THIOSULPHITO-S,S')-DECACYANODICOBALTATE(III) HYDRATE ETHANOLATE
$K_4[Co_2SSO_2(CN)_{10}].6H_2O.0{\cdot}4(C_2H_5OH)$

F.R. FRONCZEK, R.E. MARSH and W.P. SCHAEFER, 1982. J. Amer. Chem. Soc., 104, 3382-3385.

Monoclinic, C2/c, a = 36.860, b = 9.241, c = 20.772 Å, β = 124.10°, D_m = 1.83, Z = 8. Mo radiation, R = 0.052 for 3250 reflexions.

The structure contains binuclear anions, $[(NC)_5Co\text{-}S\text{-}S(O_2)\text{-}Co(CN)_5]^{4-}$; Co-S(thio) = 2.297, Co-SO_2 = 2.255(2), S-S = 2.064(3), S-O = 1.468(4), Co-N = 1.87-1.90(1) Å. One K ion and most of the solvent molecules are disordered.

SODIUM TETRAAMMINENICKEL BISTHIOSULPHATOARGENTATE(I) AMMINE
$Na_4[Ni(NH_3)_4][Ag(S_2O_3)_2]_2.0{\cdot}3NH_3$

R. STOMBERG, I.-B. SVENSSON, A.A.G. TOMLINSON and I. PERSDOTTER, 1982. Acta Chem. Scand., A36, 579-582.

Tetragonal, I4/m, a = 13.996, c = 5.709 Å, Z = 2. Cu radiation, R = 0.044 for 469 reflexions.

Atomic positions

			x	y	z
Ni	in	2(a)	0	0	0
Ag		4(d)	0	1/2	1/4
S(1)		8(h)	0.0937	0.2636	1/2
S(2)		8(h)	0.1233	0.4061	1/2
Na		8(h)	0.2930	0.1664	1/2
O(1)		8(h)	-0.0085	0.2442	1/2
O(2)		16(i)	0.1405	0.2251	0.292
N(1)		8(h)	0.3660	0.4633	1/2
0.16 N(2)		4(e)	0	0	0.364

The structure is similar to that of the compound with five NH_3 (1). It contains an anion chain along c, with tetrahedrally-coordinated Ag ions bridged by two thiosulphate groups, Ag-S = 2.596(2) Å, Ag-S-Ag = 66.7°, S-S = 2.037(3), S-O = 1.455, 1.459(7) Å. Ni has square-planar coordination, Ni-N(1) = 1.944(7) Å, with a small amount of octahedral coordination, Ni-N(2) = 2.08(6) Å. Na has 3 S at 3.10-3.41 Å and 3 O at 2.31-2.58 Å.

1. Structure Reports, 39A, 316.

AMMONIUM DEUTERIUM SELENATE
ND_4DSeO_4

RUBIDIUM DEUTERIUM SELENATE
$RbDSeO_4$

A. WAŚKOWSKA and Z. CZAPLA, 1982. Acta Cryst., B38, 2017-2020.

Orthorhombic, $P2_12_12_1$, a = 12.847, 12.887, b = 4.595, 4.599, c = 7.503, 7.515 Å, Z = 4. Mo radiation, R = 0.041, 0.061 for 1772, 1724 reflexions.

The structures of the hydrogen compounds have been described previously (1, 2). Replacement of hydrogen by deuterium changes the crystal symmetry and consequently the ferroelectric properties. The structures (Fig. 1) contain $SeO_3(OD)^-$ ions linked into chains along b by O-D...O hydrogen bonds, with the chains linked by ND_4^+ (D not located) or Rb^+ ions. Se-O = 1.592-1.629, Se-OD = 1.714(3) Å, O-Se-O = 112.1-114.1, O-Se-OD = 103.8-108.9°.

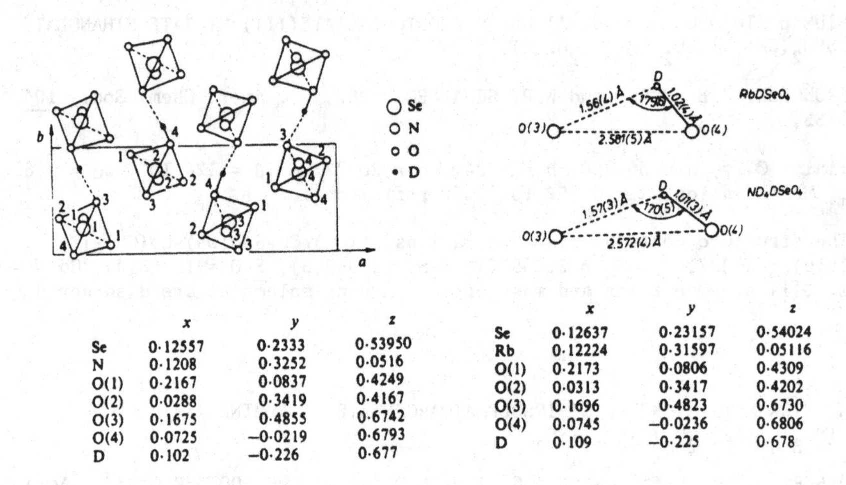

	x	y	z
Se	0·12557	0·2333	0·53950
N	0·1208	0·3252	0·0516
O(1)	0·2167	0·0837	0·4249
O(2)	0·0288	0·3419	0·4167
O(3)	0·1675	0·4855	0·6742
O(4)	0·0725	−0·0219	0·6793
D	0·102	−0·226	0·677

	x	y	z
Se	0·12637	0·23157	0·54024
Rb	0·12224	0·31597	0·05116
O(1)	0·2173	0·0806	0·4309
O(2)	0·0313	0·3417	0·4202
O(3)	0·1696	0·4823	0·6730
O(4)	0·0745	−0·0236	0·6806
D	0·109	−0·225	0·678

Fig. 1. Structures of $MDSeO_4$, M = ND_4 and Rb.

1. Structure Reports, 46A, 410; 48A, 320.
2. Ibid., 44A, 281; 46A, 366.

LITHIUM TRIHYDROGEN SELENITE (FERROELECTRIC)
$LiD_3(SeO_3)_2$

R. LIMINGA and R. TELLGREN, 1982. Acta Cryst., B38, 1551-1554.

Monoclinic, Pn, a = 6.2473, b = 7.9030, c = 5.4471 Å, β = 104.995°, Z = 2. Neutron radiation, R = 0.021 for 818 reflexions.

The structure is as previously described (1). Deuteration lengthens O...O by 0.01-0.04 Å and shortens O-D.

1. Structure Reports, 24, 393; 37A, 314; 38A, 344; 45A, 345.

AMMONIUM LITHIUM SELENATE
NH_4LiSeO_4

A. WAŚKOWSKA and R. ALLMANN, 1982. Cryst. Struct. Comm., 11, 2029-2034.

Orthorhombic, $Pbn2_1$, a = 5.278, b = 17.346, c = 5.116 Å, Z = 4. Mo radiation, R = 0.059 for 1091 reflexions.

The structure (Fig. 1) contains layers of corner-sharing SeO_4 and LiO_4 tetrahedra, linked by ammonium ions. Se-O = 1.56-1.84, Li-O = 1.86-2.11, N...O = 2.77-

3.28 Å.

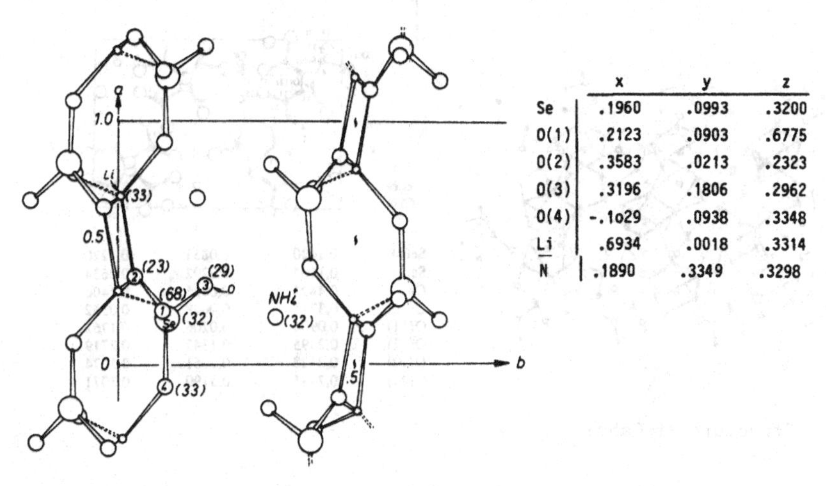

	x	y	z
Se	.1960	.0993	.3200
O(1)	.2123	.0903	.6775
O(2)	.3583	.0213	.2323
O(3)	.3196	.1806	.2962
O(4)	-.1o29	.0938	.3348
Li	.6934	.0018	.3314
N	.1890	.3349	.3298

Fig. 1. Structure of ammonium lithium selenate.

SODIUM TRIHYDROGEN SELENITE
$NaD_3(SeO_3)_2$

R.K. McMULLAN, R. THOMAS and J.F. NAGLE, 1982. J. Chem. Phys., 77, 537-547.

Paraelectric and ferroelectric phases, monoclinic, $P2_1/n$, Pn, a = 10.365, 10.314, b = 4.850, 9.663, c = 5.792, 5.768 Å, β = 91.16, 91.23°, Z = 2, 4, at 298, 173K. Neutron radiation, $R(F^2)$ = 0.038, 0.035 for 748, 1636 reflexions.

The paraelectric structure is as previously described (1), with disordered hydrogen positions. In the ferroelectric phase hydrogen positions become ordered.

1. Structure Reports, 22, 473; 33A, 389; 37A, 315; 38A, 345; 41A, 358; 43A, 284.

CALCIUM DISELENATE(IV)
$CaSe_2O_5$

C. DELAGE, A. CARPY and M. GOURSOLLE, 1982. Acta Cryst., B38, 1278-1280.

Orthorhombic, Pbca, a = 6.492, b = 14.521, c = 10.170 Å, D_m = 3.85, Z = 8. Mo radiation, R = 0.038 for 706 reflexions.

The structure (Fig. 1) contains zigzag chains along c of edge-sharing CaO_6 prisms, linked by $Se_2O_5^{2-}$ ions which consist of two SeO_3 trigonal pyramids sharing an oxygen atom. Se-O = 1.80, 1.86 (shared), 1.64-1.69(1) (terminal) Å, O-Se-O = 96-104, Se-O-Se = 124°, Ca-O = 2.34-2.54(1) Å.

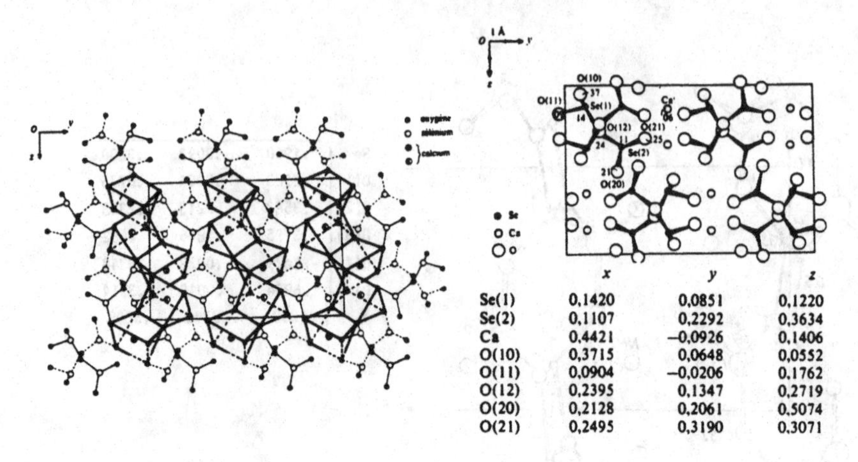

	x	y	z
Se(1)	0,1420	0,0851	0,1220
Se(2)	0,1107	0,2292	0,3634
Ca	0,4421	−0,0926	0,1406
O(10)	0,3715	0,0648	0,0552
O(11)	0,0904	−0,0206	0,1762
O(12)	0,2395	0,1347	0,2719
O(20)	0,2128	0,2061	0,5074
O(21)	0,2495	0,3190	0,3071

Fig. 1. Structure of $CaSe_2O_5$.

GOLD(III) SELENITE CHLORIDE
$Au(SeO_3)Cl$

P.G. JONES, M. KRAUSHAAR, E. SCHWARZMANN and G.M. SHELDRICK, 1982. Z. Naturforsch., 37B, 941-943.

Triclinic, $P\bar{1}$, a = 4.185, b = 9.756, c = 10.608 Å, α = 80.57, β = 88.85, γ = 87.66°, Z = 4. Mo radiation, R = 0.060 for 2273 reflexions.

The structure (Fig. 1) contains SeO_3 trigonal pyramids, linked by square-planar AuO_4, AuO_2Cl_2, and AuO_3Cl groups. Se-O = 1.70-1.72(1) Å, O-Se-O = 88-104°, Au-O = 1.97-2.00, Au-Cl = 2.25, 2.28(1) Å.

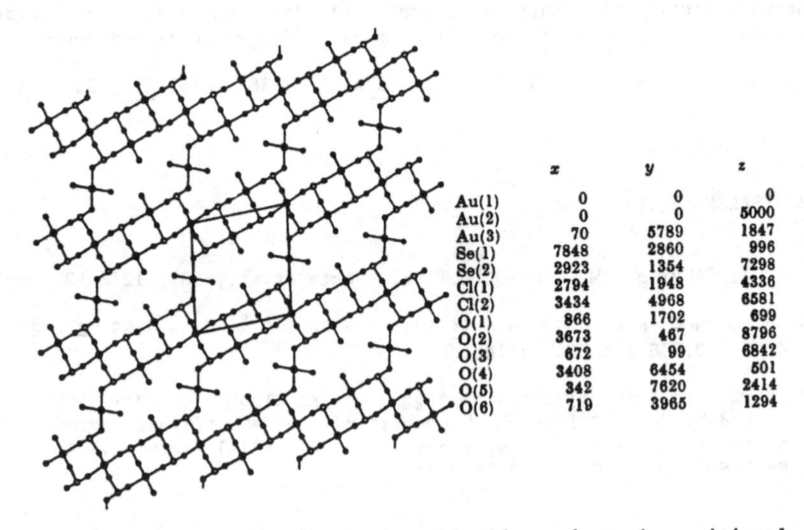

	x	y	z
Au(1)	0	0	0
Au(2)	0	0	5000
Au(3)	70	5789	1847
Se(1)	7848	2860	996
Se(2)	2923	1354	7298
Cl(1)	2794	1948	4336
Cl(2)	3434	4968	6581
O(1)	866	1702	699
O(2)	3673	467	8796
O(3)	672	99	6842
O(4)	3408	6454	501
O(5)	342	7620	2414
O(6)	719	3965	1294

Fig. 1. Structure of gold(III) selenite chloride, and atomic positional parameters (x 10^4).

SODIUM POTASSIUM TELLURITE TRIHYDRATE
$NaKTeO_3.3H_2O$

F. DANIEL, J. MORET, M. MAURIN and E. PHILIPPOT, 1982. Acta Cryst., B38, 703-706.

Trigonal, P31c, a = 6.550, c = 8.980 Å, Z = 2. R = 0.046 for 299 reflexions.

The structure (Fig. 1) contains layers parallel to (001) of edge-sharing $NaO_3(OH_2)_3$ octahedra and $KO_3(OH_2)_6$ tricapped trigonal prisms; the layers are linked by face sharing and hydrogen bonding to give a three-dimensional arrangement, with tunnels which contain trigonal pyramidal TeO_3^{2-} groups. Te-O = 1.83(1) Å, O-Te-O = 101.5°, Na-O = 2.35, 2.54, K-O = 2.86, 3.05, 3.05, O-H...O = 2.72, 2.80 Å.

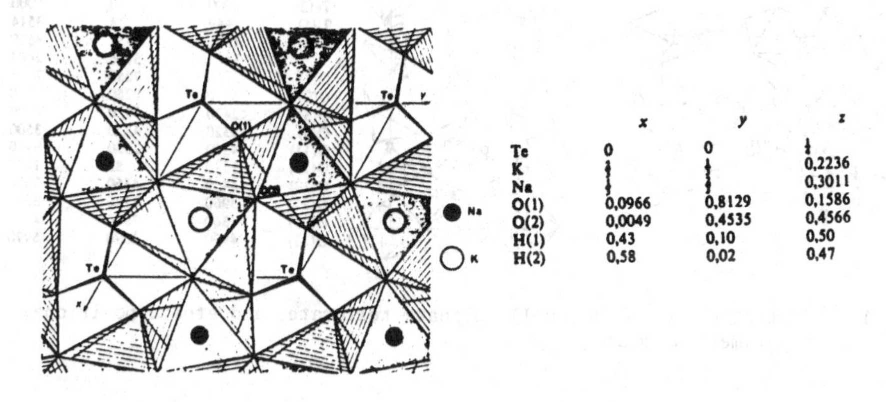

	x	y	z
Te	0	0	1/4
K	2/3	1/3	0,2236
Na	1/3	2/3	0,3011
O(1)	0,0966	0,8129	0,1586
O(2)	0,0049	0,4535	0,4566
H(1)	0,43	0,10	0,50
H(2)	0,58	0,02	0,47

Fig. 1. Structure of $NaKTeO_3.3H_2O$.

CALCIUM ORTHOTELLURATE(VI)
Ca_3TeO_6

CADMIUM ORTHOTELLURATE(VI)
Cd_3TeO_6

H.-G. BURCKHARDT, C. PLATTE and M. TRÖMEL, 1982. Acta Cryst., B38, 2450-2452.

Monoclinic, $P2_1/n$, a = 5.5782, 5.4986, b = 5.7998, 5.6383, c = 8.017, 8.0191 Å, β = 90.217, 90.00°, Dm = 4.17, 7.2, Z = 2. Mo radiation, R = 0.033, 0.032 for 625, 677 reflexions.

Atomic positions

		x	y	z
Cd_3TeO_6				
Te	2 (a)	0,0	0,0	0,0
Cd(1)	2 (b)	0,0	0,0	0,5
Cd(2)	4 (c)	0,5106	−0,0457	0,24306
O(1)	4 (c)	0,1200	0,0704	0,2231
O(2)	4 (c)	−0,2860	0,1760	0,0509
O(3)	4 (c)	0,1635	0,2810	−0,0775
Ca_3TeO_6				
Te	2 (a)	0,0	0,0	0,0
Ca(1)	2 (b)	0,0	0,0	0,5
Ca(2)	4 (c)	0,5143	−0,0525	0,2459
O(1)	4 (c)	0,1097	0,0529	0,2249
O(2)	4 (c)	−0,2820	0,1760	0,0476
O(3)	4 (c)	0,1700	0,2744	−0,0676

Cryolite-type structures (1). Te-O = 1.90-1.95 (octahedral), Ca-O = 2.27-2.32 (octahedral), 2.34-2.96 (8-coordination), Cd-O = 2.21-2.35, 2.24-3.01 Å.

1. Strukturbericht, 6, 29, 120; Structure Reports, 41A, 154.

THALLIUM(I) SULPHATE TELLURATE
$Te(OH)_6.Tl_2SO_4$

R. ZILBER, A. DURIF and M.T. AVERBUCH-POUCHOT, 1982. Acta Cryst., B38, 1554-1556.

Monoclinic, $P2_1/a$, a = 12.053, b = 7.205, c = 12.354 Å, β = 110.85°, Z = 4. Ag radiation, R = 0.055 for 1399 reflexions.

The structure (Fig. 1) contains $Te(OH)_6$ octahedra and SO_4 tetrahedra, linked by 8- and 7-coordinate Tl ions. Te-O = 1.89-1.94, S-O = 1.44-1.47, Tl-O = 2.81-3.27(1) Å.

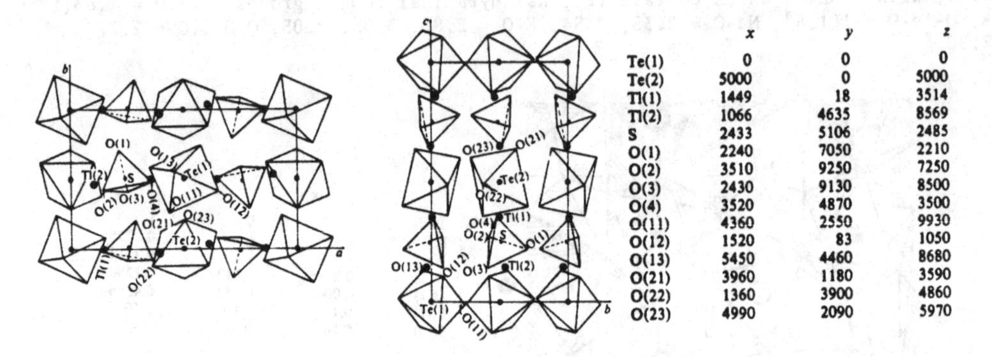

	x	y	z
Te(1)	0	0	0
Te(2)	5000	0	5000
Tl(1)	1449	18	3514
Tl(2)	1066	4635	8569
S	2433	5106	2485
O(1)	2240	7050	2210
O(2)	3510	9250	7250
O(3)	2430	9130	8500
O(4)	3520	4870	3500
O(11)	4360	2550	9930
O(12)	1520	83	1050
O(13)	5450	4460	8680
O(21)	3960	1180	3590
O(22)	1360	3900	4860
O(23)	4990	2090	5970

Fig. 1. Structure of thallium(I) sulphate tellurate, and atomic positional parameters (x 10^4).

AMMONIUM TETRAMETAPHOSPHATE-TELLURATE DIHYDRATE
$(NH_4)_4P_4O_{12}.2Te(OH)_6.2H_2O$

A. DURIF, M.T. AVERBUCH-POUCHOT and J.C. GUITEL, 1982. J. Solid State Chem., 41, 153-159.

Triclinic, $P\bar{1}$, a = 11.845, b = 8.554, c = 7.433 Å, α = 66.28, β = 95.91, γ = 76.00°, Z = 1. Ag radiation, R = 0.021 for 3713 reflexions.

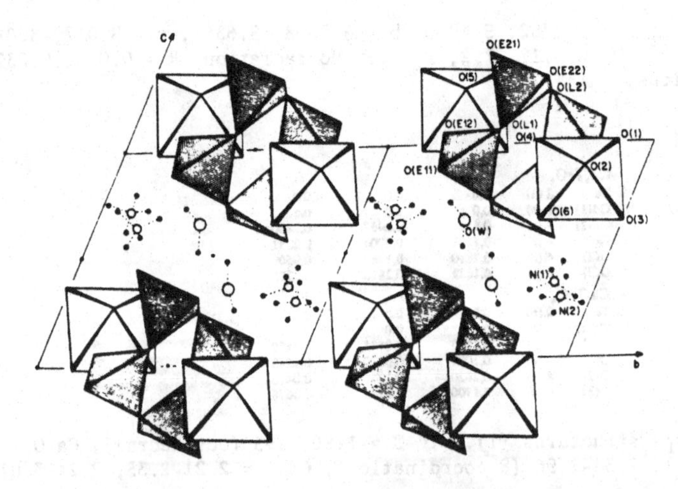

Fig. 1. Structure of $(NH_4)_4P_4O_{12}.2Te(OH)_6.2H_2O$.

The structure (Fig. 1) contains octahedral $Te(OH)_6$ groups and centrosymmetric cyclic $P_4O_{12}^{4-}$ anions, linked by hydrogen bonding which also involves the ammonium ions and water molecules. Te-O = 1.86-1.96, P-O = 1.48-1.53 (terminal), 1.54-1.61 Å (bridging), P-O-P = 124, 137°.

RUBIDIUM TELLURATE TRIMETAPHOSPHATE MONOHYDRATE
$Te(OH)_6.Rb_3P_3O_9.H_2O$

N. BOUDJADA and A. DURIF, 1982. Acta Cryst., B38, 595-597.

Monoclinic, $P2_1/a$, a = 15.56, b = 8.358, c = 13.72 Å, β = 113.27°, Z = 4. Ag radiation, R = 0.037 for 1597 reflexions.

As in related Na and K compounds (1), the structure (Fig. 1) contains cyclic P_3O_9 and octahedral TeO_6 groups, linked by 8- and 9-coordinate Rb ions. P-O = 1.61-1.63 (ring), 1.47-1.49(1) Å (terminal), P-O-P = 127-130°, Te-O = 1.91-1.93, Rb-O = 2.81-3.49 Å.

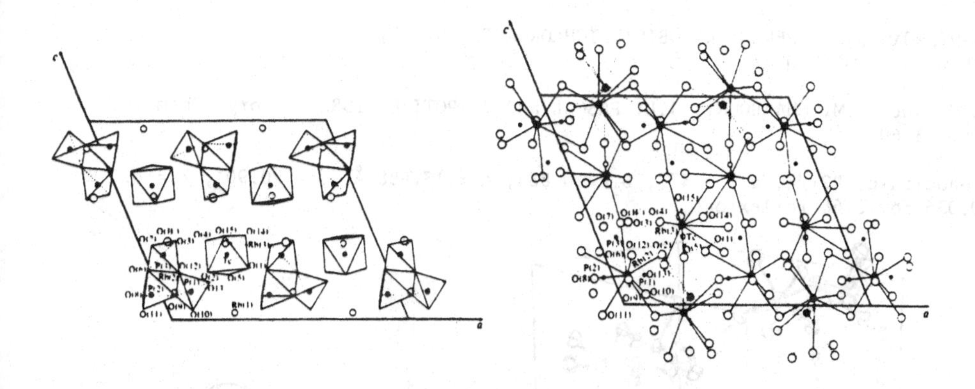

Fig. 1. Structure of $Te(OH)_6.Rb_3P_3O_9.H_2O$.

1. Structure Reports, 48A, 327.

LITHIUM PERCHLORATE TRIHYDRATE (PYROELECTRIC)
$LiClO_4.3H_2O$

J.-O. LUNDGREN, R. LIMINGA and R. TELLGREN, 1982. Acta Cryst., B38, 15-20.

Hexagonal, $P6_3mc$, a = 7.7192, c = 5.4531 Å, D_m = 1.89, Z = 2. Neutron radiation, R = 0.047 for 1284 reflexions. H parameters are (-0.0666,0.2633,0.5338).

Structure as previously described (1).

1. Strukturbericht, 3, 117, 468; Structure Reports, 33A, 450; 41A, 362; 43A, 293.

POTASSIUM PERCHLORATE
$KClO_4$

J.W. BATS and H. FUESS, 1982. Acta Cryst., B38, 2116-2120.

Orthorhombic, Pnma, a = 8.765, b = 5.620, c = 7.205 Å, at 120K, Z = 4. Mo radiation, R = 0.042, 0.026 for 1884, 1153 reflexions (two crystals), and neutron radiation, R = 0.055 for 831 reflexions.

Atomic positions

	x	y	z
K	0.18038	1/4	0.33865
Cl	0.06954	1/4	0.81160
O(1)	0.19372	1/4	0.94289
O(2)	0.08098	0.04086	0.69491
O(3)	-0.07410	1/4	0.90607

The structure is as previously described (1). Cl-O = 1.440-1.456(1) Å (corrected for libration), O-Cl-O = 108.5-110.6°, K-12 O = 2.825-3.445 Å. Deformation density corresponds to bonding electrons and oxygen lone pairs.

1. Strukturbericht, 1, 344, 372; 2, 84, 411, 414; Structure Reports, 21, 358; 43A, 293.

μ-HYDROXO-μ-OXO-PERCHLORATOBIS(TRICHLOROANTIMONY(V))
$Sb_2Cl_6(OH)(O)(ClO_4)$

C.H. BELIN, M. CHAABOUNI, J.L. PASCAL and J. POTIER, 1982. Inorg. Chem., 21, 3557-3560.

Monoclinic, $P2_1/n$, a = 9.477, b = 20.884, c = 13.588 Å, β = 90.90°, Z = 8. R = 0.036 for 3168 reflexions.

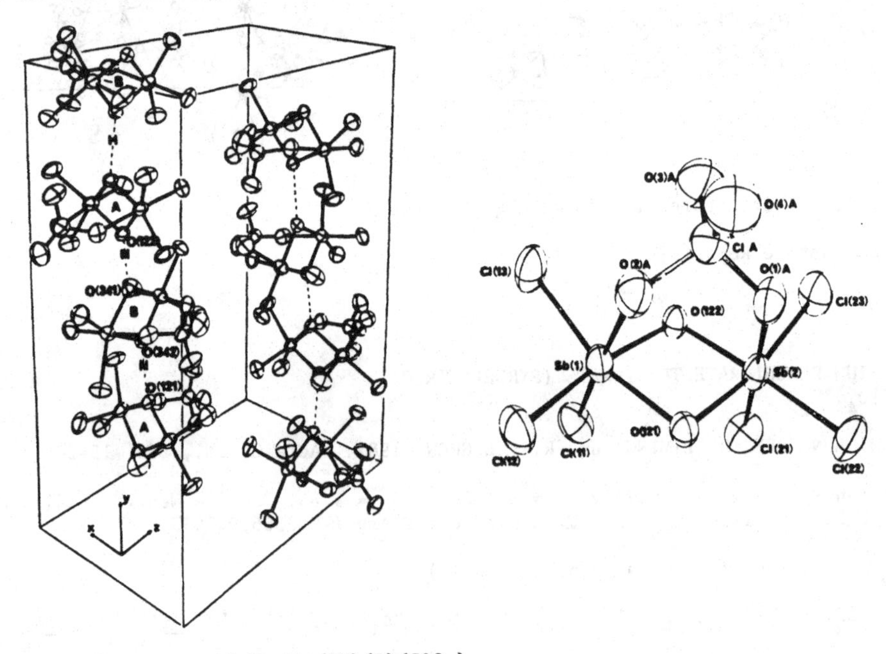

Fig. 1. Structure of $Sb_2Cl_6(OH)(O)(ClO_4)$.

The structure (Fig. 1) contains two independent molecules, each of which contains two distorted $SbCl_3O_3$ octahedra, with the Sb atoms bridged by O, OH, and a strongly-coordinated bidentate perchlorate group. Sb-Cl = 2.27-2.29, Sb-O(oxo) = 1.97, Sb-OH = 2.06, Sb-O(perchlorate) = 2.23, Cl-O = 1.48 (bridging), 1.38(1) (terminal) Å. Molecules are linked by hydrogen bonding, O-H...O = 2.65 Å

CAESIUM PERBROMATE
$CsBrO_4$

I. E. GEBERT, S.W. PETERSON, A.H. REIS and E.H. APPELMAN, 1981. J. Inorg. Nucl. Chem., 43, 3085-3089.
II. Idem, 1982. Polyhedron, 1, 567.

Tetragonal, $I4_1/a$, a = 5.751, c = 14.821 Å, Z = 4. Mo radiation, R = 0.085 for 385 reflexions. Cs in 4(a): 0,0,0; Br in 4(b): 0,0,1/2; O in 16(f): 0.0007,0.2299, 0.4389 (origin at $\bar{4}$).

The structure contains tetrahedral BrO_4^- ions linked by 12-coordinate Cs^+ ions. Br-O = 1.610(7) Å, O-Br-O = 111.0 (x 2), 108.7° (x 4), Cs-O = 3.194-3.392(7) Å.

CADMIUM BROMATE DIHYDRATE
$Cd(BrO_3)_2.2H_2O$

V.V. SATYANARAYANA MURTY, M. SESHASAYEE and B.V.R. MURTHY, 1981. Indian J. Phys., 55A, 310-315.

Orthorhombic, $P2_12_12_1$, a = 12.493, b = 6.170, c = 9.238 Å, D_m = 3.58, Z = 4. Mo radiation, R = 0.045 for 912 reflexions.

The structure (Fig. 1) contains trigonal pyramidal BrO_3^- anions linked by Cd ions, which have 7-coordination to 5 O and 2 H_2O. Mean Br-O = 1.66 Å, O-Br-O = 103°; Cd-O = 2.29-2.54 Å.

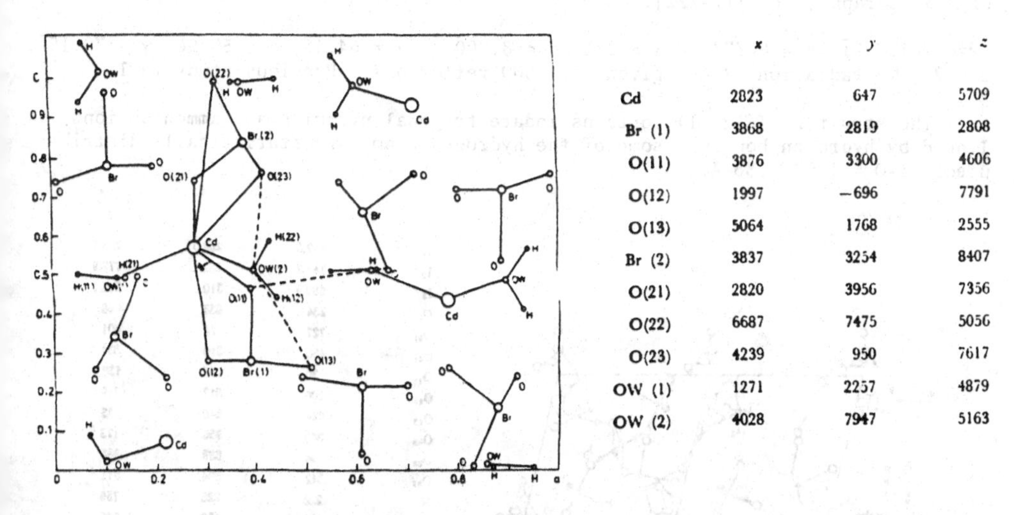

	x	y	z
Cd	2023	647	5709
Br (1)	3868	2819	2808
O(11)	3876	3300	4606
O(12)	1997	−696	7791
O(13)	5064	1768	2555
Br (2)	3837	3254	8407
O(21)	2820	3956	7356
O(22)	6687	7475	5056
O(23)	4239	950	7617
OW (1)	1271	2257	4879
OW (2)	4028	7947	5163

Fig. 1. Structure of cadmium bromate dihydrate, and atomic positional parameters (x 10^4).

TERBIUM BROMATE ENNEAHYDRATE
$Tb(BrO_3)_3.9H_2O$

J.C. GALLUCCI, R.E. GERKIN and W.J. REPPART, 1982. Cryst. Struct. Comm., 11, 1141-1145.

Hexagonal, $P6_3/mmc$, a = 11.755, c = 6.712 Å, Z = 2. Mo radiation, R = 0.024 for 441 reflexions.

Atomic positions

		x	y	z
Tb	in 2(c)	1/3	2/3	1/4
Br	6(h)	0.1302	0.2604	3/4
O(1)	12(k)	0.4200	0.8401	0.0068
O(2)	6(h)	0.2120	0.4241	1/4
O(3)	12(j)	0.2925	0.3608	3/4
0.5 O(4)	12(k)	0.0935	0.1870	0.9491
H(1)	24(ℓ)	0.3911	0.8898	-0.0433
H(2)	12(j)	0.1171	0.3715	1/4

Isostructural with the Nd, Sm, Pr, and Yb salts (1). The structure contains columns of bromate ions and columns of $Tb(H_2O)_9$ tricapped trigonal prisms, linked by hydrogen bonding. Tb-O = 2.40, 2.47, Br-O = 1.67, 1.53 (O(4) disordered), O-H...O = 2.83, 2.86 Å, O-Br-O = 104, 107°.

1. Strukturbericht, 7, 29, 135; Structure Reports, 34A, 362; 43A, 295.

AMMONIUM HYDROGEN IODATE
$NH_4IO_3.2HIO_3$

A.I. BARANOV, G.F. DOBRŽANSKIJ, V.V. ILJUKHIN, V.S. RJABKIN, Ju.N. SOKOLOV, N.I. SOROKINA and L.A. ŠUVALOV, 1981. Kristallografija, 26, 1259-1268 [Soviet Physics - Crystallography, 26, 717-722].

Triclinic, $P\bar{1}$, a = 8.374, b = 8.311, c = 8.200 Å, α = 64.45, β = 59.96, γ = 70.15°, Z = 2. Mo radiation, R not given for 3000 reflexions. Previous study in 1.

The structure (Fig. 1) contains iodate trigonal pyramids and ammonium ions linked by hydrogen bonding; some of the hydrogen atoms are statistically distributed. I-O = 1.75-1.86 Å.

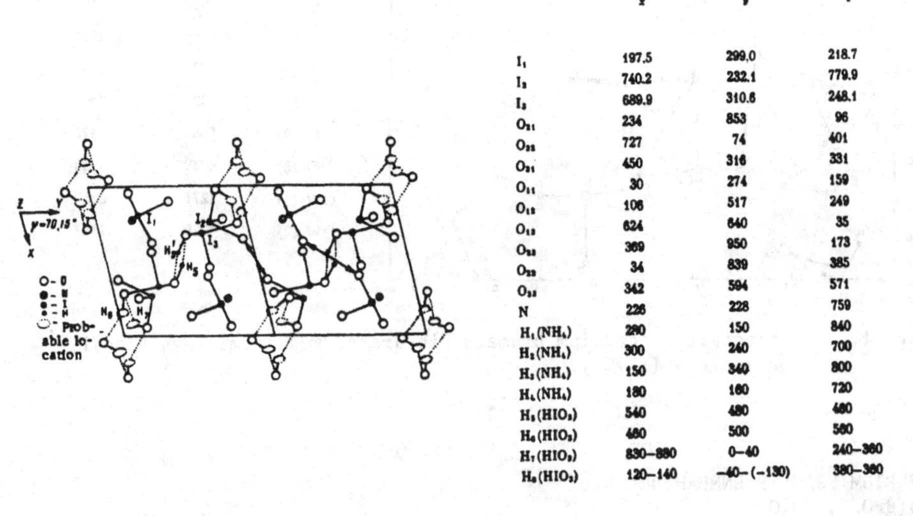

	x	y	z
I_1	197.5	299.0	218.7
I_2	740.2	232.1	779.9
I_3	689.9	310.6	248.1
O_{21}	234	853	96
O_{22}	727	74	401
O_{31}	450	316	331
O_{11}	30	274	159
O_{12}	106	517	249
O_{13}	624	640	35
O_{32}	369	950	173
O_{23}	34	839	385
O_{33}	342	594	571
N	226	228	759
$H_1(NH_4)$	280	150	840
$H_2(NH_4)$	300	240	700
$H_3(NH_4)$	150	340	800
$H_4(NH_4)$	180	160	720
$H_5(HIO_3)$	540	480	460
$H_6(HIO_3)$	460	500	560
$H_7(HIO_3)$	830–880	0–40	240–360
$H_8(HIO_3)$	120–140	−40–(−130)	380–360

Fig. 1. Structure of ammonium hydrogen iodate, and atomic positional parameters ($\times 10^3$).

1. Structure Reports, 45A, 384.

LITHIUM IODATE
γ-$LiIO_3$

R. LIMINGA, C. SVENSSON, J. ALBERTSSON and S.C. ABRAHAMS, 1982. J. Chem. Phys., 77, 4222-4226.

Orthorhombic, $Pna2_1$, a = 9.422, b = 5.861, c = 5.301 Å, at 515K, Z = 4. Powder data.

Atomic positions

	x	y	z
I	0.3179	0.0751	0.0
O(1)	0.1298	0.0556	-0.1913
O(2)	0.4117	-0.1194	-0.1618
O(3)	0.3663	0.3439	-0.1853
Li	∿0	∿0	0.0706

The structure shows similarities to that of α-$LiIO_3$ (1), with iodate rotations of about 20° and elongation of the LiO_6 octahedron, reduction in symmetry and doubling of the unit cell. A further unit cell doubling occurs at the γ-β transition (2).

1. Strukturbericht, 2, 49, 332; Structure Reports, 31A, 214; 39A, 327; 42A, 425.
2. Structure Reports, 39A, 328.

ALUMINUM HYDROGEN IODATE HYDRATE
$Al(IO_3)_3.2HIO_3.6H_2O$

H. KÜPPERS, W. SCHÄFER and G. WILL, 1982. Z. Kristallogr., 159, 231-238.

Hexagonal, $P6_3$, a = 16.126, c = 12.398 Å, D_m = 3.607, Z = 6. Mo radiation, R = 0.019 for 2416 reflexions, and neutron radiation, R = 0.022 for 709 reflexions.

Structure as previously described (1), with H atoms now located.

1. Structure Reports, 43A, 296.

POTASSIUM DIHYDROGEN SULPHATO-IODATE
$K_4H_2(S_2I_2O_{14})$

M.T. AVERBUCH-POUCHOT, 1982. J. Solid State Chem., 41, 262-265.

Monoclinic, $P2_1/n$, a = 13.84, b = 7.173, c = 7.443 Å, β = 93.16°, Z = 2. Ag radiation, R = 0.040 for 1956 reflexions.

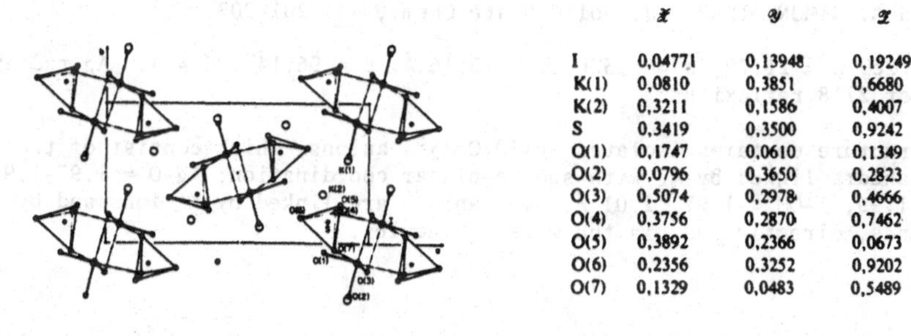

	x	y	z
I	0,04771	0,13948	0,19249
K(1)	0,0810	0,3851	0,6680
K(2)	0,3211	0,1586	0,4007
S	0,3419	0,3500	0,9242
O(1)	0,1747	0,0610	0,1349
O(2)	0,0796	0,3650	0,2823
O(3)	0,5074	0,2978	0,4666
O(4)	0,3756	0,2870	0,7462
O(5)	0,3892	0,2366	0,0673
O(6)	0,2356	0,3252	0,9202
O(7)	0,1329	0,0483	0,5489

Fig. 1. Structure of $K_4H_2(S_2I_2O_{14})$.

The structure (Fig. 1) contains $S_2I_2O_{14}^{6-}$ ions, which consist of a pair of IO_3 trigonal pyramids linked by a longer I...O bond and bridged by two SO_4 tetrahedra, I thus having distorted octahedral coordination. I-O = 1.80-1.92, I...O = 2.49-2.92, S-O = 1.47-1.50 Å. K ions have 8-coordinations, K...O = 2.69-3.11 Å; H atoms were not located.

SODIUM TETRACOBALTATOTRIPERIODATE HYDRATE
$Na_2H_2[Co_4I_3O_{18}(H_2O)_6][Co_4I_3O_{16}(OH)_2(H_2O)_6].22H_2O$

H. KONDO, A. KOBAYASHI and Y. SASAKI, 1982. Bull. Chem. Soc. Japan, 55, 2113-2117.

Monoclinic, P2/c, a = 15.172, b = 10.874, c = 19.436 Å, β = 116.58°, Z = 2. Mo radiation, R = 0.035 for 7010 reflexions.

The structure (Fig. 1) contains two complex anions with different degrees of protonation; both have the structure previously found in heteropolymolybdates with seven condensed octahedra, Co and I alternating in the peripheral sites and Co in the central site. Na is coordinated to six water molecules, and there are interanion hydrogen bonds.

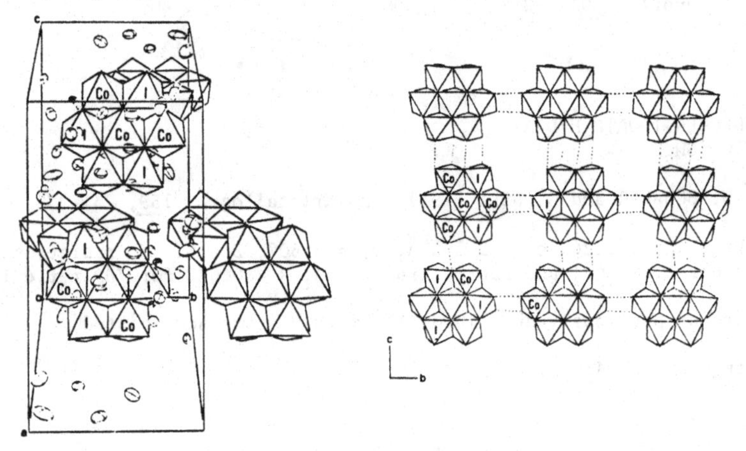

Fig. 1. Structure of the tetracobaltatotriperiodate.

POTASSIUM IODATOARGENTATE(III) HYDRATE
$K_5Ag(IO_5OH)_2.8H_2O$

R. MASSE and A. SIMON, 1982. J. Solid State Chem., 44, 201-207.

Monoclinic, Cc, a = 21.79, b = 6.320, c = 15.16 Å, β = 96.14°, Z = 4. Ag radiation, R = 0.033 for 2718 reflexions.

The structure contains isolated $Ag(IO_5OH)_2^{5-}$ anions, which consist of two $IO_5(OH)$ octahedra linked by Ag with square-planar coordination; Ag-O = 1.97-1.99, I-O = 1.82-1.99, I-OH = 1.97, 2.01 Å. The anions are linked by K^+ ions and by hydrogen bonds, directly and via the water molecules.

ZINC IODATE
$Zn(IO_3)_2$

J. LIANG and C. WANG, 1982. Huaxue Xuebao, 40, 985-993.

Monoclinic, $P2_1$, a = 5.469, b = 10.938, c (unique axis) = 5.116 Å, γ = 120°, D_m = 5.08, Z = 2.

The structure contains trigonal pyramidal IO_3^- ions (I-O = 1.81 Å), linked by octahedrally-coordinated Zn^{2+} ions.

α-QUARTZ
SiO_2

G.A. LAGER, J.D. JORGENSEN and F.J. ROTELLA, 1982. J. Appl. Phys., 53, 6751-6756.

Trigonal, $P3_121$, a = 4.9021, 4.9030, 4.9141, c = 5.3997, 5.3999, 5.4060 Å, at 13, 78, 296K, Z = 3. Neutron powder data. Si in 3(a): x,0,0, x = 0.4680, 0.4682, 0.4700; O in 6(c): x,y,z, x = 0.4124, 0.4125, 0.4131, y = 0.2712, 0.2707, 0.2677, z = -0.1163, -0.1163, -0.1189.

Structure as previously described (1), with rotation of SiO_4 tetrahedra as temperature is reduced.

1. Strukturbericht, 1, 166, 198, 776; 3, 21, 296; 4, 203; 7, 86; Structure Reports, 27, 674; 28, 119; 30A, 420; 42A, 393; 43A, 262; 44A, 297; 45A, 359, 383; 46A, 375; 48A, 334.

LITHIUM PHOSPHOSILICATE
$Li_{3.75}P_{0.25}Si_{0.75}O_4$

LITHIUM ORTHOSILICATE
Li_4SiO_4

W.H. BAUR and T. OHTA, 1982. J. Solid State Chem., 44, 50-59.

Monoclinic, $P2_1/m$, a = 5.116, 5.147, b = 6.116, 6.094, c = 5.309, 5.293 Å, β = 90.40, 90.33°, Z = 2. Mo radiation, R = 0.038, 0.033 for 478, 890 reflexions.

Atomic positions (1 = phosphosilicate, 2 = orthosilicate)

	x	y	z
	1		
$Si_{0.75}P_{0.25}$	0.6696	¼	0.3252
O(1)	0.7763	0.0334	0.1863
O(2)	0.3531	¼	0.3162
O(3)	0.7705	¼	0.6143
Li(1)	0.174	0.003	0.175
Li(2)	0.396	¼	0.703
Li(3)	0.179	¼	0.638
Li(4)	0.398	0.963	0.148
Li(5)	0.066	0.051	0.459
Li(6)	0.031	¼	0.988
	2		
Si	0.6651	¼	0.3260
O(1)	0.7745	0.0302	0.1841
O(2)	0.3479	¼	0.3149
O(3)	0.7672	¼	0.6204
Li(1)	0.1651	0.0047	0.1721
Li(2)	0.3930	¼	0.6999
Li(3)	0.1680	¼	0.6363
Li(4)	0.4010	0.9650	0.1502
Li(5)	0.0666	0.0372	0.4601
Li(6)	0.0183	¼	0.9553

OCCUPANCY FACTORS IN Li_4SiO_4 AND IN $Li_{3.75}Si_{0.75}P_{0.25}O_4$

	Disordered Li_4SiO_4 (1)	Substructure of ordered Li_4SiO_4, constrained, this work	Occupied over available sites in ordered Li_4SiO_4 (2)	Disordered $Li_{3.75}Si_{0.75}P_{0.25}O_4$, this work
Li(1)	[illegible]	0.567	[illegible] = 0.571	0.59
Li(2)	[illegible]	0.530	[illegible] = 0.571	0.20
Li(3)	[illegible]	0.414	[illegible] = 0.429	0.56
Li(4)	[illegible]	0.472	[illegible] = 0.429	0.44
Li(5)	[illegible]	0.247	[illegible] = 0.286	0.20
Li(6)	[illegible]	0.484	[illegible] = 0.429	0.21

The structure is as previously described for the Li_4SiO_4 substructure (1, 2), with partially-occupied Li sites. Si,P-O = 1.615(2), Li-O = 1.92-2.39; Si-O = 1.638, Li-O = 1.92-2.54 Å.

1. Structure Reports, 33A, 467.
2. Ibid., 45A, 360.

POTASSIUM DISILICATE

$K_6Si_2O_7$

M. JANSEN, 1982. Z. Kristallogr., 160, 127-133.

Monoclinic, $P2_1/c$, a = 6.458, b = 8.887, c = 10.879 Å, β = 125.0°, Z = 2. Mo radiation, R = 0.033 for 1073 reflexions.

Atomic positions

	x	y	z
Si	0.3267	0.6348	0.3741
K(1)	0.3949	0.2798	0.2768
K(2)	0.0227	0.3510	0.4119
K(3)	0.7922	0.0249	0.5448
O(1)	0.7384	0.4198	0.7862
O(2)	0.4975	0.2156	0.5668
O(3)	0.9257	0.3423	0.6287
O(4)	1/2	1/2	1/2

Isostructural with $K_6Co_2O_7$ (1). The structure contains a centrosymmetric disilicate anion, with a linear Si-O-Si bridge; Si-O = 1.675(2) (bridge), 1.616-1.626(4) (terminal) Å, O-Si-O = 105.9-112.3°. K ions have 7-, 7-, and 6-coordinations, K-O = 2.63-3.42 Å.

1. Structure Reports, 40A, 212.

POTASSIUM CYCLOTRISILICATE

$K_6Si_3O_9$ 3 x K_2SiO_3

R. WERTHMANN and R. HOPPE, 1981. Rev. Chim. Minér., 18, 593-607.

Monoclinic, $P2_1/c$, a = 6.270, b = 12.808, c = 16.96 Å, β = 117.38°, Z = 4. Mo radiation, R = 0.093 for 2530 reflexions.

The structure (Fig. 1) contains $Si_3O_9{}^{6-}$ anions, with rings of three SiO_4 tetrahedra, linked by K ions which have irregular coordinations. M_2XO_3 (M = Rb, Cs; X = Si, Ge) are isostructural.

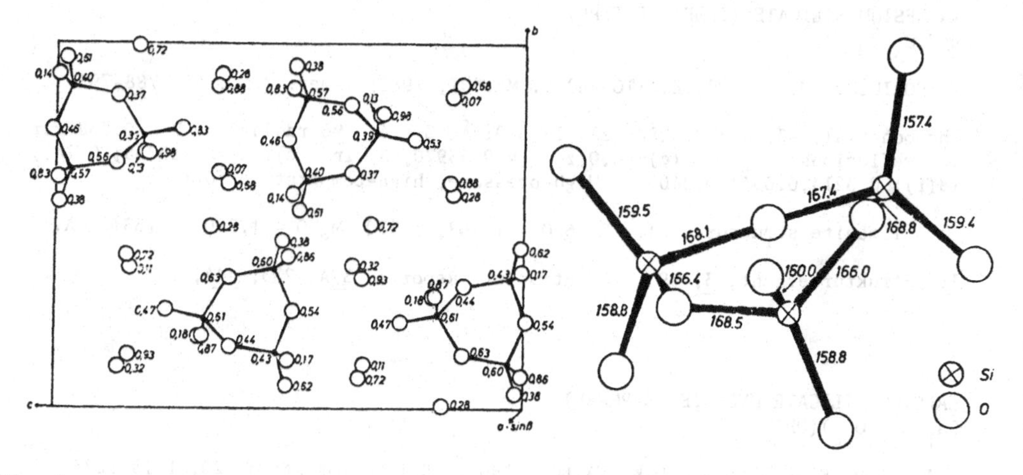

Fig. 1. Structure of $K_6Si_3O_9$ (bond lengths in Å x 10^2).

SODIUM BERYLLIUM TRISILICATE
$Na_2Be_2Si_3O_9$

D. GINDEROW, F. CESBRON and M.-C. SICHÈRE, 1982. Acta Cryst., B38, 62-66.

Orthorhombic, Pnma, a = 11.748, b = 9.415, c = 6.818 Å, Dm = 2.543, Z = 4. Mo radiation, R = 0.027 for 2023 reflexions.

The structure (Fig. 1) contains $(O_2Si)_3O_3$ rings linked by pairs of edge-sharing BeO_4 tetrahedra, with 7-coordinate Na ions in cavities in this framework. Si-O = 1.625-1.658 (ring), 1.593-1.595 (terminal), Be-O = 1.582-1.663, Na-O = 2.487-2.808(1) Å.

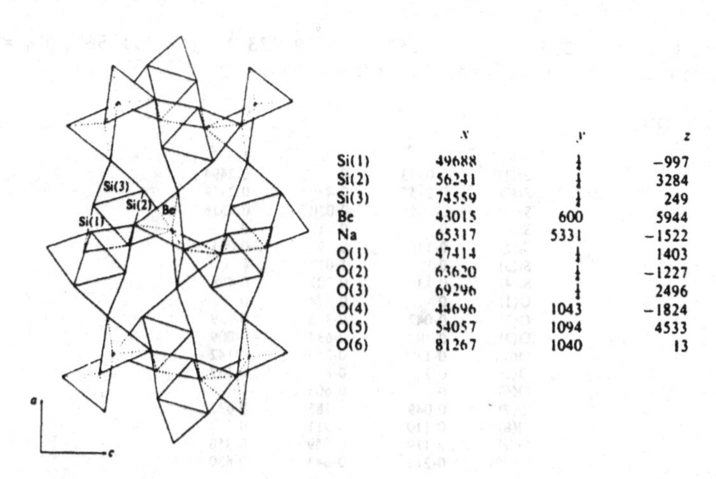

	x	y	z
Si(1)	49688	¼	−997
Si(2)	56241	¼	3284
Si(3)	74559	¼	249
Be	43015	600	5944
Na	65317	5331	−1522
O(1)	47414	¼	1403
O(2)	63620	¼	−1227
O(3)	69296	¼	2496
O(4)	44696	1043	−1824
O(5)	54057	1094	4533
O(6)	81267	1040	13

Fig. 1. Si/Be framework in $Na_2Be_2Si_3O_9$ and atomic positional parameters (x 10^5 for x, x 10^4 for y, z).

MAGNESIUM SILICATE (ILMENITE-TYPE)
$MgSiO_3$

H. HORIUCHI, M. HIRANO, E. ITO and Y. MATSUI, 1982. Amer. Min., 67, 788-793.

Rhombohedral, R$\bar{3}$, a = 4.7284, c = 13.5591 Å, Z = 6. Mo radiation, R = 0.038 for 366 reflexions. Mg in 6(c): 0,0,z, z = 0.35970; Si in 6(c): z = 0.15768; O in 18(f): 0.3214,0.0361,0.24077. High-pressure, high-temperature form.

Ilmenite structure (1). Si-6 O = 1.768, 1.830, Mg-O = 1.990, 2.163(2) Å.

1. Strukturbericht, 3, 69, 380; Structure Reports, 37A, 233; 44A, 197.

CALCIUM SILICATE HYDRATE (K-PHASE)
$Ca_7(Si_{16}O_{38})(OH)_2$

J.A. GARD, K. LUKE and H.F.W. TAYLOR, 1981. Kristallografija, 26, 1218-1223 [Soviet Physics - Crystallography, 26, 691-695].

Triclinic, P$\bar{1}$, a = 9.70, b = 9.70, c = 12.25 Å, α = 108.6, β = 78.1, γ = 120°, Z = 1. Electron diffraction.

A structure is derived, with octahedrally-coordinated Ca layers alternating with double layers of silicate tetrahedra of composition $Si_{16}O_{38}$.

STRONTIUM METASILICATE
$SrSiO_3$

I. K.-I. MACHIDA, G.-Y. ADACHI, J. SHIOKAWA, M. SHIMADA and M. KOIZUMI, 1982. Acta Cryst., B38, 386-389.

α-Form
Monoclinic, C2, a = 12.323, b = 7.139, c = 10.873 Å, β = 111.58°, Dm = 3.64, Z = 12. Mo radiation, R = 0.053 for 730 reflexions.

Atomic positions

	x	*y*	*z*
Sr(1)	0·0873	0	0·2494
Sr(2)	0·2455	0·4780	0·2458
Sr(3)	0·4128	−0·0207	0·2516
Si(1)	0	0·417	0
Si(2)	0·116	0·790	−0·001
Si(3)	0	1·075	½
Si(4)	0·130	0·702	0·498
O(1)	0	0·874	0
O(2)	0·042	0·323	0·139
O(3)	0·107	0·532	−0·009
O(4)	0·130	0·842	−0·142
O(5)	0·223	0·835	0·139
O(6)	0	0·605	½
O(7)	0·049	1·185	0·635
O(8)	0·110	0·911	0·511
O(9)	0·119	0·659	0·350
O(10)	0·213	0·643	0·630

The structure (Fig. 1) contains two Si_3O_9 rings of three SiO_4 tetrahedra, linked by three independent 8-coordinate Sr ions. Si-O = 1.48-1.85(2) Å, O-Si-O = 97-134° [the wide scatter of these distances suggests that the structure should possibly be refined in another space group, e.g. C2/c; the b-axis projection (Fig. 1) indicates that the h0ℓ reflexions with ℓ odd must be weak or absent]; Sr-O = 2.39-2.83(1) Å. Two high-pressure forms are also obtained, δ (triclinic,

$P\bar{1}$) and δ' (monoclinic, $P2_1/c$) [see II below].

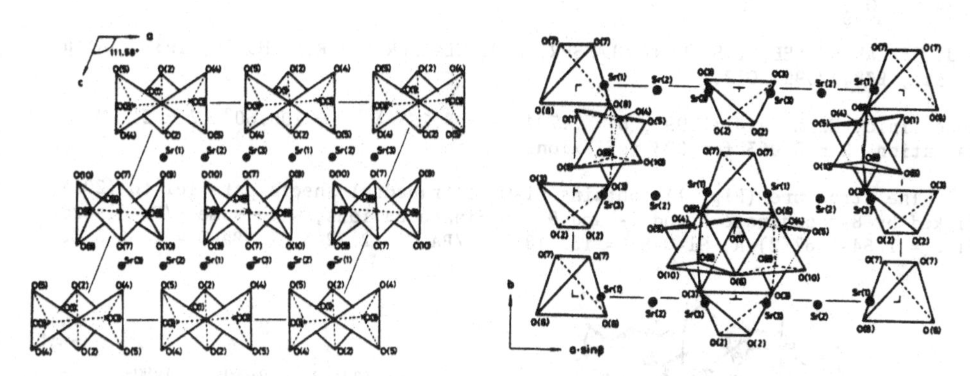

Fig. 1. Structure of α-$SrSiO_3$.

II. K.-I. MACHIDA, G.-Y. ADACHI, J. SHIOKAWA, M. SHIMADA, M. KOIZUMI, K. SUITO and A. ONODERA, 1982. Inorg. Chem., 21, 1512-1519.

δ-Form
Triclinic, $P\bar{1}$, a = 6.874, b = 6.894, c = 9.717 Å, α = 85.01, β = 110.57, γ = 104.01°, D_m = 3.87, Z = 6. Mo radiation, R = 0.043 for 1458 reflexions.

δ'-Form
Monoclinic, $P2_1/c$, a = 7.452, b = 6.066, c = 13.479 Å, β = 117.09°, D_m = 3.96, Z = 8. Mo radiation, R = 0.046 for 914 reflexions.

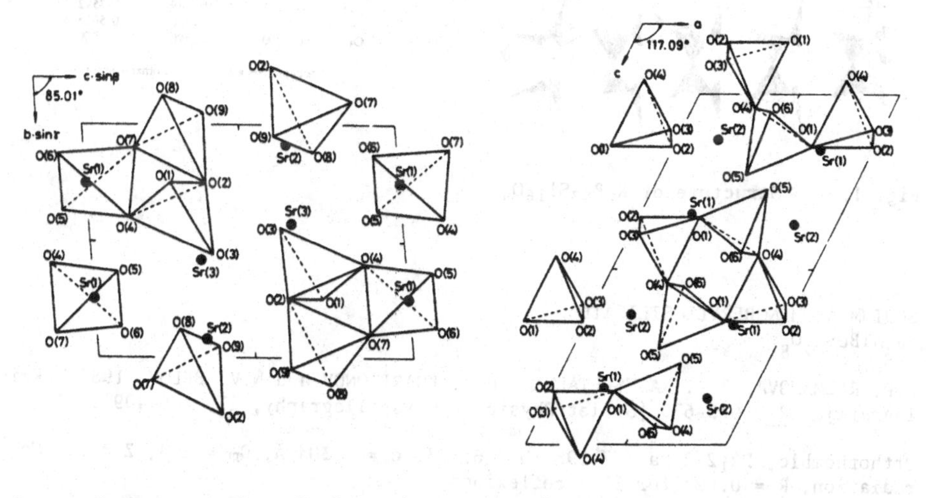

Fig. 2. Structures of δ- (left) and δ'-$SrSiO_3$ (right).

The materials are high-pressure phases. The δ-form is isostructural with δ-$CaSiO_3$ (1) and contains a cyclic trimetasilicate ion, $Si_3O_9^{6-}$ (Fig. 2), Si-O = 1.50-1.74($\bar{1}$) Å, O-Si-O = 91-130°; Sr ions have 6- and 8-coordinations, Sr-O = 2.33-2.98 Å.

The δ'-form contains a cyclic, centrosymmetric tetrametasilicate ion, $Si_4O_{12}^{8-}$ (Fig. 2), Si-O = 1.60-1.69(1) Å, O-Si-O = 103-116°; Sr ions have 8-coordinations, Sr-O = 2.41-3.06 Å.

1. Structure Reports, 41A, 430.

POTASSIUM BARIUM SILICATE
$K_2Ba_7Si_{16}O_{40}$

F.J. CERVANTES-LEE, L.S. DENT GLASSER, F.P. GLASSER and R.A. HOWIE, 1982. Acta Cryst., B38, 2099-2012.

Monoclinic, C2/m, a = 31.991, b = 7.704, c = 8.255 Å, β = 100.60°, Z = 2. Mo radiation, R = 0.063 for 993 reflexions.

The structure (Fig. 1) contains Si_2O_5 tetrahedral sheets parallel to (201), linked by 8-coordinate K and 7- and 9-coordinate Ba ions, with some K/Ba disorder. Si-O = 1.54-1.67(2) Å, Si-O-Si = 132-180°, K/Ba-O = 2.72-3.27, Ba-O = 2.66-3.22 Å.

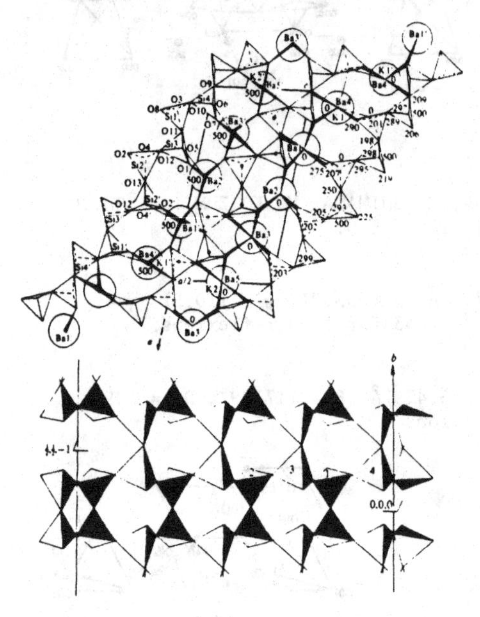

	x	y	z	Occupancy*
Ba(1)	0·14912	0·0000	0·05819	–
Ba(2)	0·26991	0·0000	0·89597	–
Ba(3)	0·38637	0·0000	0·68866	–
Ba(4)	0·0384	0·0000	0·2764	0·333
Ba(5)	0·5000	0·0000	0·5000	0·333
K(1)	0·0384	0·0000	0·2764	0·666
K(2)	0·5000	0·0000	0·5000	0·666
Si(1)	0·0739	0·2007	0·6545	–
Si(2)	0·2007	0·2070	0·4348	–
Si(3)	0·1674	0·2977	0·7531	–
Si(4)	0·0460	0·2971	0·9755	–
O(1)	0·2033	0·2187	0·8876	–
O(2)	0·1842	0·2761	0·2602	–
O(3)	0·0572	0·0000	0·6370	–
O(4)	0·1938	0·0000	0·4415	–
O(5)	0·3391	0·0000	0·1988	–
O(6)	0·4544	0·0000	–0·0369	–
O(7)	0·0839	0·2057	0·0865	–
O(8)	0·0674	0·2892	0·4823	–
O(9)	0·0000	0·2095	0·0000	–
O(10)	0·0453	0·2887	0·7796	–
O(11)	0·1227	0·1979	0·7592	–
O(12)	0·1741	0·2960	0·5628	–
O(13)	0·2500	0·2500	0·5000	–

* All occupancies are 1·0 unless otherwise stated.

Fig. 1. Structure of $K_2Ba_7Si_{16}O_{40}$.

SODIUM ALUMINOBERYLLOSILICATE
$Na_3AlBeSi_2O_8$

Z.P. RAZMANOVA, R.K. RASTSVETAEVA, Ju.A. KHARITONOV and N.V. BELOV, 1982. Kristallografija, 27, 675-679 [Soviet Physics - Crystallography, 27, 406-409].

Orthorhombic, $P2_12_12$, a = 7.198, b = 6.873, c = 7.304 Å, D_m = 2.4, Z = 2. Mo radiation, R = 0.027 for 1258 reflexions.

Atomic positions

	x	y	z
Si	0.1992	0.2648	0.2589
Al	0	0	-0.0336
Na(1)	1/2	0	0.0330
Na(2)	0.2137	0.7467	0.3242
Be	1/2	0	0.4335
O(1)	0.0890	0.3285	0.4407
O(2)	0.0416	0.2027	0.1055
O(3)	0.3385	0.0847	0.2937
O(4)	0.3037	0.4545	0.1683

The structure is as previously determined (1). Si-O = 1.608-1.651, Al-O = 1.750, Be-O = 1.626, 1.653, Na-O = 2.298-2.653(3) Å.

1. Structure Reports, 41A, 368.

LEAD SILICATE

$Pb_{11}Si_3O_{17}$ $Pb_{11}(SiO_4)(Si_2O_7)O_6$

K. KATO, 1982. Acta Cryst., B38, 57-62.

Triclinic, P$\bar{1}$, a = 22.502, b = 12.982, c = 7.313 Å, α = 92.52, β = 99.17, γ = 100.29°, Z = 4. Mo radiation, R = 0.087 for 6903 reflexions.

The structure (Fig. 1) contains SiO_4^{4-} and $Si_2O_7^{6-}$ ions, linked by 22 independent Pb^{2+} ions.

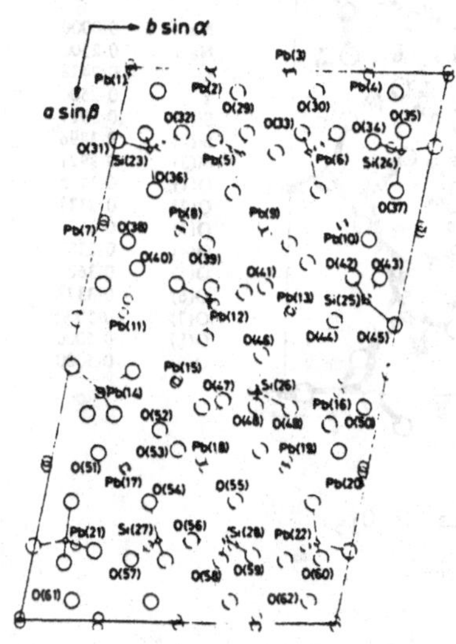

Fig. 1. Structure of $Pb_{11}Si_3O_{17}$.

CALCIUM MAGNESIUM TITANIUM ALUMINOSILICATE

$Ca(Mg,Ti)(Si,Al)_2O_6$

P. FEJDI, 1982. Silikaty, 26, 7-13.

Monoclinic, C2/c, a = 9.781, b = 8.856, c = 5.331 Å, β = 106.06°, Z = 4. R = 0.061.

Diopside structure (1), with Ti(IV) mainly in the M(1) site, but a small amount in the tetrahedral site.

1. Strukturbericht, 2, 130.

SODIUM LITHIUM YTTRIUM SILICATE
$Na_2LiYSi_6O_{15}$

R.P. GUNAWARDANE, R.A. HOWIE and F.P. GLASSER, 1982. Acta Cryst., B38, 1405-1408.

Orthorhombic, Cmca, a = 14.505, b = 17.596, c = 10.375 Å, Z = 8. Mo radiation, R = 0.082 for 1501 reflexions.

Isostructural with synthetic $Na_2Mg_2Si_6O_{15}$ (1) and with the minerals zektzerite (2) and emeleusite (3). The structure (Fig. 1) contains corrugated double silicate chains along c with a six-tetrahedra repeat. Na(2) polyhedral chains (10-coordination) and tetrahedral-octahedral (LiO_4-YO_6) chains link the silicate chains to form a three-dimensional structure. Na(1) has 9-coordination. Si-O = 1.57-1.64, Y-O = 2.21-2.25, Li-O = 2.00-2.02, Na-O = 2.39-3.36 Å, Si-O-Si = 146-160°.

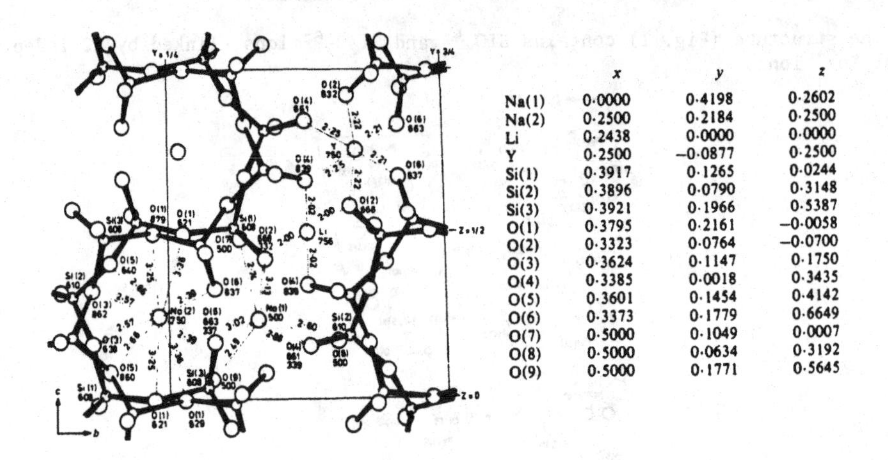

	x	y	z
Na(1)	0·0000	0·4198	0·2602
Na(2)	0·2500	0·2184	0·2500
Li	0·2438	0·0000	0·0000
Y	0·2500	−0·0877	0·2500
Si(1)	0·3917	0·1265	0·0244
Si(2)	0·3896	0·0790	0·3148
Si(3)	0·3921	0·1966	0·5387
O(1)	0·3795	0·2161	−0·0058
O(2)	0·3323	0·0764	−0·0700
O(3)	0·3624	0·1147	0·1750
O(4)	0·3385	0·0018	0·3435
O(5)	0·3601	0·1454	0·4142
O(6)	0·3373	0·1779	0·6649
O(7)	0·5000	0·1049	0·0007
O(8)	0·5000	0·0634	0·3192
O(9)	0·5000	0·1771	0·5645

Fig. 1. Structure of $Na_2LiYSi_6O_{15}$.

1. Structure Reports, 38A, 357.
2. Ibid., 44A, 317.
3. Ibid., 45A, 371.

CALCIUM NEODYMIUM SILICATE
$Ca_2Nd_8(SiO_4)_6O_2$

J. FAHEY and W.J. WEBER, 1982. Rare Earths Mod. Sci. Technol., 3, 341-344.

Hexagonal, $P6_3/m$, a = 9.5297, c = 7.0184 Å, Z = 1. Ca+Nd in 4(f); Nd in 6(h); O in 2(a). Fluorapatite structure (1).

1. Strukturbericht, 2, 99.

SAMARIUM PYROSILICATE SULPHIDE
$Sm_4S_3Si_2O_7$

T. SIEGRIST, W. PETTER and F. HULLIGER, 1982. Acta Cryst., B38, 2872-2874.

Tetragonal, $I4_1/amd$, a = 11.839, c = 13.928 Å, Z = 8. Mo radiation, R = 0.054 for 937 reflexions.

The structure (Fig. 1) contains Si_2O_7 eclipsed tetrahedra, sulphide anions, and Sm ions with 6- (3 O and 3 S) and 9-coordinations (6 O and 3 S). Si-O = 1.62-1.65(1) Å.

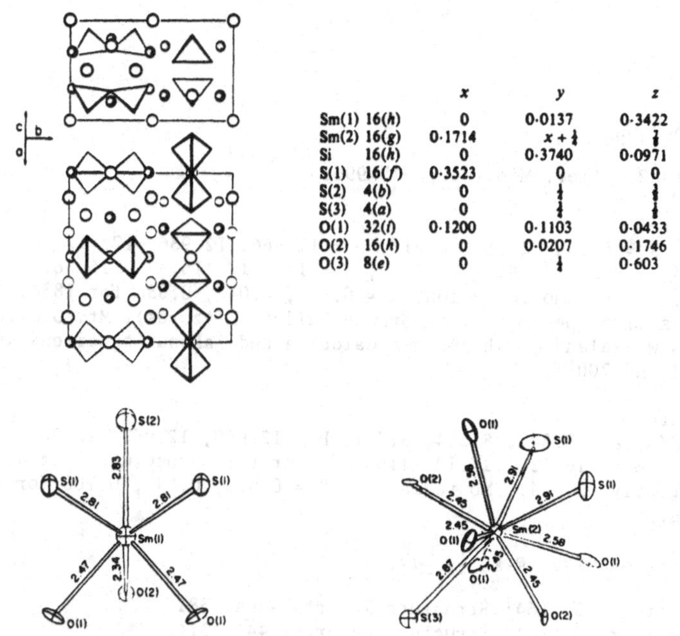

		x	y	z
Sm(1)	16(h)	0	0·0137	0·3422
Sm(2)	16(g)	0·1714	x + ¼	⅞
Si	16(h)	0	0·3740	0·0971
S(1)	16(f)	0·3523	0	0
S(2)	4(b)	0	¼	⅜
S(3)	4(a)	0	¾	⅛
O(1)	32(i)	0·1200	0·1103	0·0433
O(2)	16(h)	0	0·0207	0·1746
O(3)	8(e)	0	¼	0·603

Fig. 1. Structure of $Sm_4S_3Si_2O_7$.

SODIUM GADOLINIUM SILICATE

$NaGdSiO_4 . 0 \cdot 2NaOH$

G.D. FALLON and B.M. GATEHOUSE, 1982. Acta Cryst., B38, 919-920.

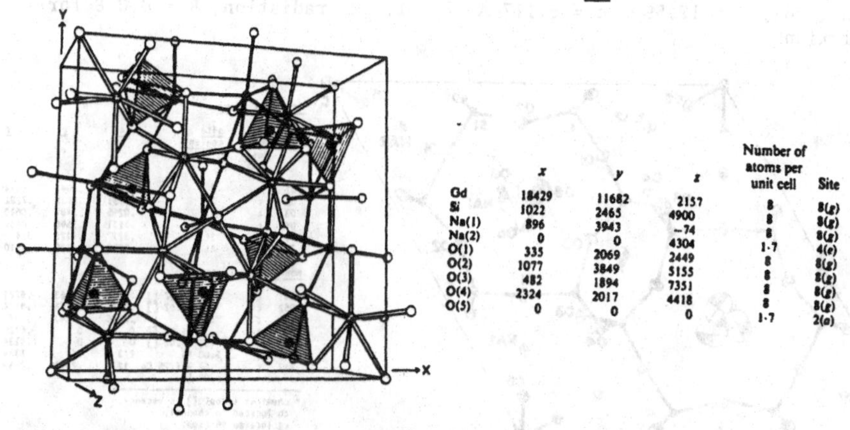

	x	y	z	Number of atoms per unit cell	Site
Gd	18429	11682	2157	8	8(g)
Si	1022	2465	4900	8	8(g)
Na(1)	896	3943	−74	8	8(g)
Na(2)	0	0	4304	1·7	4(e)
O(1)	335	2069	2449	8	8(g)
O(2)	1077	3849	5155	8	8(g)
O(3)	482	1894	7351	8	8(g)
O(4)	2324	2017	4418	8	8(g)
O(5)	0	0	0	1·7	2(a)

Fig. 1. Structure of sodium gadolinium silicate, and atomic positional parameters (x 10^5 for Gd, x 10^4 for others).

Tetragonal, $I\bar{4}$, a = 11.748, c = 5.450 Å, Z = 8. Mo radiation, R = 0.028 for 1176 reflexions.

The structure (Fig. 1) contains SiO_4 tetrahedra linked by 8-coordinate Gd and 7- and 10-coordinate Na ions; one Na site and the OH site are partially occupied. Si-O = 1.62-1.64, Gd-O = 2.30-2.57, Na-O = 2.35-3.11 Å.

ANORTHOCLASE

$(K,Na,Ca)(Al,Si)_4O_8$

G.E. HARLOW, 1982. Amer. Min., 67, 975-996.

Room temperature
Triclinic, $C\bar{1}$, a = 8.290, 8.252, 8.217, b = 12.966, 12.936, 12.917, c = 7.151, 7.139, 7.127 Å, α = 91.18, 92.11, 92.75, β = 116.31, 116.32, 116.36, γ = 90.14, 90.22, 90.24°, Z = 4. Mo radiation, R = 0.046, 0.047, 0.058 for 1834, 1697, 2141 reflexions, for specimens from Grande Caldeira (Azores), Mt. Gibele (Italy), and Kakanui (New Zealand). The Grande Caldeira and Kakanui specimens were also refined at 183 and 700°C.

High temperature
Monoclinic, C2/m, a = 8.348, 8.314, 8.321, b = 12.980, 12.973, 12.969, c = 7.158, 7.150, 7.148 Å, β = 116.11, 116.14, 116.05°, for three specimens, at 400, 510, 750°C, respectively, Z = 4. Mo radiation, R = 0.073, 0.047, 0.105 for 855, 931, 899 reflexions.

Feldspar structures (e.g. 1, 2).

1. Strukturbericht, 3, 164; Structure Reports, 46A, 381.
2. Strukturbericht, 3, 161; Structure Reports, 44A, 314.

CANCRINITE (CARBONATE)

$Na_6Ca_{1.5}Al_6Si_6O_{24}(CO_3)_{1.6}\cdot 1\cdot 75H_2O$

I. H.D. GRUNDY and I. HASSAN, 1982. Canad. Miner., 20, 239-251.

Hexagonal, $P6_3$, a = 12.590, c = 5.117 Å, Z = 1. Mo radiation, R = 0.028 for 855 reflexions.

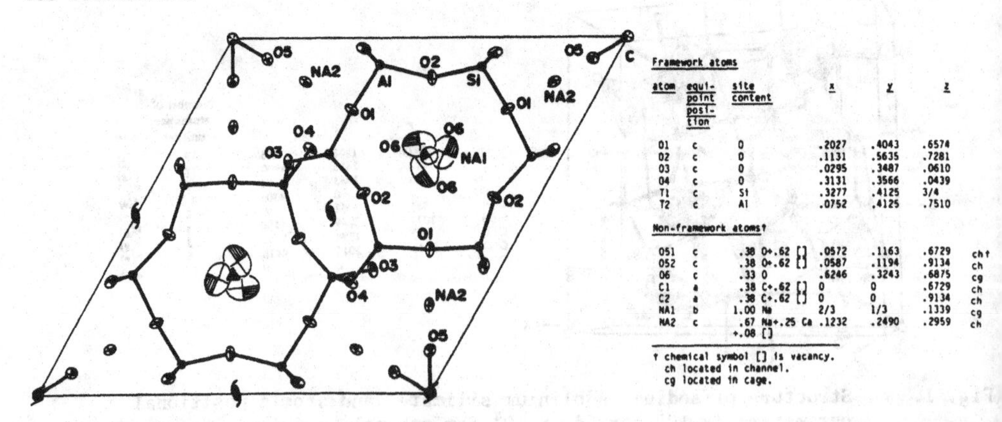

Framework atoms

atom	equipoint position	site content	x	y	z	
O1	c	O	.2027	.4043	.6574	
O2	c	O	.1131	.5635	.7281	
O3	c	O	.0295	.3487	.0610	
O4	c	O	.3131	.3566	.0439	
T1	c	Si	.3277	.4125	3/4	
T2	c	Al	.0752	.4125	.7510	
Non-framework atoms†						
O51	c	.38 O+.62 []	.0572	.1163	.6729	ch†
O52	c	.38 O+.62 []	.0587	.1194	.9134	ch
O6	c	.33 O	.6246	.3243	.6875	cg
C1	a	.38 C+.62 []	0	0	.6729	ch
C2	a	.38 C+.62 []	0	0	.9134	ch
NA1	b	1.00 Na	2/3	1/3	.1339	cg
NA2	c	.67 Na+.25 Ca +.08 []	.1232	.2490	.2959	ch

† chemical symbol [] is vacancy.
ch located in channel.
cg located in cage.

Fig. 1. Structure of cancrinite.

The structure (Fig. 1) is as previously described (1), with ordered Al/Si distribution. A superstructure (c' = 8c) arises from substitutional ordering of interchannel cations and anions. Si-O = 1.601-1.621, Al-O = 1.717-1.747(3) Å.

$Na_{7.6}Ca_{0.4}[Al_6Si_6O_{24}]CO_3.2 \cdot 2H_2O$

II. A. ÉMIRALIEV and I.I. JAMZIN, 1982. Kristallografija, 27, 51-55 [Soviet Physics - Crystallography, 27, 27-30].

Trigonal, P3 (pseudo-$P6_3$), a = 12.62, c = 5.138 Å, Z = 1. Neutron radiation, R = 0.078 for 772 reflexions.

Atomic positions

	No. of atoms	x	y	z
Al	6	0.0770	0.4114	0.7528
Si	6	0.3300	0.4104	0.7480
O_1	6	0.2014	0.4039	0.6590
O_2	6	0.1170	0.5606	0.7235
O_3	6	0.0321	0.3518	0.0581
O_4	6	0.3167	0.3565	0.9437
Na_1(Ca)	6	0.2558	0.1286	0.7901
Na_2(Ca)	2	2/3	1/3	0.1230
C	~1	0	0	0.3537
O_{CO_3}	~3	0.0604	0.1213	0.6830
O_{H_2O}	2	0.6181	0.3139	0.6287
O_{OH_m}	~0.44	0	0	0.3182

Cancrinite structure (1), with carbonate and OH^- ions on the 6_3 axis (ordered carbonate positions would reduce the symmetry to P3).

1. Strukturbericht, 3, 150, 524; Structure Reports, 19, 480; 30A, 428; 35A, 458.

CANCRINITE (BASIC)
$Na_2Al_2Si_2O_8.1 \cdot 87H_2O$ (idealized)

N. BRESCIANI PAHOR, M. CALLIGARIS, G. NARDIN and L. RANDACCIO, 1982. Acta Cryst., B38, 893-895.

Hexagonal, $P6_3$, a = 12.678, c = 5.179 Å, Z = 3. Mo radiation, R = 0.034 for 648 reflexions.

The structure (Fig. 1) contains a framework similar to those in 1 and 2, with slightly different Na sites.

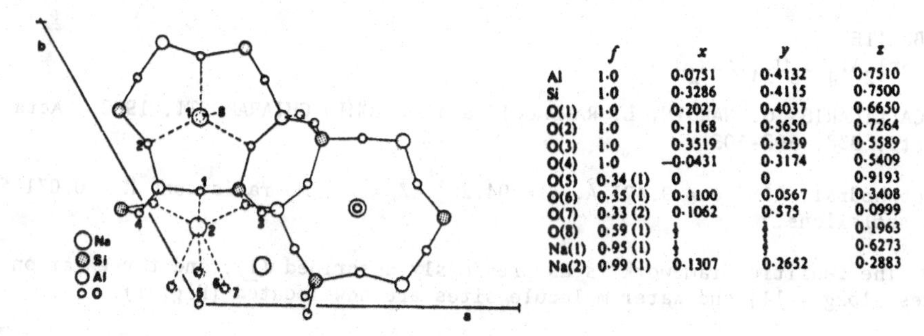

	f	x	y	z
Al	1·0	0·0751	0·4132	0·7510
Si	1·0	0·3286	0·4115	0·7500
O(1)	1·0	0·2027	0·4037	0·6650
O(2)	1·0	0·1168	0·5650	0·7264
O(3)	1·0	0·3519	0·3239	0·5589
O(4)	1·0	−0·0431	0·3174	0·5409
O(5)	0·34 (1)	0	0	0·9193
O(6)	0·35 (1)	0·1100	0·0567	0·3408
O(7)	0·33 (2)	0·1062	0·575	0·0999
O(8)	0·59 (1)	1/3	2/3	0·1963
Na(1)	0·95 (1)	1/3	2/3	0·6273
Na(2)	0·99 (1)	0·1307	0·2652	0·2883

Fig. 1. Structure of basic cancrinite (f = occupancy).

1. Structure Reports, 30A, 428.
2. Ibid., 35A, 458.

CASCANDITE
$CaScSi_3O_8(OH)$

M. MELLINI and S. MERLINO, 1982. Amer. Min., 67, 604-609.

Triclinic, $C\bar{1}$, a = 9.791, b = 10.420, c = 7.706 Å, α = 98.91, β = 102.63, γ = 84.17°, Z = 4. Mo radiation, R = 0.042 for 1526 reflexions.

The material is a pyroxenoid mineral from Baveno, Italy, with a structure related to those of pectolite, serandite, and schizolite (1). It contains two main structural units (Fig. 1): double chains of edge-sharing octahedra (M1 = Ca, M2 = Sc) and tetrahedral single chains with a repeat period of three tetrahedra. Ca-O = 2.312-2.570, Sc-O = 2.044-2.253, Si-O = 1.604-1.647(4) Å.

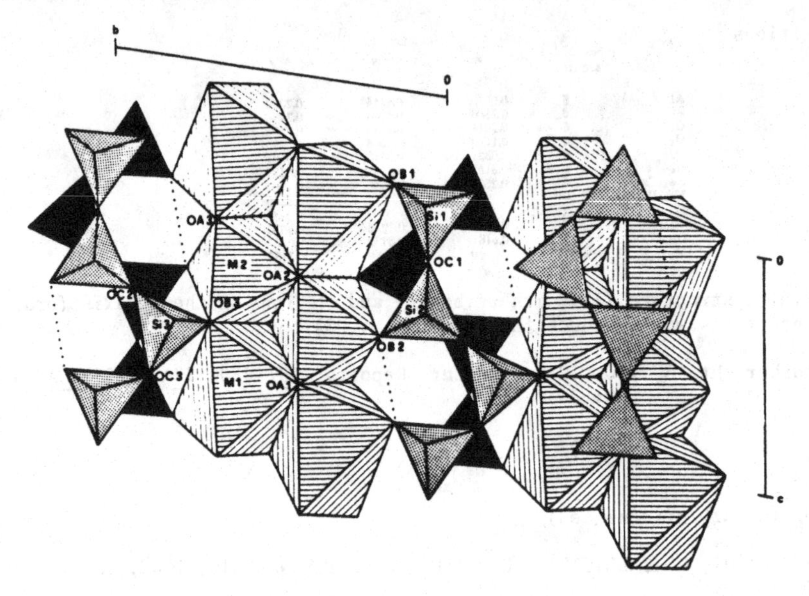

Fig. 1. Structure of cascandite.

1. Structure Reports, 20, 408; 21, 448; 32A, 461; 42A, 414; 44A, 312, 314.

CHABAZITE
$(Ca,Sr)_2Al_4Si_8O_{24}.13H_2O$

M. CALLIGARIS, G. NARDIN, L. RANDACCIO and P. COMIN CHIARAMONTI, 1982. Acta Cryst., B38, 602-605.

Rhombohedral, $R\bar{3}m$, a = 9.421 Å, α = 94.20°, Z = 1. Mo radiation, R = 0.071 for 578 reflexions.

The zeolitic framework is as previously described (1), and three cation sites along [111] and water molecule sites are now located (Fig. 1).

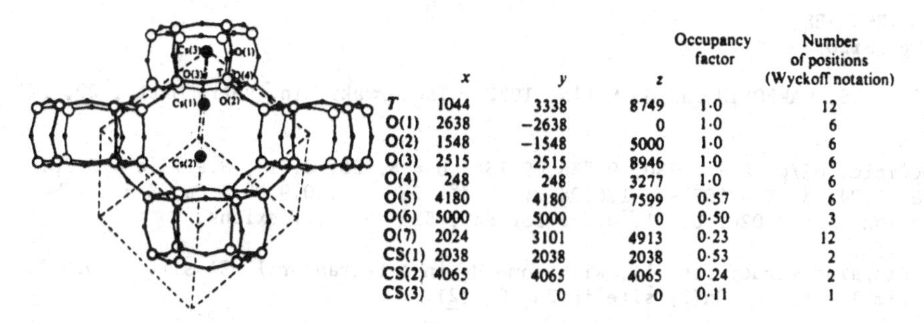

	x	y	z	Occupancy factor	Number of positions (Wyckoff notation)
T	1044	3338	8749	1·0	12
O(1)	2638	−2638	0	1·0	6
O(2)	1548	−1548	5000	1·0	6
O(3)	2515	2515	8946	1·0	6
O(4)	248	248	3277	1·0	6
O(5)	4180	4180	7599	0·57	6
O(6)	5000	5000	0	0·50	3
O(7)	2024	3101	4913	0·23	12
CS(1)	2038	2038	2038	0·53	2
CS(2)	4065	4065	4065	0·24	2
CS(3)	0	0	0	0·11	1

Fig. 1. Structure of chabazite, and atomic positional parameters (x 10^4; CS are Ca,Sr sites, O(5), O(6), O(7) are water sites).

1. Strukturbericht, 3, 151, 526; Structure Reports, 17, 564; 22, 513; 27, 691; 28, 250; 29, 401.

CHABAZITE (BARIUM and CADMIUM EXCHANGED)
$Ba_{1.8}Al_{3.8}Si_{8.2}O_{24}\cdot 9\cdot 7H_2O$, $Ca_{0.3}Sr_{0.1}Cd_{1.4}Al_{3.8}Si_{8.2}O_{24}\cdot 11\cdot 6H_2O$

M. CALLIGARIS and G. NARDIN, 1982. Zeolites, 2, 200-204.

Rhombohedral, R$\bar{3}$m, a = 9.420, 9.435 Å, α = 94.21, 94.66°, Z = 1. R = 0.097, 0.072 for 685, 555 reflexions.

Chabazite structure (1), with three main cation sites, two in the large cage along [111] and one near the 8-ring window. The site at the centre of the hexagonal prism is unoccupied.

1. Preceding report.

CHABAZITES (Na, Ca, Sr, K EXCHANGED)

A. ALBERTI, E. GALLI, G. VEZZALINI, E. PASSAGLIA and P.F. ZANAZZI, 1982. Zeolites, 2, 303-309.

Study of natural and hydrated Na-, Ca-, Sr-, and K-exchanged materials, with probable cation and water distributions.

CLINOHYPERSTHENE
$(Mg,Fe,Ca)SiO_3$

N.I. ORGANOVA, I.V. ROŽDESTVENSKAJA, I.M. MARSIJ, I.P. LAPUTINA and N.N. KURCEVA, 1982. Mineral. Ž., 4, 40-45.

Monoclinic, P2$_1$/c, a = 9.681, b = 8.932, c = 5.221 Å, β = 108.71°, Z = 8. R = 0.049.

Description of structure (1) and cation ordering.

1. Structure Reports, 40A, 283.

CLINOPYROXENES
$(Ca,Mg)_2Si_2O_6$

E. BRUNO, S. CARBONIN and G. MOLIN, 1982. Tschermaks Min. Petr. Mitt., 29, 223-240.

Monoclinic, C2/c, a = 9.750, 9.738, 9.734, b = 8.926, 8.918, 8.917, c = 5.251, 5.248, 5.247 Å, β = 105.90, 106.08, 106.33°, for 1.0, 0.9, 0.8 Ca, Z = 4. Mo radiation, R = 0.020, 0.021, 0.026 for 542, 526, 527 reflexions.

Diopside structures (1), with some Mg in a tetrahedral M2' site at 0,0.222, 1/4, similar to the Zn(2) site in $ZnSiO_3$ (2).

1. Strukturbericht, 2, 130.
2. Structure Reports, 41A, 374.

EKANITE
$ThCa_2Si_8O_{20}$

I. G. PERRAULT and J.T. SZYMAŃSKI, 1982. Canad. Miner., 20, 59-63.
II. J.T. SZYMAŃSKI, D.R. OWENS, A.C. ROBERTS, H.G. ANSELL and G.Y. CHAO, 1982. Ibid., 20, 65-75.

Tetragonal, I422, a = 7.483, c = 14.893 Å, D_m = 3.08, Z = 2. Mo radiation, R = 0.036 for 1319 reflexions.

Atomic positions

			x	y	z
Th	in	2(a)	0	0	0
Ca		4(c)	1/2	0	0
Si		16(k)	0.3335	0.2540	0.1479
O(1)		16(k)	0.2544	0.4521	0.1251
O(2)		8(j)	0.2924	0.7924	1/4
O(3)		16(k)	0.2553	0.1075	0.0818

The structure is closely related to that of the family with general composition $ThK(Na,Ca)_2Si_8O_{20}$, space group P4/mcc (1), misnamed 'ekanite', and now named steacyite. Ekanite contains Si_8O_{20} puckered sheets and steacyite discrete pseudo-cubic Si_8O_{20} arrangements. The sheets in ekanite are separated by sheets of metal ions, Th having square antiprismatic 8-coordination, Th-O = 2.405(5) Å, and Ca tetrahedral coordination, Ca-O = 2.342(5) Å, with a further 4 oxygen neighbours at Ca-O = 2.688(5) Å. Si-O = 1.585-1.640(5) Å, O-Si-O = 103.5-112.8, Si-O-Si = 147.8°.

1. Structure Reports, 31A, 225; 38A, 365.

ELBAITE
$(Na,Ca)(Al,Li,Mn)_3Al_6B_3Si_6O_{27}(O,OH)_4$

M.G. GORSKAJA, O.V. FRANK-KAMENETSKAJA, I.V. ROŽDESTVENSKAJA and V.A. FRANK-KAMENETSKIJ, 1982. Kristallografija, 27, 107-112 [Soviet Physics - Crystallography, 27, 63-66].

Rhombohedral, R3m, a = 15.802, c = 7.0861 Å, Z = 3. Mo radiation, R = 0.028 for 1404 reflexions.

Atomic positions (x 10^5)

		Occupancy factor	Composition	x	y	z
Na	3a	0.53	$Na_{0.46}K_{0.01}Ca_{0.06}\square_{0.48}$	0	0	22102
Al_1	9b	0.822	$Al_{0.74}Li_{0.18}Mn_{0.07}Fe^{3+}_{0.005}$	12157	x/2	63567
Al_2	18c	1.000	$Al_{1.00}$	29648	.25999	60662
B	9b	1.0	$B_{1.00}$	10909	2x	45078
Si	18c	1.0	$Si_{1.00}$	19168	18967	0
O_1	3a	1.0	O^{2-}, OH^-	0	0	77137
O_2	9b	1.0	O^{2-}	6034	2x	49074
O_3	9b	1.0	O^{2-}, OH^-	25949	x/2	50477
O_4	9b	1.0	O^{2-}	9454	2x	7477
O_5	9b	1.0	O^{2-}	18787	x/2	9579
O_6	18c	1.0	O^{2-}	19405	18314	77298
O_7	18c	1.0	O^{2-}	28713	28650	7560
O_8	18c	1.0	O^{2-}	20938	26963	43571

Structure as previously described (1).

1. Structure Reports, 39A, 340.

FLUOROPHLOGOPITE (MANGANOAN, 1M)
$K(Mg,Mn)_3Si_3AlO_{10}F_2$

H. TORAYA, F. MARUMO and M. HIRAO, 1982. Rep. Res. Lab. Eng. Mater., Tokyo Inst. Technol., 7, 15-21.

Monoclinic, C2/m, a = 5.371, b = 9.313, c = 10.182 Å, β = 100.13°, Z = 2. R = 0.069 for 822 reflexions.

Phlogopite structure (1), with cation disorder. T-O = 1.662, M-O = 2.085 Å.

1. Structure Reports, 27, 696; 39A, 343; 41A, 393.

FOSINAITE (ORTHORHOMBIC)
$Na_3(Ca,Ce)PSiO_7$

V.M. KRUTIK, D.Ju. PUŠČAROVSKIJ, A.P. KHOMJAKOV, E.A. POBEDIMSKAJA and N.V. BELOV, 1981. Kristallografija, 26, 1197-1203 [Soviet Physics - Crystallography, 26, 679-682].

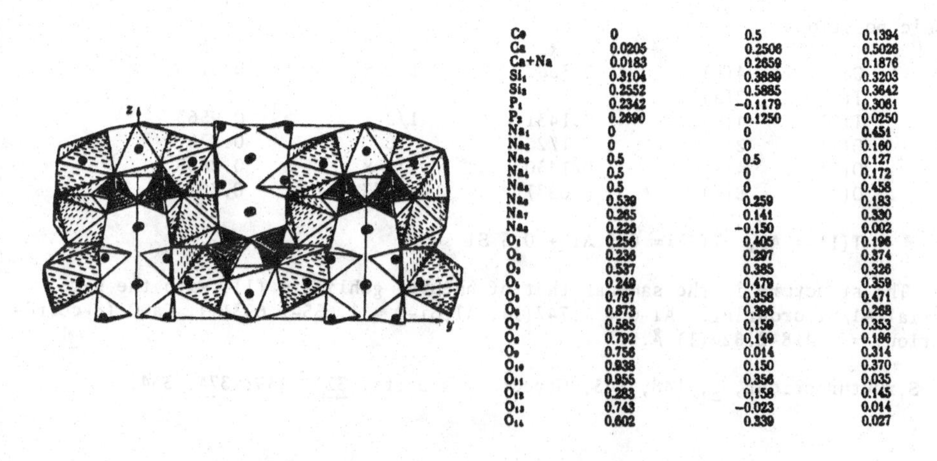

	x	y	z
Ce	0	0.5	0.1394
Ca	0.0205	0.2506	0.5026
Ca+Na	0.0183	0.2659	0.1876
Si_1	0.3104	0.3889	0.3203
Si_2	0.2552	0.5885	0.3642
P_1	0.2342	−0.1179	0.3061
P_2	0.2690	0.1250	0.0250
Na_1	0	0	0.451
Na_2	0	0	0.160
Na_3	0.5	0.5	0.127
Na_4	0.5	0	0.172
Na_5	0.5	0	0.458
Na_6	0.539	0.259	0.183
Na_7	0.265	0.141	0.330
Na_8	0.228	−0.150	0.002
O_1	0.256	0.405	0.196
O_2	0.236	0.297	0.374
O_3	0.537	0.385	0.326
O_4	0.249	0.479	0.359
O_5	0.787	0.358	0.471
O_6	0.873	0.396	0.268
O_7	0.585	0.159	0.353
O_8	0.792	0.149	0.186
O_9	0.756	0.014	0.314
O_{10}	0.938	0.150	0.370
O_{11}	0.955	0.356	0.035
O_{12}	0.283	0.158	0.145
O_{13}	0.743	−0.023	0.014
O_{14}	0.602	0.339	0.027

Fig. 1. Structure of fosinaite (Ce occupancy = 0.67).

Orthorhombic, $P2_12_12$, a = 7.234, b = 14.670, c = 12.231 Å, Z = 8. Mo radiation, R = 0.072 for 1117 reflexions.

The structure (Fig. 1) contains sheets of Ce, Ca, and Na coordination polyhedra, with interpenetrating Si_4O_{12} rings and mixed rings of PO_4 tetrahedra and Na polyhedra; the structural formula is $Na_{11}(Na,Ca)_2Ca_2Ce_{0.67}[Si_4O_{12}](PO_4)_4$. Monoclinic clinofosinaite has been described previously (1).

1. Structure Reports, 46A, 386.

GARNETS
$Ca_3(Al,Fe)_2Si_3O_{12}$ (3 specimens)

Y. TAKÉUCHI, N. HAGA, S. UMIZU and G. SATO, 1982. Z. Kristallogr., 158, 53-99.

Specimens from North Moravia, Czeckoslovakia and Kamaishi, Japan, orthorhombic, Fddd, body-centred pseudo-cubic cell, a = 11.873, 11.966, b = 11.870, 11.963, c = 11.872, 11.964 Å, α = 90.00, 89.98, β = 90.07, 90.07, γ = 90.02, 89.98°, Z = 8. Mo radiation, R = 0.029, 0.035 for 5107, 5220 reflexions.

Specimen from Munam, North Korea, triclinic, $I\bar{1}$, a = 11.920, b = 11.923, c = 11.920 Å, α = 90.03, β = 90.12, γ = 89.98°, Z = 8. Mo radiation, R = 0.030 for 2378 reflexions.

Garnet structures (1), with deviations from cubic (Ia3d) symmetry resulting from Al/Fe ordering in octahedral sites.

1. Strukturbericht, 1, 363.

GEHLENITE
$Ca_2Al_2SiO_7$

M. KIMATA and N. II, 1982. Neues Jb. Miner., Abh., 144, 254-267.

Tetragonal, $P\bar{4}2_1m$, a = 7.6770, c = 5.0594 Å, Z = 2. Mo radiation, R = 0.024 for 496 reflexions, for synthetic sample.

Atomic positions

		x	y	z
Ca	in 4(e)	0.3387	1/2-x	0.5116
T(1)	2(a)	0	0	0
T(2)	4(e)	0.1431	1/2-x	0.9563
O(1)	2(c)	1/2	0	0.1767
O(2)	4(e)	0.1430	1/2-x	0.2835
O(3)	8(f)	0.0877	0.1673	0.8072

T(1) = Al, T(2) = 0.5 Al + 0.5 Si

The structure is the same as that of natural gehlinite (1), with the same partial Al/Si ordering. Al-O = 1.747(1), Al,Si-O = 1.655-1.693(1), Ca-O (8-coordination) = 2.418-2.826(1) Å.

1. Strukturbericht, 2, 148, 543; Structure Reports, 32A, 439; 37A, 334.

GMELINITE

$Na_8Al_8Si_{16}O_{48}.22H_2O$, $(Ca,Sr)_4Al_8Si_{16}O_{48}.23H_2O$

E. GALLI, E. PASSAGLIA and P.F. ZANAZZI, 1982. Neues Jb. Miner., Mh., 145-155.

Hexagonal, $P6_3/mmc$, a = 13.756, 13.800, c = 10.048, 9.964 Å, D_m = 2.01, 2.11, Z = 1. Cu radiation, R = 0.064, 0.079 for 306, 323 reflexions.

Atomic positions

			x	y	z
Sodium-rich					
	Al/Si in	24(ℓ)	0.4419	0.1058	0.0944
	O(1)	12(k)	0.4158	0.2079	0.0605
	O(2)	12(k)	0.8524	0.4262	0.0630
	O(3)	12(j)	0.4126	0.0664	1/4
	O(4)	12(i)	0.3559	0	0
	Na(1)	4(f)	1/3	2/3	0.0743
0.33	Na(2)	12(k)	0.2246	0.1123	-0.0617
0.50	W(1)	12(j)	0.1997	0.5430	1/4
	W(2)	6(h)	0.3356	0.1678	-1/4
0.66	W(3)	12(k)	0.1544	0.0772	0.1271
Calcium-rich					
	Al/Si	24(ℓ)	0.4413	0.1054	0.0920
	O(1)	12(k)	0.4190	0.2095	0.0615
	O(2)	12(k)	0.8608	0.4304	0.0542
	O(3)	12(j)	0.4160	0.0696	1/4
	O(4)	12(i)	0.3487	0	0
0.67	Ca/Sr(1)	4(f)	1/3	2/3	0.0877
0.24	Ca/Sr(2)	12(k)	0.1986	0.0993	-0.0940
0.49	W(1)	12(j)	0.1887	0.5468	1/4
0.79	W(2)	6(h)	0.3460	0.1730	-1/4
0.77	W(3)	12(k)	0.1564	0.0782	0.1266

The structure is as previously described (1). Both specimens contain similar tetrahedral frameworks, with some differences in cation occupancies.

1. Structure Reports, 31A, 227.

GUGIAITE

$Ca_2BeSi_2O_7$

M. KIMITA and H. OHASHI, 1982. Neues Jb. Miner., Abh., 143, 210-222.

Tetragonal, $P\bar{4}2_1m$, a = 7.419, c = 4.988 Å, Z = 2. Mo radiation, R = 0.025 for 479 reflexions, for a synthetic specimen.

Atomic positions

		x	y	z
Ca in	4(e)	0.3358	1/2-x	0.5112
Be	2(a)	0	0	0
Si	4(e)	0.1469	1/2-x	0.9559
O(1)	2(c)	1/2	0	0.1654
O(2)	4(e)	0.1403	1/2-x	0.2790
O(3)	8(f)	0.0861	0.1662	0.8197

Melilite-type structure, isostructural with akermanite (1) and hardystonite (2), with completely ordered Be/Si sites. Be-4 O = 1.654, Si-O = 1.663 (bridging), 1.593, 1.617(1) Å, Si-O-Si = 135.9°, Ca-8 O = 2.355-2.784 Å.

1. Structure Reports, 48A, 342.
2. Strukturbericht, 2, 146, 542; Structure Reports, 41A, 433.

HAFNON
$HfSiO_4$

J.A. SPEER and B.J. COOPER, 1982. Amer. Min., 67, 804-808.

Tetragonal, $I4_1/amd$, a = 6.5725, c = 5.9632 Å, Z = 4. Mo radiation, R = 0.054 for 203 reflexions. Hf in 4(a): 0,3/4,1/8; Si in 4(b): 0,3/4,5/8; O in 16(h): 0,0.0655, 0.1948.

Zircon structure (1). Si-4 O = 1.620(8), Hf-8 O = 2.115, 2.260(9) Å.

1. Strukturbericht, 1, 345, 408; Structure Reports, 22, 314; 29, 411; 37A, 349; 41A, 402; 45A, 380.

HYALOTEKITE
$Pb_2Ba_2Ca_2[B_2(Si_{1.5}Be_{0.5})Si_8O_{28}]F$

P.B. MOORE, T. ARAKI and S. GHOSE, 1982. Amer. Min., 67, 1012-1020.

Triclinic, $I\bar{1}$, a = 11.310, b = 10.955, c = 10.317 Å, α = 90.43, β = 90.02, γ = 90.16°, D_m = 3.82, Z = 2. Mo radiation, R = 0.060 for 3713 reflexions. Pseudo I2/m symmetry.

The structure (Fig. 1) contains Si_4O_{12} and $B_2(Si_{1.5}Be_{0.5})O_{12}$ rings and $Ca_2Pb_4O_{26}F$ clusters. Pb and Ba are disordered in positions 0.5 Å apart, and the deviation from I2/m symmetry appears to result from distortion of PbO_7F polyhedra by the $6s^2$ lone pair, Pb-F = 2.55, 2.59, Pb-O = 2.36-2.59, 3.16-3.47 Å. Si-O = 1.58-1.64, B-O = 1.48-1.51, Si,Be-O = 1.57-1.62, Ca-F = 2.36, Ca-7 O = 2.34-2.56, Ba-F = 2.97, 3.04, Ba-8 O = 2.66-3.36 Å.

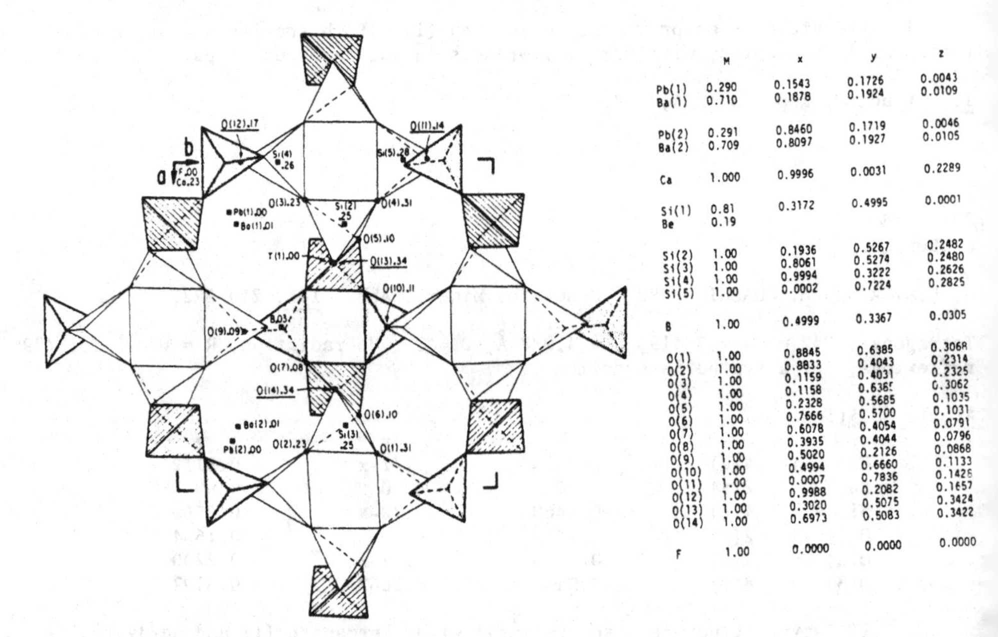

	M	x	y	z
Pb(1)	0.290	0.1543	0.1726	0.0043
Ba(1)	0.710	0.1878	0.1924	0.0109
Pb(2)	0.291	0.8460	0.1719	0.0046
Ba(2)	0.709	0.8097	0.1927	0.0105
Ca	1.000	0.9996	0.0031	0.2289
Si(1)	0.81	0.3172	0.4995	0.0001
Be	0.19			
Si(2)	1.00	0.1936	0.5267	0.2482
Si(3)	1.00	0.8061	0.5274	0.2480
Si(4)	1.00	0.9994	0.3222	0.2626
Si(5)	1.00	0.0002	0.7224	0.2825
B	1.00	0.4999	0.3367	0.0305
O(1)	1.00	0.8845	0.6385	0.3068
O(2)	1.00	0.8833	0.4043	0.2314
O(3)	1.00	0.1159	0.4031	0.2325
O(4)	1.00	0.1158	0.6365	0.3062
O(5)	1.00	0.2328	0.5685	0.1035
O(6)	1.00	0.7666	0.5700	0.1031
O(7)	1.00	0.6078	0.4054	0.0791
O(8)	1.00	0.3935	0.4044	0.0796
O(9)	1.00	0.5020	0.2126	0.0868
O(10)	1.00	0.4994	0.6660	0.1133
O(11)	1.00	0.0007	0.7836	0.1426
O(12)	1.00	0.9988	0.2082	0.1657
O(13)	1.00	0.3020	0.5075	0.3424
O(14)	1.00	0.6973	0.5083	0.3422
F	1.00	0.0000	0.0000	0.0000

Fig. 1. Structure of hyalotekite (M = occupancy).

LIZARDITE (1T)
$Mg_3Si_2O_5(OH)_4$

M. MELLINI, 1982. Amer. Min., 67, 587-598.

Trigonal, P31m, a = 5.332, c = 7.233 Å, D_m = 2.58, Z = 1. Mo radiation, R = 0.031 for 209 reflexions.

Atomic positions

		x	y	z
Si	in 2(b)	1/3	2/3	0.0766
Mg	3(c)	0.3327	0	0.4596
O(1)	2(b)	1/3	2/3	0.3000
O(2)	3(c)	0.5087	0	-0.0036
O(3)	3(c)	0.6654	0	0.5935
O(4)	1(a)	0	0	0.3088
H(3)	3(c)	0.655	0	0.709
H(4)	1(a)	0	0	0.199

The structure is essentially as previously described (1), with nearly ideal serpentine layers; the distortion of the silicate tetrahedral sheet is small, and there is no buckling of the brucite-like magnesium octahedral sheet. Si-O = 1.616, 1.646, Mg-O = 2.021-2.121(5) Å.

1. Structure Reports, 30A, 434; 33A, 473.

MAKATITE
$Na_2Si_4O_8(OH)_2.4H_2O$

H. ANNEHED, L. FÄLTH and F.J. LINCOLN, 1982. Z. Kristallogr., 159, 203-210.

Monoclinic, $P2_1/c$, a = 7.3881, b = 18.094, c = 9.5234 Å, β = 90.64°, D_m = 2.03, Z = 4. Cu radiation, R = 0.051 for 834 reflexions.

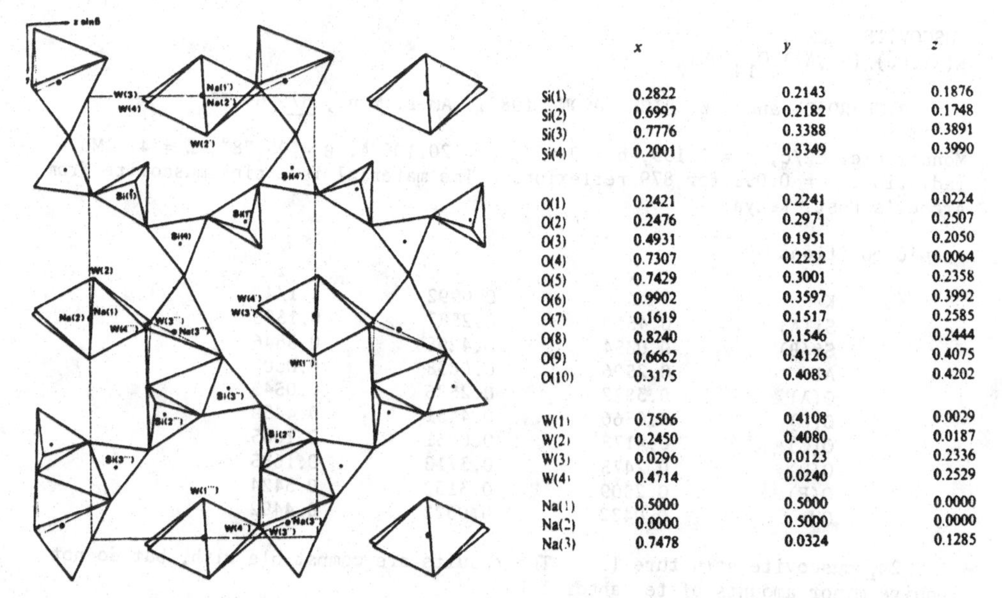

	x	y	z
Si(1)	0.2822	0.2143	0.1876
Si(2)	0.6997	0.2182	0.1748
Si(3)	0.7776	0.3388	0.3891
Si(4)	0.2001	0.3349	0.3990
O(1)	0.2421	0.2241	0.0224
O(2)	0.2476	0.2971	0.2507
O(3)	0.4931	0.1951	0.2050
O(4)	0.7307	0.2232	0.0064
O(5)	0.7429	0.3001	0.2358
O(6)	0.9902	0.3597	0.3992
O(7)	0.1619	0.1517	0.2585
O(8)	0.8240	0.1546	0.2444
O(9)	0.6662	0.4126	0.4075
O(10)	0.3175	0.4083	0.4202
W(1)	0.7506	0.4108	0.0029
W(2)	0.2450	0.4080	0.0187
W(3)	0.0296	0.0123	0.2336
W(4)	0.4714	0.0240	0.2529
Na(1)	0.5000	0.5000	0.0000
Na(2)	0.0000	0.5000	0.0000
Na(3)	0.7478	0.0324	0.1285

Fig. 1. Structure of makatite.

The structure (Fig. 1) contains corrugated $[Si_2O_4(OH)^-]_n$ layers parallel to (010), with rings of six SiO_4 tetrahedra. The layers are connected by octahedral $[Na(H_2O)_4^+]_n$ rods and NaO_5 distorted trigonal bipyramids. Si-O = 1.58-1.64(1) Å, Si-O-Si = 132-150°, Na-O = 2.33-2.56(1) Å.

MICA (CHROMIAN $2M_1$)
$(K,Na,Ca)(Al,Mg,Cr)_2(Si,Al)_4O_{10}(OH)_2$

J.D. MARTÍN-RAMOS and M. RODRÍGUEZ-GALLEGO, 1982. Miner. Mag., 46, 269-272.

Monoclinic, C2/c, a = 5.217, b = 9.045, c = 19.97 Å, β = 95.7°, Z = 4. Mo radiation, R = 0.117 for 1383 reflexions.

Atomic positions

	x	y	z
K,Na,Ca	0	0.0967	1/4
Al,Mg,Cr	0.2489	0.8260	0.0001
O(1)	0.9558	0.4386	0.0546
O(2)	0.3935	0.2529	0.0546
O(3)	0.4390	0.0934	0.1692
O(4)	0.7385	0.3234	0.1603
O(5)	0.2384	0.3611	0.1694
OH	0.9547	0.0654	0.0515
Si,Al(1)	0.9635	0.4296	0.1353
Si,Al(2)	0.4517	0.1589	0.1355

$2M_1$-mica structure (1), with no Si/Al ordering. K-O = 2.93-3.20, Al-O = 1.93-1.99, Si,Al-O = 1.61-1.64 Å.

1. Structure Reports, 37A, 339.

MUSCOVITE ($2M_1$)
$K(Al,Fe)_2(Si,Al)_4O_{10}(OH)_2$

S.M. RICHARDSON and J.W. RICHARDSON, 1982. Amer. Min., 67, 69-75.

Monoclinic, C2/c, a = 5.199, b = 9.027, c = 20.106 Å, β = 95.78°, Z = 4. Mo radiation, R = 0.099 for 879 reflexions. The material is a pink muscovite from Archer's Post, Kenya.

Atomic positions

	x	y	z
K	0	0.0992	1/4
Si(A)	0.4510	0.2587	0.1355
Si(B)	0.0354	0.4298	0.3646
Al	0.2506	0.0838	0.0002
O(A)	0.3872	0.2525	0.0543
O(B)	0.0366	0.4431	0.4459
O(C)	0.4178	0.0931	0.1685
O(D)	0.2475	0.3712	0.1685
O(E)	0.2509	0.3132	0.3424
OH	0.0422	0.0622	0.4492

$2M_1$ muscovite structure (1). The results are compatible with, but do not require, minor amounts of tetrahedral Fe^{3+}.

1. Strukturbericht, 2, 143; Structure Reports, 19, 468; 24, 483; 37A, 339; 41A, 388.

NAGASHIMALITE

$Ba_4(V,Ti)_4[Si_8B_2O_{27}](O,OH)_2Cl$

I. S. MATSUBARA and A. KATO, 1980. Mineral. J., 10, 122-130.
II. S. MATSUBARA, 1980. Ibid., 10, 131-142.

Orthorhombic, Pmmn, a = 13.937, b = 12.122, c = 7.116 Å, Dm = 4.08, Z = 2. Mo radiation, R = 0.045 for 2241 reflexions.

Isostructural with taramellite (1), the structure containing a $Si_8B_2O_{27}$ group (Fig. 1), which consists of a pair of Si_4O_{12} rings linked by a B_2O_7 group; these borosilicate groups are joined by VO_6 octahedra. Ba ions have 11- and 13-coordinations to O and Cl. Mean Si-O = 1.62, B-O = 1.48, V-O = 2.01 Å; Ba-O = 2.67-3.56, Ba-Cl = 3.15, 3.47 Å.

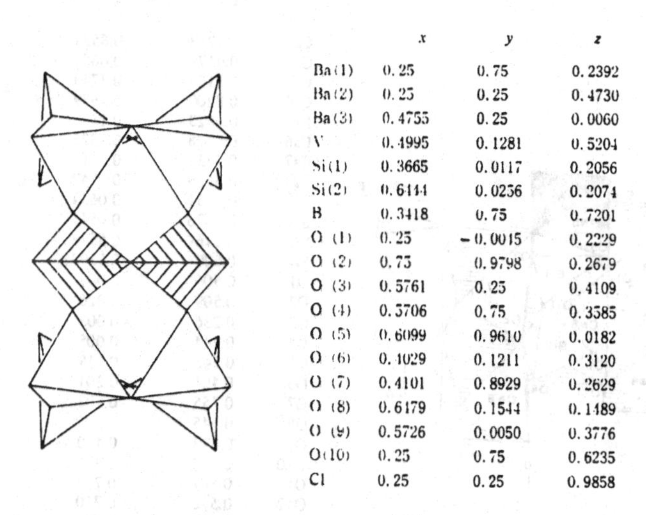

	x	y	z
Ba(1)	0.25	0.75	0.2392
Ba(2)	0.25	0.25	0.4730
Ba(3)	0.4755	0.25	0.0060
V	0.4995	0.1281	0.5204
Si(1)	0.3665	0.0117	0.2056
Si(2)	0.6444	0.0256	0.2074
B	0.3418	0.75	0.7201
O (1)	0.25	−0.0015	0.2229
O (2)	0.75	0.9798	0.2679
O (3)	0.5761	0.25	0.4109
O (4)	0.3706	0.75	0.3585
O (5)	0.6099	0.9610	0.0182
O (6)	0.4029	0.1211	0.3120
O (7)	0.4101	0.8929	0.2629
O (8)	0.6179	0.1544	0.1489
O (9)	0.5726	0.0050	0.3776
O(10)	0.25	0.75	0.6235
Cl	0.25	0.25	0.9858

Fig. 1. The borosilicate group in nagashimalite, and atomic positional parameters.

1. Structure Reports, 30A, 437; 46A, 394.

NEPHELINE HYDRATE I

$Na_3Al_3Si_3O_{12}.2H_2O$

S. HANSEN and L. FAELTH, 1982. Zeolites, 2, 162-166.

Orthorhombic, $Pna2_1$, a = 16.426, b = 15.014, c = 5.2235 Å, Z = 4. R = 0.051.

The tetrahedral framework (1) contains alternating single and double chains, with the largest channels bound by 8-ring apertures. 2 Na are each coordinated to seven framework O, and the third Na to 3 O and 3 H_2O.

1. Structure Reports, 11, 478; 17, 561; 19, 476; 35A, 454; 37A, 340; 38A, 370.

NIOCALITE
$NbCa_7(Si_2O_7)_2O_3F$

M. MELLINI, 1982. Tschermaks Min. Petr. Mitt., 30, 249-266.

Monoclinic, Pa, a = 10.863, b = 10.431, c = 7.370 Å, β = 110.1°, Z = 2. Mo radiation, R = 0.069 for 2411 reflexions (crystal probably twinned).

Niocalite is isostructural with cuspidine (1) and lovenite (2), with Nb ordering resulting in the lower symmetry space group Pa (a previous description (3) is in $P2_1$). The structure (Fig. 1) contains Si_2O_7 groups and ribbons of NbO_6, CaO_6, CaX_7, and CaX_8 polyhedra. Si-O = 1.56-1.68(1) Å, Si-O-Si = 149, 160°, Nb-O = 1.88-2.26(1), Ca-O = 2.25-3.13(1), Ca-F = 2.27-2.41 Å.

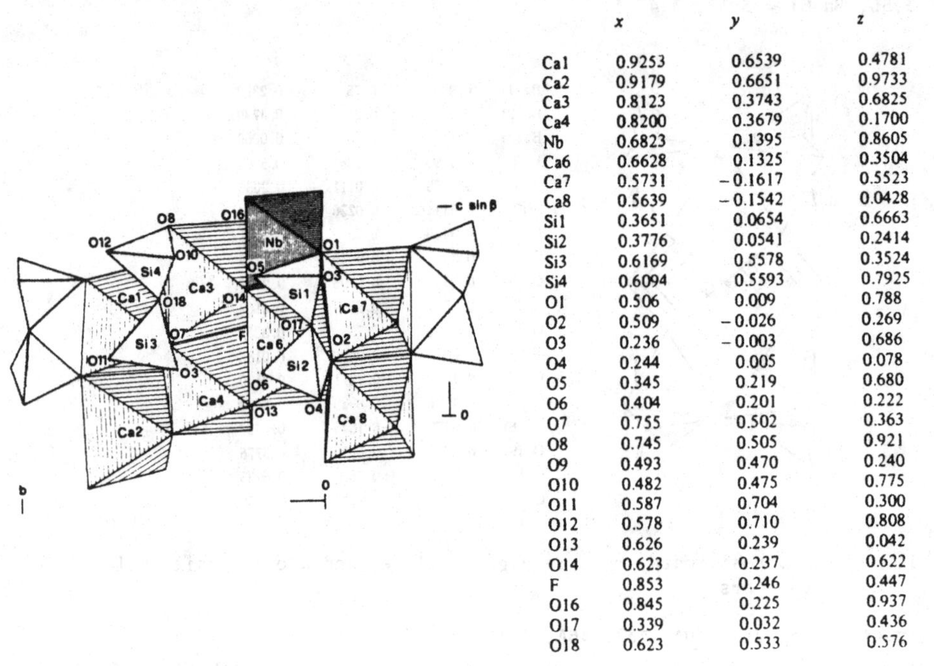

	x	y	z
Ca1	0.9253	0.6539	0.4781
Ca2	0.9179	0.6651	0.9733
Ca3	0.8123	0.3743	0.6825
Ca4	0.8200	0.3679	0.1700
Nb	0.6823	0.1395	0.8605
Ca6	0.6628	0.1325	0.3504
Ca7	0.5731	−0.1617	0.5523
Ca8	0.5639	−0.1542	0.0428
Si1	0.3651	0.0654	0.6663
Si2	0.3776	0.0541	0.2414
Si3	0.6169	0.5578	0.3524
Si4	0.6094	0.5593	0.7925
O1	0.506	0.009	0.788
O2	0.509	−0.026	0.269
O3	0.236	−0.003	0.686
O4	0.244	0.005	0.078
O5	0.345	0.219	0.680
O6	0.404	0.201	0.222
O7	0.755	0.502	0.363
O8	0.745	0.505	0.921
O9	0.493	0.470	0.240
O10	0.482	0.475	0.775
O11	0.587	0.704	0.300
O12	0.578	0.710	0.808
O13	0.626	0.239	0.042
O14	0.623	0.237	0.622
F	0.853	0.246	0.447
O16	0.845	0.225	0.937
O17	0.339	0.032	0.436
O18	0.623	0.533	0.576

Fig. 1. Structure of niocalite.

1. Structure Reports, 17, 571; 19, 461.
2. Ibid., 24, 499.
3. Ibid., 31A, 232.

ORTHOENSTATITE
$MgSiO_3$

COBALT ORTHOPYROXENE
$CoSiO_3$

ORTHOFERROSILITE
$FeSiO_3$

S. SASAKI, Y. TAKÉUCHI, K. FUJINO and S.-I. AKIMOTO, 1982. Z. Kristallogr., 158, 279-297.

Orthorhombic, Pbca, a = 18.227, 18.296, 18.427, b = 8.819, 8.923, 9.076, c = 5.179, 5.204, 5.237 Å, Z = 16. Mo radiation, R = 0.023, 0.037, 0.030 for 3438, 2930, 1984 reflexions.

Enstatite structures, as previously described (1, 2). Electron-density distributions are studied.

1. Strukturbericht, 2, 134; Structure Reports, 34A, 365; 43A, 318.
2. Structure Reports, 42A, 411.

OSUMILITE
$(K,Na)(Mg,Fe)_2(Al,Mg,Fe)_3(Si,Al)_{12}O_{30}$

K.-F. HESSE and F. SEIFERT, 1982. Z. Kristallogr., 160, 179-186.

Hexagonal, P6/mcc, a = 10.126, c = 14.319 Å, Z = 2. Mo radiation, R = 0.033 for 590 reflexions.

Atomic positions

			x	y	z
A	in	2(a)	0	0	1/4
M		4(c)	1/3	2/3	1/4
T(1)		24(m)	0.2476	0.3522	0.3920
T(2)		6(f)	1/2	1/2	1/4
O(1)		12(ℓ)	0.2841	0.4058	1/2
O(2)		24(m)	0.2833	0.2154	0.3684
O(3)		24(m)	0.3532	0.4922	0.3213

A = 0.86 K + 0.18 Na   T(1) = 10.36 Si + 1.64 Al
M = 1.44 Mg + 0.56 Fe   T(2) = 2.61 Al + 0.23 Mg + 0.16 Fe

The structure is as previously described (1), with Mg and Fe(II) in the T(2) and M positions.

1. Structure Reports, 20, 413; 34A, 388.

OXYBIOTITE (1M and $2M_1$)
$K(Mg,Fe)_3(Al,Si)_4O_{12}$ (idealized)

T. OHTA, H. TAKEDA and Y. TAKÉUCHI, 1982. Amer. Min., 67, 298-310.

1M
Monoclinic, C2/m, a = 5.320, b = 9.210, c = 10.104 Å, β = 100.102°, Z = 2. Mo radiation, R = 0.044 for 1125 reflexions.

$2M_1$
Monoclinic, C2/c, a = 5.318, b = 9.212, c = 19.976 Å, β = 95.09°, Z = 4. Mo radiation, R = 0.039 for 1676 reflexions.

Atomic positions (1M)

	x	y	z
M(1)	0	0	1/2
M(2)	0	0.3454	1/2
K	0	1/2	0
T	0.0730	0.1673	0.2228
O(1)	0.0177	0	0.1666
O(2)	0.3217	0.2315	0.1645
O(3)	0.1298	0.1697	0.3905
O(4)	0.1335	1/2	0.4006

The $2M_1$ atomic parameters are almost identical with those of 1M, when referred to a 1M setting.

The sample studied contains the two coexisting phases, which are polytypic.

The structures are similar to those of the hydrogenated structures (1), but are enriched in Fe^{3+} . Partial cation ordering is found, with Mg enriched in M(1) and Fe^{3+} mainly in M(2). The difference between the 1M and $2M_1$ structures is probably related to presence or absence of hydrogen.

1. Structure Reports, 41A, 378.

PETALITE
$LiAlSi_4O_{10}$

T. TAGAI, H. REID, W. JOSWIG and M. KOREKAWA, 1982. Z. Kristallogr., 160, 159-170.

Monoclinic, P2/a, a = 11.737, b = 5.171, c = 7.630 Å, β = 112.54°, Z = 2. Neutron radiation, R = 0.036 for 830 reflexions.

Atomic positions

	x	y	z
Li	1/4	0.2553	0
Al	1/4	0.7564	0
Si(1)	0.9980	0.5128	0.2896
Si(2)	0.1477	0.0099	0.2896
O(1)	0	1/2	1/2
O(2)	1/4	0.9653	1/2
O(3)	0.0938	0.3012	0.2704
O(4)	0.3617	0.5358	0.1342
O(5)	0.0381	0.8011	0.2518
O(6)	0.2076	0.9779	0.1353

The structure is as previously described (1), with a centrosymmetric space group and ordered cation positions. Li-O = 1.93̄3, 1.956, Al-O = 1.732, 1.742, Si-O = 1.593-1.615(2) Å.

1. Structure Reports, 26, 506; 46A, 390.

PROTOENSTATITE
$MgSiO_3$

T. MURAKAMI, Y. TAKÉUCHI and T. YAMANAKA, 1982. Z. Kristallogr., 160, 299-312.

Orthorhombic, Pbcn, a = 9.306, b = 8.892, c = 5.349 Å, at 1080°C, Z = 8. Mo radiation, R = 0.081 for 502 reflexions.

Atomic positions

	x	y	z
Mg(1)	0	0.1006	3/4
Mg(2)	0	0.2625	1/4
Si	0.2928	0.0897	0.0739
O(1)	0.1200	0.0942	0.0770
O(2)	0.3773	0.2463	0.0677
O(3)	0.3481	0.9836	0.3079

The structure is essentially as previously described for protoenstatite (1) and for $(Li,Sc,Mg)SiO_3$ (2). Mean Si-O = 1.63, Mg(1)-O = 2.12, Mg(2)-O = 2.19(1̄) Å.

1. Structure Reports, 22, 487; 23, 475; 37A, 343.
2. Ibid., 43A, 320.

SODALITES

$Li_8(Al_6Si_6O_{24})Cl_2$, $Na_8(Al_6Si_6O_{24})Cl_2$, $K_{7.6}Na_{0.4}(Al_6Si_6O_{24})Cl_2$, $Na_8(Al_6Si_6O_{24})I_2$

B. BEAGLEY, C.M.B. HENDERSON and D. TAYLOR, 1982. Miner. Mag., 46, 459-464.

Chloro-sodalites, cubic, P$\bar{4}$3n, a = 8.447, 8.879, 9.253 Å, Z = 1. Powder data. Al in 6(c): 0,1/4,1/2; Si in 6(d): 1/4,0,1/2; Cl in 2(a): 0,0,0; Li, Na, or K in 8(e): x,x,x, x = 0.1675, 0.1777, 0.1876; O in 24(i): (0.1424,0.1311,0.4108), (0.1521,0.1373,0.4391), (0.1586,0.1363,0.4786).

Iodo-sodalite, cubic, I$\bar{4}$3m, a = 9.008 Å, Z = 1. Powder data. Al + Si in 12(d): 0,1/4,1/2; I in 2(a): 0,0,0; Na in 8(c): x,x,x, x = 0.1980; O in 24(g); x,x,z, x = 0.1428, z = 0.4508.

Sodalite structures (1).

1. Strukturbericht, 2, 150, 562; Structure Reports, 32A, 490.

STELLERITE (SODIUM-EXCHANGED)

$(Na,Ca)_7Al_8Si_{28}O_{72}.25H_2O$

E. PASSAGLIA and M. SACERDOTI, 1982. Bull. Minéral., 105, 338-342.

Orthorhombic, Fmmm, a = 13.611, b = 18.227, c = 17.858 Å, Z = 2. Cu radiation, R = 0.078 for 897 reflexions.

The structure is as previously described for stellerite (1), with sodium statistically occupying sites analogous to those in barrerite (1).

1. Structure Reports, 41A, 401.

THAUMASITE

$Ca_3Si(OH)_6(SO_4)(CO_3).12H_2O$

J. ZEMANN and E. ZOBETZ, 1981. Kristallografija, 26, 1215-1217 [Soviet Physics - Crystallography, 26, 689-690].

Hexagonal, P6_3, a = 11.04, c = 10.39 Å, Z = 2. Mo radiation, R = 0.049 for 1107 reflexions.

Atomic positions

	x	y	z
Ca	0.1949	−0.0118	0.2521
Si	0	0	0.0026
C	1/3	2/3	0.4684
S	1/3	2/3	−0.0151
O(1)	0.3915	0.2292	0.2540
O(2)	0.2629	0.4036	0.2510
O(3)	0.0042	0.3411	0.0728
O(4)	0.0216	0.3482	0.4337
O(5)	0.1997	0.6220	0.4594
O(6)	0.1924	0.6235	0.0328
O(7)	0.1319	0.1248	0.1079
O(8)	0.1305	0.1251	0.3992
O(9)	1/3	2/3	0.8441

The structure is as previously described (1). Si-6 OH = 1.78(1), S-4 O = 1.47(1), C-3 O = 1.30(1) (carbonate group slightly non-planar, with C 0.09 Å out-of-plane), Ca-8 O = 2.41-2.53(1) Å.

1. Structure Reports, 18, 533; 21, 449; 37A, 344.

THOMSONITE
$NaCa_2Al_5Si_5O_{20}.6H_2O$

F. PECHAR, 1982. Cryst. Res. Technol., 17, 1141-1144.

Orthorhombic, Pncn, a = 13.124, b = 13.078, c = 6.62 Å, D_m = 2.30, Z = 4. R = 0.085 for 456 reflexions.

Structure as previously described (1). Average T-O = 1.685 Å.

1. Strukturbericht, 3, 171, 529.

TÖRNEBOHMITE
$Ln_2Al(OH)[SiO_4]_2$

J. SHEN and P.B. MOORE, 1982. Amer. Min., 67, 1021-1028.

Monoclinic, $P2_1/c$, a = 7.383, b = 5.673, c = 16.937 Å, β = 112.04°, Z = 4. Mo radiation, R = 0.033 for 2586 reflexions.

The structure (Fig. 1) contains chains of edge-sharing $Al(OH)_2O_4$ octahedra along b, linked by corner-sharing to SiO_4 tetrahedra. LnO_{10} polyhedra are between the chains. Mean Al-O = 1.90, Si-O = 1.63, Ln-O = 2.66 Å.

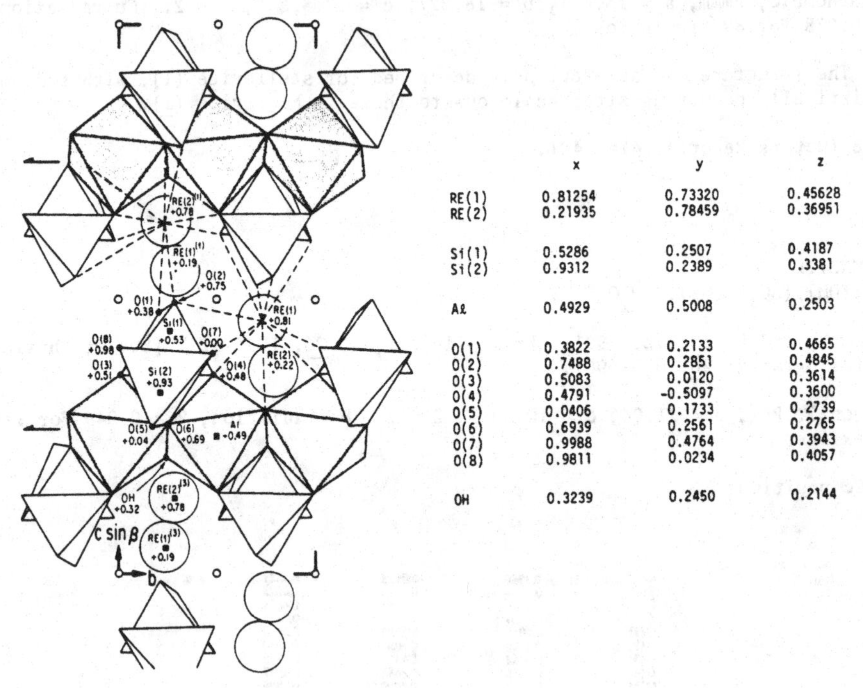

	x	y	z
RE(1)	0.81254	0.73320	0.45628
RE(2)	0.21935	0.78459	0.36951
Si(1)	0.5286	0.2507	0.4187
Si(2)	0.9312	0.2389	0.3381
Al	0.4929	0.5008	0.2503
O(1)	0.3822	0.2133	0.4665
O(2)	0.7488	0.2851	0.4845
O(3)	0.5083	0.0120	0.3614
O(4)	0.4791	-0.5097	0.3600
O(5)	0.0406	0.1733	0.2739
O(6)	0.6939	0.2561	0.2765
O(7)	0.9988	0.4764	0.3943
O(8)	0.9811	0.0234	0.4057
OH	0.3239	0.2450	0.2144

Fig. 1. Structure of törnebohmite, RE = Ln = La, Ce, Nd.

YODERITE
$Mg_2Al_{5.6}Fe_{0.4}Si_4O_{18}(OH)_2$

J.B. HIGGINS, P.H. RIBBE and Y. NAKAJIMA, 1982. Amer. Min., 67, 76-84.

Monoclinic, $P2_1/m$, a = 8.027, b = 5.805, c = 7.252 Å, β = 104.9°, D_m = 3.4, Z = 1. Ordered yoderite has satellite reflexions which indicate a commensurate antiphase structure, A-centred, a x 6b x c. Mo radiation R = 0.035 for 499 reflexions for a disordered sample (heated, no satellite reflexions), R = 0.051 for 499 reflexions for ordered sample (satellite reflexions not used).

The average structure (Fig. 1) is as previously described (1), containing chains along b of edge-sharing $A(1)O_5(OH)$ octahedra interconnected by isolated SiO_4 tetrahedra and two edge-sharing trigonal bipyramids of composition $A(2)O_4(OH)$ and $A(3)O_5$. The model for the ordered structure has 3Al + 1Mg ordered in the A(1) octahedra, 1 Al + 1Mg in A(2), and $Al_{0.84}Fe_{0.16}$ in A(3). In the disordered structure, Fe disorders over A(1), A(2), and A(3).

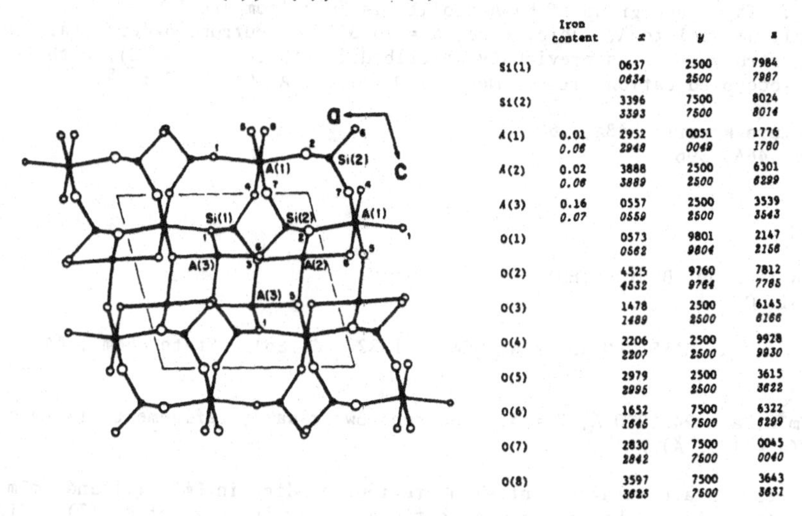

	Iron content	*x*	*y*	*z*
Si(1)		0637	2500	7984
		0634	*2500*	*7987*
Si(2)		3396	7500	8024
		3393	*7500*	*8014*
A(1)	0.01	2952	0051	1776
	0.06	*2948*	*0049*	*1780*
A(2)	0.02	3888	2500	6301
	0.06	*3889*	*2500*	*6299*
A(3)	0.16	0557	2500	3539
	0.07	*0559*	*2500*	*3543*
O(1)		0573	9801	2147
		0562	*9804*	*2156*
O(2)		4525	9760	7812
		4532	*9764*	*7785*
O(3)		1478	2500	6145
		1489	*2500*	*6166*
O(4)		2206	2500	9928
		2207	*2500*	*9930*
O(5)		2979	2500	3615
		2995	*2500*	*3622*
O(6)		1652	7500	6322
		1645	*7500*	*6299*
O(7)		2830	7500	0045
		2842	*7500*	*0040*
O(8)		3597	7500	3643
		3623	*7500*	*3631*

Fig. 1. Average structure of yoderite, and atomic positional parameters (x 10^4) in the ordered and disordered (italics) samples.

1. Structure Reports, 27, 686.

YTTRIUM OXYAPATITE

$NaY_9(SiO_4)_6O_2$

R.P. GUNAWARDANE, R.A. HOWIE and F.P. GLASSER, 1982. Acta Cryst., B38, 1564-1566.

Hexagonal, $P6_3/m$, a = 9.334, c = 6.759 Å, Z = 1. Mo radiation, R = 0.071 for 500 reflexions.

Atomic positions (Y(2) = 0.75 Y + 0.25 Na)

	No. of atoms per unit cell	*x*	*y*	*z*
O(1)	12	0·2486	0·3389	0·4380
O(2)	6	0·4868	0·3171	0·2500
O(3)	6	0·5252	0·4013	0·7500
Si	6	0·3723	0·3994	0·2500
Y(1)	6	0·2375	0·0042	0·2500
Y(2)	4	0·3333	0·6666	−0·0019
O(4)	2	0·0000	0·0000	0·2500

Fluorapatite structure (1), with oxide ion in the fluoride site and Y/Na disorder in the 4(f) site. Si-O = 1.60-1.62 Å, O-Si-O = 104-114°, Y(1)-O = 2.20-2.67, Y(2)-O = 2.32-2.78 Å.

1. Strukturbericht, 2, 99.

ZEOLITE A

I. J.M. ADAMS and D.A. HASELDEN, 1982. Chem. Comm., 822-823.
II. A.K. CHEETHAM, C.A. FYFE, J.V. SMITH and J.M. THOMAS, 1982. Ibid., 823-825.
III. A.K. CHEETHAM, M.M. EDDY, D.A. JEFFERSON and J.M. THOMAS, 1982. Nature, 299, 24-27.

I. and II. The space group of Na-A zeolite is Fm3c (compare 1).
III. Thallium zeolite A, cubic, Fm3c, a = 24.373 Å, neutron powder data. The framework structure is as previously described for Na zeolite A (2), with four partially-occupied cation sites. Mean Si-O = 1.60, Al-O = 1.73(1) Å.

1. Structure Reports, 48A, 366.
2. Ibid., 46A, 396.

ZEOLITE A (SODIUM, DEHYDRATED)
$Na_{23}Al_{23}Si_{25}O_{96}$

J.M. ADAMS, D.A. HASELDEN and A.W. HEWAT, 1982. J. Solid State Chem., 44, 245-253.

Cubic, $Fm\bar{3}c$, a = 24.5920 Å, Z = 4. Neutron powder data, refinements in $Fm\bar{3}c$ and in $Pm\bar{3}m$ (a = 12.3 Å).

The results are in agreement with previous studies in $Fm\bar{3}c$ (1) and $Pm\bar{3}m$ (2), and the data could not be interpreted satisfactorily in terms of $R\bar{3}$ (3). Si/Al ordering and crystal symmetry may depend on the conditions of preparation of the zeolite.

1. Structure Reports, 37A, 346; 46A, 396.
2. Ibid., 20, 445; 33A, 488; 43A, 326; 44A, 320.
3. Ibid., 48A, 366.

ZEOLITE A (STRONTIUM-EXCHANGED, DEHYDRATED)
$Sr_6Al_{12}Si_{12}O_{48}$ (idealized)

J.J. PLUTH and J.V. SMITH, 1982. J. Amer. Chem. Soc., 104, 6977-6982.

Cubic, $Fm\bar{3}c$, a = 24.68 Å, Z = 8. Cu radiation, R = 0.038 for 494 and 408 reflexions for two specimens.

Atomic positions

			x	y	z
96 Si	in	96(i)	0	0.0928	0.1859
96 Al		96(i)	0	0.1885	0.0914
36 Sr(1)		64(g)	0.1045	0.1045	0.1045
10 Sr(2)		64(g)	0.0758	0.0758	0.0758
1 Al*		8(b)	0	0	0
4 O*		64(g)	0.0407	0.0407	0.0407
96 O(1)		96(i)	0	0.1142	0.2462
96 O(2)		96(i)	0	0.1396	0.1416
192 O(3)		192(j)	0.0533	0.0574	0.1730

Typical zeolite structure (1), with ordered Si/Al distribution; mean Si-O = 1.599, Al-O = 1.730 Å. Sr ions lie near the centres of 6-rings, and there is no evidence for the previously-reported (2) zero-coordinate Sr. Additional electron density at the centre of the sodalite unit may be a disordered AlO_4 tetrahedron (* in table above).

1. Structure Reports, 45A, 378; 46A, 396.
2. Ibid., 44A, 318.

ZEOLITE A (COPPER, AMMONIUM EXCHANGED)
$(NH_4)_{8.2}Cu_2Al_{12}Si_{12}O_{48}(OH)_{0.2}.xH_2O$, $(NH_4)_{10}(CuOH)_2Al_{12}Si_{12}O_{48}$

H.S. LEE, W.V. CRUZ and K. SEFF, 1982. J. Phys. Chem., 86, 3562-3569.

Cubic, Pm3m, a = 12.369, 12.280 Å, Z = 1. Mo radiation, R = 0.057 and 0.053 for 330 and 260 reflexions.

Typical zeolite structure (1). In the hydrated structure one Cu is at the centre of the sodalite cavity and a second Cu is in the large cavity opposite an 8-ring; the ammonium positions are uncertain. In the dehydrated sample both Cu lie on threefold axes close to 6-ring planes of the zeolite framework; seven ammonium ions are associated with 6-rings and three occupy 8-ring sites.

1. Structure Reports, 48A, 368.

ZEOLITE Na-X (HYDRATED)

B. BEAGLEY, J. DWYER, N.P. EVMERIDES, A.I.F. HAWA and T.K. IBRAHIM, 1982. Zeolites, 2, 167-174.

Na and H_2O are located in interstitial sites.

ZEOLITES X and Y
$H_{10}Na_{11}Li_{65}Al_{86}Si_{106}O_{384}$, $H_4Na_{13}Li_{39}Al_{56}Si_{136}O_{384}$

H. HERDEN, W.D. EINICKE, R. SCHOELLNER, W.J. MORTIER, L.R. GELLENS and J.B. UYTTERHOEVEN, 1982. Zeolites, 2, 131-134.

Cubic, a = 24.71, 24.59 Å. Li ions are located at the centre of the 6-ring, inducing considerable rotation of the TO_4 tetrahedra. Na ions were not located in zeolite X, and probably share a Li site in zeolite Y.

ZEOLITE (ZSM-5)

Q. YU, W. LI, W. ZHANG, G. WEI, H. YE and B. LIN, 1982. Shiyu Xuebao, No. 3, 83-98.

Orthorhombic, Pnma, a = 20.16, b = 19.97, c = 13.34 Å. R = 0.15.

The structure contains straight elliptical channels parallel to b and Z-shaped channels parallel to a, with openings of 5.8 and 5.3 Å, respectively.

ZUNYITE

$Al_{13}Si_5O_{20}(OH)_{14}F_4Cl$

W.H. BAUR and T. OHTA, 1982. Acta Cryst., B38, 390-401.

Cubic, $F\bar{4}3m$, a = 13.8654, 13.8796 Å, for two samples, Z = 4. Mo radiation, R = 0.016, 0.020 for 904, 813 reflexions.

Atomic positions (decimal fractions x 10^5 for Si, Al, O, x 10^3 for H)

(1) Zunyite from Quartzsite, Arizona

Atom	Position	Symmetry	x	y	z
Si(1)	4(c)	$\bar{4}3m$	1/4	1/4	1/4
Si(2)	16(e)	3m	11430	11430	11430
Al(1)	4(d)	$\bar{4}3m$	3/4	3/4	3/4
Al(2)	48(h)	m	8556	8556	76670
O(1)	16(e)	3m	82478	82478	82478
O(2)	16(e)	3m	18243	18243	18243
O(h3)	24(f)	mm	27949	0	0
O(h4)	48(h)	m	17870	17870	54601
O(5)	48(h)	m	13834	13834	152
Cl	4(b)	$\bar{4}3m$	1/2	1/2	1/2
H(1a)	48(h)	m	228	228	530
H(1b)	48(h)	m	190	190	480
H(2)	24(f)	mm	336	0	0

(2) Zunyite from Zuni Mine, Colorado

Atom	Position	Symmetry	x	y	z
Si, Al(1)	4(c)	$\bar{4}3m$	1/4	1/4	1/4
Si(2)	16(e)	3m	11399	11399	11399
Al(1)	4(d)	$\bar{4}3m$	3/4	3/4	3/4
Al(2)	48(h)	m	8565	8565	76675
O(1)	16(e)	3m	82472	82472	82472
O(2)	16(e)	3m	18128	18128	18128
O(h3)	24(f)	mm	27927	0	0
O(h4)	48(h)	m	17866	17866	54558
O(5)	48(h)	m	13802	13802	119
Cl	4(b)	$\bar{4}3m$	1/2	1/2	1/2
H(1a)	48(h)	m	228	228	530
H(1b)	48(h)	m	190	190	480
H(2)	24(f)	mm	352	0	0

Si,Al(1) = 3/4 Si + 1/4 Al
O(h4) = 2/3 O + 1/3 F
H(1a) and H(1b) have occupancy = 1/3

Structure as previously described (1).

1. Strukturbericht, 3, 147, 517; Structure Reports, 24, 474; 38A, 380; 39A, 355; 40A, 313.

TABLE I

Some structural information has also been given for the following materials (listed with abbreviated 1982 reference).

Compound	Structure	Reference
Uranium dioxide, UO_2	Fluorite structure [Strukturbericht, 1, 150]. Neutron powder data	Acta Cryst., A38, 264
α-Corundum, Al_2O_3	Structure as in Strukturbericht, 1, 240; Structure Reports, 46A, 228. Neutron powder data. z(Al) = 0.35216, x(O) = 0.30624	Ibid., A38, 264, 733
KCN NaCN KOH NaOH $RbNO_3$	Structures as previously described	Ibid., A38, 274
Lead(II) fluoride, β-PbF_2	Fluorite	Ibid., A38, 729
Zeolite A	$Fm\bar{3}c$ is compatible with nmr data	Ibid., A38, 821

TABLE I

Compound	Structure	Reference
Ammonium chloride, NH_4Cl-II Scawtite, $Ca_7Si_6O_{18}CO_3.2H_2O$	Treatment of orientational disorder	Acta Cryst., B38, 1418
Potassium trifluorocuprate-(II), $KCuF_3$	Study of electron density and anharmonic vibration [structure in Structure Reports, 26, 305; 39A, 158; 45A, 159; 46A, 398]	Ibid., B38, 1422
Wustite, $Fe_{1-x}O$	Description of the modulated structure [previous study in Structure Reports, 23, 329]	Ibid., B38, 1451
Lithiophorite, (Al,Li)Mn-$O_2(OH)_2$	A superstructure of the original monoclinic structure [Structure Reports, 16, 266] is proposed	Amer. Min., 67, 817
Iron(II) manganese(II) phosphate, $(Fe,Mn)_3(PO_4)_2$	Graftonite-type with Mn preferentially in the 6-coordinate site and Fe in the 5-coordinate site. Mössbauer spectra and neutron powder data	Ibid., 67, 826
Olivines (Mg-Fe-Ni)	Cation distributions	Ibid., 67, 1206 and 1212
$In(O,N,F)_x$	Fluorite superstructures	Ann. Chim., 6, 619 (1981)
WV_2O_6	Trirutile	Ibid., 7, 275
Sr_xReO_3 $Dy_5Re_2O_{12}$ $La_6Re_4O_{18}$	As previously described [Structure Reports, 44A, 220; 45A, 265; G. BAUD, J.P. BESSE, R. CHEVALIER and M. GASPERIN, 1983. Mater. Chem., to be published]	Ibid., 7, 615
$LnVO_3$	Aragonite, vaterite, calcite, and perovskite phases	Bull. Chem. Soc. Japan, 55, 2095
$Co_3Al_2Si_3O_{12}$	Garnet	Ibid., 55, 3806
Vuorelainenite, MnV_2O_4	Spinel, a = 8.48 Å, Z = 8	Canad. Miner., 20, 281
Lead sulphate oxide, $PbSO_4.4PbO$	Disordered structure	C.R. Acad. Sci. Paris, Ser II, 293, 1053
$MnCo_2O_4$	Spinel, a = 8.28 Å, u(O) = 0.26	Ibid., 294, 427
$Co_3Al_2Si_3O_{12}$	Garnet, Ia3d, a = 11.455 Å. Si-O = 1.627 Å	Ganseki Kobutsu Kosho Gakkaishi, 76, 58
$CdMnGaO_4$	Spinel, a = 8.30 Å	Indian J. Pure Appl. Phys., 20, 765

Compound	Structure	Reference
Tetraamminelithium, $Li(NH_3)_4$	Phase II, probably $I\bar{4}3d$, a = 14.93 Å, Z = 16. Phase III, superstructure with doubled a axis	Inorg. Chem., 21, 2294
Iron(III) molybdate, $Fe_2(MoO_4)_3$	Neutron powder study at 2K [structure as in Structure Reports, 45A, 255; 46A, 267]	Ibid., 21, 4223
$Cu_{0.5}Mn_{2.5}O_4$	Spinel with Mn^{2+} in tetrahedral sites, and Cu^{2+}, Mn^{4+}, Mn^{3+} in octahedral sites	Izv. Akad. Nauk SSSR, Neorg. Mater., 18, 1060
$Zn(Cr,Fe)_2O_4$	Spinels, with Zn in tetrahedral sites, and Cr and Fe mainly in octahedral sites	J. Appl. Cryst., 15, 260
β-Aluminas	Powder diffraction study of cation distribution in Na and Ag β-aluminas	Ibid., 15, 471
α-Quartz, SiO_2 GeO_2 Si_2N_2O Ge_2N_2O α-Si_3N_4 β-Si_3N_4	Neutron powder studies at high-pressure	J. Appl. Phys., 52, 236 (1981)
$NaMnF_3$ $NaCoF_3$	Space group Pnma assigned from single-crystal polarized vibrational spectra [as in Structure Reports, 34A, 201; 41A, 414]	J. Inorg. Nucl. Chem., 43, 3143 (1981)
$LnAsO_4$, Ln = La, Ce, Pr, Nd	Monazite	J. Less-Common Metals, 83, 255
PaF_5 NpF_5	β-UF_5 α-UF_5	Ibid., 86, 75
$LiFeSnO_4$	Ca_2SnO_4 structure assumed [Structure Reports, 32A, 305; 41A, 419]	J. Mater. Sci. Lett., 1, 116
$La_4Ti_9O_{24}$	$Nd_4Ti_9O_{24}$	Ibid., 1, 312
$MnNi_2O_4$	Spinel, with Mn^{4+} in tetrahedral sites	J. Phys. C, 15, 899
Graphite-rubidium, $C_{24}Rb$	A model for the structure is proposed	J. Phys. Soc. Japan, 51, 257
Terbium oxides, δ-$Tb_{11}O_{20}$ β(2)-Tb_6O_{11} β(3)-Tb_6O_{11}	Defect fluorite structures are proposed. Electron microscope study	J. Solid State Chem., 41, 75
Bismuth lanthanum tungstate, $Bi_{2-x}La_xWO_6$ (x = 0.4-1.1)	Structure proposed. Electron microscope study	Ibid., 41, 138

TABLE I

Compound	Structure	Reference
$RbCu_3Ta_7O_{21}$ $RbCu_3Nb_7O_{21}$ $TlCu_3Ta_7O_{21}$	Structure proposed from powder data	J. Solid State Chem., 41, 221
$HNbO_3$ $HTaO_3$	Cubic perovskites, a = 3.882, 3.810 Å	Ibid., 41, 308
Cs_2KBiCl_6	Elpasolite	Ibid., 42, 130
$Ba_3CoNb_2O_6$ $Ba_3MnNb_2O_6$ $Ba_3MnTa_2O_6$	Hexagonal 3-layer perovskites	Ibid., 43, 51
Calcium ferrites, $CaFe_3O_5$, $CaFe_4O_6$, $CaFe_5O_7$	Relationship with the $CaTi_2O_4$ structure is discussed	Ibid., 43, 222
Ti_nO_{2n-1}, n = 4-9	A model is proposed to illustrate the relationship between the structures	Ibid., 43, 314
$MNbF_6$, M = Mg, Ca, Mn, Fe, Co, Ni, Zn, Cd	$MNbF_7$-type structures	Ibid., 43, 327
Vanadyl sulphate trihydrate, $VOSO_4.3H_2O$	Unit cell data for a new orthorhombic form, $P2_12_12$	Ibid., 45, 112
Vanadyl sulphate dihydrate, $VOSO_4.2H_2O$	Monoclinic	
$MLaTh(VO_4)_3$, M = Sr, Pb	Low-temperature forms with zircon structure; high-temperature monoclinic forms	Ibid., 45, 135
Zeolites	High-resolution electron microscope and nmr study	Ibid., 45, 368
$CdGa_2O_4$ $CoGa_2O_4$ $(Cd,Co)Ga_2O_4$	Spinels, a = 8.328-8.601 Å, x(O) = 0.386-0.392	Mater. Chem., 7, 675
$Cs(Ni,Cd)F_3$ (Ni:Cd = 3:1)	12R perovskite, with Ni in face-sharing octahedra forming linear Ni_3F_{12} trimers. Neutron data	Mater. Lett., 1, 49
M_xWO_3, M = Np, Pu, Am	Perovskites	Mater. Res. Bull., 17, 33
Lithium cobaltate(III) (O2 phase), $LiCoO_2$	Packing variant of the O3 phase [Structure Reports, 22, 327], with c = 2c(O3)/3	Ibid., 17, 117
Mercury(II) oxide (triclinic), HgO	Structure proposed	Ibid., 17, 179
$(Cu,Co)_3O_4$	Partially-inverted spinel	Ibid., 17, 235

Compound	Structure	Reference
$(K,Bi)_2Bi_2O_6(H_2O)$	Pyrochlore; on heating Bi^{3+} migrates from 16(c) to 16(d)	Mater. Res. Bull., 17, 309
Ruthenium pentafluoride, RuF_5 Osmium pentafluoride, OsF_5	Structures as previously determined [Structure Reports, 29, 260; 37A, 177]. Refinement of F parameters and study of magnetic interactions from neutron powder data	Ibid., 17, 315
$BaLa_4Ti_4O_{15}$	$Ba_5Ta_4O_{15}$ [Structure Reports, 26, 421; 35A, 261]	Ibid., 17, 345
Lanthanum nickelate, La_2NiO_4	K_2NiF_4-type, a = 3.9, c = 12.7 Å, z(La) = 0.362, z(O2) = 0.157. X-ray powder data	Ibid., 17, 383
$Li_{1.5}Fe_3O_4$	Spinel, with tetrahedral Fe^{3+} displaced to empty octahedral positions, and some Li in tetrahedral sites	Ibid., 17, 785
$Li_{1.7}Fe_2O_3$	α-Corundum h.c.p. transforms to c.c.p., with Li^+ distributed over two octahedral sites	
Magnetite, Fe_3O_4	Spinel, x(O) = 0.254	Ibid., 17, 1365
FeO_x, x ∿ 1.45	Spinels, x(O) = 0.253, cation vacancies in 16(d) and 8(a), with some Fe in a tetrahedrally-coordinated 48(f) site	
$(NH_4)_2PrCl_5$ $NH_4Y_2Cl_7$	K_2PrCl_5 $RbDy_2Cl_7$	Ibid., 17, 1447
Ce_2O_3	La_2O_3	Phys. Lett., 88A, 81
MMn_2O_4, M = Ni, Zn, Cu, Co	Spinels. M = Zn is a normal spinel, Co inverse, and Ni and Cu intermediate. x(O) = 0.377-0.395	Phys. Status Solidi, A, 69, K15
Vanadium difluoride, VF_2	Rutile	Port. Phys., 13, 109
NaSiON NaGeON	LiSiON [Structure Reports, 46A, 137]	Rev. Chim. Minér., 19, 701
β-Alumina (sodium) β-Alumina (silver)	β-Alumina structure, with some information on cation positions. Neutron powder data	Solid State Ionics, 6, 21
$M_{1-x}Bi_xF_{1+2x}$, M = K, Rb, x = 0.5-0.7	Fluorite, with additional F in interstitial sites	Ibid., 6, 103

TABLE I

Compound	Structure	Reference
$Ba_3ReSb\square O_9$	Tetragonal perovskite	Z. anorg. Chem., 484, 173
$A_8BB'_2M_4O_{24}$, A, B = Ba, Sr, Ca B' = Ln M = U, W	Perovskites	Ibid., 484, 177
$Cs_2LiLuCl_6$	K_2LiAlF_6 (6L) and elpasolite phases	Ibid., 485, 133
Na_3PO_4-Na_2SO_4	Na_3PO_4 (high-temperature form)	Ibid., 486, 57
$Ba_3M(II)Sb_2O_9$	Hexagonal $BaTiO_3$-type, with cation ordering	Ibid., 487, 161
$Ba_3M(III)(Pt,Ru)_2O_9$	Hexagonal $BaTiO_3$-type, with 1:2 cation ordering, sequence $(hcc)_2$	Ibid., 487, 178
$Ba_3M(II)M(V)O_9$, M(V) = Ru, Ir	Hexagonal $BaTiO_3$-type, with 1:2 cation ordering	Ibid., 487, 189
$A(II)_8B(II)B(III)_2W_4O_{24}$	Ordered perovskites with cation ordering	Ibid., 488, 159
$BaNiF_6$	$BaGeF_6$	Ibid., 489, 7
$(Ba,Sr)_2Ln_2MgW_2O_{12}$	Rhombohedral 12L perovskites, stacking sequence $(hhcc)_3$	Ibid., 489, 55
$SrNiF_5$ $CdCoF_5$	$BaGaF_5$ $CaCrF_5$	Ibid., 490, 111
Niobium pentoxide, M-Nb_2O_5	Electron microscope study indicates a disordered arrangement of blocks, rather than the linkage found in a previous X-ray study [Structure Reports, 35A, 209]	Ibid., 491, 101
$NaCaBr_3$ $MCaBr_3$, M = K, Rb, Cs $KCaBr_3$ (high-temp.) Cs_2CaBr_4 $Cs_3Ca_2Br_7$ $Cs_4Ca_3Br_{10}$ Rb_4CaBr_6 Na_6CaBr_8	Ilmenite Perovskite or deformed perovskite $TlPbI_3$ K_2NiF_4 $Rb_3Cd_2Cl_7$ $Rb_4Cd_3Cl_{10}$ K_4CdCl_6 Suzuki-phase	Ibid., 491, 301
$Ba_2WO_3F_4$ $Ba_2MoO_3F_4$ $PbWO_3F_2$ $Pb_3W_2O_6F_6$	New type $Ba_2WO_3F_4$ $SrAlF_5$ $Sr_3Fe_2F_{12}$	Ibid., 492, 63

Compound	Structure	Reference
$Ba_2M(II)M(III)F_9$, M(II) = Co, Ni, Zn, Mg; M(III) = Al, Ga, Co, Ni	Three structure types: Ba_2ZnAlF_9, a monoclinically distorted variant, and Ba_2MnFeF_9	Z. anorg. Chem., 493, 59
$M(I)M(V)Cl_6$, M(I) = Na, K, Tl, NH_4, Rb, Cs M(V) = Nb, Ta, Sb	Low-temperature forms have $CsWCl_6$-type structure; high-temperature forms have a face-centred-cubic structure	Ibid., 493, 65
$Ba_3FeIr_2O_{8.5}$	Hexagonal $BaTiO_3$	Ibid., 494, 87
$(Ba,Sr)_6(Lu,Ho)_2W_3O_{18}$	Hexagonal perovskite with cation vacancies	Ibid., 495, 89
Magnesium gallate, $MgGa_2O_4$	Spinel, Fd3m, a = 8.278 Å, x(O) = 0.256, 87% Ga in 8(a)	Z. Kristallogr., 160, 33
Manganese(II) gallate, $MnGa_2O_4$	Spinel, Fd3m, a = 8.458 Å, x(O) = 0.264, 32% Ga in 8(a)	
$Y_3FeGa_4O_{12}$ $Y_3Fe_3Ga_2O_{12}$	Garnets with Fe preferentially in octahedral and Ga in tetrahedral sites	Ibid., 161, 167
$(Fe,M)_3(PO_4)_2$, M = Mg, Ca, Ni, Zn, Cd	Graftonite	Ibid., 161, 209
$Ag_6Ge_2O_7$ $Ag_{10}Ge_4O_{13}$	$Ag_6Si_2O_7$ $Ag_{10}Si_4O_{13}$	Z. Naturforsch., 37B, 265
$CsZnCo(CN)_6$	Cubic, F$\bar{4}$3m, a = 10.336 Å. Powder data. Co(III) in 4(a); Zn in 4(b); C, N in 24(f): x ∿ 0.19, 0.30; Cs in 4(c) or 4(d) (or disordered). Perovskite-related structure	Ibid., 37B, 832
$CsFeF_4$ $CsCoF_4$ $KCoF_4$ $NaCoF_4$	β-$RbAlF_4$ β-$RbAlF_4$ $RbFeF_4$ $NaTiF_4$	Ibid., 37B, 1132
$(Co,Zn)Al_2O_4$	Spinel, with Co in tetrahedral site	Z. Phys., 127, 223 (1981)

The following compounds have been studied by electron diffraction of the vapours (listed with abbreviated 1982 reference). Bond lengths are in Å, angles in degrees.

Gold pentafluoride	82% Au_2F_{10} dimer (with four-membered Au_2F_2 ring), 18% Au_3F_{15} trimer (Au_3F_3 ring); Au octahedral coordination in each Au-F (bridging) 2.03 Au-F (terminal) 1.82, 1.89 Au-F-Au (dimer) 100 (trimer) 124	Acta Chem. Scand., A36, 705
Chlorine trifluoride oxide, F_3ClO	Distorted trigonal bipyramidal, with O, one F, and lone-pair equatorial Cl-O 1.405 Cl-F(eq) 1.603 Cl-F(ax) 1.713 O-Cl-F(eq) 109	Inorg. Chem., 21, 273
Chromyl chloride, CrO_2Cl_2	Distorted tetrahedron, C_{2v} symmetry Cr-O 1.581 Cr-Cl 2.126 O-Cr-O 108.5 Cl-Cr-Cl 113.3	Ibid., 21, 1115
(Fluoroimido)tetrafluorosulphur, $FN{=}SF_4$	Electron diffraction and microwave data. Trigonal bipyramidal at S S=N 1.520 (long) S-F(eq) 1.564 S-F(ax) 1.535, 1.615 N-F 1.357 (short) S=N-F 118	Ibid., 21, 1607
Vanadium(V) fluoride, VF_5	Trigonal bipyramid (with possible small distortions) V-F(eq) 1.71 V-F(ax) 1.73	Ibid., 21, 2690
Tungsten sulphide tetrachloride, $WSCl_4$	Square pyramid, C_{4v} symmetry W=S 2.086 W-Cl 2.277 S-W-Cl 104.2	Ibid., 21, 3280
Tungsten selenide tetrachloride, $WSeCl_4$	W=Se 2.203 W-Cl 2.284 Se-W-Cl 104.4	
Hydridogallium bis(tetrahydroborate), $HGa[(\mu\text{-}H)_2BH_2]_2$	Ga has square-pyramidal coordination. Best fit to the data is given by an asymmetrical arrangement: Ga-H (bridge) 1.76, 1.89 B-H (bridge) 1.46, 1.25	J. Chem. Soc., Dalton, 597 (1982)

Compound	Structure / Parameter	Value	Reference
Bis(difluorothiophosphoryl) ether, $O(PF_2S)_2$	Gauche conformation, but C_2 and C_s structures indistinguishable		J. Chem. Soc., Dalton, 2079 (1982)
	P-O	1.61	
	P=S	1.87	
	P-F	1.53	
	P-O-P	131	
	O-P-S	117	
	O-P-F	100	
	F-P-F	102	
Aminosulphonyl fluoride, H_2NSO_2F	S-N	1.61	J. Molec. Struct., 78, 307
	S-O	1.41	
	S-F	1.56	
	N-S-F	99	
	O-S-O	123	
Thiocarbonyl chloride, Cl_2CS	C=S	1.602	Ibid., 81, 121
	C-Cl	1.728	
	Cl-C-Cl	111.2	
Germanium dibromide, $GeBr_2$	Ge-Br	2.337	Ibid., 82, 107
	Br-Ge-Br	101.2	
Thionyl bromide, Br_2SO	S=O	1.448	Ibid., 84, 153
	S-Br	2.254	
	Br-S=O	107.4	
Thionyl tetrafluoride, OSF_4	Trigonal bipyramid, with O equatorial		J. Phys. Chem., 86, 598
	S=O	1.409	
	S-F(eq)	1.539	
	S-F(ax)	1.596	
Hydrazine, N_2H_4	N-N	1.449	Ibid., 86, 602
	N-H	1.021	
Hafnium tetrachloride, $HfCl_4$	Tetrahedral		Ž. Strukt. Khim., 22, No. 5, 65 (1981) [J. Struct. Chem., 22, 694]
	Hf-Cl	2.316	
Rhenium(VII) oxide pentafluoride, $ReOF_5$	Octahedral		Ibid., 22, No. 5, 182 (1981) [Ibid., 22, 795]
	Re=O	1.64	
	Re-F	1.81	
	O=Re-F	93	
Rubidium nitrite, $RbNO_2$ Caesium nitrite, $CsNO_2$	Previous results confirmed [Structure Reports, 46A, 407]		Ibid., 22, No. 5, 183 (1981) [Ibid., 22, 796]
	Rb-O	2.64	
	Cs-O	2.79	
	N-O	1.25	
Titanium tetrafluoride, TiF_4	Tetrahedral		Ibid., 23, No. 1, 56 [Ibid., 23, 45]
	Ti-F	1.745	
Caesium metaborate, $CsBO_2$	B-O	1.27	Ibid., 23, No. 1, 182 [Ibid., 23, 156]
	Cs-O	2.74	
	Previous study in Structure Reports, 41A, 407; 44A, 327		

Caesium sulphate, Cs_2SO_4	CsO_2SO_2Cs		Ž. Strukt. Khim., 23, No. 1, 184 [J. Struct. Chem., 23, 158]
	S-4 O	1.47	
	Cs-2 O	2.42	
Antimony pentachloride, $SbCl_5$	Trigonal bipyramid		Ibid., 23, No 2, 144 [Ibid., 23, 295]
	Sb-Cl(ax)	2.34	
	Sb-Cl(eq)	2.28	

MICROWAVE SPECTRA

Disilyl iodide, H_3Si-SiH_2I	Si-Si	2.336	Inorg. Chem., 21, 35
	Si-I	2.440	
	Si-H	1.48	
	Si-Si-I	107	
	Si-Si-H	112	
Boranediamine, $HB(NH_2)_2$	Planar, C_{2v}		J. Amer. Chem. Soc., 104, 3822
	B-H	1.193	
	B-N	1.418	
	N-H	1.003	
	N-B-N	122.0	
Disilanyl cyanide, $SiH_3.SiH_2.CN$	Si-Si	2.33	J. Chem. Phys., 76, 2210
	Si-C	1.84	
	C-N	1.16	
	Si-Si-C	107.4	
Germyl azide, H_3GeN_3	Ge-H	1.52	J. Molec. Struct., 79, 235
	Ge-N	1.82	
	N-N	1.22	
	N-N	1.13	
	Ge-N-N	123	
Germyl isocyanate, H_3GeNCO	Ge-H	1.52	
	Ge-N	1.83	
	N-C	1.17	
	C-O	1.18	
	Ge-N-C	143	

INFRARED SPECTRA

Cyanogen fluoride, F-C≡N	C-F	1.2673	J. Molec. Struct., 82, 221
	C-N	1.1571	

PAPERS REFERRED TO LATER YEARS

Many preliminary notes have not been reported, since fuller accounts will appear at a later date. The compounds studied, and abbreviated 1982 references, are listed below.

γ-Ti_3O_5 δ-Ti_3O_5	Acta Chem. Scand., A36, 207
P_4O_6	Angew. Chem., 93, 1023 (1981)
AlI_3 GaI_3 InI_3	Ibid., 94, 370

Compound	Reference
$Bi_8(AlCl_4)_2$	Angew. Chem., 94, 453
α-ThI_3 β-ThI_3	Ibid., 94, 558
Hydrogen uranyl arsenate tetrahydrate, $DUO_2AsO_4.4D_2O$	Chem. Comm., 784 (1981)
Caesium hydrogen bis(fluorosulphate), $CsH(SO_3F)_2$	Ibid., 1036 (1981)
$Na_{17}Al_5O_{16}$	Ibid., 516 (1982)
$Se_2I_4(Sb_2F_{11})_2$	Ibid., 1098 (1982)
Cycloundecasulphur, S_{11}	Ibid., 1312 (1982)
$Te_2Se_2O_8$	C.R. Acad. Sci. Paris, Ser. II, 295, 981
Silver holmium silicate, $Ag_5HoSi_4O_{12}$	Dokl. Akad. Nauk SSSR, 262, 638
Sodium heptahydroxyytterbate hydrate, $Na_4Yb(OH)_7.nH_2O$	Ibid., 262, 880
Calcium gadolinium fluoride	Ibid., 262, 883
Strontium divanadate, $Sr_2V_2O_7$	Ibid., 263, 101
Sodium zirconium germanate, $Na_2ZrGe_2O_7$	Ibid., 263, 877
Tadzhikite, $(Ca,Ln)_4(Y,Ln)_2(Ti,Fe,Al)$-$(O,OH)_2[Si_4B_4O_{22}]$	Ibid., 264, 342
Strontium hydrogarnets	Ibid., 264, 857
Caesium niobyl pyrophosphate, $CsNbOP_2O_7$	Ibid., 264, 859
Caesium tantalum phosphate triphosphate, $CsTa_2(PO_4)_2P_3O_{10}$	Ibid., 264, 862
Rare-earth gallates	Ibid., 264, 1385
$CaLn_4O_7$	Ibid., 264, 1400
$Na_2Mn_2Si_2O_7$ $Na_5(Mn,Na)_3MnSi_6O_{18}$	Ibid., 265, 76
$Fe_{0.5}H_{1.5}Na_5(PO_4)_2F_2$	Ibid., 265, 83
$(Tc_8Br_{12})Br[(H_2O)_2H]$	Ibid., 265, 1420
Lithium vanadyl phosphate	Ibid., 266, 343
Strontium osmate, $Sr[OsO_5(H_2O)].3H_2O$	Ibid., 266, 347
Rubidium tantalum phosphate pentaphosphate, $Rb_2Ta_2H(PO_4)_2P_5O_{16}$	Ibid., 266, 354
Nabaphite, $NaBaPO_4.9H_2O$	Ibid., 266, 624
Lithium hydroxoosmate(VIII), $Li_2OsO_4(OH)_2$	Ibid., 266, 628

Sodium hydroxoosmate(VIII) hydrate, $Na_2OsO_4(OH)_2.2H_2O$	Dokl. Akad. Nauk SSSR, 266, 1138
Sodium tetrathiocyclotetraphosphate hexahydrate, $Na_4P_4O_8S_4.6H_2O$	Ibid., 266, 1387
Caesium tetrathiocyclotetraphosphate	Ibid., 267, 85
α'-$Ca(Ca,Sr)SiO_4$	Ibid., 267, 641
$Na_4Sn_2Ge_5O_{16}.H_2O$	Ibid., 267, 850
Barium chlorosilicate	Ibid., 267, 1125
ClF_3, BrF_3 (gases)	Ibid., 267, 1143
Zeolite ZSM-39	Nature, 294, 340 (1981)
Ice-Ih	Ibid., 294, 432 (1981)
$Ca_2CoSi_2O_7$	Naturwissenschaften, 69, 40
Garnet	Ibid., 69, 141
$Cs_2Cu_2Si_8O_{19}$	Ibid., 69, 142
$Ba_7Nb_4Ti_2\square O_{21}$	Ibid., 69, 445
Rubidium hexabromotellurate(IV), Rb_2TeBr_6	Z. Kristallogr., 159, 1
Sodalite, $Ca_8[Al_{12}O_{24}](WO_4)_2$	Ibid., 159, 30
$Pb_2(NO_3)(PO_4).H_2O$	Ibid., 159, 31
Gryolite, $Ca_{13}[Si_8O_{20}]_3(OH)_2.22H_2O$	Ibid., 159, 34
Okenite, $Ca_9[Si_6O_{15}]_3.26H_2O$	Ibid., 159, 37
$MgS_2O_3.6H_2O$	Ibid., 159, 42
$CoSeO_4.6H_2O$ $NiSeO_4.6H_2O$	Ibid., 159, 44
$RbLiSO_4$ (phase II)	Ibid., 159, 48
$Cs[Br(ICN)_2]$ $Rb[I(ICN)_2]$	Ibid., 159, 50
Dodecasilicate-1H	Ibid., 159, 52
$K_2W_2O_7$	Ibid., 159, 55
Parabustamite (2M), $(Ca,Mn)_3Si_3O_9$	Ibid., 159, 58
Osumilite	Ibid., 159, 59 [This volume, p. 343]
Cs_2KHoF_6 Rb_2NaHoF_6	Ibid., 159, 62
$RbOH$ $RbOH.H_2O$	Ibid., 159, 63
$NdGaGe_2O_7$	Ibid., 159, 64

$SmGaGeO_5$ — Z. Kristallogr., 159, 65

$Ln_2AlSi_2O_8(OH)$ — Ibid., 159, 66

$LiBGeO_4$ — Ibid., 159, 67

Forsterite, Mg_2SiO_4 — Ibid., 159, 73

Gebhardite, $Pb_8OCl_6(As_2O_5)_2$ — Ibid., 159, 75

$Li_3Na_2GaO_4$ — Ibid., 159, 81

$NaZr_2(Si,P)_3O_{12}$ — Ibid., 159, 82

$SeOBr_2$ — Ibid., 159, 84

$NaNiAsO_4$ — Ibid., 159, 93

Cs_2DyCl_5
K_2PrCl_5 — Ibid., 159, 94

$NH_3.nHF$ — Ibid., 159, 96

$Pb_5Ge_3O_{11}$
$Pb_3Ge_2O_7$ — Ibid., 159, 100

$KLiSO_4$ — Ibid., 159, 115

$KNa_2(GaSiO_4)_3$ — Ibid., 159, 119

$(NH_4)_2S_2O_3$ — Ibid., 159, 123

Zeolite mineral, $K_{0.7}CaSi_3Al_3O_{12}.4{\cdot}5H_2O$ — Ibid., 159, 125

$PbRe_2O_6$ — Ibid., 159, 135

Zeolite N, $NaAlSiO_4.1{\cdot}35H_2O$ — Ibid., 160, 313

$Mg[UO_2(OH)CrO_4]_2.12H_2O$ — Ž. Neorg. Khim., 26, 532 (1981) [Russian J. Inorg. Chem., 26, 288 (1981)]

$[UO_2(H_2O)_5].B_{12}H_{12}.6H_2O$ — Ibid., 27, 2343

ADDITIONAL PAPERS

The following reports were prepared too late for inclusion in the main text.

trans-TETRAZENE-(2)
N_4H_4 H_2N-N=N-NH_2

M. VEITH and G. SCHLEMMER, 1982. Z. anorg. Chem., 494, 7-19.

Triclinic, $P\bar{1}$, a = 10.23, b = 7.12, c = 4.19 Å, α = 102.0, β = 90.0, γ = 106.5°, at -90°C, Z = 4. Cu radiation, R = 0.12 for 508 reflexions (twinned crystal).

The structure (Fig. 1) contains four molecules lying on centres of symmetry, linked by N-H...N hydrogen bonds (N...N = 3.03-3.42 Å). N=N = 1.21(2), N-N = 1.43(1) Å, N-N-N = 109°.

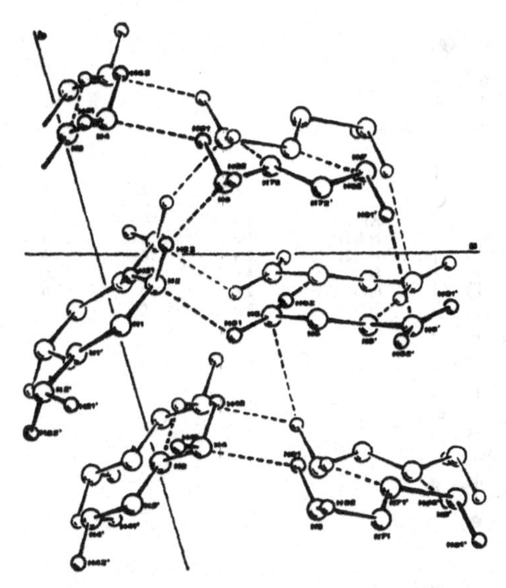

Fig. 1. Structure of N_4H_4.

PHOSPHORUS OXYNITRIDE
PON

B. BONDARS, T. MILLERS, A. VITOLA and J. VILKS, 1982. Latv. PSR Zinat. Akad. Vestis, Kim. Ser., No. 4, 498.

Trigonal, $P3_121$, a = 4.748, c = 5.237 Å, Z = 3. R = 0.10.

α-Quartz structure (1). [Compare 2.]

1. Strukturbericht, 1, 166.
2. Structure Reports, 44A, 324.

NEODYMIUM ALUMINUM OXYNITRIDE
Nd_2AlO_3N

R. MARCHAND, R. PASTUSZAK, Y. LAURENT and G. ROULT, 1982. Rev. Chim. Minér., 19, 684-689.

Tetragonal, I4mm, a = 3.705, c = 12.530 Å, Z = 2. Neutron powder data.

The structure is similar to that of K_2NiF_4 (1), but with an ordered O/N arrangement which lowers the symmetry; it contains AlO_5N octahedra linked by two independent 9-coordinated Nd ions (Fig. 1). Al-O = 1.86 (x 4), 2.09, Al-N = 2.13, Nd-O/N = 2.28-2.65(1) Å.

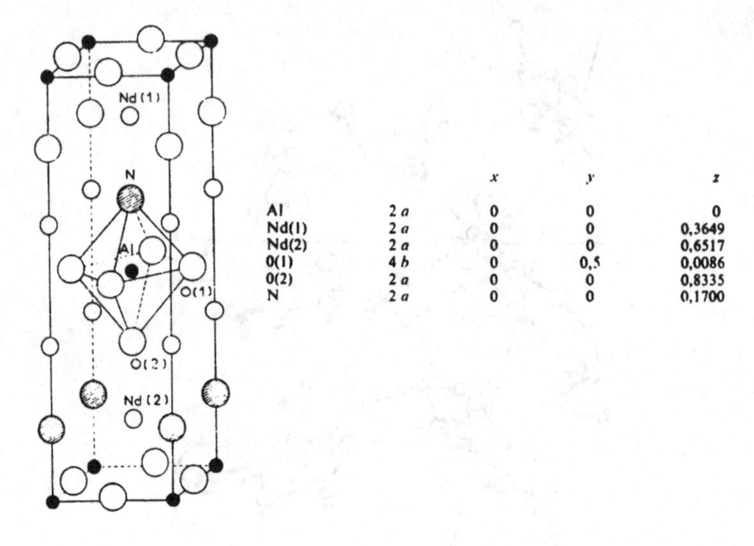

		x	y	z
Al	2a	0	0	0
Nd(1)	2a	0	0	0,3649
Nd(2)	2a	0	0	0,6517
O(1)	4b	0	0,5	0,0086
O(2)	2a	0	0	0,8335
N	2a	0	0	0,1700

Fig. 1. Structure of Nd_2AlO_3N.

1. Structure Reports, 17, 332; 19, 323.

DI-μ3-TELLURIDO-TRIS(TRICARBONYLIRON)
$Fe_3Te_2(CO)_9$

H. SCHUMANN, M. MAGERSTAEDT and J. PICKARDT, 1982. J. Organometal. Chem., 240, 407-411.

Triclinic, $P\bar{1}$, a = 9.525, b = 13.143, c = 6.947 Å, α = 95.27, β = 112.78, γ = 81.44°, Z = 2. Mo radiation, R = 0.036 for 2079 reflexions.

The molecule contains three $Fe(CO)_3$ groups and two triply bridging Te atoms. The Fe_3Te_2 cluster is a square-pyramid with one Fe at the apex. Fe-Te = 2.524-2.550 Å.

STRONTIUM AZIDE
$Sr(N_3)_2$

CALCIUM AZIDE
$Ca(N_3)_2$

H. KRISCHNER and G. KELZ, 1982. Z. anorg. Chem., 494, 203-206.

Orthorhombic, Fddd, a = 11.802, 11.338, b = 11.511, 11.038, c = 6.108, 5.938 Å, Z = 8. Powder data. Sr or Ca in 8(a): 1/8,1/8,1/8; N(1) in 32(h): x,y,z, x = -0.0015, -0.0026, y = 0.0567, 0.0571, z = 0.7700, 0.7756; N(2) in 16(e): x,1/8,1/8, x = 0.4988, 0.5027 (origin at $\bar{1}$).

Structures as previously described for the Sr compound (1). N-N = 1.18, 1.17(1), Sr-N = 2.63, 2.75, Ca-N = 2.51, 2.64 Å, N-N-N = 178, 174°.

1. Structure Reports, 11, 357; 33A, 168.

POTASSIUM CADMIUM AZIDE MONOHYDRATE
$KCd(N_3)_3.H_2O$

POTASSIUM CADMIUM AZIDE
$K_2Cd(N_3)_4$

W. CLEGG, H. KRISCHNER, A.I. SARACOGLU and G.M. SHELDRICK, 1982. Z. Kristallogr., 161, 307-313.

$KCd(N_3)_3.H_2O$, orthorhombic, Pnma, a = 11.991, b = 3.712, c = 17.292 Å, Z = 4. Mo radiation, R = 0.074 for 1193 reflexions.

$K_2Cd(N_3)_4$, monoclinic, C2/m, a = 14.272, b = 3.787, c = 8.887 Å, β = 92.83°, Z = 2. Mo radiation, R = 0.063 for 794 reflexions.

Both structures (Fig. 1) contain edge-sharing $Cd(N_3)_6$ octahedra; K ions have (7 N + 2 O) and 8 N coordinations.

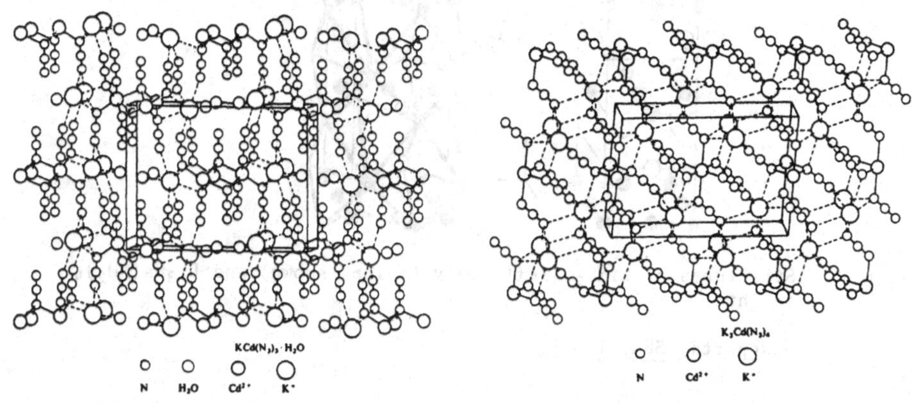

Fig. 1. Structures of $KCd(N_3)_3.H_2O$ (left) and $K_2Cd(N_3)_4$ (right).

THALLIUM CADMIUM AZIDE
$Tl_8Cd_3(N_3)_{14}$

H. KRISCHNER, C. KRATKY and H.E. MAIER, 1982. Z. Kristallogr., 161, 225-229.

Triclinic, P$\bar{1}$, a = 8.603, 8.552, b = 10.532, 10.487, c = 11.767, 11.614 Å, α = 99.92, 100.31, β = 107.68, 107.50, γ = 113.51, 113.47°, at 298, 100K, Z = 1. Mo radiation R = 0.087 for 3302 reflexions at 100K.

The structure contains $Cd(N_3)_6$ octahedra which share two nitrogen corners; Tl^+ ions are surrounded by eight azide groups. Azide groups are linear, mean N-N = 1.17(5) Å. Cd-N = 2.30-2.43, Tl-N = 2.77-3.40 Å.

LITHIUM SODIUM AMIDE
$Li_3Na(NH_2)_4$

H. JACOBS and B. HARBRECHT, 1982. J. Less-Common Metals, 85, 87-95.

Tetragonal, I$\bar{4}$, a = 5.072, 5.057, c = 11.478, 11.432 Å, at 293, 180K, Z = 2. Mo radiation, R = 0.052, 0.042 for 338, 192 reflexions at 293, 180K.

Atomic positions (at 293K)

			x	y	z
Na	in	2(c)	0	1/2	1/4
Li(1)		2(a)	0	0	0
Li(2)		4(f)	0	1/2	0.0087
N		8(g)	0.2396	0.2380	0.1007
H(1)		8(g)	0.365	0.290	0.110
H(2)		8(g)	0.233	0.139	0.168

The structure is similar to that of $LiNH_2$ (1), with Na replacing the Li ion in the 2(c) site (Fig. 1). Li-4N = 2.07, 2.09, 2.19, Na-4N = 2.49(1) Å.

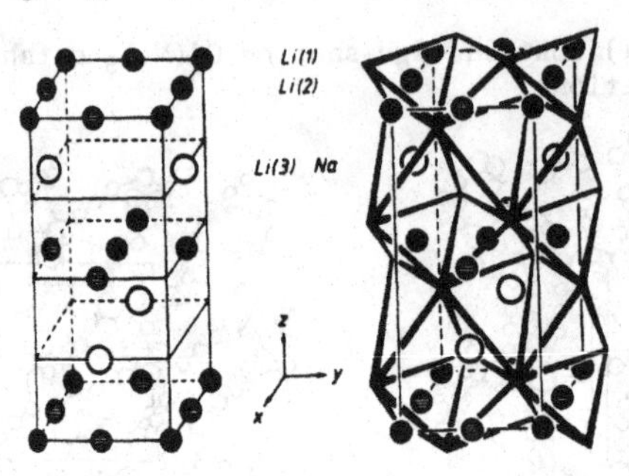

Fig. 1. Structures of $LiNH_2$ (left, only Li ions shown) and $Li_3Na(NH_2)_4$ (right).

1. Structure Reports, 38A, 188.

CAESIUM MAGNESIUM AMIDE
$Cs_2Mg(NH_2)_4$

H. JACOBS, J. BIRKENBEUL and D. SCHMITZ, 1982. J. Less-Common Metals, 85, 79-86.

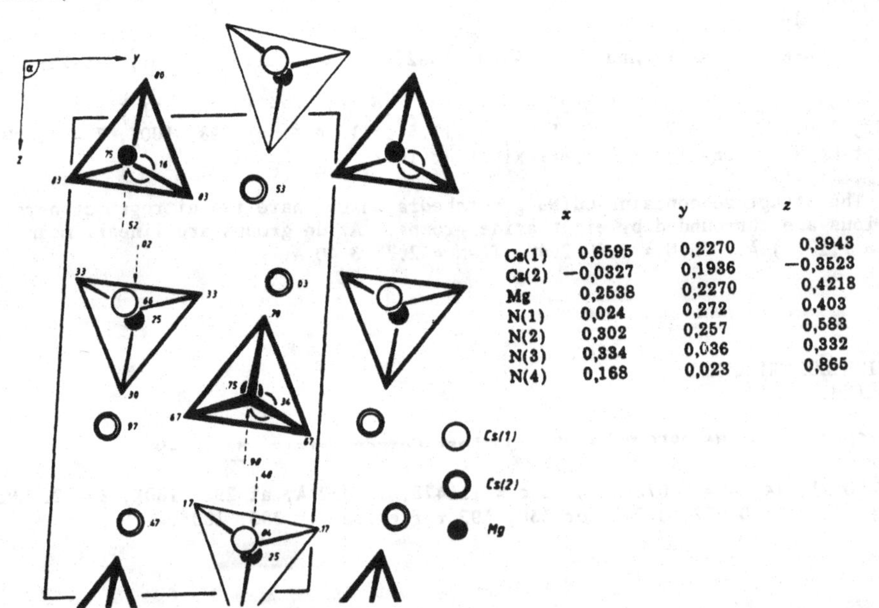

	x	y	z
Cs(1)	0,6595	0,2270	0,3943
Cs(2)	−0,0327	0,1936	−0,3523
Mg	0,2538	0,2270	0,4218
N(1)	0,024	0,272	0,403
N(2)	0,302	0,257	0,583
N(3)	0,334	0,036	0,332
N(4)	0,168	0,023	0,865

Fig. 1. Structure of $Cs_2Mg(NH_2)_4$.

Monoclinic, $P2_1/n$, a (unique axis) = 9.446, b = 7.028, c = 12.373 Å, α = 94.81°, D_m = 2.86, Z = 4. Mo radiation, R = 0.081 for 898 reflexions.

The structure (Fig. 1) is closely related to that of β-K_2SO_4 (1) and contains tetrahedral $Mg(NH_2)_4^{2-}$ ions linked by 9- and 11-coordinate Cs^+ ions. Mg-N = 2.09-2.12(1), Cs-N = 3.17-4.10(1) Å.

1. Strukturbericht, 2, 86, 423; Structure Reports, 22, 447; 33A, 367; 38A, 330; 48A, 304.

CAESIUM BARIUM AMIDE
$CsBa(NH_2)_3$

H. JACOBS, J. BIRKENBEUL and J. KOCKELKORN, 1982. J. Less-Common Metals, 85, 71-78.

Orthorhombic, Pnma, a = 9.602, b = 4.228, c = 15.481 Å, D_m = 3.35, Z = 4. Ag radiation, R = 0.055 for 777 reflexions.

Atomic positions

	x	y	z
Cs	0.5416	1/4	0.8276
Ba	0.3333	1/4	0.4322
N(1)	0.472	1/4	0.602
N(2)	0.721	1/4	0.227
N(3)	0.325	1/4	0.003

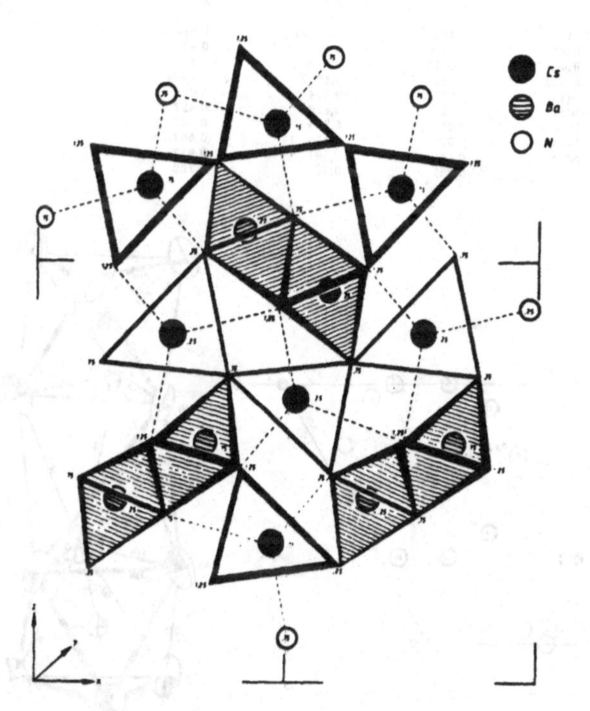

Fig. 1. Structure of $CsBa(NH_2)_3$.

Isostructural with NH_4CdCl_3 (1), the structure (Fig. 1) containing double chains of edge-sharing BaN_6 octahedra, linked by 9-coordinate Cs ions. Ba-N = 2.70-2.95, Cs-N = 3.40-3.61, 4.27(1) Å.

1. Strukturbericht, 6, 13, 79; 7, 19, 115; Structure Reports, 44A, 160.

ALKALI-METAL LANTHANON AMIDES
$K_3Y(NH_2)_6$, $K_3Yb(NH_2)_6$, $Rb_3Eu(NH_2)_6$, $Rb_3Y(NH_2)_6$, $Rb_3Yb(NH_2)_6$

H. JACOBS and J. KOCKELKORN, 1982. J. Less-Common Metals, 85, 97-110.

All except Rb_3Yb, rhombohedral, R32, a = 11.620, 11.565, 12.124, 12.066, c = 13.347, 13.244, 13.553, 13.465 Å, Z = 6. Ag radiation, R = 0.058, 0.068, 0.048, 0.060 for 721, 836, 356, 318 reflexions.

Rb_3Yb, monoclinic, $P2_1/n$, a = 11.413, b = 12.035, c = 8.186 Å, β = 94.55°, Z = 4. Ag radiation, R = 0.045 for 1083 reflexions.

Atomic positions

K_3Y (similar values for the other rhombohedral structures)

		x	y	z
Y	in 6(c)	0	0	0.2468
K(1)	9(d)	0.4354	0	0
K(2)	9(e)	0.6138	0	1/2
N(1)	18(f)	0.2850	0.4826	0.0209
N(2)	18(f)	0.1484	0.9672	0.1444

Rb_3Yb

	x	y	z
Yb	0,24364	0,26021	0,5118
Rb(1)	0,1960	0,0691	0,8795
Rb(2)	0,0111	0,6420	0,2312
Rb(3)	0,3015	0,4677	0,1684
N(1)	0,077	0,854	0,026
N(2)	0,123	0,416	0,427
N(3)	0,276	0,709	0,269
N(4)	0,071	0,156	0,557
N(5)	0,141	0,601	0,917
N(6)	0,261	0,328	0,783

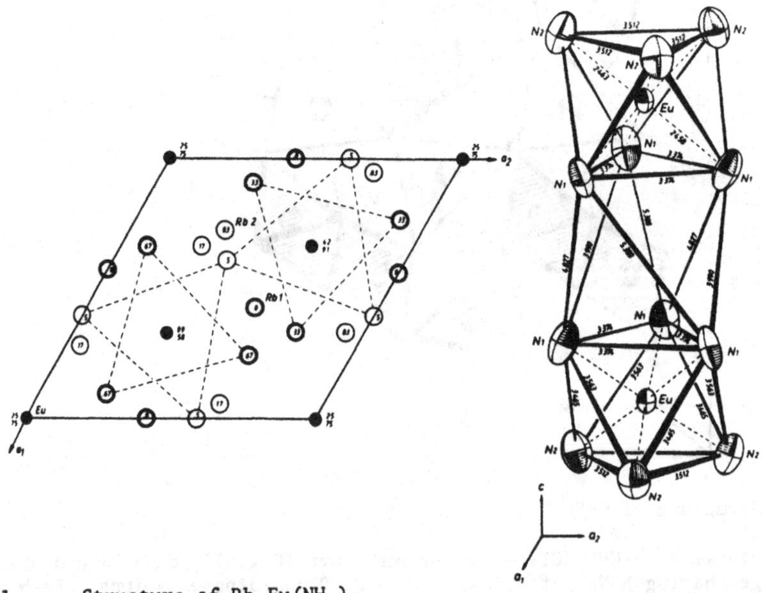

Fig. 1. Structure of $Rb_3Eu(NH_2)_6$.

The rhombohedral phases (Fig. 1) contain $Ln(NH_2)_6$ octahedra stacked along c, linked by K or Rb ions. $Rb_3Yb(NH_2)_6$ has a closely related structure, with octahedral coordination for Yb.

SULPHUR - CYCLOHEXASULPHUR-1,3-DIIMIDE
$2S_8.S_6(NH)_2$

H. GARCIA-FERNANDEZ, M. GASPÉRIN and R. FREYMANN, 1982. C.R. Acad. Sci. Paris, Ser. II, 295, 1109-1112.

Monoclinic, $P2_1/c$, a = 10.883, b = 10.730, c = 10.674 Å, β = 95.67°, D_m = 1.98, Z = 2. R = 0.066 for 2568 reflexions.

β-S_8 structure (1) with statistical distribution of NH groups (see also 2).

1. Structure Reports, 42A, 149.
2. Ibid., 48A, 125.

BARIUM ALUMINUM FLUORIDES
α-$BaAlF_5$

I. R. DOMESLE and R. HOPPE, 1982. Z. anorg. Chem., 495, 16-26.

Orthorhombic, $P2_12_12_1$, a = 13.710, b = 5.604, c = 4.930 Å, Z = 4. Mo radiation, R = 0.048 for 1203 reflexions.

Atomic positions

	x	y	z
Ba	-0.0938	0.0888	0.0199
Al	-0.1650	0.5971	0.4367
F(1)	-0.2733	0.1482	0.8147
F(2)	-0.4195	0.2154	0.1170
F(3)	-0.1055	0.3354	0.5497
F(4)	-0.0796	0.5991	0.1570
F(5)	-0.2490	0.4122	0.2251

Isostructural with $BaGaF_5$ (1); Al-F = 1.77-1.87, Ba-F = 2.62-3.41 Å. High-temperature β- and γ-forms are monoclinic.

$Ba_3Al_2F_{12}$ 0.5 x $Ba_6F_4[Al_4F_{20}]$

II. R. DOMESLE and R. HOPPE, 1982. Z. anorg. Chem., 495, 27-38.

Orthorhombic, Pnnm, a = 10.187, b = 9.869, c = 9.502 Å, D_m = 4.79, Z = 4. Mo radiation, R = 0.073 for 1958 reflexions.

The structure contains tetrameric $[Al_4F_{20}]^{8-}$ (Fig. 1) and F^- anions, linked by Ba ions with 10-, (10 + 3)-, and (11 + 2)-coordinations. Al-F = 1.82, 1.86 (bridging), 1.75-1.82 Å (terminal). $Ba_3Ti_2F_{10}O_2$ is probably isostructural.

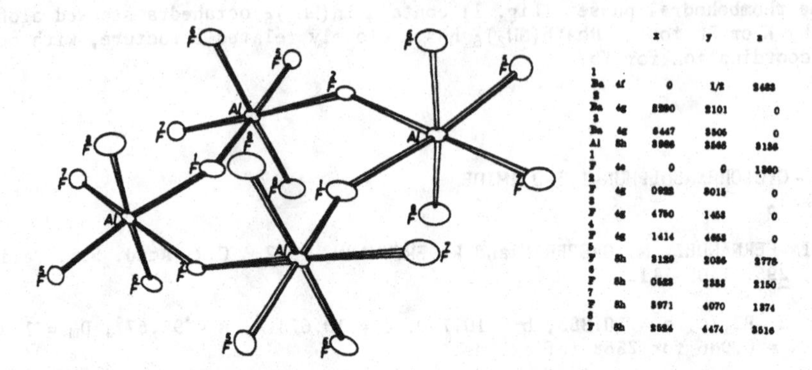

		x	y	z
Ba 1	4f	0	1/2	2483
Ba 2	4g	2290	2101	0
Ba 3	4g	6447	3506	0
Al	8h	3906	3565	3136
F 1	4e	0	0	1360
F 2	4g	0922	8015	0
F 3	4g	4760	1453	0
F 4	4g	1414	4653	0
F 5	8h	3120	2086	2778
F 6	8h	0523	2333	2150
F 7	8h	3971	4070	1374
F 8	8h	2524	4474	3516

Fig. 1. The $[Al_4F_{20}]^{8-}$ anion and atomic positional parameters (x 10^4) in $Ba_3Al_2F_{12}$.

1. Structure Reports, 45A, 149.

STRONTIUM HEXAFLUOROSILICATE DIHYDRATE
$SrSiF_6.2H_2O$

N.I. GOLOVASTIKOV and N.V. BELOV, 1982. Kristallografija, 27, 1084-1086 [Soviet Physics - Crystallography, 27, 649-650].

Monoclinic, $P2_1/n$, a = 10.760, b = 5.892, c = 9.462 Å, γ = 99.46°, Z = 4. Mo radiation, R = 0.049 for 1366 reflexions.

The structure (Fig. 1) contains SiF_6 octahedra linked by pairs of edge-shared $SrF_5(H_2O)_3$ dodecahedra. Si-F = 1.68-1.74, Sr-F = 2.45-2.52, Sr-O = 2.59-2.68 Å.

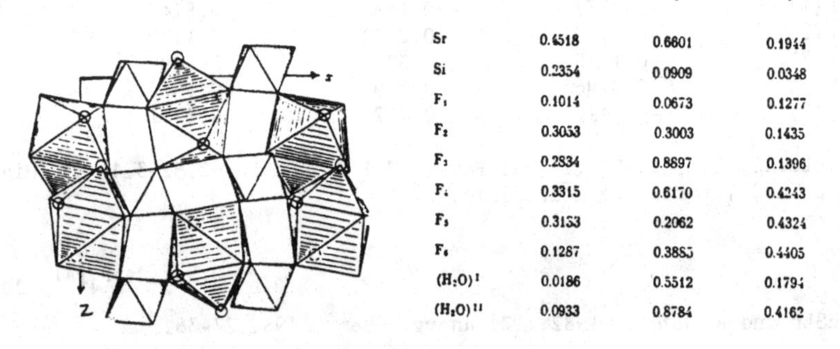

	x	y	z
Sr	0.4518	0.6601	0.1944
Si	0.2354	0.0909	0.0348
F_1	0.1014	0.0673	0.1277
F_2	0.3053	0.3003	0.1435
F_3	0.2834	0.8897	0.1396
F_4	0.3315	0.6170	0.4243
F_5	0.3153	0.2062	0.4324
F_6	0.1287	0.3883	0.4405
$(H_2O)^I$	0.0186	0.5512	0.1794
$(H_2O)^{II}$	0.0933	0.8784	0.4162

Fig. 1. Structure of strontium hexafluorosilicate dihydrate.

AMMONIUM TETRAFLUOROANTIMONATE(III) THALLIUM(I) TETRAFLUOROANTIMONATE(III)
NH_4SbF_4 $TlSbF_4$

V.E. OVČINNIKOV, A.A. UDOVENKO, L.P. SOLOV'EVA, L.M. VOLKOVA and R.L. DAVIDOVIČ, 1982. Koord. Khim., 8, 697-701.

Ammonium salt, monoclinic, $P2_1/b$, a = 8.184, b = 16.407, c = 6.947 Å, γ = 104.32°, D_m = 3.16, Z = 8. Mo radiation, R = 0.048 for 1688 reflexions.

Thallium salt, monoclinic, $P2_1/b$, a = 7.441, b = 8.343, c = 7.061 Å, γ = 98.98°, D_m = 6.13, Z = 4. Mo radiation, R = 0.058 for 558 reflexions.

Both structures (Fig. 1) contain distorted trigonal bipyramidal $SbEF_4$ groupings (E = equatorial lone pair) linked by longer Sb...F interactions which give five and six fluorine coordinations for Sb in the ammonium and thallium salts, respectively. Sb-F = 1.93-1.97 (equatorial), 1.97-2.18 (axial) Å, F(ax)-Sb-F(ax) = 154-157°, Sb...F = 2.45-2.80 Å. NH_4 ions have 8 or 9 F neighbours at 2.75-3.29 Å, and Tl has 10 F at 2.72-3.40 Å.

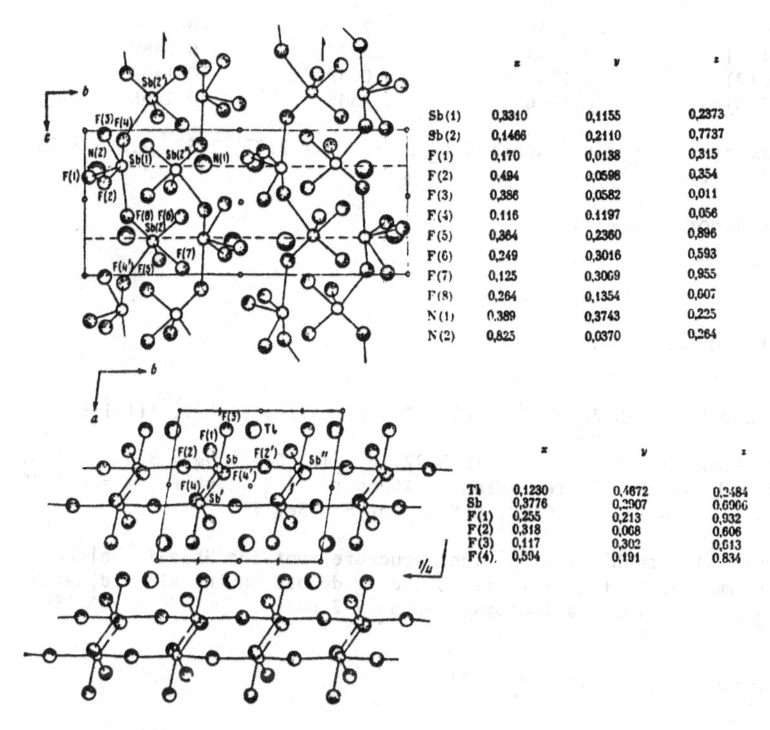

	x	y	z
Sb(1)	0,3310	0,1155	0,2373
Sb(2)	0,1466	0,2110	0,7737
F(1)	0,170	0,0138	0,315
F(2)	0,494	0,0598	0,354
F(3)	0,386	0,0582	0,011
F(4)	0.116	0.1197	0,056
F(5)	0,364	0,2360	0,896
F(6)	0,249	0,3016	0,593
F(7)	0,125	0,3069	0,955
F(8)	0,264	0,1354	0,607
N(1)	0.389	0,3743	0,225
N(2)	0,825	0,0370	0,264

	x	y	z
Tl	0,1230	0,4672	0,2484
Sb	0,3776	0,2907	0,6906
F(1)	0,255	0,213	0,932
F(2)	0,318	0,068	0,606
F(3)	0,117	0,302	0,613
F(4)	0,594	0,191	0.834

Fig. 1. Structures of NH_4SbF_4 (top) and $TlSbF_4$ (bottom).

POTASSIUM CADMIUM FLUORIDE (PHASE III)
$KCdF_3$

M. HIDAKA and S. HOSOGI, 1982. J. Phys., Fr., 43, 1227-1232.

Orthorhombic, Pbnm, a = 6.103, b = 6.103, c = 8.660 Å, Z = 4. Mo radiation, R = 0.098 (film data, crystal consisted of several domains).

Atomic positions

	x	y	z
K	-0.0058	0.0260	1/4
Cd	1/2	0	0
F(1)	0.0603	0.4839	1/4
F(2)	-0.2864	0.2855	0.0267

Distorted perovskite structure, as for $NaNiF_3$ (1).

1. Structure Reports, 44A, 336.

BISMUTH(III) CHLORIDE
$BiCl_3$

H. BARTL, 1982. Z. Anal. Chem., 312, 17-18.

Orthorhombic, $Pn2_1a$, a = 7.67, b = 9.16, c = 6.18 Å, Z = 4. Neutron radiation, R = 0.04 for 550 reflexions.

Atomic positions

	x	y	z
Bi	-0.0466	-0.25	-0.2280
Cl(1)	0.0564	-0.2517	0.3489
Cl(2)	0.1747	-0.0649	-0.1454
Cl(3)	0.1766	-0.4348	-0.1455

Structure as previously described (1). Bi-Cl = 2.47, 2.52, 2.53; 3.23, 3.23, 3.24; 3.42, 3.43(1) Å.

1. Structure Reports, 37A, 193.

ANTIMONY SELENOIODIDE
SbSeI

G.P. VOUTSAS and P.J. RENTZEPERIS, 1982. Z. Kristallogr., 161, 111-118.

Orthorhombic, Pnam, a = 8.6862, b = 10.3927, c = 4.1452 Å, D_m = 5.81, Z = 4. Mo radiation, R = 0.054 for 821 reflexions. Atoms in 4(c): x,y,1/4, x = 0.1185, 0.8347, 0.5157, y = 0.1281, 0.0482, 0.8269 for Sb, Se, I.

Isostructural with BiSCl (1). The structure contains $SbSe_3I_4$ polyhedra linked by a common Se to form infinite $Sb_2Se_4I_8$ double chains along c, with further linking of chains via I atoms. Sb-Se = 2.601, 2.796, Sb-I = 3.150, 3.822(1) Å.

1. Structure Reports, 13, 206; 46A, 217.

AMMONIUM HEXAISOTHIOCYANATOMOLYBDATE(III) MONOHYDRATE MONOHYDROCHLORIDE
$(NH_4)_3Mo(NCS)_6.H_2O.HCl$

S. LIU, A. ZHENG and J. HUANG, 1982. Kexue Tongbao, 27, 666-669.

Orthorhombic, Pnam, a = 16.179, b = 10.001, c = 14.529 Å, D_m = 1.57, Z = 4. Mo radiation, R = 0.095 for 1089 reflexions.

The structure contains octahedral $Mo(NCS)_6^{3-}$ anions, NH_4^+ ions, and H_2O and HCl linked by an O-H...Cl hydrogen bond [presumably H_3O^+ and Cl^-]. Mo-N = 2.071-2.102(6), N-C = 1.104-1.149(12), C-S = 1.621-1.680(12), NH_4^+...S = 3.44-3.82 Å.

DIAMMINENICKEL THIOCYANATE
$Ni(NH_3)_2(SCN)_2$

NICKEL THIOCYANATE
$Ni(SCN)_2$

E. DUBLER, A. RELLER and H.R. OSWALD, 1982. Z. Kristallogr., 161, 265-277.

$Ni(NH_3)_2(SCN)_2$, monoclinic, C2/m, a = 8.698, b = 7.991, c = 5.619 Å, β = 105.91°, D_m = 1.84, Z = 2. Mo radiation, R = 0.024 for 524 reflexions.

$Ni(SCN)_2$, monoclinic, C2/m, a = 10.476, b = 3.628, c = 6.165 Å, β = 106.89°, D_m = 2.56, Z = 2. Mo radiation, R = 0.042 for 611 reflexions.

Atomic positions

$Ni(NH_3)_2(SCN)_2$	x	y	z
Ni	0	0	0
S	0.2601	0	0.3496
C	0.1911	0	0.5929
N(1)	0.1435	0	0.7644
N(2)	0	0.2604	0
$Ni(SCN)_2$			
Ni	0	0	0
S	0.3815	0	0.7362
C	0.2350	0	0.7870
N	0.1347	0	0.8282

The diammine structure contains $Ni(SCN)_2Ni$ chains along c, with NH_3 ligands completing octahedral coordination at Ni (Fig. 1). Similar chains in $Ni(SCN)_2$ are linked by additional S bridges into layers (Fig. 1). Ni-S = 2.557, 2.511, Ni-N = 2.00-2.08 Å.

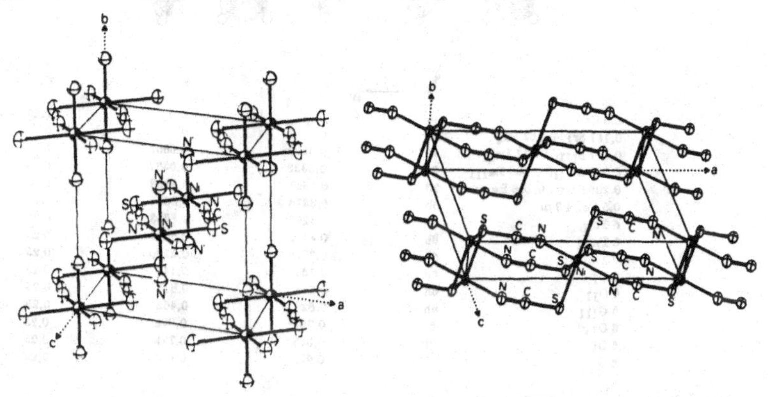

Fig. 1. Structures of $Ni(NH_3)_2(SCN)_2$ (left) and $Ni(SCN)_2$ (right).

RUTILE
TiO_2

W. GONSCHOREK and R. FELD, 1982. Z. Kristallogr., 161, 1-5.

Crystal data as in 1. Neutron radiation, R = 0.033 for 134 reflexions. x(O) = 0.3048.

1. This volume, p. 160.

BARIUM STRONTIUM CALCIUM THULIUM OXIDE
$BaSrCaTm_{22}O_{36}$

J. KRÜGER and H. MÜLLER-BUSCHBAUM, 1982. Z. anorg. Chem., 494, 103-108.

Hexagonal, $P6_3/m$, a = 17.600, c = 3.353 Å, Z = 1. R = 0.09 for 840 reflexions.

The structure (Fig. 1) is related to that of $SrCa_2Lu_{10}O_{18}$ (1), with a complicated and disordered distribution of alkaline earth metals in tunnels. Tm-O = 2.15-2.36 (6-coordination), other M-O = 2.45-3.70 Å.

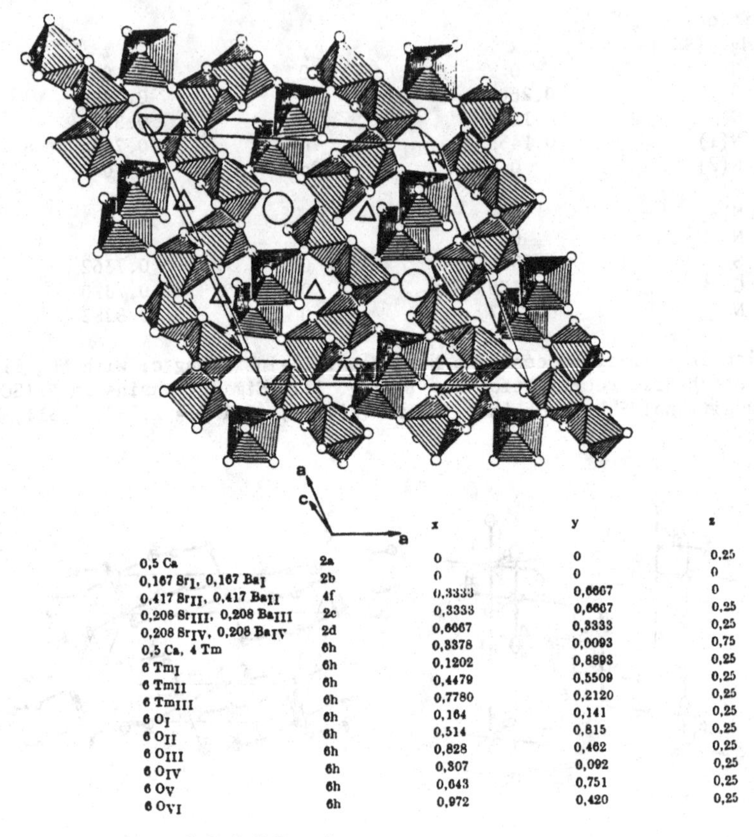

		x	y	z
0,5 Ca	2a	0	0	0,25
0,167 Sr_I, 0,167 Ba_I	2b	0	0	0
0,417 Sr_{II}, 0,417 Ba_{II}	4f	0,3333	0,6667	0
0,208 Sr_{III}, 0,208 Ba_{III}	2c	0,3333	0,6667	0,25
0,208 Sr_{IV}, 0,208 Ba_{IV}	2d	0,6667	0,3333	0,25
0,5 Ca, 4 Tm	6h	0,3378	0,0093	0,75
6 Tm_I	6h	0,1202	0,8893	0,25
6 Tm_{II}	6h	0,4479	0,5509	0,25
6 Tm_{III}	6h	0,7780	0,2120	0,25
6 O_I	6h	0,164	0,141	0,25
6 O_{II}	6h	0,514	0,815	0,25
6 O_{III}	6h	0,828	0,462	0,25
6 O_{IV}	6h	0,307	0,092	0,25
6 O_V	6h	0,643	0,751	0,25
6 O_{VI}	6h	0,972	0,420	0,25

Fig. 1. Structure of $BaSrCaTm_{22}O_{36}$.

1. Structure Reports, 42A, 293.

SUBJECT INDEX

This index contains the names of the substances printed at the heads of the reports, and some additional entries. Greek letter and numerical prefixes, and prefixes such as cis, trans, etc. are disregarded in fixing the alphabetical order.

METALS FORMULA INDEX

The entries are in alphabetical order by formula.

AUTHOR INDEX

Names beginning with a separated prefix are listed before single-word names beginning with the same letters; accents are omitted.